CISM Courses and Lectures

The series presents lecture notes, monographs, edited works and proceedings in the field of Mechanics, Engineering, Computer Science and Applied Mathematics.
Purpose of the series is to make known in the international scientific and technical community results obtained in some of the activities organized by CISM, the International Centre for Mechanical Sciences.

International Centre for Mechanical Sciences

Courses and Lectures Vol. 541

For further volumes:
www.springer.com/series/76

Holm Altenbach · Victor A. Eremeyev
Editors

Generalized Continua from the Theory to Engineering Applications

Springer

Editors

Holm Altenbach
Otto Von Guericke University
Magdeburg, Germany

Victor A. Eremeyev
South Federal University and South
Scientific Center of Rasci
Rostov on Don, Russian Federation

ISSN 0254-1971
ISBN 978-3-7091-1370-7 ISBN 978-3-7091-1371-4 (eBook)
DOI 10.1007/978-3-7091-1371-4
Springer Wien Heidelberg New York Dordrecht London

All contributions have been typeset by the authors
Printed in Italy

Printed on acid-free paper

Springer is part of Springer Science+Business Media (www.springer.com)

PREFACE

The course Generalized continua - from the theory to engineering applications *brought together doctoral students, young researcher, senior researchers, and practicing engineers. The need of generalized continua models is coming from the practice. Complex material behavior sometimes cannot be presented by the classical Cauchy continua.*

Generalized Continua are in the focus of scientists from the end of the 19th century. A first summary was given in 1909 by the Cosserat brothers. After World War II a true renaissance in this field occurred with a publication of Ericksen & Truesdell in 1958. Further developments were connected with the fundamental contributions of, among others, Kröner (Germany), Aero and Palmov (Soviet Union), Nowacki (Poland), Eringen (USA), and Maugin (France).

The Mechanics of Generalised Continua is an established research topic since the end of the 50s - early 60s of the last century. The starting point was the monograph of the Cosserat brothers from 1909 Théorie des corps déformables *and some previous works of such famous scientists like Lord Kelvin. All these contributions were focussed on the fact that in a continuum one has to define translations and rotations independently (or in other words, one has to establish force and moment actions as it was done by Euler).*

The reason for the revival of generalized continua is that some effects of the mechanical behavior of solids and fluids could not be explained by the available classical models. Examples of this are the turbulence of a fluid or the behavior of solids with a significant and very complex microstructure. Since the suggested models satisfy all requirements from Continuum Thermomechanics (the balance laws were formulated and the general representations of the constitutive equations were suggested) the scientific community accepted for a while but missed real applicative developments.

Indeed, for practical applications the developed models were not useful. The reason for this was a gap between the formulated constitutive equations and the possibilities to identify the material parameters. As often the case one had much more parameters compared to classical models.

During the last ten years the situation has drastically changed. More and more researches emerged, being kindled by the partly for-

gotten models since now one has available much more computational possibilities and very complex problems can be simulated numerically. In addition, with the increased attention paid to a large number of materials with complex microstructure and a deeper understanding of the meaning of the material parameters (scale effects) the identification becomes much more well founded. We have thus contributions describing the micro- and macrobehaviors, new existence and uniqueness theorems, the formulation of multi-scale problems, etc, and now it is time to ponder again the state of matter and to discuss new trends and applications. In addition, generalized continua models are not included in the actual BSc or MSc programs.

At present the attention of the scientists in this field is focussed on the most recent research items

- new models,
- application of well-known models to new problems,
- micro-macro aspects,
- computational effort, and
- possibilities to identify the constitutive equations

The new research directions were discussed during the course from the point of view of modeling and simulation, identification, and numerical methods. The following lectures were presented:

- On the Roots of Continuum Mechanics in Differential Geometry - A Review - by Paul Steinmann
- Cosserat Media by Holm Altenbach & Victor A. Eremeyev
- Cosserat-type Shells by Holm Altenbach & Victor A. Eremeyev
- Cosserat-type Rods by Holm Altenbach, Mircea Bîrsan & Victor A. Eremeyev
- Micromorphic Media by Samuel Forest
- Electromagnetism and Generalized Continua by Gérard A. Maugin
- Computational Methods for Generalised Continua by René de Borst

Finally the lecturers should acknowledge the German Research Foundation supporting the Course by the Grant No. AL 341/40-1.

Holm Altenbach and Victor A. Eremeyev

CONTENTS

On the Roots of Continuum Mechanics in Differential Geometry - A Review -
by P. Steinmann 1

1 Introduction **1**

2 Differential Geometry **5**
2.1 Overview 5
2.2 Manifolds 6
2.3 Connection 9
2.4 Parallel Transport 12
2.5 Torsion 13
2.6 Curvature 18
2.7 Metric 26
2.8 Metric Curvature 32

3 Continuum Mechanics **36**
3.1 Kinematics 36
3.2 Distortion 40
3.3 Integrability 41
3.4 Elasticity 44
3.5 Elastoplasticity 47

4 Summary **61**

Bibliography **61**

Cosserat Media
by H. Altenbach & V.A. Eremeyev 65

1 Introduction **65**
1.1 Elements of Rigid Body Dynamics 67
2.1 Elements of Mechanics of Elastic Rods 74

2 **Kinematics of Cosserat Continuum** **78**

3 **Forces and Couples, Stress and Couple Stress Tensors in Micropolar Continua** **80**
3.1 Forces and Couples 80
3.2 Eulers Laws of Motion 82
3.3 Stress Tensor and Couple Stress Tensor 83
3.4 Principal Stresses in Micropolar Continua 87
3.5 Equations of Motion 89
3.6 Boundary-Value Problems 90

4 **Constitutive Equations of Cosserat Continua** **93**
4.1 General Principles Restricting the Constitutive Equations 93
4.2 Natural Lagrangian Strain Measures of Cosserat Continuum 95
4.3 Vectorial Parameterizations of Strain Measures . . 101
4.4 Kinetic Constitutive Equations 102
4.5 Material Symmetry Group 104
4.6 Non-Linear Micropolar Isotropic Solids 108
4.7 Physically Linear Micropolar Solids 110
4.8 Linear Micropolar Isotropic Solids 111
4.9 Constraints 113
4.10 Constitutive Inequalities 115
4.11 Micropolar Fluid 119
4.12 Some Sources of Cosserat's Constitutive Equations 120

Bibliography **122**

Cosserat-Type Shells
by H. Altenbach & Victor A. Eremeyev 131

1 **Introduction** **131**

2 **Cosserat Surface** **137**
2.1 Kinematics 137

2.2 Strain Energy Density of an Elastic Cosserat Surface 138
2.3 Principle of Virtual Work and the Equilibrium Conditions 139

3 Micropolar Shells **140**
3.1 Kinematics 141
3.2 Principle of Virtual Work and Boundary-Value Problems 143
3.3 On the Constitutive Equations 145
3.4 Compatibility Conditions 146
3.5 Variational Statements 147
3.6 Linear Theory of Micropolar Shells 149
3.7 Constitutive Restrictions for Micropolar Shells . . 152
3.8 Strong Ellipticity Condition and Acceleration Waves 155
3.9 Principle Peculiarities of the Micropolar Shell Theory 157

4 Theories of Shells and Plates by Reduction of the Three-Dimensional Micropolar Continuum **158**
4.1 Basic Equations of Three-Dimensional Linear Cosserat Continuum 158
4.2 Transition to the Two-Dimensional Equilibrium Equations: Eringen's Approach 160
4.3 Transition to the Two-Dimensional Equilibrium Equations: Other Reduction Procedures 162

5 Conclusions and Discussion **163**

Bibliography **165**

Cosserat-Type Rods
by H. Altenbach, M. Bîrsan & Victor A. Eremeyev. 179

1 Introduction **179**

2 Kinematical Model of Directed Curves 181

3 Governing Equations of the Non-Linear Theory 184

4 Constitutive Equations for Thermoelastic Porous Rods 189
4.1 Free Energy Function 190
4.2 Structure of Constitutive Tensors 190

5 Linearized Equations of Directed Rods 192
5.1 Boundary-Initial-Value Problems 193
5.2 Uniqueness of Solution 194
5.3 Existence Results in the Dynamical Theory 196

6 Statical Theory for Rods 200
6.1 Inequalities of Korn-Type for Cosserat Rods . . . 201
6.2 Existence of Solution 204
6.3 Analysis of Pure Traction Problems 206

7 Equations for Straight Rods 208
7.1 Decoupling of the Problem 209
7.2 Solution of Simple Problems 210

8 Derivation of Rods Equations from the Three-Dimensional Equations 213

9 Identification of Constitutive Coefficients for Thermo-Elastic Porous Orthotropic Rods 216
9.1 Bending and Extension of Orthotropic Rods . . . 217
9.2 Torsion of Orthotropic Rods 218
9.3 Shear Vibrations of an Orthotropic Rod 219
9.4 Problem of Thermal Deformation 221
9.5 Extension of Porous Thermo-Elastic Rods 222

10 Extended Thermodynamic Theory for Rods with Two Temperature Fields 225
10.1 Basic Laws 225

10.2 Constitutive Equations 227

10 Non-Homogeneous Rods and Composite Beams Analyzed by the Direct Approach 232

11.1 Constitutive Coefficients for Non-Homogeneous Rods 235

11.2 Effective Stiffness Properties of Composite Beams 240

12 Conclusions 244

Bibliography 244

Micromorphic Media
by S. Forest. 249

1 Introduction 249

1.1 Scope of this chapter 249

1.2 Notations 250

2 Micromorphic Continua 251

2.1 Kinematics of Micromorphic Media 251

2.2 Principle of Virtual Power 253

2.3 Elastoviscoplasticity of Micromorphic Media . . . 255

3 From a Heterogeneous Cauchy Material to a Homogeneous Equivalent Micromorphic Medium 258

3.1 Definition of the Micromorphic Degrees of Freedom 260

3.2 Higher Order Strain Measures 262

3.3 Polynomial Ansatz 263

3.4 Identification of Generalised Effective Elastic Moduli 265

3.5 Validation of the Extended Homogenisation Method 273

4 Homogenization of Micromorphic Media 274

4.1 Multiscale Asymptotic Expansion Method 274

4.2 Application to Polycrystalline Plasticity 290

Bibliography **296**

Electromagnetism and Generalized Continua
by G.A. Maugin 301

1 Introduction and Historical Perspective **301**

2 Electromagnetic Sources in Galilean Invariant Continuum Physics **306**
2.1 Maxwell's Equations 306
2.2 Ponderomotive Force and Couple in a Continuum 309

3 Deformable Magnetized Bodies with Magnetic Microstructure **312**
3.1 Model of Interactions 312
3.2 Statement of Global Balance Laws 315
3.3 Approach via the Principle of Virtual Power . . . 319
3.4 Hamiltonian Variational Formulation 321
3.5 Ferrimagnetic and Antiferromagnetic Materials . . 323
3.6 Analogy with Cosserat Continua 325
3.7 Reduction to a Model without Microstructure (Paramagnetic and Soft-Ferromagnetic Bodies) 325

4 Deformable Dielectrics with Electric-Polarization Microstructure **326**
4.1 Model of Interactions 326
4.2 Approach via the Principle of Virtual Power . . . 328
4.3 Hamiltonian Variational Principle 329
4.4 Antiferroelectric Materials 330
4.5 Analogy with Cosserat Continua 330
4.6 Reduction to a Model without Microstructure . . 331
4.7 Remark on Electric Quadrupoles 331

5 Dynamical Couplings between Deformation and Electromagnetic Microstructure **332**

5.1 Introductory Note: Resonance Coupling between Wave Modes 332
5.2 The Case of Magnetoelasticity in Ferromagnets . 335
5.3 The Case of Electroelasticity in Ferroelectrics . . 343

6 Configurational Forces in Presence of an Electromagnetic Microstructure 345
6.1 Definition 345
6.2 Reminder of a Purely Mechanical Case 346
6.3 The Ferroelectric Case 349
6.4 The Ferromagnetic Case 351

7 Conclusive Remark 353

A Reminder of Basic Equations of Generalized Mechanical Continua 353

Bibliography 355

Computational Methods for Generalised Continua
by R. deBorst. 361

1 Introduction 361

2 Isotropic Elasticity-Based Damage 362

3 Stability, Ellipticity, and Mesh Sensitivity 366
3.1 Stability and Ellipticity 366
3.2 Mesh Sensitivity 370

4 Non-Local and Gradient Damage Models 372
4.1 Non-local Damage Models 372
4.2 Gradient Damage Models 373

5 Cosserat Elasto-Plasticity 377
5.1 Cosserat Elasticity 377
5.2 Cosserat Plasticity 379
5.3 A Return-Mapping Algorithm 382

5.4 Consistent Tangent Operator 383

6 Non-Local and Gradient Plasticity **383**
6.1 Non-Local Plasticity 383
6.2 Gradient Plasticity 385

Bibliography **387**

On the Roots of Continuum Mechanics in Differential Geometry – A Review –

Paul Steinmann
Chair of Applied Mechanics, University of Erlangen-Nuremberg, Germany
E-mail: Paul.Steinmann@ltm.uni-erlangen.de

Abstract The aim of this contribution is to illustrate the roots of the geometrically nonlinear kinematics of (generalized) continuum mechanics in differential geometry. Firstly several relevant concepts from differential geometry, such as connection, parallel transport, torsion, curvature, and metric (in index notation) for holonomic and anholonomic coordinate transformations are reiterated. The notation and the selection of these topics are essentially motivated by their relation to the geometrically nonlinear kinematics of continuum mechanics. Then, secondly, the kinematics are considered from the point of view of nonlinear coordinate transformations and nonlinear point transformations, respectively. Together with the discussion on the integrability conditions for the (first-order) distortions, the concept of dislocation density tensors is introduced. After touching on the possible interpretations of nonlinear elasticity using concepts from differential geometry, a detailed discussion of the kinematics of multiplicative elastoplasticity is given. The discussion culminates in a comprehensive set of twelve different types of dislocation density tensors. Potentially, these can be used to model densities of geometrically necessary dislocations and the accompanying hardening in crystalline materials. Continuum elastoplasticity formulations of this kind fall into the class of generalized (gradient-type) plasticity models.

1 Introduction

The kinematics of geometrically nonlinear continuum mechanics is deeply rooted in differential geometry. An appreciation thereof is thus particularly illuminating. This is especially true for some generalized models of continuum mechanics, for example, gradient crystal plasticity. Here, the amount of accumulated dislocations (point defects in an otherwise perfect crystalline lattice) is typically deemed responsible for the state of hardening that the crystalline material displays. Thereby, the total amount of arrested dislocations is decomposed into *statistically stored dislocations* (SSD) and *geometrically necessary dislocations* (GND). The former are then assumed responsible for isotropic hardening. The latter are necessary to support the plastic part of the deformation and form an (additional) obstacle to further dislocation flow. Geometrically necessary dislocations may be subdivided further into dislocations responsible for a macroscopically stress free curvature of the crystal lattice, and dislocations responsible for macroscopic residual stresses, both after the removal of external loads. The illuminating relation between the stress free curvature of the crystal lattice and the part of the dislocation density that is geometrically necessary to support this curvature was establishes by Nye (1953). Both contributions to the geometrically necessary dislocations, i.e. those resulting in a stress free curvature of the crystal lattice and those resulting in residual stresses, constitute additional contributions to the hardening of the crystalline material. Thus, geometrically necessary dislocations obviously have to be taken into account when modelling of plasticity to describe the hardening behaviour more realistically and thus more accurately.

A consideration of the continuum version of geometrically necessary dislocations, i.e. the *dislocation density tensor*, in a thermodynamically consistent modelling framework inevitably results in a form of gradient crystal plasticity, see Steinmann (1996), Menzel and Steinmann (2000). The dislocation density tensor, however, is intimately related to one of the key concepts in non-Riemann differential geometry, i.e. the third-order torsion tensor as introduced by Cartan (1922). For anholonomic coordinates, as in the case of crystal plasticity, the Cartan torsion coincides moreover with the so-called anholonomic object of differential geometry. The important relation between the continuum description of dislocation density and a non-Riemann geometry was discovered by Kondo (1952) and Bilby et al. (1955); Bilby and Smith (1956). Prior to this, differential geometry was instrumental in the development of general relativity and the theory of gravitation, see Misner et al. (1998). Important contributions to the elaboration of differential geometry in this context have been made by Schouten (1954, 1989).

Kröner (1958) proposed a geometrically linear continuum theory of residual stresses based on the concept of dislocation densities. Motivated by insights into differential geometry, the corresponding extension to the geometrically nonlinear case was developed by Kröner and Seeger (1959) and Kröner (1960). It turned out that the interplay between continuum mechanics and differential geometry is extremely helpful: firstly, rather involved relations of the geometrically nonlinear kinematics of continuum mechanics such as the connection between the dislocation density and the St. Venant compatibility conditions for the strains could be clarified; and secondly, generalized continuum formulations that consider more general (point) defects besides dislocations, such as the distribution of quasi-dislocations caused, e.g., by inhomogeneous temperature distributions, electric or magnetic fields, vacancies, interstitial atoms and the like in the crystal lattice, are motivated by the existence of other, more involved, types of differential geometries, see e.g. the contributions by Anthony (1970a,b, 1971). A comprehensive account of the geometrically linearized version of the continuum theory of general defects in crystal lattices is found in de Wit (1981). Further interesting contributions to the continuum theory of dislocations are e.g. by Kondo (1964), Noll (1967), and Kröner (1981).

After the prolific developments in the 1950's to 1970's the topic became somewhat dormant, but since the 1990's there has been a renewed interest. This had to do with, on the one hand, the intense research on possibilities to overcome the pathological dependencies on the discretization that computational solutions, mainly based on the finite element method, displayed for the simulation of inelastic materials with a softening response. The incorporation (in one way or another) of gradients of the inelastic variables into the modelling has a regularizing effect that results in discretization-independent simulations, see e.g. Liebe and Steinmann (2001). On the other hand, the continuing trend towards miniaturization made clear that the inelastic response of a material especially is length scale (size) dependent. Again, size dependence can be included into the modelling by incorporating gradients of the inelastic variables. For an overview of a variety of possibilities to arrive at a generalized model of plasticity see, e.g., Hirschberger and Steinmann (2009); the micromorphic approach has recently been advocated strongly by Forest (2009) and Grammenoudis and Tsakmakis (2010).

However, purely phenomenological approaches for generalized models of continuum mechanics are somewhat unsatisfying if a clear link to the underlying physics is lacking. The plasticity of crystalline materials is a notable exception, as the mechanisms of plasticity are well understood to depend on the concepts of dislocations and dislocation flow. The flow of dislocations causes the plastic deformation process while the accumulating arrest of single dislocations represents an obstacle that has to be overcome if ongoing flow of dislocations is to occur. To better capture the underlying physics of crystalline material was the main motivation for the proposal in Steinmann (1996) to include the dislocation density tensor as an additional argument in the free energy density. As a consequence a gradient-type crystal plasticity formulation emerges. Subsequently, many more or less related formulations considering specific versions of dislocation density tensors were pursued, among them the important contributions by, e.g., Le and Stumpf (1996), Acharya and Bassani (2000), Cermelli and Gurtin (2001), Gurtin (2002), Svendsen (2002), Becker (2006), Reddy et al. (2008), Clayton et al. (2006) (and many more). Other aspects such as the gauge theory of dislocations as treated, e.g., by Lazar and Hehl (2010) or nonsingular stress and strain fields of dislocations and disclinations embedded in gradient elasticity, see Lazar and Maugin (2005) are exciting topics of current research activities.

It is the aim of this contribution to review and highlight the roots of the kinematics of this type of generalized crystal plasticity using relevant concepts of differential geometry. A comprehensive and as clear as possible exposition of relevant concepts from differential geometry, alone an interesting field in itself, serves as a strong guide for the sound formulation of physically based continuum theories. The interplay between materials science and mathematical underpinning results in a very powerful and fruitful approach. It is the hope that in this fashion the way may be paved to more complex continuum models that take into account, e.g., disclinations and further types of distributed (point) defects.

This contribution is decomposed into two major sections: In Sect. 2, the essential concepts from *differential geometry* are reviewed. Thereby, in the spirit of deduction, the present exposition is clearly in reverse to the historical developments in which, starting with the idea of an Euclidean space, the complexity was increased step by step resulting eventually in the treatment of general affine spaces. Thus, after giving an overview of various geometries of spaces in Sect. 2.1, some aspects of general manifolds are treated in Sect. 2.2. The linear connection, the concept of parallel transport and the torsion are then touched upon in Sects. 2.3, 2.4 and 2.5. Section 2.6 highlights the general concept of curvature. The previous concepts are equipped with more structure by introducing the metric in Sect. 2.7. The implications of the metric on the curvature are considered in Sect. 2.8. Section 3 applies the previously outlined concepts from differential geometry to the kinematics of *continuum mechanics*. To this end, Sect. 3.1 recalls the underlying ideas of continuum kinematics while Sect. 3.2 investigates the formulation of the distortion. In Sect. 3.3 the integrability of the distortion into a compatible vector field is analysed. The kinematics of elasticity are studied subsequently in terms of concepts from differential geometry in Sect. 3.4. Finally, in Sect. 3.5, as the main outcome of this review, the case of (crystal) elastoplasticity is treated along the same lines, and, in particular, a set of twelve different dislocation density tensors is proposed.

2 Differential Geometry

This section is intended to give a concise but self contained exposition of the here relevant basics of differential geometry as needed in the following section to discuss the kinematics of (generalized) continuum mechanics.

2.1 Overview

Differential geometry deals with the geometry of spaces, which may be characterized essentially in terms of only a few fundamental objects that will be discussed in detail in the following sections, i.e.:

- Connection $\mathcal{L}^I{}_{JK}$
  $\rightarrow$ Torsion $\mathcal{T}^I{}_{JK} = \mathcal{L}^I{}_{[JK]}$
- Curvature $\mathcal{R}^I{}_{JKL}$
- Metric $\mathcal{M}_{IJ}$

These objects then allow for the classification of (affine) geometries as outlined in Table 1. A geometry with vanishing *torsion* is called *symmetric*, a geometry with vanishing *curvature* is called *flat* (or equivalently a geometry with *teleparallelism*), and a geometry with vanishing covariant derivative of the metric with respect to the connection is called *metric*, it thus possesses a *metric connection*.

An *Euclidean space* is defined as a symmetric, flat and metric geometry; a symmetric, non-flat but metric geometry defines a *Riemann space*; a non-symmetric but flat and metric geometry defines a *Cartan space*, a *Riemann-Cartan space* is defined as a non-symmetric and non-flat but metric geometry; finally a *general affine space* may be defined as a non-symmetric, non-flat and non-metric geometry.

It is interesting to note that all of these geometries have corresponding counterparts in the kinematics of various continuum theories: the kinematics of elasticity may be considered a flat Riemann geometry, i.e. simply an Euclidean geometry; a Riemann geometry describes, e.g., the kinematics of (a somewhat exotic) continuum-disclination-based elastoplasticity; a Cartan geometry describes, e.g., the kinematics of (well-accepted) continuum-dislocation-based elastoplasticity; the kinematics of (an again exotic) continuum-disclination- and continuum-dislocation-based elastoplasticity may be regarded a Riemann-Cartan geometry; and the kinematics of the continuum version of even more general (point) defects such as the distribution of quasi-dislocations caused, e.g., by inhomogeneous temperature distributions, electric or magnetic fields, vacancies, interstitial atoms and the like may finally be considered within a general affine geometry that is

Table 1. Classification of affine geometries of spaces based on three fundamental attributes (symmetric, flat, and metric) from differential geometry.

	Symmetric	Flat	Metric
Euclid	yes	yes	yes
Riemann	yes	no	yes
Cartan	no	yes	yes
Riemann & Cartan	no	no	yes
General Affine	no	no	no

essentially characterized by a non-metric connection, see Anthony (1971) and more recently Clayton (2011).

2.2 Manifolds

Central to the following discussions is the notion of a manifold. Thereby, the key idea of a manifold is to allow for general coordinate systems and corresponding transformations between these coordinate systems, see e.g. the discussion in Marsden and Hughes (1994). Correspondingly and more formal is the following

Definition:
A smooth n_{dm}-dimensional manifold is a set $\mathcal{M}$ such that for each point $\mathcal{P} \in \mathcal{M}$ there is a subset $\mathcal{U}$ of $\mathcal{M}$ containing $\mathcal{P}$, and a one-to-one mapping called chart (coordinate system) $\{\mathcal{X}^I\}$ from $\mathcal{U}$ onto an open set in $\mathbb{R}^{n_{dm}}$. Multiple charts may be needed to cover the manifold. Coordinate transformations $\{\mathcal{X}^I\} \to \{\chi^i\}$ (on a region of $\mathcal{M}$) are infinitely differentiable, i.e. C^∞. A collection of charts covering $\mathcal{M}$ is called an atlas. □

As a simple example for a manifold consider either a circle or a sphere that can only be covered by at least two charts. Thus the corresponding atlas also consists of at least two charts. Abstracting from of our usual idea of a (Euclidean) space a manifold may also be considered as a generalized space. Thus less formal is the alternative

Definition:
A system that is assigned to n_{dm} variables $\mathcal{X}^1, \mathcal{X}^2, \cdots, \mathcal{X}^{n_{dm}}$ is a point $\mathcal{P}$ of an n_{dm}-dimensional manifold $\mathcal{M}$. The n_{dm} numbers $\mathcal{X}^1, \mathcal{X}^2, \cdots, \mathcal{X}^{n_{dm}}$ are the coordinates of the point $\mathcal{P}$. The set of all points $\mathcal{P}$ then defines the manifold $\mathcal{M}$. □

To illuminate this viewpoint consider as specific examples: (i) a mechanical system with n_{dm} generalized coordinates $\mathcal{X}^1, \mathcal{X}^2, \cdots, \mathcal{X}^{n_{dm}}$, (ii) the set of ellipsoids with $n_{dm} = 3$ half-axes $\mathcal{X}^1, \mathcal{X}^2, \mathcal{X}^3$, or (iii) as the most basic case simply the ordinary n_{dm}-dimensional Euclidean space.

Differentials Let a chart (coordinate system) consist of n_{dm} coordinates

$$\mathcal{X}^1, \mathcal{X}^2, \cdots, \mathcal{X}^{n_{dm}} := \{\mathcal{X}^I\}. \tag{1}$$

Then a coordinate transformation from the n_{dm} coordinates $\{\mathcal{X}^I\}$ to a new n_{dm}-dimensional set of coordinates $\{\chi^i\}$ is given by the (one-to-one) mapping

$$\chi^i = y^i(\{\mathcal{X}^J\}) \quad \text{with} \quad \mathcal{X}^J = \mathcal{Y}^J(\{\chi^i\}). \tag{2}$$

Consequently the chain rule allows to work out the transformation behaviour of coordinate differentials simply as

$$\mathrm{d}\chi^i = \frac{\partial y^i}{\partial \mathcal{X}^J}\,\mathrm{d}\mathcal{X}^J =: \mathcal{F}^i{}_J\,\mathrm{d}\mathcal{X}^J \quad \text{and} \quad \mathrm{d}\mathcal{X}^J = \frac{\partial \mathcal{Y}^J}{\partial \chi^i}\,\mathrm{d}\chi^i =: f^J{}_i\,\mathrm{d}\chi^i. \tag{3}$$

Please note that it is by purpose that the notation for coordinate mappings and their Jacobians resembles notation typically used in the kinematics of continuum mechanics, see Sect. 3.1. Thus to unify terminology, coordinates $\mathcal{X}^I$ and χ^i will also be addressed as material and spatial coordinates, respectively.

Gradients Consider next a (scalar-valued) field that depends on either of the n_{dm}-dimensional coordinate systems

$$\vartheta = \Theta(\{\mathcal{X}^J\}) = \theta(\{\chi^i\}) \circ y^i(\{\mathcal{X}^J\}). \tag{4}$$

Then the total differential involves the gradient of the field with respect to the coordinates

$$\mathrm{d}\vartheta = \frac{\partial \Theta}{\partial \mathcal{X}^J}\,\mathrm{d}\mathcal{X}^J = \frac{\partial \theta}{\partial \chi^i}\,\mathrm{d}\chi^i. \tag{5}$$

Thus by either using the chain rule or by incorporating the coordinate differentials as derived in Eq. (3) the transformation of gradients follows as

$$\begin{aligned} \frac{\partial \Theta}{\partial \mathcal{X}^J} &= \frac{\partial \theta}{\partial \chi^i}\frac{\partial y^i}{\partial \mathcal{X}^J} = \frac{\partial \theta}{\partial \chi^i}\mathcal{F}^i{}_J, \\ \frac{\partial \theta}{\partial \chi^i} &= \frac{\partial \Theta}{\partial \mathcal{X}^J}\frac{\partial \mathcal{Y}^J}{\partial \chi^i} = \frac{\partial \Theta}{\partial \mathcal{X}^J}f^J{}_i. \end{aligned} \tag{6}$$

In conclusion it shall be recognized carefully that differentials and gradients obey different transformation behaviours upon a change of coordinates.[1]

Co- and Contravariant Transformations We may next attach n_{dm}-dimensional tupel $\mathcal{V}^J(\{\mathcal{X}^K\})$ and $\mathcal{V}_J(\{\mathcal{X}^K\})$ to each point $\mathcal{P}$ of $\mathcal{M}$. $\mathcal{V}^J$ are denoted the contravariant coefficients (of a vector) while $\mathcal{V}_J$ are the covariant coefficients (of a covector), both evaluated at point $\mathcal{P}$ with coordinates $\{\mathcal{X}^K\}$. Obviously these have to be distinguished by their transformation behavior upon a change of coordinates:

Contravariant coefficients (of a vector) transform like differentials

$$v^i = \mathcal{F}^i{}_J \mathcal{V}^J \quad \text{and} \quad \mathcal{V}^J = f^J{}_i v^i, \tag{7}$$

whereas covariant coefficients (of a covector) transform like gradients

$$\mathcal{V}_J = v_i \mathcal{F}^i{}_J \quad \text{and} \quad v_i = \mathcal{V}_J f^J{}_i. \tag{8}$$

Tensors Sloppily speaking tensors are objects with multiple indices that respect the following

Definition:
Coefficients of tensors change in a 'proper way' with coordinate transformations. □

As an example the previously introduced vectors and covectors may be regarded as first-order tensors with transformation properties

$$u^i = \mathcal{F}^i{}_J \mathcal{U}^J \quad \text{and} \quad u_i = f^J{}_i \mathcal{U}_J. \tag{9}$$

Consequently four different types[2] of (simple) second-order tensors may be constructed from dyadic products of first-order tensors and may be distin-

[1] Recall that the coordinate basis in a manifold corresponding to the coordinate system $\{\mathcal{X}^I\}$ is denoted by $\boldsymbol{\partial}_{\mathcal{X}^I}$, whereas the dual basis is denoted by $\mathbf{d}\mathcal{X}^I$, see Marsden and Hughes (1994). Then the coordinate representation of a vector reads $\boldsymbol{V}^\sharp = \mathcal{V}^I(\boldsymbol{\partial}_{\mathcal{X}^I})$, the coordinate representation of a covector (one-form) correspondingly follows as $\boldsymbol{V}^\flat = \mathcal{V}_I \mathbf{d}\mathcal{X}^I$. It is only in an Euclidean space parameterized by curvilinear coordinates $\{\mathcal{X}^I\}$ that the coordinate and dual basis coincide with the co- and contravariant base vectors $\boldsymbol{G}_I = \boldsymbol{\partial}_{\mathcal{X}^I}$ and $\boldsymbol{G}^I = \mathbf{d}\mathcal{X}^I$ that in turn may be related to the orthonormal Cartesian base vectors $\boldsymbol{E}^A$ and $\boldsymbol{E}_A$, respectively, see Sect. 3.1.

[2] Fully contravariant, fully covariant, contra-covariant, and co-contravariant, the latter two collectively being referred to as mixedvariant, types of second-order tensors may be distinguished.

guished by their transformation behaviour

$$
\begin{array}{lllllll}
t^{ij} & := & u^i v^j & = & \mathcal{F}^i_{\ K} \mathcal{U}^K \mathcal{V}^L \mathcal{F}^j_{\ L} & =: & \mathcal{F}^i_{\ K} \mathcal{T}^{KL} \mathcal{F}^j_{\ L}, \\
t_{ij} & := & u_i v_j & = & f^K_{\ i} \mathcal{U}_K \mathcal{V}_L f^L_{\ j} & =: & f^K_{\ i} \mathcal{T}_{KL} f^L_{\ j}, \\
t^i_{\ j} & := & u^i v_j & = & \mathcal{F}^i_{\ K} \mathcal{U}^K \mathcal{V}_L f^L_{\ j} & =: & \mathcal{F}^i_{\ K} \mathcal{T}^K_{\ L} f^L_{\ j}, \\
t_i^{\ j} & := & u_i v^j & = & f^K_{\ i} \mathcal{U}_K \mathcal{V}^L \mathcal{F}^j_{\ L} & =: & f^K_{\ i} \mathcal{T}_K^{\ L} \mathcal{F}^j_{\ L}.
\end{array}
\tag{10}
$$

Clearly these transformations do also hold for general second-order tensors that are constructed from a sum of simple second-order tensors. The extension to higher-order tensors follows the same pattern and is thus straightforward.

Affine Tangent Space In general no vectors are defined in a manifold $\mathcal{M}$. However a n_{dm}-dimensional vector space (the tangent space $T_{\mathcal{P}}\mathcal{M}$), satisfying the axioms of an affine vector space[3], may be attached to each point $\mathcal{P}$ of an n_{dm}-dimensional manifold $\mathcal{M}$. It thus follows from the

Definition:
The tangent space $T_{\mathcal{P}}\mathcal{M}$ consists of all vectors $\mathcal{V}^I$ emanating from $\mathcal{P}$. □

Moreover at each point $\mathcal{P}$ a (covariant) basis of the affine tangent space denoted by $\boldsymbol{\partial}_{X^I}$ with $I = 1 \cdots n_{dm}$ may be introduced.

As elementary but already specialized examples consider 1-dimensional curves and 2-dimensional surfaces embedded into the Euclidean ambient space: Then for a parameter curve $X^I = X^I(t)$ the 1-dimensional tangent space follows from the assignment $\mathrm{d}X^I \leftrightarrow \mathrm{d}\boldsymbol{X} = \mathrm{d}X^I \boldsymbol{G}_I$. Likewise the 2-dimensional tangent space of the surface is given by its tangent plane spanned by $\boldsymbol{G}_1$ and $\boldsymbol{G}_2$. However, in general a manifold and its tangent space do not necessitate the concept of an embedding Euclidean space.

2.3 Connection

Partial Derivatives Based on the transformation rule for contravariant first-order tensors and the chain rule the partial derivatives (PD) of vectors with respect to the coordinates are computed as

$$
\begin{array}{lllllllll}
v^i & = & \mathcal{F}^i_{\ J} \mathcal{V}^J & \rightarrow & v^i_{\ ,k} & = & \mathcal{F}^i_{\ J} \mathcal{V}^J_{\ ,L} f^L_{\ k} & + & \underline{\mathcal{F}^i_{\ M,L} \mathcal{V}^M f^L_{\ k}}, \\
\mathcal{V}^I & = & f^I_{\ j} v^j & \rightarrow & \mathcal{V}^I_{\ ,K} & = & f^I_{\ j} v^j_{\ ,l} \mathcal{F}^l_{\ K} & + & \underline{f^I_{\ m,l} v^m \mathcal{F}^l_{\ K}}.
\end{array}
\tag{11}
$$

[3] In an affine vector space addition of vectors and multiplication of vectors with scalars are defined.

Likewise, based on the transformation rule for covariant first-order tensors and the chain rule the partial derivatives of covectors with respect to the coordinates follow as

$$\begin{aligned} v_i &= f^J{}_i \mathcal{V}_J &\rightarrow\quad v_{i,k} &= f^J{}_i \mathcal{V}_{J,L} f^L{}_k &+\quad \underline{\mathcal{V}_M f^M{}_{i,k}}, \\ \mathcal{V}_I &= \mathcal{F}^j{}_I v_j &\rightarrow\quad \mathcal{V}_{I,K} &= \mathcal{F}^j{}_I v_{j,l} \mathcal{F}^l{}_K &+\quad \underline{v_m \mathcal{F}^m{}_{I,K}}. \end{aligned} \tag{12}$$

It is obvious from the discussion in the preceding section and the representation in Eq. (10) that the underlined terms conflict with the transformation rules for second-order tensors. As a result it may be stated that the partial derivative of a vector or a covector does not result in a second-order tensor.

Covariant Derivatives Thus the challenge is to find a correction to the partial derivative of a vector or a covector so as to reinstall the transformation behavior of second-order tensors. As a result an alternative derivative with respect to the coordinates (indicated by a vertical bar $|$) is sought for vectors that transforms as

$$v^i{}_{|k} \doteq \mathcal{F}^i{}_J \mathcal{V}^J{}_{|L} f^L{}_k \quad \text{and} \quad \mathcal{V}^I{}_{|K} \doteq f^I{}_j v^j{}_{|l} \mathcal{F}^l{}_K. \tag{13}$$

Likewise a corresponding derivative for covectors is sought with the following transformation behaviour

$$v_{i|k} \doteq f^J{}_i \mathcal{V}_{J|L} f^L{}_k \quad \text{and} \quad \mathcal{V}_{I|K} \doteq \mathcal{F}^j{}_I v_{j|l} \mathcal{F}^l{}_K. \tag{14}$$

If such derivatives may be found the resulting operation shall be called *covariant derivative* (CD). A suited ansatz to solve the above problem is to introduce third-order objects $\mathcal{L}^I{}_{KL}$ and $\ell^i{}_{jk}$, the so-called linear (or affine) *connection*. Then the connection allows to reinstall the transformation behavior of the covariant derivatives of vectors and covectors provided the connection satisfies the following non tensorial transformation properties[4]

$$\begin{aligned} \mathcal{F}^i{}_{M,L} f^L{}_k &= \mathcal{F}^i{}_J \mathcal{L}^J{}_{ML} f^L{}_k &-\quad \ell^i{}_{nk} \mathcal{F}^n{}_M, \\ f^I{}_{m,l} \mathcal{F}^l{}_K &= f^I{}_j \ell^j{}_{ml} \mathcal{F}^l{}_K &-\quad \mathcal{L}^I{}_{NK} f^N{}_m. \end{aligned} \tag{15}$$

[4] A special case occurs whenever the covariant and the partial derivative coincide for a particular coordinate system, which is only possible for a Cartesian coordinate system in Euclidean space, i.e. in a flat manifold. Then the connection in the Cartesian coordinates vanishes identically and the connection in the transformed coordinates consequently reads

$$\ell^i{}_{jk} = F^i{}_A f^A{}_{j,k} \quad \text{and} \quad \mathcal{L}^I{}_{JK} = f^I{}_a F^a{}_{J,K}.$$

Connections of this type are also denoted as *integrable connections*, the reason for this terminology becoming clear only after the concept of curvature has been introduced.

By rearrangement these non tensorial transformation properties of the connection may also be stated equivalently as

$$
\begin{aligned}
f^{M}{}_{i,k} &= f^{M}{}_{n}\ell^{n}{}_{ik} - \mathcal{L}^{M}{}_{JL}f^{J}{}_{i}f^{L}{}_{k}, \\
\mathcal{F}^{m}{}_{I,K} &= \mathcal{F}^{m}{}_{N}\mathcal{L}^{N}{}_{IK} - \ell^{m}{}_{jl}\mathcal{F}^{j}{}_{I}\mathcal{F}^{l}{}_{K}.
\end{aligned} \tag{16}
$$

By combining the transformation behaviour of the connection in Eq. (15) with that of the partial derivative of a vector in Eq. (11) the covariant derivative of a vector is eventually given by

$$
v^{i}{}_{|j} = v^{i}{}_{,j} + \ell^{i}{}_{mj}v^{m} \quad \text{and} \quad \mathcal{V}^{I}{}_{|J} = \mathcal{V}^{I}{}_{,J} + \mathcal{L}^{I}{}_{MJ}\mathcal{V}^{M}. \tag{17}
$$

Please observe that the position for the running index m or M, repectively, and thus the precise arrangement of indices in all later expressions that involve the connection varies in the literature, however once defined as in the above it only matters to consequently stick to this convention in the sequel. Likewise the covariant derivative of a covector follows from inserting the transformation in Eq. (16) into Eq. (12) to render

$$
v_{i|j} = v_{i,j} - v_{m}\ell^{m}{}_{ij} \quad \text{and} \quad \mathcal{V}_{I|J} = \mathcal{V}_{I,J} - \mathcal{V}_{M}\mathcal{L}^{M}{}_{IJ}. \tag{18}
$$

Then the covariant derivatives of the four types of (simple) second-order tensors follow from the product rule applied to their dyadic representation

$$
\begin{aligned}
\mathcal{T}^{IJ}{}_{|K} &= \mathcal{T}^{IJ}{}_{,K} + \mathcal{L}^{I}{}_{MK}\mathcal{T}^{MJ} + \mathcal{L}^{J}{}_{MK}\mathcal{T}^{IM}, \\
\mathcal{T}^{I}{}_{J|K} &= \mathcal{T}^{I}{}_{J,K} + \mathcal{L}^{I}{}_{MK}\mathcal{T}^{M}{}_{J} - \mathcal{L}^{M}{}_{JK}\mathcal{T}^{I}{}_{M}, \\
\mathcal{T}_{IJ|K} &= \mathcal{T}_{IJ,K} - \mathcal{L}^{M}{}_{IK}\mathcal{T}_{MJ} - \mathcal{L}^{M}{}_{JK}\mathcal{T}_{IM}, \\
\mathcal{T}_{I}{}^{J}{}_{|K} &= \mathcal{T}_{I}{}^{J}{}_{,K} - \mathcal{L}^{M}{}_{IK}\mathcal{T}_{M}{}^{J} + \mathcal{L}^{J}{}_{MK}\mathcal{T}_{I}{}^{M}.
\end{aligned} \tag{19}
$$

Again these expressions do also hold for general second-order tensors that are constructed from a sum of simple second-order tensors. The covariant derivatives of higher-order tensors (and objects) follow likewise, e.g. for third-order objects as occurring in the sequel one finds

$$
\begin{aligned}
\mathcal{T}^{I}{}_{JL|K} &= \mathcal{T}^{I}{}_{JL,K} + \mathcal{L}^{I}{}_{MK}\mathcal{T}^{M}{}_{JL} - \mathcal{L}^{M}{}_{JK}\mathcal{T}^{I}{}_{ML} - \mathcal{L}^{M}{}_{LK}\mathcal{T}^{I}{}_{JM}, \\
\mathcal{T}_{IJL|K} &= \mathcal{T}_{IJL,K} - \mathcal{L}^{M}{}_{IK}\mathcal{T}_{MJL} - \mathcal{L}^{M}{}_{JK}\mathcal{T}_{IML} - \mathcal{L}^{M}{}_{LK}\mathcal{T}_{IJM}.
\end{aligned} \tag{20}
$$

Based on its definition the covariant derivative obeys a number of important rules, for example:

- The CD of scalars coincides with the PD of scalars,

- The CD obeys the distribution rule,
- The CD obeys the Leibniz (product) rule.

Proof:
The first and second rule are obvious, the proof of the last rule is based on the application of the partial derivative to the contraction of a vector and a covector into a scalar (whereby opposite but otherwise identical indices follow the Einstein summation rule)

$$[\mathcal{V}^I \mathcal{V}_I]_{,J} = \mathcal{V}^I{}_{,J} \mathcal{V}_I + \mathcal{V}^I \mathcal{V}_{I,J}. \tag{21}$$

Since based on the first rule the partial and the covariant derivatives of scalars coincide it also holds that

$$[\mathcal{V}^I \mathcal{V}_I]_{|J} = [\mathcal{V}^I{}_{,J} + \mathcal{L}^I{}_{MJ} \mathcal{V}^M] \mathcal{V}_I + \mathcal{V}^I [\mathcal{V}_{I,J} - \mathcal{V}_M \mathcal{L}^M{}_{IJ}] \doteq [V^I V_I]_{,J}. \tag{22}$$

Comparing the two results in Eqs. (21) and (22) and noting that $\mathcal{L}^I{}_{MJ} \mathcal{V}^M \mathcal{V}_I \equiv \mathcal{V}^I \mathcal{V}_M \mathcal{L}^M{}_{IJ}$ concludes the proof. □

2.4 Parallel Transport

It shall be observed that in general tangent spaces $T_\mathcal{P}\mathcal{M}$ and cotangent spaces $T^*_\mathcal{P}\mathcal{M}$ at different points $\mathcal{P}$ of a manifold $\mathcal{M}$ are not connected. However, if a covariant derivative of vectors $\mathcal{V}^I|_K = \mathcal{V}^I{}_{,K} + \mathcal{L}^I{}_{JK} \mathcal{V}^J$ and covectors $\mathcal{V}_J|_K = \mathcal{V}_{J,K} - \mathcal{V}_I \mathcal{L}^I{}_{JK}$ based on a linear connection $\mathcal{L}^I{}_{JK}$ is introduced, the notion of *parallel transport* may be defined. Thus the bundle $T\mathcal{M}$ of tangent spaces $T_\mathcal{P}\mathcal{M}$ and the bundle $T^*\mathcal{M}$ of cotangent spaces $T^*_\mathcal{P}\mathcal{M}$ constitute affinely connected spaces.

Thereby the motivation for the notion of parallel transport $\mathrm{p}\mathcal{V}^I$ of a vector $\mathcal{V}^I$ is as follows: The comparison of two vectors $\mathcal{V}^I(\{\mathcal{X}^J + \mathrm{d}\mathcal{X}^J\})$ and $\mathcal{V}^I(\{\mathcal{X}^J\})$ in two different (infinitesimal close) tangent spaces attached to $\{\mathcal{X}^J\}$ and $\{\mathcal{X}^J + \mathrm{d}\mathcal{X}^J\}$ necessitates first a parallel (back) transport of $\mathcal{V}^I(\{\mathcal{X}^J + \mathrm{d}\mathcal{X}^J\})$ to $\{\mathcal{X}^J\}$. Thereby this parallel transport is assumed proportional to $\mathcal{V}^K$ and $\mathrm{d}\mathcal{X}^J$, i.e. $\mathrm{p}\mathcal{V}^I := -\mathcal{L}^I{}_{KJ} \mathcal{V}^K \mathrm{d}\mathcal{X}^J$, the minus sign (and the sequence of indices) being convention. The argument holds likewise for covectors. From these considerations we may derive the

Definition:
The transport along a parameter curve $\mathcal{X}^J(t)$ of a vector $\mathcal{V}^I$ that is attached to a manifold is called parallel if the covariant derivative (or rather the *covariant differential* $\mathrm{D}\mathcal{V}^I$) of $\mathcal{V}^I$ vanishes

$$\mathcal{V}^I{}_{|J} = \mathcal{V}^I{}_{,J} + \mathcal{L}^I{}_{KJ} \mathcal{V}^K = 0 \quad \text{with} \quad \mathrm{D}\mathcal{V}^I := \mathcal{V}^I{}_{|J} \, \mathrm{d}\mathcal{X}^J = 0. \tag{23}$$

Thus for a parallel transport the change of the vector in the direction of the parameter curve (directional derivative) follows as

$$\mathrm{d}\mathcal{V}^I := \mathcal{V}^I{}_{,J}\,\mathrm{d}\mathcal{X}^J \equiv -\mathcal{L}^I{}_{KJ}\mathcal{V}^K\,\mathrm{d}\mathcal{X}^J =: \mathrm{p}\mathcal{V}^I. \tag{24}$$

The notation $\mathrm{p}\mathcal{V}^I$ for the parallel transport of $\mathcal{V}^I$ is motivated by simply rotating the common notation for a differential $\mathrm{d}\mathcal{V}^I$ upside down. □

As a conclusion it may be stated that for a covariant derivative the change of a vector $\mathcal{V}^I$ due to its partial derivative with respect to the coordinates, i.e.

$$\mathrm{d}\mathcal{V}^I := \mathcal{V}^I{}_{,J}\,\mathrm{d}\mathcal{X}^J = \mathcal{V}^I(\{\mathcal{X}^J + \mathrm{d}\mathcal{X}^J\}) - \mathcal{V}^I(\{\mathcal{X}^J\}) \tag{25}$$

is corrected by the contribution of the parallel transport

$$\mathrm{p}\mathcal{V}^I := -\mathcal{L}^I{}_{KJ}\mathcal{V}^K\,\mathrm{d}\mathcal{X}^J \tag{26}$$

to render the covariant differential

$$\mathrm{D}\mathcal{V}^I := \mathrm{d}\mathcal{V}^I - \mathrm{p}\mathcal{V}^I = \left[\mathcal{V}^I{}_{,J} + \mathcal{L}^I{}_{KJ}\mathcal{V}^K\right]\mathrm{d}\mathcal{X}^J. \tag{27}$$

It shall be noted that the same arguments hold likewise for covectors to render eventually

$$\mathrm{D}\mathcal{V}_J := \mathrm{d}\mathcal{V}_J - \mathrm{p}\mathcal{V}_J = \left[\mathcal{V}_{J,K} - \mathcal{V}_I\mathcal{L}^I{}_{JK}\right]\mathrm{d}\mathcal{X}^K. \tag{28}$$

2.5 Torsion

Transformation of Connection As a motivation for the introduction of the *torsion* remember that the linear (or affine) connections $\mathcal{L}^I{}_{KL}$ and $\ell^i{}_{jk}$ do not transform like a tensor, but according to Eqs. (15) and (16) transforms rather like

$$\begin{aligned} \ell^i{}_{jk} &= \mathcal{F}^i{}_I \;\; \mathcal{L}^I{}_{JK} \;\; f^J{}_j \;\; f^K{}_k \;+\; \mathcal{F}^i{}_I \;\; f^I{}_{j,k}, \\ \mathcal{L}^I{}_{JK} &= f^I{}_i \;\; \ell^i{}_{jk} \;\; \mathcal{F}^j{}_J \;\; \mathcal{F}^k{}_K \;+\; f^I{}_i \;\; \mathcal{F}^i{}_{J,K}. \end{aligned} \tag{29}$$

Observe that it is the second term in each line that conflicts with a tensorial transformation behaviour. By resorting to the following easy to proof relations for the partial derivatives of the tangent maps

$$\mathcal{F}^i{}_I f^I{}_{j,k} = -\mathcal{F}^i{}_{J,K} f^J{}_j f^K{}_k \quad \text{and} \quad f^I{}_i \mathcal{F}^i{}_{J,K} = -f^I{}_{j,k}\mathcal{F}^j{}_J\mathcal{F}^k{}_K \tag{30}$$

it is useful in the sequel to express the transformation of the connections also alternatively as

$$\begin{aligned} \ell^i{}_{jk} &= \mathcal{F}^i{}_I \; \mathcal{L}^I{}_{JK} \; f^J{}_j \; f^K{}_k \; - \; \mathcal{F}^i{}_{J,K} \; f^J{}_j \; f^K{}_k, \\ \mathcal{L}^I{}_{JK} &= f^I{}_i \; \ell^i{}_{jk} \; \mathcal{F}^j{}_J \; \mathcal{F}^k{}_K \; - \; f^I{}_{j,k} \; \mathcal{F}^j{}_J \; \mathcal{F}^k{}_K. \end{aligned} \tag{31}$$

It shall be observed carefully that these transformations of the connections are valid for *holonomic* as well as *anholonomic* coordinate transformations. Here holonomic and anholonomic refers to the *integrability* and *non-integrability* of the tangent map $\mathcal{F}^i{}_I$ (or likewise $f^I{}_i$) into a map $x^i = y^i(X^I)$ (or likewise $X^I = \mathcal{Y}^I(x^i)$).

Holonomic Transformation It is obvious from the previous discussion that the connections $\mathcal{L}^I{}_{KL}$ and $\ell^i{}_{jk}$ do not transform like third-order tensors. Under a holonomic change of coordinates, however, due to the symmetry of the second partial derivatives contained in the second terms of Eq. (29) its (right) skew symmetric contribution does

$$\begin{aligned} \mathcal{F}^i{}_{[J,K]} &= 0 \;\rightarrow\; \ell^i{}_{[jk]} = \mathcal{F}^i{}_I \; \mathcal{L}^I{}_{[JK]} \; f^J{}_j \; f^K{}_k, \\ f^I{}_{[j,k]} &= 0 \;\rightarrow\; \mathcal{L}^I{}_{[JK]} = f^I{}_i \; \ell^i{}_{[jk]} \; \mathcal{F}^j{}_J \; \mathcal{F}^k{}_K. \end{aligned} \tag{32}$$

Here, skew symmetry in an index pair is denoted by square brackets, i.e. for example $\mathcal{F}^i{}_{[J,K]} := [\mathcal{F}^i{}_{J,K} - \mathcal{F}^i{}_{K,J}]/2$.

Now as a new object, the skew symmetric part of the connection is called the (Cartan) *torsion* or rather the *torsion tensor*

$$\mathcal{T}^I{}_{JK} := \mathcal{L}^I{}_{[JK]} \quad \text{and} \quad t^i{}_{jk} := \ell^i{}_{[jk]}. \tag{33}$$

The meaning of the torsion can be highlighted by considering the situation sketched in Fig. 1, compare also to Schouten (1989). The parallel transport of two coordinate differentials $\mathrm{d}X^I$ and $\mathrm{d}\mathcal{Y}^I$ along each other results in a pentagon formed by $\mathrm{d}X^I$ and $\mathrm{d}\mathcal{Y}^I$ together with the parallel transported coordinate differentials

$$\mathrm{d}X^I_{(\circ)\rightarrow(\circ\circ)} = \mathrm{d}X^I - \mathcal{L}^I{}_{JK}\,\mathrm{d}X^J\,\mathrm{d}\mathcal{Y}^K \tag{34}$$

and

$$\mathrm{d}\mathcal{Y}^I_{(\bullet)\rightarrow(\bullet\bullet)} = \mathrm{d}\mathcal{Y}^I - \mathcal{L}^I{}_{JK}\,\mathrm{d}\mathcal{Y}^J\,\mathrm{d}X^K. \tag{35}$$

From the situation sketched in Fig. 1 it is thus clear that torsion measures the closure gap

$$\mathrm{d}X^I + \mathrm{d}\mathcal{Y}^I_{(\bullet)\rightarrow(\bullet\bullet)} - \mathrm{d}\mathcal{Y}^I - \mathrm{d}X^I_{(\circ)\rightarrow(\circ\circ)} = \tag{36}$$

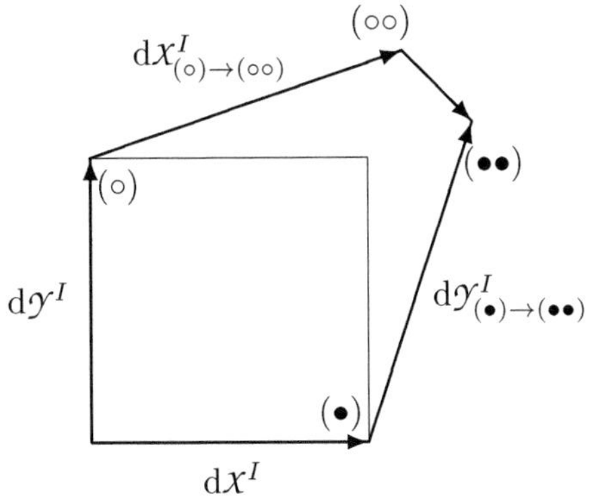

Figure 1. In a space with torsion parallel transport of coordinate differentials along each other results in a pentagon.

$$\mathcal{L}^I{}_{JK}\,\mathrm{d}X^J\,\mathrm{d}\mathcal{Y}^K - \mathcal{L}^I{}_{JK}\,\mathrm{d}\mathcal{Y}^J\,\mathrm{d}X^K = 2\mathcal{T}^I{}_{JK}\,\mathrm{d}X^J\,\mathrm{d}\mathcal{Y}^K.$$

As a result infinitesimal parallelograms constructed from coordinate differentials do only exist in spaces with vanishing torsion.

Anholonomic Transformation Recall that the connections $\mathcal{L}^I{}_{KL}$ and $\ell^i{}_{jk}$ do not transform like third-order tensors. Under an anholonomic change of coordinates its (right) skew symmetric contribution thus transforms as

$$\begin{aligned} \mathcal{F}^i{}_{[J,K]} \neq 0 \;\rightarrow\; \ell^i{}_{[jk]} + a^i{}_{jk} &= \mathcal{F}^i{}_I\;\mathcal{T}^I{}_{JK}\;f^J{}_j\;f^K{}_k, \\ f^I{}_{[j,k]} \neq 0 \;\rightarrow\; \mathcal{L}^I{}_{[JK]} + \mathcal{A}^I{}_{JK} &= f^I{}_i\;t^i{}_{jk}\;\mathcal{F}^j{}_J\;\mathcal{F}^k{}_K, \end{aligned} \tag{37}$$

whereby the torsion in the holonomic coordinates follows the standard definition

$$\mathcal{T}^I{}_{JK} = \mathcal{L}^I{}_{[JK]} \quad \text{and} \quad t^i{}_{jk} = \ell^i{}_{[jk]}. \tag{38}$$

The additional contribution appearing in the transformation due to the lack of integrability is called the *anholonomic object*:

$$a^i{}_{jk} := \mathcal{F}^i{}_{[J,K]}f^J{}_j f^K{}_k \quad \text{and} \quad \mathcal{A}^I{}_{JK} := f^I{}_{[j,k]}\mathcal{F}^j{}_J\mathcal{F}^k{}_K. \tag{39}$$

It shall be noted that in the above expressions either the coordinates χ^i in the first row of Eq. (37) or the coordinates X^I in the second row of Eq. (37) are anholonomic. Based on the anholonomic object and the representation in Eq. (37) the torsion in a space that is equipped with anholonomic coordinates follows from the

Definition:
The torsion in an anholonomic space with either anholonomic coordinate χ^i or anholonomic coordinates $\mathcal{X}^I$, respectively, is given as

$$t^i{}_{jk} := \ell^i{}_{[jk]} + a^i{}_{jk} \quad \text{and} \quad \mathcal{T}^I{}_{JK} := \mathcal{L}^I{}_{[JK]} + \mathcal{A}^I{}_{JK}. \tag{40}$$

The situation is highlighted in Fig. 2. □

It will be shown in the sequel, that the anholonomic objects may be associated with *dislocation density tensors*. Thereby, quite like in the definition of the various stress measures in nonlinear continuum mechanics, Piola-type anholonomic objects corresponding to two-point description dislocation density tensors together with Cauchy-type anholonomic objects follow from the

Definition:
The Piola-type anholonomic object corresponding to the two-point description dislocation density tensor is given by

$$\mathcal{D}^i{}_{JK} := \mathcal{F}^i{}_{[J,K]} \quad \text{and} \quad d^I{}_{jk} := f^I{}_{[j,k]}. \tag{41}$$

Consequently the anholonomic object introduced previously is of Cauchy-type and results from a convection (push-forward/pull-back) by the corre-

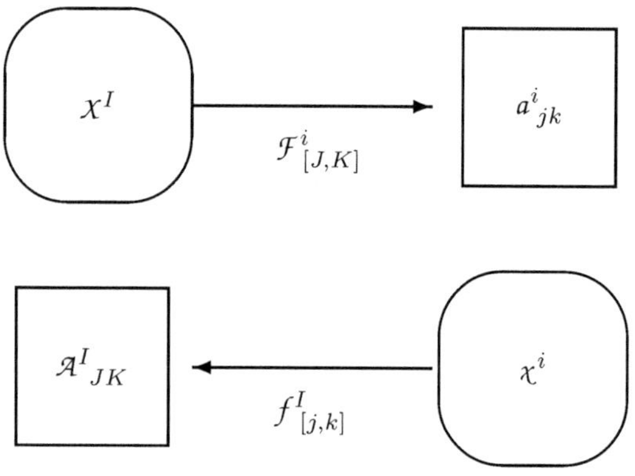

Figure 2. The anholonomic object characterizes the non-integrability of the tangent map, i.e. the transformation of coordinate differentials. In the top figure χ^i are anholonomic while $\mathcal{X}^I$ are holonomic; in the bottom figure the situation is reversed, i.e. $\mathcal{X}^I$ are anholonomic and χ^i are holonomic.

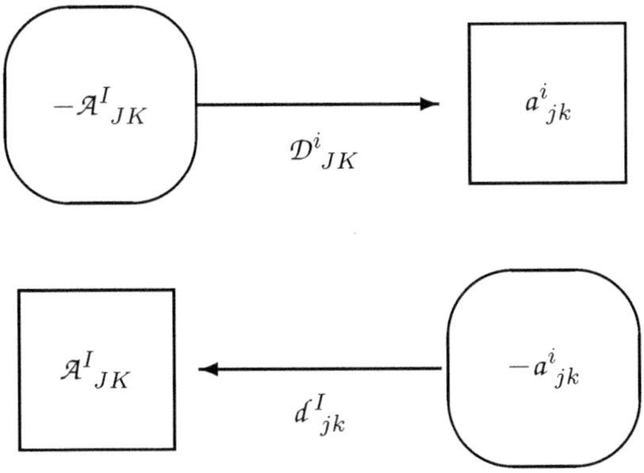

Figure 3. The anholonomic object characterizes the non-integrability of the tangent map, i.e. the transformation of coordinate differentials. In the top figure χ^i are anholonomic while $\mathcal{X}^I$ are holonomic; in the bottom figure the situation is reversed, i.e. $\mathcal{X}^I$ are anholonomic and χ^i are holonomic.

sponding tangent map

$$a^i{}_{jk} := \mathcal{D}^i{}_{JK} f^J{}_j f^K{}_k \quad \text{and} \quad \mathcal{A}^I{}_{JK} := d^I{}_{jk} \mathcal{F}^j{}_J \mathcal{F}^k{}_K. \tag{42}$$

The situation is highlighted in Fig. 3. □

Finally, Piola-Kirchhoff-type anholonomic objects may be regarded either as the pull-back/push-forward of the Piola-type or the Cauchy-type anholonomic objects due to the

Definition:
The Piola-Kirchhoff-type anholonomic object follows from the convection (pull-back/push-forward) of the Piola-type anholonomic object by the corresponding tangent map

$$-\mathcal{A}^I{}_{JK} = f^I{}_i \mathcal{D}^i{}_{JK} \quad \text{and} \quad - a^i{}_{jk} = \mathcal{F}^i{}_I d^I{}_{jk}. \tag{43}$$

It coincides with the definition of the previously introduced Cauchy-type anholonomic object if the following *anholonomic partial derivatives* are defined

$$f^I{}_i \mathcal{F}^i{}_{[J,K]} =: -f^I{}_{[j,k]} \mathcal{F}^j{}_J \mathcal{F}^k{}_K \quad \text{and} \quad \mathcal{F}^i{}_I f^I{}_{[j,k]} =: -\mathcal{F}^i{}_{[J,K]} f^J{}_j f^K{}_k. \tag{44}$$

Recall that in the above expressions either the coordinates χ^i or the coordinates $\mathcal{X}^I$ are anholonomic. The situation is highlighted in Fig. 3. □

From the above definitions it is clear that the terminology *Cauchy-type* and *Piola-Kirchhoff-type* is used interchangeably if instead of the tangent map $\mathcal{F}^i_{\ I} : \mathrm{d}\mathcal{X}^I \mapsto \mathrm{d}\chi^i$ the (inverse) tangent map $f^I_{\ i} : \mathrm{d}\chi^i \mapsto \mathrm{d}\mathcal{X}^I$ is considered.

2.6 Curvature

The notion of *curvature* or rather the *curvature tensor* is central to the differential geometry of manifolds. Formally the curvature tensor is introduced by the following

Definition:
Based on the linear connection a fourth-order object, the curvature tensor, is defined as:

$$\mathcal{R}^I_{\ JKL} := \mathcal{L}^I_{\ JL,K} - \mathcal{L}^I_{\ JK,L} + \mathcal{L}^I_{\ MK}\mathcal{L}^M_{\ JL} - \mathcal{L}^I_{\ ML}\mathcal{L}^M_{\ JK}. \tag{45}$$

The tensorial transformation properties of the curvature tensor will be demonstrated later. □

From its definition the curvature tensor obeys the following skew symmetries:

$$\mathcal{R}^I_{\ JKL} = 2\mathcal{L}^I_{\ J[L,K]} + 2\mathcal{L}^I_{\ M[K}\mathcal{L}^M_{\ JL]} = \mathcal{R}^I_{\ J[KL]}. \tag{46}$$

Note carefully that here, in contrast to most of the literature on differential geometry, the notation for skew symmetrization of the two indices in the term quadratic in the connection is used in the following format

$$2\mathcal{L}^I_{\ M[K}\mathcal{L}^M_{\ JL]} := \mathcal{L}^I_{\ MK}\mathcal{L}^M_{\ JL} - \mathcal{L}^I_{\ ML}\mathcal{L}^M_{\ JK}. \tag{47}$$

This somewhat less heavy notation is here preferred over the traditional $\mathcal{L}^I_{\ M[K}\mathcal{L}^M_{\ |J|L]}$. Less formal and more operational is the alternative

Definition:
The curvature tensor determines the change of a vector $\mathcal{V}^I$ for a parallel transport along infinitesimal closed curves as

$$\Delta\mathcal{V}^I = \mathcal{R}^I_{\ JKL}\mathcal{V}^J\,\mathrm{d}\mathcal{X}^K\,\mathrm{d}\mathcal{Y}^L. \tag{48}$$

Thus, upon transporting a vector $\mathcal{V}^I$ parallel along infinitesimal closed curves if suffers a change $\Delta\mathcal{V}^I$ that depends on the curvature tensor, i.e. on the curvature of the manifold.

Moreover, the curvature tensor also determines the skew symmetric contribution to the second covariant derivatives of a vector

$$2\mathcal{V}^I_{\ |[KL]} = -\mathcal{R}^I_{\ JKL}\mathcal{V}^J - 2\mathcal{V}^I_{\ |M}\mathcal{T}^M_{\ KL}. \tag{49}$$

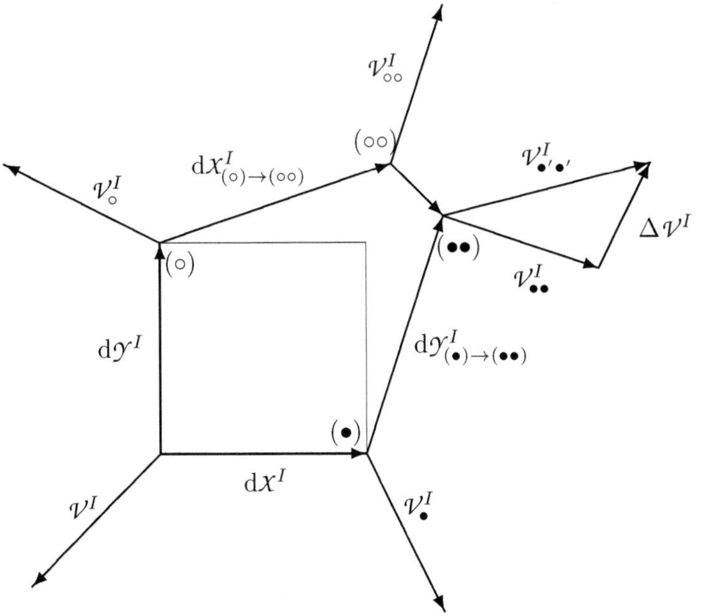

Figure 4. Parallel transport of a vector $\mathcal{V}^I$ along an infinitesimal closed curve. Due to the curvature $\mathcal{V}^I$ suffers a change $\Delta\mathcal{V}^I$. The closed curve consists of the coordinate differentials $\mathrm{d}\mathcal{X}^I$ and $\mathrm{d}\mathcal{Y}^I$ together with their parallel transport along each other $\mathrm{d}\mathcal{Y}^I_{(\bullet)\to(\bullet\bullet)}$ and $\mathrm{d}\mathcal{X}^I_{(\circ)\to(\circ\circ)}$ and the resulting closure gap, compare Fig. 1.

Observe that the second covariant derivatives of a vector involves in particular the torsion. □

Note that similar operational definitions of the curvature tensor hold in terms of covectors. The concrete representation and a proof are however left to the reader.

In the sequel both, the change of a vector $\mathcal{V}^I$ for a *parallel transport along infinitesimal closed curves* and the *skew symmetric contribution to the second covariant derivatives* of a vector shall be investigated.

Parallel Transport Along Infinitesimal Closed Curves By referring to Fig. 4 the proof of $\Delta\mathcal{V}^I = \mathcal{R}^I{}_{JKL}\mathcal{V}^J\,\mathrm{d}\mathcal{X}^K\,\mathrm{d}\mathcal{Y}^L$ may be sketched in nine steps:

1. Parallel transport of $\mathcal{V}^I$ to $(\bullet)$ and connection $\mathcal{L}^I{}_{JK\bullet}$ at $(\bullet)$,
2. Parallel transport of $\mathcal{V}^I_\bullet$ to $(\bullet\bullet)$,

3. Retain terms up to quadratic order,
4. Parallel transport of $\mathcal{V}^I$ to ($\circ$) and connection $\mathcal{L}^I{}_{JK\circ}$ at ($\circ$),
5. Parallel transport of $\mathcal{V}^I_\circ$ to ($\circ\circ$),
6. Retain terms up to quadratic order,
7. Parallel transport from ($\circ\circ$) to ($\bullet\bullet$), retain terms up to quadratic order,
8. Substract,
9. Express Result in terms of the curvature tensor.

Proof:
These steps shall now be outlined in more detail:

1. Parallel transport of $\mathcal{V}^I$ to ($\bullet$) and connection $\mathcal{L}^I{}_{JK\bullet}$ at ($\bullet$):

$$\mathcal{V}^I_\bullet = \mathcal{V}^I - \mathcal{L}^I{}_{JK}\mathcal{V}^J \,\mathrm{d}\mathcal{X}^K \quad \text{and} \quad \mathcal{L}^I{}_{JK\bullet} = \mathcal{L}^I{}_{JK} + \mathcal{L}^I{}_{JK,L}\,\mathrm{d}\mathcal{X}^L.$$

2. Parallel transport of $\mathcal{V}^I_\bullet$ to ($\bullet\bullet$):

$$\mathcal{V}^I_{\bullet\bullet} = \mathcal{V}^I_\bullet - \mathcal{L}^I{}_{JK\bullet}\mathcal{V}^J_\bullet \left[\mathrm{d}\mathcal{Y}^K - \mathcal{L}^K{}_{OP}\,\mathrm{d}\mathcal{Y}^O\,\mathrm{d}\mathcal{X}^P\right] =$$

$$\left[\mathcal{V}^I - \mathcal{L}^I{}_{JK}\mathcal{V}^J\,\mathrm{d}\mathcal{X}^K\right] - \left[\mathcal{L}^I{}_{JK} + \mathcal{L}^I{}_{JK,L}\,\mathrm{d}\mathcal{X}^L\right]\times$$
$$\left[\mathcal{V}^J - \mathcal{L}^J{}_{MN}\mathcal{V}^M\,\mathrm{d}\mathcal{X}^N\right]\times\left[\mathrm{d}\mathcal{Y}^K - \mathcal{L}^K{}_{OP}\,\mathrm{d}\mathcal{Y}^O\,\mathrm{d}\mathcal{X}^P\right].$$

3. Retain terms up to quadratic order in the coordinate differentials:

$$\mathcal{V}^I_{\bullet\bullet} = \mathcal{V}^I - \mathcal{L}^I{}_{JK}\mathcal{V}^J\,\mathrm{d}\mathcal{X}^K - \mathcal{L}^I{}_{JK}\mathcal{V}^J\,\mathrm{d}\mathcal{Y}^K - \mathcal{L}^I{}_{JK,L}\,\mathrm{d}\mathcal{X}^L\mathcal{V}^J\,\mathrm{d}\mathcal{Y}^K$$

$$+\mathcal{L}^I{}_{JK}\mathcal{L}^J{}_{MN}\mathcal{V}^M\,\mathrm{d}\mathcal{X}^N\,\mathrm{d}\mathcal{Y}^K + \mathcal{L}^I{}_{JK}\mathcal{L}^K{}_{OP}\mathcal{V}^J\,\mathrm{d}\mathcal{Y}^O\,\mathrm{d}\mathcal{X}^P.$$

4. Parallel transport of $\mathcal{V}^I$ to ($\circ$) and connection $\mathcal{L}^I{}_{JK\circ}$ at ($\circ$):

$$\mathcal{V}^I_\circ = \mathcal{V}^I - \mathcal{L}^I{}_{JK}\mathcal{V}^J\,\mathrm{d}\mathcal{Y}^K \quad \text{and} \quad \mathcal{L}^I{}_{JK\circ} = \mathcal{L}^I{}_{JK} + \mathcal{L}^I{}_{JK,L}\,\mathrm{d}\mathcal{Y}^L.$$

5. Parallel transport of $\mathcal{V}^I_\circ$ to ($\circ\circ$):

$$\mathcal{V}^I_{\circ\circ} = \mathcal{V}^I_\circ - \mathcal{L}^I{}_{JK\circ}\mathcal{V}^J_\circ\left[\mathrm{d}\mathcal{X}^K - \mathcal{L}^K{}_{OP}\,\mathrm{d}\mathcal{X}^O\,\mathrm{d}\mathcal{Y}^P\right] =$$

$$\left[\mathcal{V}^I - \mathcal{L}^I{}_{JK}\mathcal{V}^J\,\mathrm{d}\mathcal{Y}^K\right] - \left[\mathcal{L}^I{}_{JK} + \mathcal{L}^I{}_{JK,L}\,\mathrm{d}\mathcal{Y}^L\right]\times$$
$$\left[\mathcal{V}^J - \mathcal{L}^J{}_{MN}\mathcal{V}^M\,\mathrm{d}\mathcal{Y}^N\right]\times\left[\mathrm{d}\mathcal{X}^K - \mathcal{L}^K{}_{OP}\,\mathrm{d}\mathcal{X}^O\,\mathrm{d}\mathcal{Y}^P\right].$$

6. Retain terms up to quadratic order in the coordinate differentials:

$$\mathcal{V}^I_{\circ\circ} = \mathcal{V}^I - \mathcal{L}^I{}_{JK}\mathcal{V}^J\,\mathrm{d}\mathcal{Y}^K - \mathcal{L}^I{}_{JK}\mathcal{V}^J\,\mathrm{d}\mathcal{X}^K - \mathcal{L}^I{}_{JK,L}\,\mathrm{d}\mathcal{Y}^L\mathcal{V}^J\,\mathrm{d}\mathcal{X}^K$$

$$+\mathcal{L}^I{}_{JK}\mathcal{L}^J{}_{MN}\mathcal{V}^M\,\mathrm{d}\mathcal{y}^N\,\mathrm{d}\mathcal{x}^K+\mathcal{L}^I{}_{JK}\mathcal{L}^K{}_{OP}\mathcal{V}^J\,\mathrm{d}\mathcal{x}^O\,\mathrm{d}\mathcal{y}^P.$$

7. Subtract:

$$\mathcal{V}^I_{\circ\circ}-\mathcal{V}^I_{\bullet\bullet}=-2\mathcal{L}^I{}_{J[K,L]}\,\mathrm{d}\mathcal{y}^L\mathcal{V}^J\,\mathrm{d}\mathcal{x}^K$$

$$+2\mathcal{L}^I{}_{J[K}\mathcal{L}^J{}_{MN]}\mathcal{V}^M\,\mathrm{d}\mathcal{y}^N\,\mathrm{d}\mathcal{x}^K+2\mathcal{L}^I{}_{JK}\mathcal{L}^K{}_{[OP]}\mathcal{V}^J\,\mathrm{d}\mathcal{x}^O\,\mathrm{d}\mathcal{y}^P.$$

8. Parallel transport from ($\circ\circ$) to ($\bullet\bullet$), retain terms up to quadratic order in the coordinate differentials:

$$\mathcal{V}^I_{\bullet'\bullet'}=\mathcal{V}^I_{\circ\circ}-\mathcal{L}^I{}_{JK}\mathcal{V}^J\left[2\mathcal{L}^K{}_{[OP]}\,\mathrm{d}\mathcal{x}^O\,\mathrm{d}\mathcal{y}^P\right].$$

9. Subtract:

$$\mathcal{V}^I_{\bullet'\bullet'}-\mathcal{V}^I_{\bullet\bullet}=-2\mathcal{L}^I{}_{J[K,L]}\,\mathrm{d}\mathcal{y}^L\mathcal{V}^J\,\mathrm{d}\mathcal{x}^K+2\mathcal{L}^I{}_{J[K}\mathcal{L}^J{}_{MN]}\mathcal{V}^M\,\mathrm{d}\mathcal{y}^N\,\mathrm{d}\mathcal{x}^K.$$

Thus, in summary the change of the vector $\mathcal{V}^I$ may be expressed in terms of the curvature tensor as defined in Eq. (46)

$$\Delta\mathcal{V}^I:=\mathcal{V}^I_{\bullet'\bullet'}-\mathcal{V}^I_{\bullet\bullet}=:\mathcal{R}^I{}_{JKL}\mathcal{V}^J\,\mathrm{d}\mathcal{x}^K\,\mathrm{d}\mathcal{y}^L. \tag{50}$$

Clearly, from the above derivation and in accordance with the definition in Eq. (46) the curvature tensor is eventually recognized as

$$\frac{1}{2}\mathcal{R}^I{}_{JKL}=\mathcal{L}^I{}_{J[L,K]}+\mathcal{L}^I{}_{M[K}\mathcal{L}^M{}_{JL]}. \tag{51}$$

This concludes the proof. □

Skew Symmetric Contribution to Second Covariant Derivatives The second covariant derivative of a vector is computed as the covariant derivative of the (mixedvariant) second-order tensor represented by $\mathcal{V}^I{}_{|J}$, see the definition in Eq. (19).2

$$\mathcal{V}^I{}_{|JK}=\mathcal{V}^I{}_{|J,K}+\mathcal{L}^I{}_{MK}\mathcal{V}^M{}_{|J}-\mathcal{L}^M{}_{JK}\mathcal{V}^I{}_{|M}. \tag{52}$$

Involving next the covariant derivative of a vector in Eq. (17) inflates the above expression to

$$\begin{aligned}\mathcal{V}^I{}_{|JK}=\mathcal{V}^I{}_{,JK}\quad &+\quad \mathcal{L}^I{}_{MJ}\mathcal{V}^M{}_{,K}+\mathcal{L}^I{}_{MJ,K}\mathcal{V}^M\\ &+\quad \mathcal{L}^I{}_{MK}\mathcal{V}^M{}_{,J}+\mathcal{L}^I{}_{MK}\mathcal{L}^M{}_{NJ}\mathcal{V}^N\\ &-\quad \mathcal{L}^M{}_{JK}\mathcal{V}^I{}_{|M}.\end{aligned} \tag{53}$$

Likewise, changing the sequence of the indices JK renders the corresponding result

$$\begin{aligned} \mathcal{V}^I{}_{|KJ} = \mathcal{V}^I{}_{,KJ} &+ \mathcal{L}^I{}_{MK}\mathcal{V}^M{}_{,J} + \mathcal{L}^I{}_{MK,J}\mathcal{V}^M \\ &+ \mathcal{L}^I{}_{MJ}\mathcal{V}^M{}_{,K} + \mathcal{L}^I{}_{MJ}\mathcal{L}^M{}_{NK}\mathcal{V}^N \\ &- \mathcal{L}^M{}_{KJ}\mathcal{V}^I{}_{|M}. \end{aligned} \tag{54}$$

Finally, substracting the two results in Eqs. (53) and (54) and taking into account the symmetry of the second partial derivatives renders the skew symmetric contribution to the second covariant derivative of a vector in terms of the curvature and the torsion

$$\mathcal{V}^I{}_{|[JK]} = \underbrace{\left[\mathcal{L}^I{}_{N[J,K]} + \mathcal{L}^I{}_{M[K}\mathcal{L}^M{}_{NJ]}\right]}_{-\mathcal{R}^I{}_{NJK}/2} \mathcal{V}^N - \mathcal{T}^M{}_{JK}\mathcal{V}^I{}_{|M}. \tag{55}$$

A similar result may be derived for the second covariant derivative of a covector.

Transformation of the Curvature Tensor Due to the tensor property of the curvature the following convection or rather pull-back ($\mathcal{Y}$)/push-forward ($\mathcal{y}$) relations hold in the case of holonomic coordinate transformations:

Curvature($\mathcal{Y}$(connection))=$\mathcal{Y}$(curvature(connection))

curvature($\mathcal{y}$(Connection))=$\mathcal{y}$(Curvature(Connection))

As an example the pull-back of the spatial curvature expressed in terms of the spatial connection equals the material curvature expressed in terms of the pull-back of the spatial connection. The corresponding relation holds if spatial and material objects are exchanged. However, in the case of anholonomic coordinate transformations extra contributions in terms of the anholonomic object arise.

The tensorial transformation of the curvature tensor upon changing the coordinate system between holonomic coordinates $\mathcal{X}^I$ and anholonomic coordinates χ^i is stated as

$$\mathcal{R}^I{}_{JKL} = f^I{}_i r^i{}_{jkl} \mathcal{F}^j{}_J \mathcal{F}^k{}_K \mathcal{F}^l{}_L. \tag{56}$$

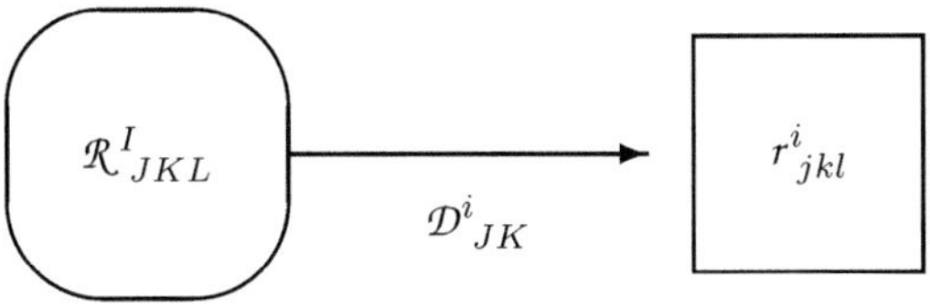

Figure 5. Transformation of the curvature tensor for the case of holonomic $\mathscr{X}^I$ and anholonomic $\mathscr{x}^i$.

Thereby, for holonomic $\mathscr{X}^I$ and anholonomic $\mathscr{x}^i$ the curvature tensor $\mathscr{R}^I{}_{JKL}$ follows the standard definition in Eqs. (45) and (46) whereas the curvature tensor $r^i{}_{jkl}$ involves extra contributions in terms of the connection and the anholonomic object

$$r^i{}_{jkl} = 2\ell^i{}_{j[l,k]} + 2\ell^i{}_{m[k}\ell^m{}_{jl]} + 2\ell^i{}_{jm}a^m{}_{lk}. \tag{57}$$

The situation is depicted in Fig. 5.

Likewise, the tensorial transformation of the curvature tensor upon changing the coordinate system between holonomic coordinates $\mathscr{x}^i$ and anholonomic coordinates $\mathscr{X}^I$ is stated as

$$r^i{}_{jkl} = \mathscr{F}^i{}_I \mathscr{R}^I{}_{JKL} f^J{}_j f^K{}_k f^L{}_l. \tag{58}$$

Then, for holonomic $\mathscr{x}^i$ and anholonomic $\mathscr{X}^I$ the curvature tensor $r^i{}_{jkl}$ follows the standard definition corresponding to Eqs. (45) and (46) whereas the curvature tensor $\mathscr{R}^I{}_{JKL}$ involves extra contributions in terms of the connection and the anholonomic object

$$\mathscr{R}^I{}_{JKL} = 2\mathscr{L}^I{}_{J[L,K]} + 2\mathscr{L}^I{}_{M[K}\mathscr{L}^M{}_{JL]} + 2\mathscr{L}^I{}_{JM}\mathscr{A}^M{}_{LK}. \tag{59}$$

The situation is depicted in Fig. 6.

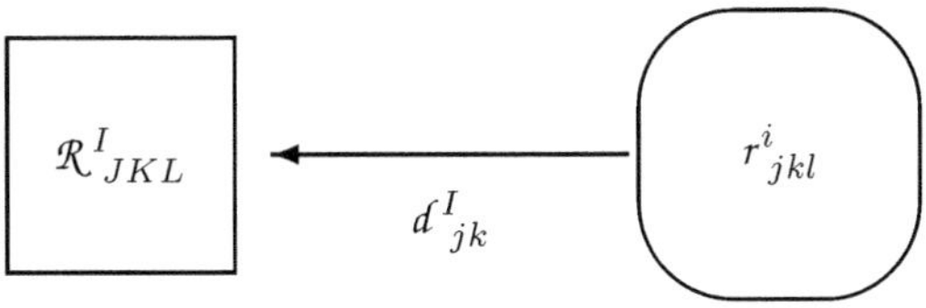

Figure 6. Transformation of the curvature tensor for the case of holonomic $\mathscr{x}^i$ and anholonomic $\mathscr{X}^I$.

Proof:
To proof the above assertions in Eqs. (57) and (59) the transformation of the connection according to Eq. (29) has to be inserted into the standard definition of the curvature tensor in Eq. (45) or (46).

As an example the case of holonomic $\mathcal{X}^I$ and anholonomic χ^i shall be considered in detail. To start with, the transformation of the connection reads as

$$\mathcal{L}^I{}_{JL} = f^I{}_i \ell^i{}_{jl} \mathcal{F}^j{}_J \mathcal{F}^l{}_L + f^I{}_i \mathcal{F}^i{}_{J,L}.$$

Computing the partial derivative of the connection as needed in the definition of the curvature renders the lengthy expression

$$\begin{aligned}
\mathcal{L}^I{}_{JL,K} &= f^I{}_{i,K} \ell^i{}_{jl} \mathcal{F}^j{}_J \mathcal{F}^l{}_L \\
&+ f^I{}_i \ell^i{}_{jl,k} \mathcal{F}^j{}_J \mathcal{F}^k{}_K \mathcal{F}^l{}_L \\
&+ f^I{}_i \ell^i{}_{jl} \mathcal{F}^j{}_{J,K} \mathcal{F}^l{}_L \\
&+ f^I{}_i \ell^i{}_{jl} \mathcal{F}^j{}_J \mathcal{F}^l{}_{L,K} \\
&+ f^I{}_{i,K} \mathcal{F}^i{}_{J,L} \\
&+ f^I{}_i \mathcal{F}^i{}_{J,LK}.
\end{aligned}$$

Unfortunately, upon skew symmetrization in L and K only one term drops out so far

$$\begin{aligned}
\mathcal{L}^I{}_{J[L,K]} &= f^I{}_{i,[K} \ell^i{}_{jl} \mathcal{F}^j{}_J \mathcal{F}^l{}_{L]} \\
&+ \underline{f^I{}_i \ell^i{}_{j[l,k]}} \mathcal{F}^j{}_J \mathcal{F}^k{}_K \mathcal{F}^l{}_L \\
&+ f^I{}_i \ell^i{}_{jl} \mathcal{F}^j{}_{J,[K} \mathcal{F}^l{}_{L]} \\
&+ f^I{}_i \ell^i{}_{jl} \mathcal{F}^j{}_J \mathcal{F}^l{}_{[L,K]} \\
&+ f^I{}_{i,[K} \mathcal{F}^i{}_{J,L]}.
\end{aligned}$$

Observe that the underlined term is already part of the sought for curvature tensor $r^i{}_{jkl}$. Next the term of the curvature quadratic in the connection shall be computed. To this end the transformation of the connection is recalled once again with the right set of indices

$$\begin{aligned}
\mathcal{L}^I{}_{MK} &= f^I{}_i \ell^i{}_{mk} \mathcal{F}^m{}_M \mathcal{F}^k{}_K + f^I{}_i \mathcal{F}^i{}_{M,K}, \\
\mathcal{L}^M{}_{JL} &= f^M{}_m \ell^m{}_{jl} \mathcal{F}^j{}_J \mathcal{F}^l{}_L + f^M{}_m \mathcal{F}^m{}_{J,L}.
\end{aligned}$$

Then multiplication of the two representations of the connection in the above results in the multi-term expression

$$\begin{aligned} & \mathcal{L}^I{}_{MK}\mathcal{L}^M{}_{JL} \\ = \; & f^I{}_i \ell^i{}_{mk} \mathcal{F}^m{}_M \mathcal{F}^k{}_K f^M{}_n \ell^n{}_{jl} \mathcal{F}^j{}_J \mathcal{F}^l{}_L \\ + \; & f^I{}_i \ell^i{}_{mk} \mathcal{F}^m{}_M \mathcal{F}^k{}_K f^M{}_n \mathcal{F}^n{}_{J,L} \\ + \; & f^I{}_i \mathcal{F}^i{}_{M,K} f^M{}_m \ell^m{}_{jl} \mathcal{F}^j{}_J \mathcal{F}^l{}_L \\ + \; & f^I{}_i \mathcal{F}^i{}_{M,K} f^M{}_m \mathcal{F}^m{}_{J,L}. \end{aligned}$$

Here many terms may be simplified by taking out multiplications of the tangent map by its inverse and by substituting partial derivatives of the tangent map by those of its inverse in the spirit of Eq. (44):

$$\begin{aligned} & \mathcal{L}^I{}_{MK}\mathcal{L}^M{}_{JL} \\ = \; & f^I{}_i \underline{\ell^i{}_{mk} \ell^m{}_{jl}} \mathcal{F}^j{}_J \mathcal{F}^k{}_K \mathcal{F}^l{}_L \\ + \; & f^I{}_i \ell^i{}_{jl} \mathcal{F}^j{}_{J,L} \mathcal{F}^l{}_K \\ - \; & f^I{}_{i,K} \ell^i{}_{jl} \mathcal{F}^j{}_J \mathcal{F}^l{}_L \\ - \; & f^I{}_{i,K} \mathcal{F}^i{}_{J,L}. \end{aligned}$$

Observe that the underlined term will be another part of the sought for curvature tensor $r^i{}_{jkl}$. Upon skew symmetrization in L and K no term drops out:

$$\begin{aligned} \mathcal{L}^I{}_{M[K}\mathcal{L}^M{}_{JL]} = \; & f^I{}_i \underline{\ell^i{}_{m[k} \ell^m{}_{jl]}} \mathcal{F}^j{}_J \mathcal{F}^k{}_K \mathcal{F}^l{}_L \\ - \; & f^I{}_i \ell^i{}_{jl} \mathcal{F}^j{}_{J,[K} \mathcal{F}^l{}_{L]} \\ - \; & f^I{}_{i,[K} \ell^i{}_{jl} \mathcal{F}^j{}_J \mathcal{F}^l{}_{L]} \\ - \; & f^I{}_{i,[K} \mathcal{F}^i{}_{J,L]}. \end{aligned}$$

However, if we combine the above results so as to produce the curvature tensor $\mathcal{R}^I{}_{JKL}$ many terms drop out and the resulting expression reads as

$$\frac{1}{2}\mathcal{R}^I{}_{JKL} := \mathcal{L}^I{}_{J[L,K]} + \mathcal{L}^I{}_{M[K}\mathcal{L}^M{}_{JL]}$$

$$= f^I{}_i \left[\ell^i{}_{j[l,k]} + \ell^i{}_{m[k}\ell^m{}_{jl]} + \ell^i{}_{jm}\mathcal{F}^m{}_{[L,K]}f^L{}_l f^K{}_k \right] \mathcal{F}^j{}_J \mathcal{F}^k{}_K \mathcal{F}^l{}_L.$$

Inserting finally the definition of the anholonomic object

$$a^m{}_{lk} := \mathcal{F}^m{}_{[L,K]} f^L{}_l f^K{}_k$$

into the above result concludes the proof. □

2.7 Metric

The *metric* is an important object that introduces more structure into a (differential) manifold as may be seen from the

Definition:
If a n_{dm}-dimensional (differentiable) manifold $\mathcal{M}$ is equipped with a symmetric field of metric coefficients $\mathcal{M}_{IJ}(\mathcal{X}^1, \mathcal{X}^2, \cdots, \mathcal{X}^{n_{dm}})$ such that the arclength of a parameter curve $\mathcal{X}^I = \mathcal{X}^I(t)$ between parameter values t_a and t_b is given by

$$S(t_b) - S(t_a) = \int_{t_a}^{t_b} \sqrt{\dot{\mathcal{X}}^I \mathcal{M}_{IJ} \dot{\mathcal{X}}^J}\, \mathrm{d}t \tag{60}$$

the manifold $\mathcal{M}$ is a metric space. Its tangent space $T_{\mathcal{P}}\mathcal{M}$ at $\mathcal{P}$ is an Euclidean (tangent) space. □

Thereby the metric shall obey the following properties:

- $\mathcal{M}_{IJ} = \mathcal{M}_{JI} = \mathcal{M}_{(IJ)}$ with $\mathcal{M}_{[IJ]} = 0$ symmetric
- $\mathcal{V}^I \mathcal{M}_{IJ} \mathcal{V}^J > 0 \quad \forall \{\mathcal{V}^K\} \neq \{0\}$ positive definite
- $\mathcal{M}_{IJ}$ transforms as 2nd-order tensor, i.e. $m_{kl} = f^I{}_k \mathcal{M}_{IJ} f^J{}_l$

The first property is obvious since any skew symmetric contributions would not contribute to a quadratic form as needed for the determination of the length. The second property is specific to the later application to (three-dimensional) continuum mechanics, relativity and general relativity allow also for indefinite metrics, see, e.g., Misner et al. (1998). Finally the proof of the third property is straightforward from the transformation behaviour of the coordinate differentials, see Eq. (3):

$$\mathrm{d}S^2 = \mathrm{d}\mathcal{X}^I \mathcal{M}_{IJ}\, \mathrm{d}\mathcal{X}^J = \mathrm{d}x^k f^I{}_k \mathcal{M}_{IJ} f^J{}_l\, \mathrm{d}x^l = \mathrm{d}x^k m_{kl}\, \mathrm{d}x^l. \tag{61}$$

Metric Connection As an immediate consequence the introduction of a metric allows to formulate the

Ricci Postulate:
In a *metric geometry* the covariant derivative of the covariant metric coefficients vanishes, i.e.

$$\mathcal{M}_{IJ|K} = \mathcal{M}_{IJ,K} - \mathcal{L}^{M}{}_{IK}\mathcal{M}_{MJ} - \mathcal{L}^{M}{}_{JK}\mathcal{M}_{IM} = 0. \tag{62}$$

From resolving the above equality the partial derivative of the metric may thus alternatively be stated in terms of the (left) symmetric part of the *fully covariant connection*

$$\mathcal{M}_{IJ,K} = \mathcal{L}_{JIK} + \mathcal{L}_{IJK} = 2\mathcal{L}_{(IJ)K}. \tag{63}$$

Note that the covariant metric has been used to lower the contravariant indices of the connection[5]. The above representation will be useful in the sequel. The underlying connection is denoted a *metric connection.* □

As a consequence a metric connection is additively decomposed into the *Riemann part of the connection* and the *contortion*:

$$\mathcal{L}^{I}{}_{JK} = \underbrace{\mathcal{M}^{I}{}_{JK}}_{Riemann} + \underbrace{\mathcal{K}^{I}{}_{JK}}_{Contortion} . \tag{64}$$

Proof:
The proof follows directly by stating the Ricci postulate three times upon cyclic permutation of the indices and subsequent addition of the resulting expressions to render:

$$\begin{array}{rcrcrcrcl}
\mathcal{M}_{IJ|K} &=& \mathcal{M}_{IJ,K} &-& \mathcal{L}^{M}{}_{IK}\mathcal{M}_{MJ} &-& \mathcal{L}^{M}{}_{JK}\mathcal{M}_{IM} &=& 0\\
-\mathcal{M}_{JK|I} &=& -\mathcal{M}_{JK,I} &+& \mathcal{L}^{M}{}_{JI}\mathcal{M}_{MK} &+& \mathcal{L}^{M}{}_{KI}\mathcal{M}_{JM} &=& 0\\
\mathcal{M}_{KI|J} &=& \mathcal{M}_{KI,J} &-& \mathcal{L}^{M}{}_{KJ}\mathcal{M}_{MI} &-& \mathcal{L}^{M}{}_{IJ}\mathcal{M}_{KM} &=& 0\\
\hline\hline
\end{array}$$

$$\mathcal{M}_{IJ,K} - \mathcal{M}_{JK,I} + \mathcal{M}_{KI,J} + 2\mathcal{M}_{JM}\mathcal{L}^{M}{}_{[KI]} - 2\mathcal{M}_{KM}\mathcal{L}^{M}{}_{[IJ]} - 2\mathcal{M}_{IM}\mathcal{L}^{M}{}_{(JK)} = 0$$

Then by dividing by 2 and by adding and subtracting the terms

$$\mathcal{M}_{IM}\mathcal{L}^{M}{}_{JK} = \mathcal{M}_{IM}\mathcal{L}^{M}{}_{(JK)} + \mathcal{M}_{IM}\mathcal{L}^{M}{}_{[JK]}$$

---

[5] The matrix arrangement of the contravariant metric coefficients $[\mathcal{M}^{IJ}]$ follows from the (algebraic) inversion of the corresponding matrix arrangement of the covariant metric coefficients $[\mathcal{M}_{IJ}]$. The contravariant metric coefficients $\mathcal{M}^{IJ}$ may be used to raise covariant indices.

the result follows immediately as

$$\begin{aligned} \mathcal{M}_{IM}\mathcal{L}^{M}{}_{JK} &= \frac{1}{2}\left[\mathcal{M}_{IJ,K} - \mathcal{M}_{JK,I} + \mathcal{M}_{KI,J}\right] \\ &+ \mathcal{M}_{IM}\mathcal{L}^{M}{}_{[JK]} + \mathcal{M}_{JM}\mathcal{L}^{M}{}_{[KI]} - \mathcal{M}_{KM}\mathcal{L}^{M}{}_{[IJ]}. \end{aligned}$$

This concludes the proof. □

Thereby the Riemann part of the connection or in short simply the *Riemann connection* follows from the

Definition:
The Riemann connection is exclusively computed from the metric coefficients:

$$\mathcal{M}_{IJK} := \frac{1}{2}\left[\mathcal{M}_{IJ,K} - \mathcal{M}_{JK,I} + \mathcal{M}_{KI,J}\right]. \tag{65}$$

As will be demonstrated below the Riemann connection is a metric connection. □

It corrresponds to the Christoffel symbols in Euclidean space and obeys the transformation behavior of the connection, thus it is not a third-order tensor. It may be shown by inspection that, based on the symmetry of the metric coefficients, its (right) skew symmetric contribution vanishes:

$$\mathcal{M}_{I[JK]} = 0. \tag{66}$$

Likewise the contortion is given in the

Definition:
The contortion depends (linearly) on the metric coefficients and in particular on the (Cartan) torsion tensor:

$$\mathcal{K}_{IJK} := \mathcal{T}_{IJK} + \mathcal{T}_{JKI} - \mathcal{T}_{KIJ}. \tag{67}$$

The contortion inherits from the torsion the transformation properties of a tensor, thus it is also denoted the *contortion tensor.* □

It may be shown by inspection, based on the symmetry of the metric coefficients and the skew symmetry of the torsion tensor, that the (left) symmetric contribution to the contortion vanishes:

$$\mathcal{K}_{(IJ)K} = 0. \tag{68}$$

Here, symmetry in an index pair is denoted by round brackets, i.e. for example $\mathcal{K}_{(IJ)K} := [\mathcal{K}_{IJK} + \mathcal{K}_{JIK}]/2$. Thus the conjecture in the above that the Riemann connection is also a metric connection is obviously proven by

$$\mathcal{M}_{IJ,K} \doteq 2\mathcal{L}_{(IJ)K} = 2\mathcal{M}_{(IJ)K} + 2\mathcal{K}_{(IJ)K} \equiv 2\mathcal{M}_{(IJ)K}. \tag{69}$$

Observe that as a consequence of the decomposition of the connection into Riemann connection and contortion tensor together with the (right) symmetry of the Riemann connection

$$\mathcal{L}_{IJK} = \mathcal{M}_{IJK} + \mathcal{K}_{IJK} \quad \text{and} \quad \mathcal{M}_{I[JK]} = 0 \tag{70}$$

the (right) skew symmetric contribution of the contortion tensor equals the torsion tensor

$$\mathcal{T}_{IJK} := \mathcal{L}_{I[JK]} = \mathcal{K}_{I[JK]}. \tag{71}$$

It is also interesting to note that due to its left skew symmetry the contortion tensor does neither contribute to the projection of the parallel transport of a vector

$$\mathcal{V}_I \,\mathrm{p}\mathcal{V}^I = -\mathcal{V}^L \mathcal{V}^K \mathcal{L}_{LKJ} \,\mathrm{d}\mathcal{X}^J = -\mathcal{V}^L \mathcal{V}^K \mathcal{M}_{LKJ} \,\mathrm{d}\mathcal{X}^J \tag{72}$$

nor to the projection of the parallel transport of a covector

$$\mathcal{V}^J \,\mathrm{p}\mathcal{V}_J = \mathcal{V}^J \mathcal{V}^N \mathcal{L}_{NJK} \,\mathrm{d}\mathcal{X}^K = \mathcal{V}^J \mathcal{V}^N \mathcal{M}_{NJK} \,\mathrm{d}\mathcal{X}^K. \tag{73}$$

Length As a consequence it may easily been proven that in a metric space the (quadratic) length $\mathcal{V}^2 = \mathcal{V}^I \mathcal{M}_{IJ} \mathcal{V}^J$ of a vector $\mathcal{V}^I$ is preserved upon an infinitesimal parallel transport along a parameter curve $\mathcal{X}^K(t)$ since firstly

$$\mathrm{p}[\mathcal{V}^2] = \mathcal{M}_{IJ,K} \mathcal{V}^I \mathcal{V}^J \,\mathrm{d}\mathcal{X}^K + \mathcal{M}_{IJ} \,\mathrm{p}\mathcal{V}^I \mathcal{V}^J + \mathcal{M}_{IJ} \mathcal{V}^I \,\mathrm{p}\mathcal{V}^J. \tag{74}$$

Using next the definition of parallel transport then renders

$$\mathrm{p}[\mathcal{V}^2] = \left[\mathcal{M}_{IJ,K} - \mathcal{L}^M{}_{IK} \mathcal{M}_{MJ} - \mathcal{L}^M{}_{JK} \mathcal{M}_{IM}\right] \mathcal{V}^I \mathcal{V}^J \,\mathrm{d}\mathcal{X}^K. \tag{75}$$

Including finally the definition of a metric connection thus results in

$$\mathrm{p}[\mathcal{V}^2] = \left[\mathcal{M}_{IJ,K} - 2\mathcal{L}_{(IJ)K}\right] \mathcal{V}^I \mathcal{V}^J \,\mathrm{d}\mathcal{X}^K \equiv 0. \tag{76}$$

Observe again that the contortion does not contribute, obviously this result is in agreement with Eqs. (72) and (73).

Reduction of Torsion and Contortion in 3d In later sections the previous concepts of manifolds and differential geometry shall be applied to the geometrically nonlinear kinematics of continuum mechanics, thus all considerations may be restricted to three-dimensional metric spaces. Then, in 3d, and due to its (right) skew symmetry the third-order fully covariant torsion tensor

$$\mathcal{T}_{IMN} = \mathcal{T}_{I[MN]} \tag{77}$$

may equivalently be represented by the second-order mixed-variant torsion tensor

$$\mathcal{T}_I{}^J := -\mathcal{T}_{I[OP]}\mathcal{E}^{OPJ} = -\mathcal{T}_{IOP}\mathcal{E}^{OPJ}. \tag{78}$$

Here, $\mathcal{E}^{OPJ}$ denotes the third-order, fully skew symmetric permutation symbol [6]. Note that the scaling by the factor $\frac{1}{2}$ as common for the notion of two-forms (see, e.g., Lazar and Hehl (2010)) has been omitted to obtain an easier comparison to the curl operator that will be introduced later in Sect. 3.3. With the properties of the permutation symbol the inverse relation (now including the factor $\frac{1}{2}$) follows immediately as

$$\mathcal{T}_{IMN} = -\frac{1}{2}\mathcal{T}_I{}^J\mathcal{E}_{JMN}. \tag{79}$$

Likewise, in 3d due to its (left) skew symmetry the third-order fully covariant contortion tensor

$$\mathcal{K}_{IJN} = \mathcal{K}_{[IJ]N} \tag{80}$$

may equivalently be represented by the second-order mixed-variant contortion tensor

$$\mathcal{K}^M{}_N := -\frac{1}{2}\mathcal{E}^{MOP}\mathcal{K}_{[OP]N} = -\frac{1}{2}\mathcal{E}^{MOP}\mathcal{K}_{OPN}. \tag{81}$$

Here the definition follows the usual scaling by the factor $\frac{1}{2}$ known from the notion of two-forms, see Lazar and Hehl (2010). Again with the properties

[6] The co- and contravariant permutation symbols satisfy the following relations upon a single contraction

$$\mathcal{E}^{IJM}\mathcal{E}_{MKL} = \delta^I{}_K\delta^J{}_L - \delta^I{}_L\delta^J{}_K \quad \text{and} \quad \mathcal{E}_{IJM}\mathcal{E}^{MKL} = \delta_I{}^K\delta_J{}^L - \delta_I{}^L\delta_J{}^K.$$

Likewise, upon a double contraction the co- and contravariant permutation symbols satisfy

$$\mathcal{E}^{IMN}\mathcal{E}_{MNJ} = 2\delta^I{}_J \quad \text{and} \quad \mathcal{E}_{IMN}\mathcal{E}^{MNJ} = 2\delta_I{}^J.$$

Finally, a triple contraction of the co- and contravariant permutation symbols renders

$$\mathcal{E}^{IJK}\mathcal{E}_{IJK} = 6 \quad \text{and} \quad \mathcal{E}_{IJK}\mathcal{E}^{IJK} = 6.$$

of the permutation symbol the inverse relation follows as

$$\mathcal{K}_{IJN} = -\mathcal{E}_{IJM}\mathcal{K}^{M}{}_{N}. \tag{82}$$

Recall that the (right) skew symmetry part of the contortion tensor coincides with the torsion tensor

$$\mathcal{K}_{I[JK]} = \mathcal{T}_{IJK}. \tag{83}$$

Thus, as a consequence, in 3d the second-order mixed-variant torsion and contortion tensors are related by:

$$\mathcal{T}_{I}{}^{J} = \mathcal{K}^{M}{}_{M}\delta_{I}{}^{J} - \mathcal{K}^{J}{}_{I}. \tag{84}$$

Finally, the inverse relation reads

$$\mathcal{K}^{J}{}_{I} = \frac{1}{2}\mathcal{T}_{M}{}^{M}\delta_{I}{}^{J} - \mathcal{T}_{I}{}^{J}. \tag{85}$$

It turns out that this is the analogue in differential geometry to the relation between the stress free curvature of a crystalline lattice and the density of dislocations geometrically necessary to support this irreversible deformation as discovered by Nye (1953).

Reduction of Anholonomic Objects in 3d Moreover it is observed that, due to their (right) skew symmetry, the (Cauchy- and Piola-type) anholonomic objects

$$a^{i}{}_{op} = \mathcal{D}^{i}{}_{OP}f^{O}{}_{o}f^{P}{}_{p} \quad \text{and} \quad \mathcal{A}^{I}{}_{OP} = d^{I}{}_{op}\mathcal{F}^{o}{}_{O}\mathcal{F}^{p}{}_{P} \tag{86}$$

may equivalently be represented in 3d by second-order tensors

$$a^{ij} := -a^{i}{}_{op}e^{opj} \quad \text{and} \quad \mathcal{D}^{iJ} := -\mathcal{D}^{i}{}_{OP}\mathcal{E}^{OPJ}, \tag{87}$$

$$\mathcal{A}^{IJ} := -\mathcal{A}^{I}{}_{OP}\mathcal{E}^{OPJ} \quad \text{and} \quad d^{Ij} := -d^{I}{}_{op}e^{opj}.$$

with the inverse relation

$$a^{i}{}_{mn} = -\frac{1}{2}a^{ij}e_{jmn} \quad \text{and} \quad \mathcal{D}^{i}{}_{MN} = -\frac{1}{2}\mathcal{D}^{iJ}\mathcal{E}_{JMN}, \tag{88}$$

$$\mathcal{A}^{I}{}_{MN} = -\frac{1}{2}\mathcal{A}^{IJ}\mathcal{E}_{JMN} \quad \text{and} \quad d^{I}{}_{mn} = -\frac{1}{2}d^{Ij}e_{jmn}.$$

Thus, the second-order anholonomic objects are related by

$$a^{ij} = -\mathcal{D}^{i}{}_{OP}f^{O}{}_{o}f^{P}{}_{p}e^{opj} = -j\mathcal{D}^{i}{}_{OP}\mathcal{E}^{OPJ}\mathcal{F}^{j}{}_{J} = j\mathcal{D}^{iJ}\mathcal{F}^{j}{}_{J}, \tag{89}$$

$$\mathcal{A}^{IJ} = -d^I_{\ op}\mathcal{F}^o_{\ O}\mathcal{F}^p_{\ P}\mathcal{E}^{OPJ} = -\mathcal{J}d^I_{\ op}e^{opj}f^J_{\ j} = \mathcal{J}d^{Ij}f^J_{\ j}.$$

Here the transformation behaviour of the third-order permutation symbol upon a change of coordinates has been involved

$$f^O_{\ o}f^P_{\ p}e^{opj} = j\mathcal{E}^{OPJ}\mathcal{F}^j_{\ J} \quad \text{and} \quad \mathcal{F}^o_{\ O}\mathcal{F}^p_{\ P}\mathcal{E}^{OPJ} = \mathcal{J}e^{opj}f^J_{\ j}, \tag{90}$$

whereby $j := \det[f^I_{\ i}]$ and $\mathcal{J} := \det[\mathcal{F}^i_{\ I}]$ denote the Jacobian determinants of the matrix arrangement of the tangent maps $f^I_{\ i}$ and $\mathcal{F}^i_{\ I}$, respectively. Note that the relations between the second-order Cauchy- and Piola-type anholonomic objects in Eq. (89) correspond to Piola transformations well-established in geometrically nonlinear continuum mechanics.

2.8 Metric Curvature

The presence of a metric has considerable impact on the possible representations of the curvature tensor, in particular a *fully covariant version of the curvature tensor* follows from the

Definition:
The fully covariant curvature tensor is obtained by lowering the first index

$$\mathcal{R}_{IJKL} := \mathcal{M}_{IM}\mathcal{R}^M_{\ \ JKL}. \tag{91}$$

Thus from the definition of the mixedvariant curvature tensor it follows:

$$\mathcal{R}_{IJKL} = 2\mathcal{M}_{IM}\mathcal{L}^M_{\ \ J[L,K]} + 2\mathcal{L}_{IM[K}\mathcal{L}^M_{\ \ JL]}. \tag{92}$$

As a result of the subsequent step by step proof the fully covariant curvature tensor is eventually obtained explicitly as

$$\mathcal{R}_{IJKL} = 2\mathcal{L}_{IJ[L,K]} - 2\mathcal{L}_{MI[K}\mathcal{L}^M_{\ \ JL]}. \tag{93}$$

Observe the different sequence of indices in the term quadratic in the connection and its difference in sign for the mixed- and the covariant curvature tensors, respectively. □

Proof:
The proof of the explicit representation for the covariant curvature follows in four steps:

Step 1: Application of the product rule to the first term in Eq. (92) and index lowering

$$\mathcal{M}_{IM}\mathcal{L}^M_{\ \ J[L,K]} = \mathcal{L}_{IJ[L,K]} - \mathcal{M}_{IM,[K}\mathcal{L}^M_{\ \ JL]}.$$

Step 2: Exploitation of the connection being a metric connection

$$\mathcal{M}_{IM,K} = \mathcal{L}_{MIK} + \mathcal{L}_{IMK}.$$

Step 3: Insert result of step 2 into the second term of the result in step 1

$$\mathcal{M}_{IM,[K}\mathcal{L}^{M}{}_{JL]} = \left[\mathcal{L}_{MI[K} + \mathcal{L}_{IM[K}\right]\mathcal{L}^{M}{}_{JL]}.$$

Step 4: Insert the result of steps 1 to 3 into the definition in Eq. (92)

$$\mathcal{M}_{IM}\mathcal{L}^{M}{}_{J[L,K]} + \mathcal{L}_{IM[K}\mathcal{L}^{M}{}_{JL]} = \mathcal{L}_{IJ[L,K]} - \mathcal{L}_{MI[K}\mathcal{L}^{M}{}_{JL]}.$$

This concludes the proof. □

It shall be noted further that the fully covariant curvature tensor obeys the following left and right (minor) skew symmetries

$$\mathcal{R}_{IJKL} = \mathcal{R}_{IJ[KL]} = \mathcal{R}_{[IJ][KL]}. \tag{94}$$

Proof:
The skew symmetry in the index pair KL is a direct result from the definition of the curvature tensor in Eq. (93). The assertion of the skew symmetry in the index pair IJ can be proven easily in two steps:

Step 1: Exploit the connection being metric together with the symmetry of second partial derivatives

$$\mathcal{M}_{IJ,KL} = \mathcal{L}_{JIK,L} + \mathcal{L}_{IJK,L} = \mathcal{L}_{JIL,K} + \mathcal{L}_{IJL,K},$$

$$\rightarrow \qquad \mathcal{L}_{IJ[L,K]} = -\mathcal{L}_{JI[L,K]}.$$

Step 2: Rearranging the skew symmetry in the indices of the quadratic term

$$2\mathcal{L}_{MI[K}\mathcal{L}^{M}{}_{JL]} = \mathcal{L}_{MIK}\mathcal{L}^{M}{}_{JL} - \mathcal{L}_{MIL}\mathcal{L}^{M}{}_{JK} = -2\mathcal{L}_{MJ[K}\mathcal{L}^{M}{}_{IL]}.$$

This concludes the proof. □

Reduction of Curvature Tensor in 3d Since the coefficients of the forth-order fully covariant curvature tensor based on a metric connection obey the (minor) skew symmetries

$$\mathcal{R}_{IJKL} = \mathcal{R}_{IJ[KL]} = \mathcal{R}_{[IJ][KL]}. \tag{95}$$

they may likewise be represented in 3d by the second-order contravariant *Einstein tensor*

$$\mathcal{E}^{MN} := \frac{1}{4}\mathcal{E}^{MIJ}\mathcal{R}_{IJKL}\mathcal{E}^{KLN}. \tag{96}$$

Vice versa the forth-order curvature tensor may be extracted from the second-order Einstein tensor

$$\mathcal{R}_{IJKL} = \mathcal{E}_{IJM}\mathcal{E}^{MN}\mathcal{E}_{NKL}. \tag{97}$$

Furthermore the corresponding (covariant) axial vector of $\mathcal{E}^{MN}$ follows as

$$\mathcal{E}_I = -\frac{1}{2}\mathcal{E}^{MN}\mathcal{E}_{MNI} = \frac{1}{4}\mathcal{R}_{IJKL}\mathcal{E}^{JKL}. \tag{98}$$

Raising finally the index renders in addition its contravariant version

$$\mathcal{E}^I = \mathcal{M}^{IM}\mathcal{E}_M = \frac{1}{4}\mathcal{R}^I{}_{JKL}\mathcal{E}^{JKL}. \tag{99}$$

Detailed representations equivalent to the skew part of $\mathcal{E}^{MN}$ are therefore[7]

$$\begin{aligned} \mathcal{E}_I = \frac{1}{4}\mathcal{R}_{IJKL}\mathcal{E}^{JKL} &= \frac{1}{2}\left[\mathcal{T}_I{}^M{}_{|M} - 2\mathcal{T}_{IMN}\mathcal{T}^{MN}\right], \\ \mathcal{E}^I = \frac{1}{4}\mathcal{R}^I{}_{JKL}\mathcal{E}^{JKL} &= \frac{1}{2}\left[\mathcal{T}^{IM}{}_{|M} - 2\mathcal{T}^I{}_{MN}\mathcal{T}^{MN}\right]. \end{aligned} \tag{100}$$

Thus, as can be read of this representation, the Einstein tensor is in particular symmetric for vanishing torsion, since its axial vector then vanishes identically, i.e.

$$\mathcal{E}_I = 0 \quad \text{and} \quad \mathcal{E}^I = 0. \tag{101}$$

However, Eq. (101) is also true for a flat geometry with vanishing curvature tensor, since then $\mathcal{R}^I{}_{JKL} = 0$ identically. The above result is called the (first or algebraic) *Bianchi identity* (not to be confused with a number of differential Bianchi identities, see Schouten (1954, 1989)).

Proof:
The proof of this assertion follows in three steps:

Step 1: Write down the curvature tensor as being defined before

$$\begin{aligned} \mathcal{R}_{IJKL} &= 2\mathcal{L}_{IJ[L,K]}\underline{-2\mathcal{L}_{MI[K}\mathcal{L}^M{}_{JL]}}, \\ \mathcal{R}^I{}_{JKL} &= 2\mathcal{L}^I{}_{J[L,K]}\underline{+2\mathcal{L}^I{}_{M[K}\mathcal{L}^M{}_{JL]}}. \end{aligned}$$

[7] Note that here the covariant derivative is based on the connection $\mathcal{L}^I{}_{JK}$. Thus terms as, e.g., $\mathcal{T}^M{}_{I|M}$ do not coincide with the divergence in Euclidean space unless $\mathcal{L}^I{}_{JK}$ coincides with the appropriate Christoffel symbols. This distinction is of considerable importance when discussing the geometrically nonlinear kinematics of continuum mechanics in Sect. 3.1.

Step 2: Relate the partial derivative of the connection to its covariant derivative

$$
\begin{aligned}
2\mathcal{L}_{IJ[L,K]} &= 2\mathcal{L}_{IJ[L|K]} \underline{+2\mathcal{L}_{MI[K}\mathcal{L}^M{}_{JL]}} - 2\mathcal{L}_{IM[K}\mathcal{L}^M{}_{JL]} - 2\mathcal{L}_{IJM}\mathcal{L}^M{}_{[KL]}, \\
2\mathcal{L}^I{}_{J[L,K]} &= 2\mathcal{L}^I{}_{J[L|K]} \underline{-2\mathcal{L}^I{}_{M[K}\mathcal{L}^M{}_{JL]}} - 2\mathcal{L}^I{}_{M[K}\mathcal{L}^M{}_{JL]} - 2\mathcal{L}^I{}_{JM}\mathcal{L}^M{}_{[KL]}.
\end{aligned}
$$

Step 3: Combine the results in step 1 and step 2 and note that the underlined terms cancel out to produce an alternative representation of the curvature tensor in terms of the covariant rather than the partial dervative

$$
\begin{aligned}
\mathcal{R}_{IJKL} &= 2\mathcal{L}_{IJ[L|K]} - 2\mathcal{L}_{IM[K}\mathcal{L}^M{}_{JL]} - 2\mathcal{L}_{IJM}\mathcal{T}^M{}_{KL}, \\
\mathcal{R}^I{}_{JKL} &= 2\mathcal{L}^I{}_{J[L|K]} - 2\mathcal{L}^I{}_{M[K}\mathcal{L}^M{}_{JL]} - 2\mathcal{L}^I{}_{JM}\mathcal{T}^M{}_{KL}.
\end{aligned}
$$

The proof is concluded by projecting the last result with $\mathcal{E}^{JKL}$ and exploiting $\mathcal{E}^{JKL}{}_{|M} = 0$ (being true since $\mathcal{E}^{JKL}$ is scaled by $\det[\mathcal{M}_{IJ}]$). □

Riemann Curvature Tensor In a Riemann geometry, since there is no torsion, the fully covariant curvature tensor or rather the Riemann curvature tensor is computed from the Riemann connection solely. The Riemann connection in turn depends only on the metric, see Eq. (65). Besides the (minor) skew symmetries discussed in the above it turns out that it obeys the following additional (major) symmetry for exchanging index pairs

$$
\mathcal{M}_{[IJ][KL]} = \mathcal{M}_{[KL][IJ]}. \tag{102}
$$

Thus, as an immediate result, the corresponding Einstein tensor associated to the Riemann curvature tensor is always symmetric.

Proof:
The proof of the (major) symmetry assertion follows directly in two steps:

Step 1: Exploit the structure of the Riemann connection, and take into account the symmetry of the metric and that of the second partial derivatives

$$
2\mathcal{M}_{IJ[K,L]} = -\mathcal{M}_{JK,IL} + \mathcal{M}_{KI,JL} + \mathcal{M}_{JL,IK} - \mathcal{M}_{LI,JK} = 2\mathcal{M}_{KL[I,J]},
$$

$$
2\mathcal{M}_{KL[I,J]} = -\mathcal{M}_{LI,KJ} + \mathcal{M}_{IK,LJ} + \mathcal{M}_{LJ,KI} - \mathcal{M}_{JK,LI} = 2\mathcal{M}_{IJ[K,L]}.
$$

Step 2: Take the vanishing torsion of the Riemann connection into account

$$
2\mathcal{M}_{MI[K}\mathcal{M}^M{}_{JL]} = \mathcal{M}_{M(IK)}\mathcal{M}^M{}_{(JL)} - \mathcal{M}_{M(IL)}\mathcal{M}^M{}_{(JK)} = 2\mathcal{M}_{MK[I}\mathcal{M}^M{}_{LJ]},
$$

$$
2\mathcal{M}_{MK[I}\mathcal{M}^M{}_{LJ]} = \mathcal{M}_{M(KI)}\mathcal{M}^M{}_{(LJ)} - \mathcal{M}_{M(KJ)}\mathcal{M}^M{}_{(LI)} = 2\mathcal{M}_{MI[K}\mathcal{M}^M{}_{JL]}.
$$

This concludes the proof. □

3 Continuum Mechanics

In this section the previously discussed concepts of differential geometry shall be applied to the geometrically nonlinear kinematics of (generalized) continuum mechanics. Thereby the aim is eventually to investigate the underlying structure of multiplicative elastoplasticity as a Cartan space and to derive corresponding measures of dislocation density. These could then be used in the modelling of the hardening due to geometrically necessary dislocations in crystalline materials.

3.1 Kinematics

The kinematics of (nonlinear) continuum mechanics and their relation to the previously introduced concepts of coordinates and their transformations within differential geometry are sketched in Fig. 7. Thereby this representation is in particular inspired by the exposition in Haupt (2000), Chapter 1. Further details may also be found, e.g., in Marsden and Hughes (1994)

Spatial and Material Coordinates Thereby, a *physical body* $\mathscr{B}$ consists in a set of *physical points* $\mathscr{P}$ (think of individual atoms or molecules)

$$\mathscr{B} = \{\mathscr{P}\}. \tag{103}$$

Then at time t the *assignment* of the physical points $\mathscr{P}$ to *spatial coordinates*, i.e. triplets $\{x^i(t)\} \in \mathbb{R}^3$ may be stated by the help of the map χ_t as

$$\{\mathscr{P}, t\} \mapsto \{x^1(t), x^2(t), x^3(t)\} = \chi(\mathscr{P}, t) = \chi_t(\mathscr{P}). \tag{104}$$

Accordingly, the *placement* of the whole physical body $\mathscr{B}$ into the *spatial configuration* $B_t \subset \mathbb{R}^3$ reads at time t as

$$\{\mathscr{B}, t\} \mapsto B_t = \chi(\mathscr{B}, t) = \chi_t(\mathscr{B}). \tag{105}$$

Alternatively, by the map χ_0 at time $t = t_0 = 0$, the physical points $\mathscr{P}$ may be assigned to *material coordinates*, i.e. triplets $\{X^I = x^i(0)\} \in \mathbb{R}^3$

$$\{\mathscr{P}, 0\} \mapsto \{X^1, X^2, X^3\} = \chi(\mathscr{P}, 0) = \chi_0(\mathscr{P}). \tag{106}$$

Thus, the placement of the physical body $\mathscr{B}$ into the *material configuration* $B_0 \subset \mathbb{R}^3$ follows at time $t = t_0 = 0$ as

$$\{\mathscr{B}, 0\} \mapsto B_0 = \chi(\mathscr{B}, 0) = \chi_0(\mathscr{B}). \tag{107}$$

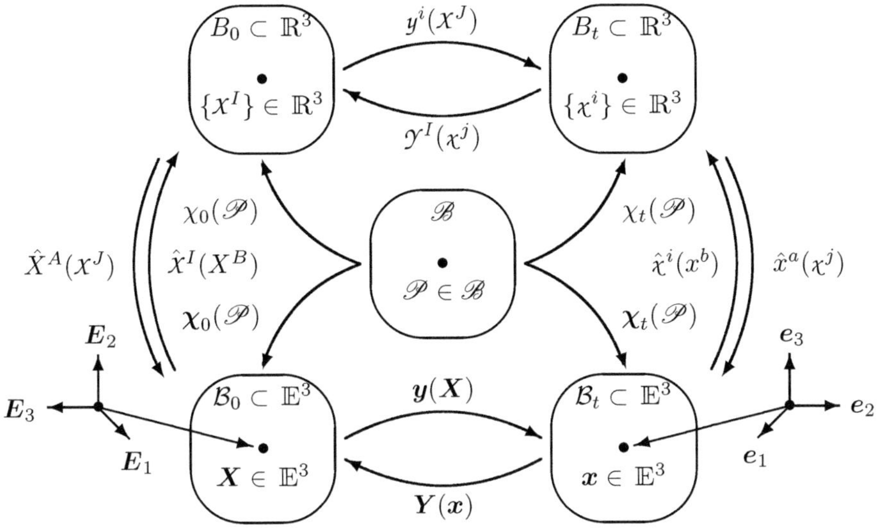

Figure 7. Kinematics of geometrically nonlinear continuum mechanics. *Physical body* $\mathscr{B}$ consisting of *physical points* $\mathscr{P}$, their assignment to material and spatial *coordinates* in $\mathbb{R}^3$ and *position vectors* in $\mathbb{E}^3$, and its *placement* into material and spatial *configurations*. The material and spatial coordinates are related by an invertible and continuously differentiable *coordinate transformation*, whereas the material and spatial position vectors are related by an invertible and continuously differentiable *point transformation*.

The relation between the material and the spatial coordinates is then given in terms of an invertible and continuously differentiable *coordinate transformation*

$$\{X^J\} \mapsto x^i = y^i(X^J) \quad \text{and} \quad \{x^i\} \mapsto X^I = \mathcal{Y}^I(x^j). \tag{108}$$

It shall be noted that these coordinate transformations may be regarded as an example of Eq. (2).

Spatial and Material Position Vectors For the present purposes a continuum body is always embedded into the ordinary three-dimensional ambient space[8] that is mathematically described by an Euclidean space $\mathbb{E}^3$. Then, at time t, the physical points $\mathscr{P}$ may be associated by the map χ_t to *spatial position vectors* $\boldsymbol{x}(t) \in \mathbb{E}^3$

$$\{\mathscr{P}, t\} \mapsto \boldsymbol{x}(t) = \chi(\mathscr{P}, t) = \chi_t(\mathscr{P}). \tag{109}$$

Accordingly, the placement of the whole physical body $\mathscr{B}$ into the spatial configuration $\mathcal{B}_t \subset \mathbb{E}^3$ reads at time t as

$$\{\mathscr{B}, t\} \mapsto \mathcal{B}_t = \chi(\mathscr{B}, t) = \chi_t(\mathscr{B}). \tag{110}$$

Likewise, by the map χ_0 at time $t = t_0 = 0$, the physical points $\mathscr{P}$ may be assigned to *material position vectors* $\boldsymbol{X} = \boldsymbol{x}(0) \in \mathbb{E}^3$

$$\{\mathscr{P}, t\} \mapsto \boldsymbol{X} = \chi(\mathscr{P}, 0) = \chi_0(\mathscr{P}). \tag{111}$$

Thus the placement of the physical body $\mathscr{B}$ into material configuration $\mathcal{B}_0 \subset \mathbb{E}^3$ reads at time $t_0 = t = 0$

$$\{\mathscr{B}, 0\} \mapsto \mathcal{B}_0 = \chi(\mathscr{B}, 0) = \chi_0(\mathscr{B}). \tag{112}$$

Finally the material and spatial position vectors are related by an invertible and continuously differentiable *point transformation*

$$\boldsymbol{X} \mapsto \boldsymbol{x} = \boldsymbol{y}(\boldsymbol{X}) \quad \text{and} \quad \boldsymbol{x} \mapsto \boldsymbol{X} = \boldsymbol{Y}(\boldsymbol{x}). \tag{113}$$

Rectilinear versus Curvilinear Coordinates In the embedding ambient (flat) Euclidean space *Cartesian coordinates* X^A and x^a and corresponding orthonormal base vectors $\boldsymbol{E}_A$ and $\boldsymbol{e}_a$ may be used, thus the material and spatial position vectors may be written as $\boldsymbol{X} = X^A \boldsymbol{E}_A$ and $\boldsymbol{x} = x^a \boldsymbol{e}_a$, respectively[9]. Then the relation between the *rectilinear* (Cartesian) coordinates X^A and x^a and the *curvilinear* coordinates X^J and χ^j is stated as

$$X^A = \hat{X}^A(X^J) \text{ and } X^I = \hat{X}^I(X^B), \quad x^a = \hat{x}^a(\chi^j) \text{ and } \chi^i = \hat{\chi}^i(x^b). \tag{114}$$

[8] In his excellent overview article de Wit (1981) distinguishes spaces from geometries: *..., it is a characteristic feature of defect theory that these geometries always exist in Euclidean space. Hence, in this paper we are not dealing with curved space, but with curved geometry in flat space.*. Thus the Euclidean embedding of the continuum body does not preclude to work with non-Euclidean geometries in order to describe distributed defects.

[9] Note that indices A, B, C, D or a, b, c, d are here reserved for Cartesian coordinates and Cartesian base vectors, respectively. The curvilinear coordinates and accompanying base vectors are equipped with indices I, J, K, L or i, j, k, l, respectively.

The corresponding *Jacobians* of these coordinate mappings are denoted by

$$J^A{}_J = \hat{X}^A{}_{,J} \quad \text{and} \quad \jmath^I{}_B = \hat{\mathcal{X}}^I{}_{,B}, \qquad \jmath^a{}_j = \hat{x}^a{}_{,j} \quad \text{and} \quad j^i{}_b = \hat{\chi}^i{}_{,b}. \tag{115}$$

Finally the *Christoffel symbols* associated with the curvilinear coordinates in the material and spatial configuration are computed based on the above concepts as

$$\mathcal{N}^I{}_{JK} = \jmath^I{}_A J^A{}_{J,K} \quad \text{and} \quad n^i{}_{jk} = j^i{}_a \jmath^a{}_{j,k} \tag{116}$$

Thereby the use of the Christoffel symbols is motivated as follows: In Euclidean space *tangent vectors* $\boldsymbol{g}_i := \boldsymbol{x}_{,i} = x^a{}_{,i}\boldsymbol{e}_a$ to *coordinate lines* $\boldsymbol{x} = \boldsymbol{x}(\chi^i)$ with χ^j =const and χ^k =const together with *normal vectors* $\boldsymbol{g}^j := \nabla_x \chi^j = \chi^j{}_{,b}\boldsymbol{e}^b$ to *coordinate surfaces* $\chi^j(\boldsymbol{x})$ =const may be defined if curvilinear coordinates $\{\chi^k\}$ are utilized, thus

$$\boldsymbol{g}_i = \jmath^a{}_i \boldsymbol{e}_a \quad \text{and} \quad \boldsymbol{g}^j = j^j{}_b \boldsymbol{e}^b. \tag{117}$$

Moreover co- and contravariant metric coefficients (that serve among other purposes to lower and raise indices) are defined as

$$g_{ij} = \boldsymbol{g}_i \cdot \boldsymbol{g}_j = \jmath^a{}_i \delta_{ab} \jmath^b{}_j \quad \text{and} \quad g^{ij} = \boldsymbol{g}^i \cdot \boldsymbol{g}^j = j^i{}_a \delta^{ab} j^j{}_b. \tag{118}$$

Then the Christoffel symbols allow essentially to express the partial derivatives of the base vectors as

$$\boldsymbol{g}_{j,k} = n^i{}_{jk} \boldsymbol{g}_i \quad \text{and} \quad \boldsymbol{g}^i{}_{,k} = -n^i{}_{jk} \boldsymbol{g}^j. \tag{119}$$

In an Euclidean space the Christoffel symbols take the role of the connection. Thus covariant differentiation of vectors and covectors in Euclidean space based on the Christoffel symbols, i.e. Euclidean covariant differentiation (denoted by a semicolon ;), is defined as

$$v^i{}_{;k} = v^i{}_{,k} + n^i{}_{jk} v^j \quad \text{and} \quad v_{j;k} = v_{j,k} - v_i n^i{}_{jk}. \tag{120}$$

Consequently, the spatial gradient of, e.g., a vector field reads

$$\nabla_x \boldsymbol{v} = v^i{}_{;j} \boldsymbol{g}_i \otimes \boldsymbol{g}^j = \jmath^a{}_i v^i{}_{;j} j^j{}_b \boldsymbol{e}_a \otimes \boldsymbol{e}^b = v_{i;j} \boldsymbol{g}^i \otimes \boldsymbol{g}^j = j^i{}_a v_{i;j} j^j{}_b \boldsymbol{e}^a \otimes \boldsymbol{e}^b. \tag{121}$$

In summary the Euclidean covariant derivatives in terms of the curvilinear coordinates may be related to the ordinary partial derivative in terms of the rectilinear (Cartesian) coordinates by

$$v^a{}_{,b} = \jmath^a{}_i v^i{}_{;j} j^j{}_b \quad \text{and} \quad v_{a,b} = j^i{}_a v_{i;j} j^j{}_b. \tag{122}$$

Finally the Euclidean covariant derivative, e.g., of the covariant metric coefficients vanishes identically

$$g_{ij;k} = g_{ij,k} - 2n_{(ij)k} \equiv 0 \quad \text{with} \quad n_{ijk} = g_{im} n^m{}_{jk} = \jmath^a{}_i \delta_{ab} \jmath^b{}_{j,k}. \tag{123}$$

3.2 Distortion

First-Order Distortions The coefficients of the first-order distortions are introduced as the Jacobians of the curvilinear and the rectilinear coordinate transformations, respectively, as

$$\mathcal{F}^i{}_J = y^i{}_{,J} \qquad \mathcal{f}^I{}_j = \mathcal{Y}^I{}_{,j} \quad \text{and} \quad F^a{}_B = y^a{}_{,B} \qquad f^A{}_b = Y^A{}_{,b}. \tag{124}$$

The corresponding *first-order distortion* (two-point) tensors follow accordingly as the first-order gradients of the point transformation

$$\boldsymbol{F} = \nabla_X \boldsymbol{y} = \underbrace{y^a{}_{,B}}_{F^a{}_B} \boldsymbol{e}_a \otimes \boldsymbol{E}^B = \underbrace{y^i{}_{,J}}_{\mathcal{F}^i{}_J} \boldsymbol{g}_i \otimes \boldsymbol{G}^J, \tag{125}$$

$$\boldsymbol{f} = \nabla_x \boldsymbol{Y} = \underbrace{Y^A{}_{,b}}_{f^A{}_b} \boldsymbol{E}_A \otimes \boldsymbol{e}^b = \underbrace{\mathcal{Y}^I{}_{,j}}_{\mathcal{f}^I{}_j} \boldsymbol{G}_I \otimes \boldsymbol{g}^j. \tag{126}$$

The first-order distortion tensors $\boldsymbol{F}$ and $\boldsymbol{f}$ are also denoted as (first-order) *deformation gradients.*

Second-Order Distortions The coefficients of the second-order distortions are introduced as the *Hessians* of the curvilinear and the rectilinear coordinate transformations, respectively, as

$$\mathcal{G}^i{}_{JK} = \mathcal{F}^i{}_{J,K} \quad \mathcal{g}^I{}_{jk} = \mathcal{f}^I{}_{j,k} \quad \text{and} \quad G^a{}_{BC} = F^a{}_{B,C} \quad g^A{}_{bc} = f^I{}_{b,c}. \tag{127}$$

The (Euclidean) *total covariant derivative* (denoted by a double dot :) simultaneously takes into account the material and spatial tangent and normal vectors of the material and spatial curvilinear coordinate systems, for more details see for example Ericksen (1960). The total covariant derivatives of the first-order distortions in terms of the curvilinear coordinates are thus computed as

$$\mathsf{G}^i{}_{JK} = \mathcal{F}^i{}_{J:K} \quad \text{with} \quad \mathcal{F}^i{}_{J:K} = \mathcal{F}^i{}_{J,K} + n^i{}_{jk}\mathcal{F}^j{}_J\mathcal{F}^k{}_K - \mathcal{F}^i{}_I\mathcal{N}^I{}_{JK}, \tag{128}$$

$$\mathsf{g}^I{}_{jk} = \mathcal{f}^I{}_{j:k} \quad \text{with} \quad \mathcal{f}^I{}_{j:k} = \mathcal{f}^I{}_{j,k} + \mathcal{N}^I{}_{JK}\mathcal{f}^J{}_j\mathcal{f}^K{}_k - \mathcal{f}^I{}_i n^i{}_{jk}.$$

Note the sans serif font used to distinguish the total covariant derivative of the first-order distortion $\mathsf{G}^i{}_{JK} = \mathcal{F}^i{}_{J:K}$ from the corresponding partial derivative $\mathcal{G}^i{}_{JK} = \mathcal{F}^i{}_{J,K}$. The corresponding *second-order distortion* (two-point) tensors follow accordingly as the second-order gradients of the point transformation

$$\boldsymbol{G} = \nabla_X \boldsymbol{F} = \underbrace{F^a{}_{B,C}}_{G^a{}_{BC}} \boldsymbol{e}_a \otimes \boldsymbol{E}^B \otimes \boldsymbol{E}^C = \underbrace{\mathcal{F}^i{}_{J:K}}_{\mathsf{G}^i{}_{JK}} \boldsymbol{g}_i \otimes \boldsymbol{G}^J \otimes \boldsymbol{G}^K, \tag{129}$$

$$\boldsymbol{g} = \nabla_x \boldsymbol{f} = \underbrace{f^A{}_{b,c}}_{g^A{}_{bc}} \boldsymbol{E}_A \otimes \boldsymbol{e}^b \otimes \boldsymbol{e}^c = \underbrace{f^I{}_{j:k}}_{\mathrm{g}^I{}_{jk}} \boldsymbol{G}_I \otimes \boldsymbol{g}^j \otimes \boldsymbol{g}^k.$$

The second-order distortion tensors $\boldsymbol{G}$ and $\boldsymbol{g}$ are also denoted as second-order deformation gradients.

3.3 Integrability

It is of considerable interest to relate the conditions for the integrability of the first-order distortion into a compatible vector field (the deformation) to the previously discussed concepts of differential geometry, in particular to the various anholonomic objects. To this end it is useful to first recall some necessary aspects of tensor calculus in three dimensions:

Permutation Tensors In the tangent space to the material configuration the material version of the permutation tensor has the following representation in terms of rectilinear and curvilinear coordinates, respectively

$$\boldsymbol{E} := E^{ABC} \boldsymbol{E}_A \otimes \boldsymbol{E}_B \otimes \boldsymbol{E}_C = \mathcal{E}^{IJK} \boldsymbol{G}_I \otimes \boldsymbol{G}_J \otimes \boldsymbol{G}_K. \tag{130}$$

The permutation symbol E^{ABC} takes values 1, -1 and 0, for even permutations of A, B, C, odd permutations of A, B, C, and else. Then the material permutation symbols in curvilinear and rectilinear coordinates are related as

$$\mathcal{E}^{IJK} = \sqrt[-1]{G} E^{ABC} \delta^I_A \delta^J_B \delta^K_C = \sqrt[-1]{G} E^{IJK}, \tag{131}$$

whereby $G = \det[G_{IJ}]$ denotes the determinant of the matrix arrangement of the metric coefficients $G_{IJ} := \boldsymbol{G}_I \cdot \boldsymbol{G}_J$ related to the curvilinear coordinates in the material configuration. Since the covariant base vectors follow as $\boldsymbol{G}_I = J^A{}_I \boldsymbol{E}_A$, the metric coefficients may also be expressed as $G_{IJ} = J^A{}_I \delta_{AB} J^B{}_J$ and thus it holds also that $G = \det^2[J^A{}_I]$. Due to the transformation properties of the base vectors it then also holds that

$$E^{ABC} = \mathcal{E}^{IJK} J^A{}_I J^B{}_J J^C{}_K = \sqrt[-1]{G} E^{IJK} J^A{}_I J^B{}_J J^C{}_K. \tag{132}$$

In conclusion the permutation symbol $\mathcal{E}^{IJK}$ takes values $\sqrt[-1]{G}$, $-\sqrt[-1]{G}$ and 0, for even permutations of I, J, K, odd permutations of I, J, K, and else. In analogy, the spatial permutation tensor in the tangent space to the spatial configuration reads in terms of rectilinear and curvilinear coordinates, respectively, as

$$\boldsymbol{e} := e^{abc} \boldsymbol{e}_a \otimes \boldsymbol{e}_b \otimes \boldsymbol{e}_c = e^{ijk} \boldsymbol{g}_i \otimes \boldsymbol{g}_j \otimes \boldsymbol{g}_k \tag{133}$$

Clearly, here the spatial permutation symbols in curvilinear and rectilinear coordinates are related in terms of $g = \det[g_{ij}]$, i.e. the determinant of the matrix arrangement of the metric coefficients $g_{ij} := \boldsymbol{g}_i \cdot \boldsymbol{g}_j$ related to the curvilinear coordinates in the spatial configuration

$$\mathscr{e}^{ijk} = \sqrt[-1]{g} e^{abc} \delta^i_a \delta^j_b \delta^k_c = \sqrt[-1]{g} e^{ijk}. \tag{134}$$

Finally the spatial and material permutation tensors are convected as

$$\boldsymbol{e} = j[\boldsymbol{F} \overline{\otimes} \boldsymbol{F}] : \boldsymbol{E} \cdot \boldsymbol{F}^t \quad \text{and} \quad \boldsymbol{E} = J[\boldsymbol{f} \overline{\otimes} \boldsymbol{f}] : \boldsymbol{e} \cdot \boldsymbol{f}^t, \tag{135}$$

whereby $J = \det \boldsymbol{F}$ and $j = \det \boldsymbol{f}$ denote the determinants of the deformation gradients with $J = \sqrt{g/G}$ and $j = \sqrt{G/g}$, and $\overline{\otimes}$ denotes a non-standard dyadic product that is highlighted in the corresponding index notation

$$e^{abc} = jE^{ABC} F^a{}_A F^b{}_B F^c{}_C \quad \text{and} \quad \mathscr{e}^{ijk} = \mathscr{j} \mathscr{E}^{IJK} \mathscr{F}^i{}_I \mathscr{F}^j{}_J \mathscr{F}^k{}_K. \tag{136}$$

Here $\mathscr{J} = \det[\mathscr{F}^i{}_I]$ and $\mathscr{j} = \det[\mathscr{f}^I{}_i]$, respectively.

Curl Operators Next, the material *curl operator* with respect to the material coordinates as applied to the deformation gradient $\boldsymbol{F}$ allows the following coordinate representation

$$\text{Curl}\boldsymbol{F} := D^{aC} \boldsymbol{e}_a \otimes \boldsymbol{E}_C = \mathscr{D}^{iK} \boldsymbol{g}_i \otimes \boldsymbol{G}_K, \tag{137}$$

whereby the coefficients of Curl$\boldsymbol{F}$ are given in terms of the permutation symbols expressed in material rectilinear and curvilinear coordinates, respectively

$$D^{aC} = -F^a{}_{A,B} E^{ABC} \quad \text{and} \quad \mathscr{D}^{iK} = -\mathscr{F}^i{}_{I:J} \mathscr{E}^{IJK} \equiv -\mathscr{F}^i{}_{I,J} \mathscr{E}^{IJK}. \tag{138}$$

Note that expressed in curvilinear coordinates the Euclidean total covariant derivative may be substituted by the ordinary partial derivative due to the (right) symmetry of the spatial and material Christoffel symbols. The relation of the coefficients of Curl$\boldsymbol{F}$ to the (Piola-type) anholonomic object in Eq. (41).1, respectively to the second-order dislocation density tensor in Eq. (86).2 is obvious from the above equations.

Likewise, the spatial curl operator with respect to the spatial coordinates as applied to the deformation gradient $\boldsymbol{f}$ allows the following representation

$$\text{curl}\boldsymbol{f} := d^{Ac} \boldsymbol{E}_A \otimes \boldsymbol{e}_c = \mathscr{d}^{Ik} \boldsymbol{G}_I \otimes \boldsymbol{g}_k, \tag{139}$$

whereby here the coefficients of curl$\boldsymbol{f}$ are given in terms of the permutation symbols expressed in spatial rectilinear and curvilinear coordinates, respectively

$$d^{Ac} = -f^A{}_{a,b} e^{abc} \quad \text{and} \quad \mathscr{d}^{Ik} = -\mathscr{f}^I{}_{i:j} \mathscr{e}^{ijk} \equiv -\mathscr{f}^I{}_{i,j} \mathscr{e}^{ijk}. \tag{140}$$

Finally, it may be shown by the chain rule applied to, e.g. $\boldsymbol{f}\cdot\boldsymbol{F}=\boldsymbol{I}$ that the spatial and material curl operators are convected as

$$\mathrm{Curl}\boldsymbol{F}=-J\boldsymbol{F}\cdot\mathrm{curl}\boldsymbol{f}\cdot\boldsymbol{f}^t \quad\text{and}\quad \mathrm{curl}\boldsymbol{f}=-j\boldsymbol{f}\cdot\mathrm{Curl}\boldsymbol{F}\cdot\boldsymbol{F}^t. \tag{141}$$

Integrability of First-Order Distortions In rectilinear (Cartesian) coordinates the condition for the first-order distortions $F^a{}_B$ and $f^A{}_b$, respectively, to be integrable into a vector field $y^a(X^B)$ and $Y^A(x^b)$, respectively, derive simply from the symmetry of the second partial derivatives, i.e.

$$F^a{}_{[B,C]}\doteq 0 \quad\text{and}\quad f^A{}_{[b,c]}\doteq 0. \tag{142}$$

Since in curvilinear coordinates the Euclidean total covariant derivative (indicated by a double dot :) of $\mathcal{F}^i{}_J$ and $f^I{}_j$, i.e. $\mathcal{F}^i{}_{J:K}$ and $f^I{}_{j:k}$, may be considered a transformation of the $F^a{}_{B,C}$ or $f^A{}_{b,c}$, respectively, i.e.

$$F^a{}_{B,C}=j^a{}_i\mathcal{F}^i{}_{J:K}\mathcal{J}^J{}_B\mathcal{J}^K{}_C \quad\text{and}\quad f^A{}_{b,c}=J^A{}_I f^I{}_{j:k}j^j{}_b j^k{}_c, \tag{143}$$

and since this is a similarity transformation in the indices JK and jk, respectively, that preserves the symmetry and skew symmetry properties, the integrability conditions for the first-order distortions in curvilinear coordinates may be stated likewise as

$$\mathcal{F}^i{}_{[J:K]}\doteq 0 \quad\text{and}\quad f^I{}_{[j:k]}\doteq 0. \tag{144}$$

Recall that the Euclidean total covariant derivatives of the first-order distortion $\mathcal{F}^i{}_J$ and $f^I{}_j$, respectively, follow as

$$\mathcal{F}^i{}_{J:K}=\mathcal{F}^i{}_{J,K}+n^i{}_{mk}\mathcal{F}^m{}_J\mathcal{F}^k{}_K-\mathcal{F}^i{}_M\mathcal{N}^M{}_{JK}, \tag{145}$$

$$f^I{}_{j:k}=f^I{}_{j,k}+\mathcal{N}^I{}_{MK}f^M{}_j f^K{}_k-f^I{}_m n^m{}_{jk}.$$

Since, however, the spatial and material Christoffel symbols are (right) symmetric, the integrability condition for the first-order distortion expressed in curvilinear coordinates may eventually be expressed in terms of the ordinary partial derivative

$$\mathcal{F}^i{}_{[J:K]}=\mathcal{F}^i{}_{[J,K]}=:\mathcal{D}^i{}_{JK}\doteq 0 \quad\text{and}\quad f^I{}_{[j:k]}=f^I{}_{[j,k]}=:d^I{}_{jk}\doteq 0. \tag{146}$$

This relates the (Piola-type) anholonomic objects $\mathcal{D}^i{}_{JK}$ and $d^I{}_{jk}$ in Eq. (41), or rather the third-order dislocation density tensors to the integrability conditions in curvilinear coordinates. Finally from the preceding discussion on the coordinate representation of the permutation tensor and the curl operator applied to the deformation gradient, the integrability condition may alternatively be stated as

$$\boldsymbol{D} := \mathrm{Curl}\boldsymbol{F} \doteq \boldsymbol{0} \quad \text{and} \quad \boldsymbol{d} := \mathrm{curl}\boldsymbol{f} \doteq \boldsymbol{0}. \tag{147}$$

Clearly, it is again the second-order (Piola-type) dislocation density tensors $\boldsymbol{D}$ and $\boldsymbol{d}$ that may be used to express the integrability condition of the first-order distortion.

3.4 Elasticity

Next, the previous concepts of differential geometry, in particular aspects related to compatibility, shall be applied to the kinematics of (nonlinear) elasticity. Thereby it shall be noted that the material and spatial configurations of elasticity ought to be compatible. Then, two cases may be considered: firstly the material configuration is assumed compatible and the condition on the deformation for the spatial configuration to also be compatible is sought; secondly the situation is reversed: the spatial configuration is assumed compatible and the condition on the (inverse) deformation for the material configuration to be compatible is sought.

Integrability of Material Metric For a spatial configuration of elasticity to be compatible distance determination and parallel transport ought to be of Euclidean type, thus Cartesian coordinates may be selected with the spatial metric coefficients coinciding with the Kronecker-delta and a vanishing spatial connection

$$m_{ab} = \delta_{ab} \quad \text{and} \quad l^a{}_{bc} = 0. \tag{148}$$

The convection of the spatial metric and the spatial connection by the deformation map then renders the *Cauchy-Green tensor* and an *integrable connection*[10] in the material configuration

$$\mathcal{M}_{IJ} = \mathcal{C}_{IJ} := F^a{}_I \delta_{ab} F^b{}_J \quad \text{and} \quad \mathcal{L}^I{}_{JK} = f^I{}_a F^a{}_{J,K}. \tag{149}$$

Furthermore, it may easily be shown that the fully covariant material connection $\mathcal{L}_{IJK} = \mathcal{C}_{IM} \mathcal{L}^M{}_{JK}$ is metric with respect to $\mathcal{C}_{IJ}$, i.e.

$$\mathcal{C}_{IJ,K} = F^a{}_{I,K} \delta_{ab} F^b{}_J + F^a{}_I \delta_{ab} F^b{}_{J,K} = 2\mathcal{L}_{(IJ)K}. \tag{150}$$

Consequently, and due to the integrability of $F^a{}_I$, i.e. due to $\mathcal{L}^I{}_{JK} = \mathcal{L}^I{}_{(JK)}$ (compare Eq. (66)) the material connection $\mathcal{L}_{IJK}$ coincides exclusively with the material Riemann connection

$$\mathcal{M}_{IJK} = \mathcal{C}_{IJK} := [\mathcal{C}_{IJ,K} - \mathcal{C}_{JK,I} + \mathcal{C}_{KI,J}]/2 \tag{151}$$

based on the metric $\mathcal{C}_{IJ}$ while the corresponding material contortion

$$\mathcal{K}_{IJK} \equiv 0 \tag{152}$$

vanishes identically:

$$\begin{array}{rcccccccccc} \mathcal{C}_{IJ,K} & = & + & F^a{}_J & \delta_{ab} & F^b{}_{I,K} & + & F^a{}_I & \delta_{ab} & F^b{}_{J,K} \\ -\mathcal{C}_{JK,I} & = & - & F^a{}_K & \delta_{ab} & F^b{}_{J,I} & - & F^a{}_J & \delta_{ab} & F^b{}_{K,I} \\ \mathcal{C}_{KI,J} & = & + & F^a{}_I & \delta_{ab} & F^b{}_{K,J} & + & F^a{}_K & \delta_{ab} & F^b{}_{I,J} \\ \hline\hline \end{array}$$

$$\mathcal{C}_{IJK} = F^a{}_I \delta_{ab} F^b{}_{(J,K)} \quad \text{and} \quad \mathcal{K}_{IJK} \equiv 0. \tag{153}$$

Next, since the material connection is integrable, its associated material Riemann curvature tensor vanishes identically

$$\mathcal{M}_{IJKL} = \mathcal{C}_{IJKL}(\mathcal{C}) := 2\mathcal{C}_{IJ[L,K]} - 2\mathcal{C}_{MI[K} \mathcal{C}^M{}_{JL]} \equiv 0. \tag{154}$$

---

[10] An integrable connection results in a vanishing curvature tensor due to the symmetry of the second partial derivatives:

$$\begin{aligned} F^a{}_{J,[LK]} &= [F^a{}_I \mathcal{L}^I{}_{J[L}]_{,K]} \\ &= F^a{}_I [\mathcal{L}^I{}_{J[L,K]} + f^I{}_b F^b{}_{M,[K} \mathcal{L}^M{}_{JL]}] \\ &= F^a{}_I [\mathcal{L}^I{}_{J[L,K]} + \mathcal{L}^I{}_{M[K} \mathcal{L}^M{}_{JL]}] \\ &= F^a{}_I \mathcal{R}^I{}_{JKL}/2 \\ &\equiv 0. \end{aligned}$$

In the context of plasticity an integrable connection is also sometimes denoted as *crystal connection*, see, e.g., Clayton (2011).

This finding coincides of course with the postulated Euclidean character of the spatial configuration

$$0 \equiv c_{abcd}(\delta) = c_{abcd}(y(\mathcal{C})) = y(\mathcal{C}_{IJKL}(\mathcal{C})). \tag{155}$$

Thus in the material configuration a vanishing Riemann curvature tensor based on a Riemann connection assures integrability of the material metric into a compatible deformation map, i.e. into a compatible spatial configuration.

In summary the kinematics of elasticity may be considered a flat Riemann geometry over the material configuration B_0. A flat Riemann geometry is of course an Euclidean geometry. Concluding, flatness assures integrability, i.e. the Cauchy-Green tensor may be derived from a vector valued deformation map. This is the nonlinear version of the *St. Venant compatibility conditions* expressed in terms of the material metric.

Integrability of Spatial Metric Next the situation is reversed, i.e. for the material configuration of elasticity to be compatible distance determination and parallel transport ought to be of Euclidean type, thus Cartesian coordinates may be selected with the material metric coefficients agreeing with the Kronecker-delta and a null material connection

$$M_{AB} = \delta_{AB} \quad \text{and} \quad L^{A}{}_{BC} = 0. \tag{156}$$

Then the convection of the material metric and the material connection by the (inverse) deformation map render the *Finger tensor* and an integrable connection in the spatial configuration

$$m_{ij} = c_{ij} := f^{A}{}_{i}\delta_{AB}f^{B}{}_{j} \quad \text{and} \quad \ell^{i}{}_{jk} = F^{i}{}_{A}f^{A}{}_{j,k}. \tag{157}$$

It thus follows that the fully covariant spatial connection $\ell_{ijk} = c_{im}\ell^{m}{}_{jk}$ is metric with respect to c_{ij}, i.e.

$$c_{ij,k} = f^{A}{}_{i,k}\delta_{AB}f^{B}{}_{j} + f^{A}{}_{i}\delta_{AB}f^{B}{}_{j,k} = 2\ell_{(ij)k}. \tag{158}$$

As a consequence, and due to the integrability of $f^{A}{}_{i}$, i.e. due to $\ell^{i}{}_{jk} = \ell^{i}{}_{(jk)}$ (refer again to Eq. (66)), the spatial connection ℓ_{ijk} coincides with the Riemann connection

$$m_{ijk} = c_{ijk} := [c_{ij,k} - c_{jk,i} + c_{ki,j}]/2 \tag{159}$$

based on the metric c_{ij} while the corresponding spatial contortion

$$k_{ijk} \equiv 0 \tag{160}$$

vanishes identically:

$$\begin{array}{rcrccccccc} c_{ij,k} & = & + & f^A{}_j & \delta_{AB} & f^B{}_{i,k} & + & f^A{}_i & \delta_{AB} & f^B{}_{j,k} \\ -c_{jk,i} & = & - & f^A{}_k & \delta_{AB} & f^B{}_{j,i} & - & f^A{}_j & \delta_{AB} & f^B{}_{k,i} \\ c_{ki,j} & = & + & f^A{}_i & \delta_{AB} & f^B{}_{k,j} & + & f^A{}_k & \delta_{AB} & f^B{}_{i,j} \\ \hline\hline \end{array}$$

$$c_{ijk} = f^A{}_i \delta_{AB} f^B{}_{(j,k)} \quad \text{and} \quad k_{ijk} \equiv 0. \tag{161}$$

Moreover, since the spatial connection is integrable, its associated spatial Riemann curvature tensor vanishes identically

$$m_{ijkl} = c_{ijkl}(c) := 2c_{ij[l,k]} - 2c_{mi[k}c^m{}_{jl]} \equiv 0. \tag{162}$$

Clearly, this finding coincides with the Euclidean character of the material configuration

$$0 \equiv C_{ABCD}(\delta) = C_{ABCD}(Y(c)) = Y(c_{ijkl}(c)). \tag{163}$$

Thus in the spatial configuration a vanishing Riemann curvature tensor based on a Riemann connection assures integrability of the spatial metric into a compatible (inverse) deformation map, i.e a compatible material configuration.

In summary the kinematics of elasticity may alternatively be considered a flat Riemann geometry (i.e. an Euclidean geometry) over the spatial configuration $\mathcal{B}_t$. To conclude flatness assures integrability, i.e. the Finger tensor may be derived from a vector valued (inverse) deformation map. This is again the nonlinear version of the St. Venant compatibility conditions.

3.5 Elastoplasticity

Plastic Part of the Deformation A well-accepted assumption in the kinematics of elastoplastic crystalline materials is the *multiplicative decomposition* of the deformation gradient into a plastic and an elastic contribution

$$\boldsymbol{F} = \bar{\boldsymbol{F}} \cdot \widetilde{\boldsymbol{F}}. \tag{164}$$

Besides the material and the spatial configuration this decomposition also introduces the so-called *intermediate configuration*. In the sequel quantities referring to the plastic part of the deformation either in the material or in the intermediate configuration shall be denoted by a tilde $(\tilde{\bullet})$. The intermediate configuration is characterized by an unchanged inclination of the crystalline lattice as compared to the material configuration and is deemed accessible by locally relaxing the stress state. As a consequence the so-called *isoclinic* intermediate configuration of elastoplasticity is generally incompatible and can not be patched together to a single compatible configuration in the Euclidean space.

However, due to the above conceptual definition distance determination and parallel transport in the intermediate configuration of elastoplasticity are of Euclidean type, thus Cartesian-like coordinates may be selected with the metric coefficients coinciding with the Kronecker-delta and a vanishing connection[11]

$$\tilde{m}_{\alpha\beta} = \delta_{\alpha\beta} \quad \text{and} \quad \tilde{l}^{\alpha}{}_{\beta\gamma} = 0. \tag{165}$$

The plastic convection of the metric and the connection then render the *plastic Cauchy-Green tensor* and an *integrable plastic connection* in the material configuration

$$\tilde{\mathfrak{M}}_{IJ} = \tilde{C}_{IJ} := \tilde{F}^{\alpha}{}_{I}\delta_{\alpha\beta}\tilde{F}^{\beta}{}_{J} \quad \text{and} \quad \tilde{\mathcal{L}}^{I}{}_{JK} = \tilde{f}^{I}{}_{\alpha}\tilde{F}^{\alpha}{}_{J,K}. \tag{166}$$

It is easy to check that the fully covariant plastic connection $\tilde{\mathcal{L}}_{IJK} = \tilde{C}_{IM}\tilde{\mathcal{L}}^{M}{}_{JK}$ is indeed a metric connection with respect to $\tilde{C}_{IJ}$, since

$$\tilde{C}_{IJ,K} = \tilde{F}^{\alpha}{}_{I,K}\delta_{\alpha\beta}\tilde{F}^{\beta}{}_{J} + \tilde{F}^{\alpha}{}_{I}\delta_{\alpha\beta}\tilde{F}^{\beta}{}_{J,K} = 2\tilde{\mathcal{L}}_{(IJ)K}. \tag{167}$$

Thus, the plastic connection $\tilde{\mathcal{L}}_{IJK}$ splits into the plastic Riemann connection

$$\tilde{\mathfrak{M}}_{IJK} = \tilde{C}_{IJK} := [\tilde{C}_{IJ,K} - \tilde{C}_{JK,I} + \tilde{C}_{KI,J}]/2 \tag{168}$$

based on $\tilde{C}_{IJ}$ and the corresponding plastic contortion

$$\tilde{\mathfrak{K}}_{IJK} := \tilde{\mathcal{L}}_{IJK} - \tilde{\mathfrak{M}}_{IJK}. \tag{169}$$

This may be verified directly by cyclic permutation of indices in Eq. (167) and subsequent addition of the resulting expressions

$$\begin{array}{rcccccccccc}
\tilde{C}_{IJ,K} & = & + & \tilde{F}^{\alpha}{}_{J} & \delta_{\alpha\beta} & \tilde{F}^{\beta}{}_{I,K} & + & \tilde{F}^{\alpha}{}_{I} & \delta_{\alpha\beta} & \tilde{F}^{\beta}{}_{J,K} \\
-\tilde{C}_{JK,I} & = & - & \tilde{F}^{\alpha}{}_{K} & \delta_{\alpha\beta} & \tilde{F}^{\beta}{}_{J,I} & - & \tilde{F}^{\alpha}{}_{J} & \delta_{\alpha\beta} & \tilde{F}^{\beta}{}_{K,I} \\
\tilde{C}_{KI,J} & = & + & \tilde{F}^{\alpha}{}_{I} & \delta_{\alpha\beta} & \tilde{F}^{\beta}{}_{K,J} & + & \tilde{F}^{\alpha}{}_{K} & \delta_{\alpha\beta} & \tilde{F}^{\beta}{}_{I,J} \\
\hline\hline
\end{array}$$

$$\tilde{C}_{IJK} = \tilde{F}^{\alpha}{}_{I}\delta_{\alpha\beta}\tilde{F}^{\beta}{}_{(J,K)} - \tilde{F}^{\alpha}{}_{J}\delta_{\alpha\beta}\tilde{F}^{\beta}{}_{[K,I]} + \tilde{F}^{\alpha}{}_{K}\delta_{\alpha\beta}\tilde{F}^{\beta}{}_{[I,J]} \tag{170}$$

and consequently, due to Eq. (169) the plastic contortion reads

$$\tilde{\mathfrak{K}}_{IJK} = \tilde{F}^{\alpha}{}_{I}\delta_{\alpha\beta}\tilde{F}^{\beta}{}_{[J,K]} + \tilde{F}^{\alpha}{}_{J}\delta_{\alpha\beta}\tilde{F}^{\beta}{}_{[K,I]} - \tilde{F}^{\alpha}{}_{K}\delta_{\alpha\beta}\tilde{F}^{\beta}{}_{[I,J]}. \tag{171}$$

[11] Note that greek indices $\alpha, \beta, \gamma, \delta$ are here reserved for the Cartesian-like coordinates and base vectors in the intermediate configuration. Here *Cartesian-like* refers to rectilinear but anholonomic coordinates.

Moreover, since the plastic (material) connection is integrable, its associated curvature tensor vanishes identically

$$\tilde{\mathcal{R}}_{IJKL} = 2\tilde{\mathcal{L}}_{IJ[L,K]} - 2\tilde{\mathcal{L}}_{MI[K}\tilde{\mathcal{L}}^M{}_{JL]} \equiv 0. \tag{172}$$

Then, since the plastic connection decomposes into the plastic Riemann connection and the plastic contortion, the corresponding curvature tensors

$$\tilde{\mathcal{C}}_{IJKL} := 2\tilde{\mathcal{C}}_{IJ[L,K]} - 2\tilde{\mathcal{C}}_{MI[K}\tilde{\mathcal{C}}^M{}_{JL]} \neq 0, \tag{173}$$

and

$$\tilde{\mathcal{K}}_{IJKL} := 2\tilde{\mathcal{K}}_{IJ[L,K]} - 2\tilde{\mathcal{K}}_{MI[K}\tilde{\mathcal{K}}^M{}_{JL]} \neq 0 \tag{174}$$

establish a relation between the (plastic) incompatibilities of the plastic Cauchy-Green tensor $\tilde{C}_{IJ}$ contained in $\tilde{\mathcal{C}}_{IJKL}$ on the one hand and the dislocation density contained in $\tilde{\mathcal{K}}_{IJKL}$ on the other hand

$$\tilde{\mathcal{C}}_{IJKL} = -\tilde{\mathcal{K}}_{IJKL} + 2\tilde{\mathcal{K}}_{MI[K}\tilde{\mathcal{C}}^M{}_{JL]} + 2\tilde{\mathcal{C}}_{MI[K}\tilde{\mathcal{K}}^M{}_{JL]}. \tag{175}$$

Recall that $\tilde{\mathcal{C}}_{IJKL}$ obeys major symmetries in the index pairs IJ and KL. The relation between incompatibilities of the strain and the (geometrically necessary) dislocation density as displayed in Eq. (175) is a key result in the geometrically linear theory of continuous dislocations that may be established comparatively straightforward. However, for the geometrically nonlinear case it would not have been easily discovered without the tools from differential geometry.

Thus Eq. (175) is a result of the major symmetric part of $\tilde{\mathcal{R}}_{IJKL} = 0$, the consequence of the major skew symmetric part of $\tilde{\mathcal{R}}_{IJKL} = 0$ or, equivalently, of the right hand part of Eq. (175) is discussed in the following: Associated with the incompatibility of the intermediate configuration, and referring to Eqs. (43).1 and (39).1, are the third-order Piola-Kirchhoff- and Cauchy-type dislocation densities (anholonomic objects)

$$\tilde{\mathcal{A}}^I{}_{JK} = -\tilde{f}^I{}_\alpha \tilde{F}^\alpha{}_{[J,K]} \quad \text{and} \quad \tilde{a}^\alpha{}_{\beta\gamma} = \tilde{F}^\alpha{}_{[J,K]}\tilde{f}^J{}_\beta \tilde{f}^K{}_\gamma. \tag{176}$$

These dislocation densities are related to the corresponding torsions as computed from the connections in Eqs. (166).2 and (165).2

$$\tilde{\mathcal{A}}^I{}_{JK} = -\tilde{\mathcal{L}}^I{}_{[JK]} = -\tilde{\mathcal{T}}^I{}_{JK} \quad \text{and} \quad \tilde{a}^\alpha{}_{\beta\gamma} = \tilde{t}^\alpha{}_{\beta\gamma} - \tilde{l}^\alpha{}_{[\beta\gamma]} = \tilde{t}^\alpha{}_{\beta\gamma}. \tag{177}$$

Clearly, due to the representation of the material plastic connection in Eq. (166).2 its torsion coincides with the negative of the plastic (Piola-Kirchhoff-type) anholonomic object in Eq. (176).1; likewise due to the representation of the torsion in an anholonomic space in Eq. (40).1 it coincides with the plastic (Cauchy-type) anholonomic object in Eq. (176).2 due to the vanishing plastic connection in Eq. (165).2. Then, the (contravariant) second-order Piola-Kirchhoff- and Cauchy-type dislocation density tensors are expressed in the ambient Euclidean space as

$$\widetilde{\boldsymbol{A}} = -\widetilde{\boldsymbol{f}} \cdot \widetilde{\boldsymbol{D}} \quad \text{and} \quad \widetilde{\boldsymbol{a}} = \widetilde{\boldsymbol{D}} \cdot \text{cof}\widetilde{\boldsymbol{f}} \quad \text{with} \quad \widetilde{\boldsymbol{D}} = \text{Curl}\widetilde{\boldsymbol{F}}. \tag{178}$$

Finally the continuity equation of continuum dislocation theory states that *dislocation lines* may not end within the volume, i.e. the plastic dislocation density is source free (solenoidal)

$$\text{Div}\widetilde{\boldsymbol{D}} = -\widetilde{\boldsymbol{F}} \cdot [\text{Div}\widetilde{\boldsymbol{A}} + \widetilde{\boldsymbol{f}} \cdot \nabla_X \widetilde{\boldsymbol{F}} : \widetilde{\boldsymbol{A}}] = \boldsymbol{0} \quad \text{since} \quad \widetilde{\boldsymbol{D}} = \text{Curl}\widetilde{\boldsymbol{F}}. \tag{179}$$

The solenoidal character of the plastic (Piola-type) dislocation density is directly related to the algebraic Bianchy identity corresponding to Eq. (101) as may be seen from the following

Proof:
According to Eq. (87).2 the coefficients of the second-order dislocation density tensor follow from those of the third-order dislocation density tensor as

$$\tilde{D}^{\alpha L} = -\tilde{D}^{\alpha}{}_{JK} \mathcal{E}^{JKL}.$$

Then the divergence of the second-order dislocation density tensor involves the total covariant derivative in Euclidean ambient space in terms of the material Christoffel symbol $\mathcal{N}^I{}_{JK}$ (the Christoffel symbol related to the coordinates in the intermediate configuration $\tilde{n}^{\alpha}{}_{\beta\gamma} \equiv 0$ vanishes due to the selection of Cartesian-like coordinates, see Eq. (165).1)

$$\tilde{D}^{\alpha L}{}_{:L} = -\tilde{D}^{\alpha}{}_{JK:L} \mathcal{E}^{JKL} \equiv -\tilde{D}^{\alpha}{}_{JK,L} \mathcal{E}^{JKL}.$$

Here, firstly $\mathcal{E}^{JKL}{}_{:L} = 0$ (holds true since $\mathcal{E}^{JKL}$ is scaled by $\det[G_{IJ}]$) and secondly the right symmetry of the material Christoffel symbols have been exploited to derive at a result in terms of the partial rather than the total covariant derivative of the dislocation density. Next, $\tilde{D}^{\alpha}{}_{JK}$ is re-expressed in terms of the connection in Eq. (166).2 to render

$$\tilde{D}^{\alpha}{}_{JK} = \tilde{F}^{\alpha}{}_{I} \tilde{\mathcal{L}}^{I}{}_{[JK]}.$$

Then, application of the product rule gives[12]

$$\begin{aligned}\tilde{D}^{\alpha L}{}_{:L} &= -[\tilde{F}^{\alpha}{}_{I}\tilde{\mathcal{L}}^{I}{}_{[JK]}]_{,L}\mathcal{E}^{JKL}\\ &= -\tilde{F}^{\alpha}{}_{I}[\tilde{\mathcal{L}}^{I}{}_{[JK],L} + \tilde{f}^{I}{}_{\beta}\tilde{F}^{\beta}{}_{M,L}\tilde{\mathcal{L}}^{M}{}_{[JK]}]\mathcal{E}^{JKL}.\end{aligned}$$

Reassigning further the skew symmetries in the indices JKL due to the triple contraction with the permutation symbol results in an expression containing the plastic curvature tensor

$$\tilde{D}^{\alpha L}{}_{:L} = \tilde{F}^{\alpha}{}_{I}[\tilde{\mathcal{L}}^{I}{}_{J[L,K]} + \tilde{\mathcal{L}}^{I}{}_{M[K}\tilde{\mathcal{L}}^{M}{}_{JL]}]\mathcal{E}^{JKL}.$$

Finally, the axial vector $\tilde{\mathcal{E}}^{I}$ of the plastic Einstein tensor and thus the algebraic Bianchi identity is recognized from the triple contraction of the curvature tensor with the permutation symbol

$$\tilde{D}^{\alpha L}{}_{:L} = \frac{1}{2}\tilde{F}^{\alpha}{}_{I}\tilde{\mathcal{R}}^{I}{}_{JKL}\mathcal{E}^{JKL} = 2\tilde{F}^{\alpha}{}_{I}\tilde{\mathcal{E}}^{I} = 0.$$

Summarizing, since elastoplasticity may be considered a flat but non-symmetric geometry over the material configuration with $\tilde{\mathcal{R}}^{I}{}_{JKL} = 0$ the skew part of $\tilde{\mathcal{R}}^{I}{}_{JKL} = 0$ coincides with the corresponding algebraic Bianchi identity, compare Eq. (101). □

Elastic Part of the Deformation The the well-accepted multiplicative decomposition of the deformation gradient in Eq. (164) allows an alternative representation in terms of the inverse deformation gradient and its corresponding decomposition into an elastic and a plastic part

$$\boldsymbol{f} = \tilde{\boldsymbol{f}} \cdot \bar{\boldsymbol{f}}. \tag{180}$$

In the sequel quantities referring to the elastic part of the deformation either in the spatial or in the intermediate configuration shall be denoted by an overbar $(\bar{\bullet})$. Recall that the isoclinic intermediate configuration of elastoplasticity is incompatible.

However, due to its isoclinic character the intermediate configuration of elastoplasticity is endowed with distance determination and parallel transport of Euclidean type, thus Cartesian-like coordinates may be selected

[12] Sticking to the total covariant derivative produces alternatively

$$\tilde{D}^{\alpha L}{}_{:L} = -\tilde{F}^{\alpha}{}_{I}[\tilde{\mathcal{L}}^{I}{}_{[JK]:L} + \tilde{f}^{I}{}_{\beta}\tilde{F}^{\beta}{}_{M:L}\tilde{\mathcal{L}}^{M}{}_{[JK]}]\mathcal{E}^{JKL}.$$

By resorting to Eqs. (165).2 and (176).1 and using the relation between the third and second-order dislocation density tensors the divergence statement in Eq. (179) is recovered.

again with the metric coefficients equal to the Kronecker-delta and a zero connection

$$\bar{M}_{\alpha\beta} = \delta_{\alpha\beta} \quad \text{and} \quad \bar{L}^{\alpha}{}_{\beta\gamma} = 0. \tag{181}$$

The elastic convection of the metric and the connection then render the *elastic Finger tensor* and an *integrable elastic connection* in the spatial configuration

$$\bar{m}_{ij} = \bar{c}_{ij} := \bar{f}^{\alpha}{}_{i}\delta_{\alpha\beta}\bar{f}^{\beta}{}_{j} \quad \text{and} \quad \bar{l}^{i}{}_{jk} = \bar{F}^{i}{}_{\alpha}\bar{f}^{\alpha}{}_{j,k}. \tag{182}$$

It is again easy to proof that the fully covariant elastic connection $\bar{l}^{i}{}_{jk} = \bar{c}_{im}\bar{l}^{m}{}_{jk}$ is indeed a metric connection with respect to $\bar{c}_{ij}$, due to

$$\bar{c}_{ij,k} = \bar{f}^{\alpha}{}_{i,k}\delta_{\alpha\beta}\bar{f}^{\beta}{}_{j} + \bar{f}^{\alpha}{}_{i}\delta_{\alpha\beta}\bar{f}^{\beta}{}_{j,k} = 2\bar{l}_{(ij)k}. \tag{183}$$

Therefore the elastic connection $\bar{l}_{ijk}$ decomposes into the elastic Riemann connection

$$\bar{m}_{ijk} = \bar{c}_{ijk} := [\bar{c}_{ij,k} - \bar{c}_{jk,i} + \bar{c}_{ki,j}]/2 \tag{184}$$

based on $\bar{c}_{ij}$ and the associated elastic contortion

$$\bar{k}_{ijk} := \bar{l}_{ijk} - \bar{m}_{ijk}. \tag{185}$$

The verification follows easily from rewriting Eq. (183) three times upon cyclic permutation of indices and subsequent addition to render

$$\begin{array}{rcrlllrlll}
\bar{c}_{ij,k} & = & + & \bar{f}^{\alpha}{}_{j} & \delta_{\alpha\beta} & \bar{f}^{\beta}{}_{i,k} & + & \bar{f}^{\alpha}{}_{i} & \delta_{\alpha\beta} & \bar{f}^{\beta}{}_{j,k} \\
-\bar{c}_{jk,i} & = & - & \bar{f}^{\alpha}{}_{k} & \delta_{\alpha\beta} & \bar{f}^{\beta}{}_{j,i} & - & \bar{f}^{\alpha}{}_{j} & \delta_{\alpha\beta} & \bar{f}^{\beta}{}_{k,i} \\
\bar{c}_{ki,j} & = & + & \bar{f}^{\alpha}{}_{i} & \delta_{\alpha\beta} & \bar{f}^{\beta}{}_{k,j} & + & \bar{f}^{\alpha}{}_{k} & \delta_{\alpha\beta} & \bar{f}^{\beta}{}_{i,j} \\
\hline\hline
\end{array} \tag{186}$$

$$\bar{c}_{ijk} = \bar{f}^{\alpha}{}_{i}\delta_{\alpha\beta}\bar{f}^{\beta}{}_{(j,k)} - \bar{f}^{\alpha}{}_{j}\delta_{\alpha\beta}\bar{f}^{\beta}{}_{[k,i]} + \bar{f}^{\alpha}{}_{k}\delta_{\alpha\beta}\bar{f}^{\beta}{}_{[i,j]} \tag{187}$$

and consequently by Eq. (185) the elastic contortion is computed by

$$\bar{k}_{ijk} = \bar{f}^{\alpha}{}_{i}\delta_{\alpha\beta}\bar{f}^{\beta}{}_{[j,k]} + \bar{f}^{\alpha}{}_{j}\delta_{\alpha\beta}\bar{f}^{\beta}{}_{[k,i]} - \bar{f}^{\alpha}{}_{k}\delta_{\alpha\beta}\bar{f}^{\beta}{}_{[i,j]}. \tag{188}$$

Moreover, since the elastic (spatial) connection is integrable, its associated curvature tensor vanishes identically

$$\bar{r}_{ijkl} = 2\bar{l}_{ij[l,k]} - 2\bar{l}_{mi[k}\bar{l}^{m}{}_{jl]} \equiv 0. \tag{189}$$

Here, since the elastic connection decomposes into the elastic Riemann connection and the elastic contortion, the corresponding curvature tensors

$$\bar{c}_{ijkl} := 2\bar{c}_{ij[l,k]} - 2\bar{c}_{mi[k}\bar{c}^{m}{}_{jl]} \neq 0, \tag{190}$$

and

$$\bar{\kappa}_{ijkl} := 2\bar{\kappa}_{ij[l,k]} - 2\bar{\kappa}_{mi[k}\bar{\kappa}^m{}_{jl]} \neq 0 \tag{191}$$

establish a relation between the (elastic) incompatibilities of the elastic Finger tensor $\bar{c}_{ij}$ as contained in $\bar{c}_{ijkl}$ on the one hand and the dislocation density contained in $\bar{\kappa}_{ijkl}$ on the other hand

$$\bar{c}_{ijkl} = -\bar{\kappa}_{ijkl} + 2\bar{\kappa}_{mi[k}\bar{c}^m{}_{jl]} + 2\bar{c}_{mi[k}\bar{\kappa}^m{}_{jl]}. \tag{192}$$

Observe that $\bar{c}_{ijkl}$ obeys major symmetry in the index pairs ij and kl, thus Eq. (192) is a result of the major symmetric part of $\bar{r}_{ijkl} = 0$. The consequence of its major skew symmetric part or, likewise, of the right hand side of Eq. (192) is highlighted in the sequel.

Associated with the incompatibility of the intermediate configuration, and referring to Eqs. (43).2 and (39).2, are the third-order dislocation densities (anholonomic objects)

$$\bar{a}^i{}_{jk} = -\bar{F}^i{}_\alpha \bar{f}^\alpha{}_{[j,k]} \quad \text{and} \quad \bar{A}^\alpha{}_{\beta\gamma} = \bar{f}^\alpha{}_{[j,k]}\bar{F}^j{}_\beta \bar{F}^k{}_\gamma. \tag{193}$$

These dislocation densities are related to the corresponding torsions as computed from the connections in Eqs. (182).2 and (181).2

$$\bar{a}^i{}_{jk} = -\bar{l}^i{}_{[jk]} = -\bar{t}^i{}_{jk} \quad \text{and} \quad \bar{A}^\alpha{}_{\beta\gamma} = \bar{T}^\alpha{}_{\beta\gamma} - \bar{L}^\alpha{}_{[\beta\gamma]} = \tilde{T}^\alpha{}_{\beta\gamma}. \tag{194}$$

Obviously, due to the representation of the spatial elastic connection in Eq. (182).2 its torsion coincides with the negative of the elastic (Piola-Kirchhoff-type) anholonomic object in Eq. (193).1; likewise due to the representation of the torsion in an anholonomic space in Eq. (40).2 it coincides with the elastic (Cauchy-type) anholonomic object in Eq. (193).2 due to the vanishing elastic connection in Eq. (181).2. Then, the (contravariant) second-order Piola-Kirchhoff- and Cauchy-type dislocation density tensors are expressed in the ambient Euclidean space as

$$\bar{\boldsymbol{a}} = -\bar{\boldsymbol{F}} \cdot \bar{\boldsymbol{d}} \quad \text{and} \quad \bar{\boldsymbol{A}} = \bar{\boldsymbol{d}} \cdot \mathrm{cof}\bar{\boldsymbol{F}} \quad \text{with} \quad \bar{\boldsymbol{d}} = \mathrm{curl}\bar{\boldsymbol{f}}. \tag{195}$$

Finally the continuity equation of continuum dislocation theory states that the dislocation lines may not end within the volume, i.e. the elastic dislocation density is source free (solenoidal)

$$\mathrm{div}\bar{\boldsymbol{d}} = -\bar{\boldsymbol{f}} \cdot [\mathrm{div}\bar{\boldsymbol{a}} + \bar{\boldsymbol{F}} \cdot \nabla_x \bar{\boldsymbol{f}} : \bar{\boldsymbol{a}}] = \boldsymbol{0} \quad \text{since} \quad \bar{\boldsymbol{d}} = \mathrm{curl}\bar{\boldsymbol{f}}. \tag{196}$$

The solenoidal character of the plastic (Piola-type) dislocation density is directly related to the algebraic Bianchi identity according to Eq. (101) as may be seen from the following

Proof:
According to Eq. (87).4 the coefficients of the second-order dislocation density tensor follow from its third-order counterpart as

$$\bar{d}^{\alpha l} = -\bar{d}^{\alpha}{}_{jk} e^{jkl}.$$

Then the divergence of the second-order dislocation density tensor involves the total covariant derivative in Euclidean ambient space in terms of the spatial Christoffel symbol $n^{i}{}_{jk}$ (the Christoffel symbol related to the coordinates in the intermediate configuration $\bar{N}^{\alpha}{}_{\beta\gamma} \equiv 0$ vanishes due to the selection of Cartesian-like coordinates, see Eq. (181).1)

$$\bar{d}^{\alpha l}{}_{:l} = -\bar{d}^{\alpha}{}_{jk:l} e^{jkl} \equiv -\bar{d}^{\alpha}{}_{jk,l} e^{jkl}.$$

Here, firstly $e^{jkl}{}_{:l} = 0$ and secondly the right symmetry of the spatial Christoffel symbol have been exploited to derive at a result in terms of the partial rather than the total covariant derivative of the dislocation density. Next, $\bar{d}^{\alpha}{}_{jk}$ is re-expressed in terms of the connection in Eq. (182).2 to render

$$\bar{d}^{\alpha}{}_{jk} = \bar{f}^{\alpha}{}_{i} \bar{\ell}^{i}{}_{[jk]}.$$

Then, application of the product rule gives

$$\begin{aligned} \bar{d}^{\alpha l}{}_{:l} &= -[\bar{f}^{\alpha}{}_{i} \bar{l}^{i}{}_{[jk]}]_{,l} e^{jkl} \\ &= -\bar{f}^{\alpha}{}_{i} [\bar{\ell}^{i}{}_{[jk],l} + \bar{F}^{i}{}_{\beta} \bar{f}^{\beta}{}_{m,l} \bar{\ell}^{m}{}_{[jk]}] e^{jkl}. \end{aligned}$$

Reassigning further the skew symmetries in the indices jkl due to the triple contraction with the permutation symbol results in an expression containing the elastic curvature tensor in the spatial configuration

$$\bar{d}^{\alpha l}{}_{:l} = \bar{f}^{\alpha}{}_{i} [\bar{\ell}^{i}{}_{j[l,k]} + \bar{\ell}^{i}{}_{m[k} \bar{\ell}^{m}{}_{jl]}] e^{jkl}.$$

Finally, the axial vector $\bar{e}^{i}$ of the elastic Einstein tensor and thus the associated algebraic Bianchi identity is recognized from the triple contraction of the elastic curvature tensor with the permutation symbol

$$\bar{d}^{\alpha l}{}_{:l} = \frac{1}{2} \bar{f}^{\alpha}{}_{i} \bar{r}^{i}{}_{jkl} e^{jkl} = 2 \bar{f}^{\alpha}{}_{i} \bar{e}^{i} = 0.$$

Summarizing, since elastoplasticity may be considered a flat but non-symmetric geometry over the spatial configuration with $\bar{r}^{i}{}_{jkl} = 0$ the skew part of $\bar{r}^{i}{}_{jkl} = 0$ coincides with the corresponding algebraic Bianchi identity, compare Eq. (101). □

Dislocation Densities in the Intermediate Configuration Due to the multiplicative decomposition and the compatibility of the total deformation the plastic Cauchy-type and the elastic Piola-Kirchhoff-type *third-order* dislocation densities as introduced in Eqs. (176).2 and (193).2 together with the corresponding torsions in the intermediate configuration are related by the identities

$$\tilde{a}^{\alpha}{}_{\beta\gamma} \equiv \bar{A}^{\alpha}{}_{\beta\gamma} \quad \text{and} \quad \tilde{t}^{\alpha}{}_{\beta\gamma} \equiv \bar{T}^{\alpha}{}_{\beta\gamma}. \tag{197}$$

Proof:
The multiplicative decomposition of the deformation gradient reads

$$F^{i}{}_{J} = \bar{F}^{i}{}_{\alpha} \tilde{F}^{\alpha}{}_{J}.$$

Consequently the material gradient of the deformation gradient, i.e. the second-order distorsion expands into

$$F^{i}{}_{J,K} = \bar{F}^{i}{}_{\alpha,K} \tilde{F}^{\alpha}{}_{J} + \bar{F}^{i}{}_{\alpha} \tilde{F}^{\alpha}{}_{J,K} = -\bar{F}^{i}{}_{\alpha} \bar{f}^{\alpha}{}_{j,k} F^{j}{}_{J} F^{k}{}_{K} + \bar{F}^{i}{}_{\alpha} \tilde{F}^{\alpha}{}_{J,K}.$$

Here the relation $\bar{F}^{i}{}_{\alpha,K} = -\bar{F}^{i}{}_{\beta} \bar{f}^{\beta}{}_{j,K} \bar{F}^{j}{}_{\alpha}$ has been used. Multiplication with $\tilde{f}^{J}{}_{\beta} \tilde{f}^{K}{}_{\gamma}$ and skew symmetrization in the indices JK then discovers the relation between the elastic and plastic Piola-type dislocation densities

$$\bar{d}^{\alpha}{}_{jk} \bar{F}^{j}{}_{\beta} \bar{F}^{k}{}_{\gamma} = \tilde{D}^{\alpha}{}_{JK} \tilde{f}^{J}{}_{\beta} \tilde{f}^{K}{}_{\gamma}.$$

As a consequence the relation between the elastic Piola-Kirchhoff-type and the elastic Cauchy-type third-order dislocation densities then follows as

$$\bar{A}^{\alpha}{}_{\beta\gamma} = \tilde{a}^{\alpha}{}_{\beta\gamma}$$

Clearly this relation also holds for the corresponding torsions. □

Alternatively consider the

Proof:
The multiplicative decomposition of the inverse deformation gradient reads

$$f^{I}{}_{j} = \tilde{f}^{I}{}_{\alpha} \bar{f}^{\alpha}{}_{j}.$$

Consequently the spatial gradient of the deformation gradient expands into

$$f^{I}{}_{j,k} = \tilde{f}^{I}{}_{\alpha,k} \bar{f}^{\alpha}{}_{j} + \tilde{f}^{I}{}_{\alpha} \bar{f}^{\alpha}{}_{j,k} = -\tilde{f}^{I}{}_{\alpha} \tilde{F}^{\alpha}{}_{J,K} f^{J}{}_{j} f^{K}{}_{k} + \tilde{f}^{I}{}_{\alpha} \bar{f}^{\alpha}{}_{j,k}.$$

Here the relation $\tilde{f}^{I}{}_{\alpha,k} = -\tilde{f}^{I}{}_{\beta} \tilde{F}^{\beta}{}_{J,k} \tilde{f}^{J}{}_{\alpha}$ has been used. Multiplication with $\bar{F}^{j}{}_{\beta} \bar{F}^{k}{}_{\gamma}$ and skew symmetrization in the indices jk then discovers the relation between the plastic and elastic Piola-type dislocation densities

$$\tilde{D}^{\alpha}{}_{JK} \tilde{f}^{J}{}_{\beta} \tilde{f}^{K}{}_{\gamma} = \bar{d}^{\alpha}{}_{jk} \bar{F}^{j}{}_{\beta} \bar{F}^{k}{}_{\gamma}.$$

As a consequence the relation between the plastic Cauchy-type and the elastic Piola-Kirchhoff-type dislocation densities then follows as

$$\tilde{a}^{\alpha}{}_{\beta\gamma} = \bar{A}^{\alpha}{}_{\beta\gamma}.$$

Obviously this relation also holds for the corresponding torsions. □

As a consequence the plastic Cauchy-type and the elastic Piola-Kirchhoff-type *second-order* dislocation densities in the intermediate configuration as expressed in the ambient Euclidean space are likewise related by the identity

$$\widetilde{\boldsymbol{a}} \equiv \bar{\boldsymbol{A}}. \tag{198}$$

Proof:
The second-order plastic and elastic Piola-type dislocation densities in Euclidean ambient space are computed from their third-order counterparts

$$\tilde{D}^{\alpha L} = -\tilde{D}^{\alpha}{}_{JK}\mathcal{E}^{JKL} \quad \text{and} \quad \bar{d}^{\alpha l} = -\bar{d}^{\alpha}{}_{jk}e^{jkl}.$$

Then the convection to the intermediate configuration according to Eqs. (178).2 and (195).2 and by considering the inverse of the relation between the third-order tensors as stated in Eqs. (176).2 and (193).2 renders

$$\tilde{a}^{\alpha\delta} = -\tilde{j}\tilde{a}^{\alpha}{}_{\beta\gamma}\tilde{F}^{\beta}{}_{J}\tilde{F}^{\gamma}{}_{K}\mathcal{E}^{JKL}\tilde{F}^{\delta}{}_{L} \quad \text{and} \quad \bar{A}^{\alpha\delta} = -\bar{J}\bar{A}^{\alpha}{}_{\beta\gamma}\bar{f}^{\beta}{}_{j}\bar{f}^{\gamma}{}_{k}e^{jkl}\bar{f}^{\delta}{}_{l}.$$

Here, $\tilde{j} = \det \widetilde{\boldsymbol{f}}$ and $\bar{J} = \det \bar{\boldsymbol{F}}$ denote the Jacobian determinants of the (inverse) plastic and the elastic part of the deformation gradient, respectively. Thus, due to the relation between the permutation tensors $\boldsymbol{E}$, $\widetilde{\boldsymbol{e}} \equiv \bar{\boldsymbol{E}}$ and $\boldsymbol{e}$ in the material, the intermediate and the spatial configuration, i.e.

$$\widetilde{\boldsymbol{e}} := \tilde{j}[\widetilde{\boldsymbol{F}}\overline{\otimes}\widetilde{\boldsymbol{F}}] : \boldsymbol{E}\cdot\widetilde{\boldsymbol{F}}^{t} \equiv \bar{J}[\bar{\boldsymbol{f}}\overline{\otimes}\bar{\boldsymbol{f}}] : \boldsymbol{e}\cdot\bar{\boldsymbol{f}}^{t} =: \bar{\boldsymbol{E}} \tag{199}$$

together with the result as stated in Eq. (197).1 for the third-order dislocation densities the equality of the second-order plastic and elastic Cauchy-type dislocation densities per unit volume in the intermediate configuration in Eq. (198) follows identically with

$$\tilde{a}^{\alpha\delta} \equiv \bar{A}^{\alpha\delta}.$$

Clearly this relation also holds for the corresponding second-order torsions. □

Anholonomic Curl Operators It proves useful to introduce besides the curl operators as defined in Euclidean space with respect to the material and spatial coordinates in Eqs. (137) and (139) also curl operators with respect to the anholonomic coordinates in the intermediate configuration, compare Menzel and Steinmann (2007). Thus, with the Cartesian base vectors $\bar{\boldsymbol{E}}_\gamma$ and the tangent vectors $\bar{\boldsymbol{G}}_\gamma$ in the intermediate configuration denoted by an overbar, the *elastic curl operator* as applied to the elastic part of the deformation gradient is introduced as

$$\mathrm{C\bar{u}rl}\bar{\boldsymbol{F}} := \bar{D}^{a\gamma}\boldsymbol{e}_a \otimes \bar{\boldsymbol{E}}_\gamma = \bar{\mathcal{D}}^{i\gamma}\boldsymbol{g}_i \otimes \bar{\boldsymbol{G}}_\gamma. \tag{200}$$

Here, the coefficients follow essentially the standard definition for the (material) curl operator, however, since the intermediate configuration is incompatible, the gradient with respect to the anholonomic coordinates in the intermediate configuration is expanded as the (elastic) *anholonomic gradient* $(\bullet)_{,\beta} := (\bullet)_{,b}\bar{F}^b{}_\beta$ to render

$$\bar{D}^{a\gamma} = -\bar{F}^a{}_{\alpha,b}\bar{F}^b{}_\beta\bar{E}^{\alpha\beta\gamma} \quad \text{and} \quad \bar{\mathcal{D}}^{i\gamma} = -\bar{\mathcal{F}}^i{}_{\alpha:j}\bar{\mathcal{F}}^j{}_\beta\bar{\mathcal{E}}^{\alpha\beta\gamma}. \tag{201}$$

Furthermore, with the Cartesian base vectors $\widetilde{\boldsymbol{e}}_\gamma$ and the tangent vectors $\widetilde{\boldsymbol{g}}_\gamma$ in the intermediate configuration[13] denoted by a tilde, the *plastic curl operator* as applied to the plastic part of the deformation gradient is introduced as

$$\mathrm{c\tilde{u}rl}\widetilde{\boldsymbol{f}} := \tilde{d}^{A\gamma}\boldsymbol{E}_A \otimes \widetilde{\boldsymbol{e}}_\gamma = \tilde{d}^{I\gamma}\boldsymbol{G}_I \otimes \widetilde{\boldsymbol{g}}_\gamma. \tag{202}$$

Here, the coefficients follow essentially the standard definition for the (spatial) curl operator, however, since the intermediate configuration is incompatible, the gradient with respect to the anholonomic coordinates in the intermediate configuration is expanded as the (plastic) anholonomic gradient $(\bullet)_{,\beta} := (\bullet)_{,B}\tilde{f}^B{}_\beta$ to render

$$\tilde{d}^{A\gamma} = -\tilde{f}^A{}_{\alpha,B}\tilde{f}^B{}_\beta\tilde{e}^{\alpha\beta\gamma} \quad \text{and} \quad \tilde{d}^{I\gamma} = -\tilde{f}^I{}_{\alpha:J}\tilde{f}^J{}_\beta\tilde{e}^{\alpha\beta\gamma}. \tag{203}$$

These definitions together with the relation between the permutation tensors $\boldsymbol{E}$, $\widetilde{\boldsymbol{e}} \equiv \bar{\boldsymbol{E}}$ and $\boldsymbol{e}$ in the material, the intermediate and the spatial configuration in Eq. (199) allow to introduce the convection of the spatial into the elastic and the material into the plastic curl operators, respectively

$$\mathrm{C\bar{u}rl}\bar{\boldsymbol{F}} = -\bar{J}\bar{\boldsymbol{F}} \cdot \mathrm{curl}\bar{\boldsymbol{f}} \cdot \bar{\boldsymbol{f}}^t \quad \text{and} \quad \mathrm{c\tilde{u}rl}\widetilde{\boldsymbol{f}} = -\tilde{j}\widetilde{\boldsymbol{f}} \cdot \mathrm{Curl}\widetilde{\boldsymbol{F}} \cdot \widetilde{\boldsymbol{F}}^t. \tag{204}$$

[13] Note that the notation for the base and tangent vectors in the intermediate configuration is doubled for notational consistency, i.e. $\bar{\boldsymbol{E}}_\gamma \equiv \widetilde{\boldsymbol{e}}_\gamma$ and $\bar{\boldsymbol{G}}_\gamma \equiv \widetilde{\boldsymbol{g}}_\gamma$ hold.

The above relations will be exploited in the sequel to (i) introduce further versions of dislocation density tensors and to (ii) identify relations between these and the previously defined dislocation density tensors.

Furthermore as a consequence of the multiplicative decomposition of the deformation gradient in Eq. (164), the definition of the elastic curl operator and the convection of the permutation tensors allow to decompose the material curl of the deformation gradient *additively* into elastic and plastic contributions[14]

$$\mathrm{Curl}\boldsymbol{F} = \mathrm{C\bar{u}rl}\bar{\boldsymbol{F}} \cdot \mathrm{cof}\widetilde{\boldsymbol{F}} + \bar{\boldsymbol{F}} \cdot \mathrm{Curl}\widetilde{\boldsymbol{F}} \doteq \boldsymbol{0}. \tag{205}$$

Likewise the multiplicative decomposition of the inverse deformation gradient in Eq. (180) together with the definition of the plastic curl operator and the convection of the permutation tensors allow to decompose the spatial curl of the inverse deformation gradient *additively* into plastic and elastic contributions

$$\mathrm{curl}\boldsymbol{f} = \mathrm{c\tilde{u}rl}\widetilde{\boldsymbol{f}} \cdot \mathrm{cof}\bar{\boldsymbol{f}} + \widetilde{\boldsymbol{f}} \cdot \mathrm{curl}\bar{\boldsymbol{f}} \doteq \boldsymbol{0}. \tag{206}$$

The additive decomposition of the material and spatial curl as applied to the total deformation gradient is in agreement with the equality of the plastic and elastic Cauchy-type dislocation densities in Eq. (198) as may be demonstrated in the following.

Dislocation Densities Based on Anholonomic Curl Based on the anholonomic elastic curl operator an additional (contravariant) Piola-type dislocation density may be introduced in the ambient Euclidean space as

$$\bar{\boldsymbol{D}} := \mathrm{C\bar{u}rl}\bar{\boldsymbol{F}}. \tag{207}$$

Observe that $\bar{\boldsymbol{D}}$ is a two-point tensor deriving from an anholonomic elastic curl that maps from the intermediate to the spatial configuration. As such

[14] The additive decomposition of the material curl of the deformation gradient into elastic and plastic parts follows in detail from the step by step calculation

$$\begin{aligned}
\mathrm{Curl}\boldsymbol{F} &= -\nabla_X(\bar{\boldsymbol{F}} \cdot \widetilde{\boldsymbol{F}}) : \boldsymbol{E} \\
&= -\nabla_x \bar{\boldsymbol{F}} : [\widetilde{\boldsymbol{F}} \overline{\otimes} \boldsymbol{F}] : \boldsymbol{E} - \bar{\boldsymbol{F}} \cdot \nabla_X \widetilde{\boldsymbol{F}} : \boldsymbol{E} \\
&= -[\nabla_x \bar{\boldsymbol{F}} \cdot \bar{\boldsymbol{F}}] : [\widetilde{\boldsymbol{F}} \overline{\otimes} \widetilde{\boldsymbol{F}}] : \boldsymbol{E} + \bar{\boldsymbol{F}} \cdot \mathrm{Curl}\widetilde{\boldsymbol{F}} \\
&= -\tilde{J}[\nabla_x \bar{\boldsymbol{F}} \cdot \bar{\boldsymbol{F}}] : [\widetilde{\boldsymbol{F}} \overline{\otimes} \widetilde{\boldsymbol{F}}] : [\widetilde{\boldsymbol{f}} \overline{\otimes} \widetilde{\boldsymbol{f}}] : \bar{\boldsymbol{E}} \cdot \widetilde{\boldsymbol{f}}^t + \bar{\boldsymbol{F}} \cdot \mathrm{Curl}\widetilde{\boldsymbol{F}} \\
&= -[\nabla_x \bar{\boldsymbol{F}} \cdot \bar{\boldsymbol{F}}] : \bar{\boldsymbol{E}} \cdot \mathrm{cof}\widetilde{\boldsymbol{F}} + \bar{\boldsymbol{F}} \cdot \mathrm{Curl}\widetilde{\boldsymbol{F}} \\
&= \mathrm{C\bar{u}rl}\bar{\boldsymbol{F}} \cdot \mathrm{cof}\widetilde{\boldsymbol{F}} + \bar{\boldsymbol{F}} \cdot \mathrm{Curl}\widetilde{\boldsymbol{F}}.
\end{aligned}$$

this situation is different from the previously introduced elastic Piola-type dislocation density $\bar{\boldsymbol{d}} = \text{curl}\bar{\boldsymbol{f}}$ that is a two-point tensor mapping from the spatial to the intermediate configuration and that derives from the ordinary spatial curl. However, due to the convection of curl operators in Eq. (204).1 these two Piola-type dislocation densities are related by

$$\bar{\boldsymbol{D}} = -\bar{\boldsymbol{F}} \cdot \bar{\boldsymbol{d}} \cdot \text{cof}\bar{\boldsymbol{F}}. \tag{208}$$

Therefore, the (elastic) Piola-Kirchhoff- and Cauchy-type dislocation densities $\bar{\boldsymbol{A}}$ and $\bar{\boldsymbol{a}}$ as introduced in Eq. (195) are related to $\bar{\boldsymbol{D}}$ by

$$\bar{\boldsymbol{A}} = -\bar{\boldsymbol{f}} \cdot \bar{\boldsymbol{D}} \quad \text{and} \quad \bar{\boldsymbol{a}} = \bar{\boldsymbol{D}} \cdot \text{cof}\bar{\boldsymbol{f}}. \tag{209}$$

Likewise, based on the anholonomic plastic curl operator a further (contravariant) Piola-type dislocation density may be introduced in the ambient Euclidean space as

$$\widetilde{\boldsymbol{d}} = \widetilde{\text{curl}}\widetilde{\boldsymbol{f}}. \tag{210}$$

Note that $\widetilde{\boldsymbol{d}}$ is a two-point tensor deriving from an anholonomic plastic curl that maps from the intermediate to the material configuration. As such this situation is different from the previously introduced plastic Piola-type dislocation density $\widetilde{\boldsymbol{D}} = \text{Curl}\widetilde{\boldsymbol{F}}$ that is a two-point tensor mapping from the material to the intermediate configuration and that derives from the ordinary material curl. However, due to the convection of curl operators in Eq. (204).2 these two Piola-type dislocation densities are related by

$$\widetilde{\boldsymbol{d}} = -\widetilde{\boldsymbol{f}} \cdot \widetilde{\boldsymbol{D}} \cdot \text{cof}\widetilde{\boldsymbol{f}}. \tag{211}$$

Thus, the (plastic) Piola-Kirchhoff- and Cauchy-type dislocation densities $\widetilde{\boldsymbol{A}}$ and $\widetilde{\boldsymbol{a}}$ as introduced in Eq. (178) are related to $\widetilde{\boldsymbol{d}}$ by

$$\widetilde{\boldsymbol{a}} = -\widetilde{\boldsymbol{F}} \cdot \widetilde{\boldsymbol{d}} \quad \text{and} \quad \widetilde{\boldsymbol{A}} = \widetilde{\boldsymbol{d}} \cdot \text{cof}\widetilde{\boldsymbol{F}}. \tag{212}$$

Summarizing, as depicted in Fig. 8 the multiplicative decomposition of the deformation gradient into incompatible elastic and plastic contributions in Eq. (164) (or likewise Eq. (180)) results in the introduction of twelve (total, elastic and plastic) dislocation density tensors of Piola- , Piola-Kirchhoff- and Cauchy-type, respectively. Thereby the six Piola-type dislocation density tensors in a) are defined as the application of the appropriate curl operator to the corresponding contribution to the deformation gradient. Accordingly, the six Piola-Kirchhoff- and Cauchy-type dislocation density tensors in b) and c) are either a result from appropriate pull-back/push-forward operations or a result from the convection of curl operators, respectively.

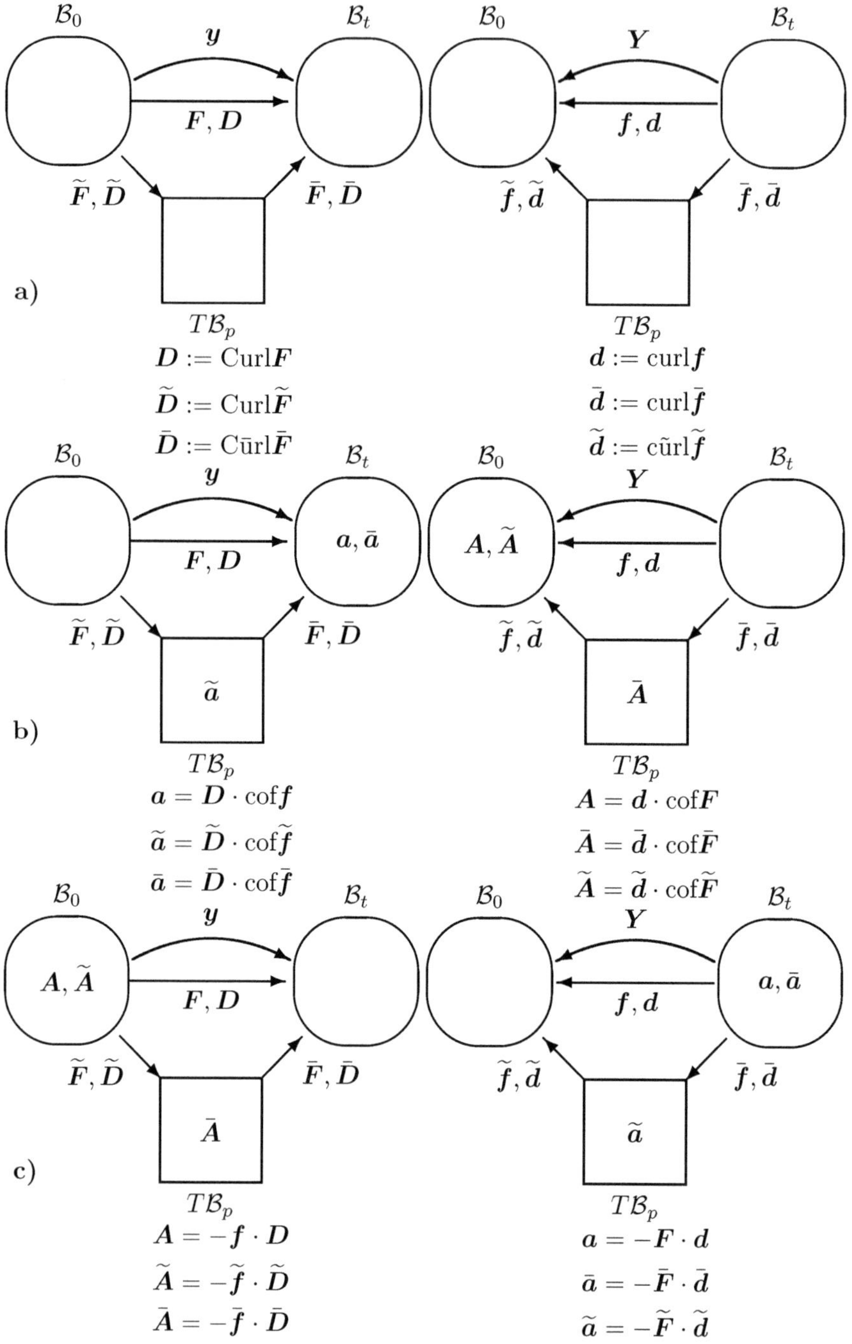

Figure 8. **a)** Piola-type dislocation density tensors; **b)** Cauchy-type dislocation density tensors; **c)** Piola-Kirchhoff-type dislocation density tensors.

From Eqs. (205) and (206) the previously derived equality of the elastic and plastic dislocation densities in the intermediate configuration finally follows in the format

$$\underbrace{\bar{\boldsymbol{f}} \cdot \bar{\boldsymbol{D}}}_{-\bar{\boldsymbol{A}}} + \underbrace{\widetilde{\boldsymbol{D}} \cdot \operatorname{cof} \widetilde{\boldsymbol{f}}}_{\widetilde{a}} = \boldsymbol{0} \quad \text{and} \quad \underbrace{\widetilde{\boldsymbol{F}} \cdot \widetilde{\boldsymbol{d}}}_{-\widetilde{a}} + \underbrace{\bar{\boldsymbol{d}} \cdot \operatorname{cof} \bar{\boldsymbol{F}}}_{\bar{\boldsymbol{A}}} = \boldsymbol{0}. \tag{213}$$

Obviously this is nothing but the statement of compatibility of the total deformation gradient as stated in Eqs. (205) and (206).

4 Summary

The foundations of geometrically nonlinear continuum mechanics in differential geometry are considered a key for the sound formulation of generalized models. Here this aspect has been demonstrated especially for the prominent example of multiplicative (crystal) elastoplasticity, whereby the dislocation density tensor and its variants are related to the torsion of the underlying Cartan space or rather the corresponding anholonomic object. Considering these kinematic quantities in a thermodynamically consistent formulation results typically in a gradient-type crystal plasticity, see Steinmann (1996). Intriguing interrelationships between concepts of differential geometry and physically relevant statements such as the algebraic Bianchi identity on the one hand and the solenoidal character of the dislocation density (corresponding to the continuity equation of continuum dislocation theory) on the other hand are paradigmatic to demonstrate the strength of this approach. However, differential geometry is much richer in that it allows also for more complex geometries such as Riemann-Cartan spaces and general affine spaces. Thus the combination of differential geometry and continuum mechanics opens the door to further, more sophisticated models of generalized continuum models that shall be explored elsewhere.

Bibliography

A. Acharya, and J.L. Bassani. Lattice Incompatibility and a Gradient Theory of Crystal Plasticity. *J. Mech. Phys. Solids* 48:1565–1595, 2000.

K.H. Anthony. Die Reduktion von nichteuklidischen geometrischen Objekten in eine euklidische Form und physikalische Deutung der Reduktion durch Eigenspannungszustnde in Kristallen. *Arch. Rat. Mech. Anal.* 37:161–180, 1970.

K.H. Anthony. Die Theorie der Disklinationen. *Arch. Rat. Mech. Anal.* 39:43–88, 1970.

K.H. Anthony. Die Theorie der nichtmetrischen Spannungen in Kristallen. *Arch. Rat. Mech. Anal.* 40:50–78, 1971.

M. Becker. *Incompatibility and Instability Based Size Effects in Crystals and Composites at Finite Elastoplastic Strains.* Bericht Nr.: I-18, Institut für Mechanik, Lehrstuhl I, Stuttgart, 2006.

B.A. Bilby, R. Bullough, and E. Smith. Continuous Distributions of Dislocations: A New Application of the Methods of Non-Riemannian Geometry. *Proc. Roy. Soc.* A231:263–237, 1955.

B.A. Bilby, and E. Smith. Continuous Distributions of Dislocations III. *Proc. Roy. Soc.* A236:481–505, 1956.

E. Cartan. Sur une Généralisation de la Notion de Courbure de Riemann et les Espace à Torsion. *Comptes Rend.* 174:593–575, 1922.

P. Cermelli, and M. Gurtin. On the Characterization of Geometrically Necessary Dislocations in Finite Plasticity. *J. Mech. Phys. Solids* 49:1539–1568, 2001.

J.D. Clayton, D.L. McDowell, and D.J. Bammann. Modeling Dislocations and Disclinations with Finite Micropolar Elastoplasticity. *Int. J. Plast.* 22:210–256, 2006.

J.D. Clayton. *Nonlinear Mechanics of Crystals.* Springer, Dordrecht etc., 2011.

J.L. Ericksen. Tensor Fields. In *Handbuch der Physik III/1.* Springer, Berlin etc., 1960.

S. Forest. The Micromorphic Approach for Gradient Elasticity, Viscoplasticity and Damage. *ASCE J. Engng. Mech.* 135:117–131, 2009.

P. Grammenoudis, and Ch. Tsakmakis. Plastic Intermediate Configuration and Related Spatial Differential Operators in Micromorphic Plasticity. *Math. Mech. Solids* 15:515–538, 2010.

M. Gurtin. A Gradient Theory of Single-Crystal Viscoplasticity that Accounts for Geometrically Necessary Dislocations. *J. Mech. Phys. Solids* 50:5–32, 2002.

P. Haupt. *Continuum Mechanics and Theory of Materials.* Springer, Berlin etc., 2000.

B. Hirschberger, and P. Steinmann. Classification of Concepts in Thermodynamically Consistent Generalized Plasticity. *ASCE J. Engng. Mech.* 135:156–170, 2009.

K. Kondo. On the Geometrical and Physical Foundation of the Theory of Yielding. *Proc. Japan Nat. Congress Appl. Mech.* 2:41–47, 1952.

K. Kondo. On the Analytical and Physical Foundations of the Theory of Dislocations and Yielding by the Differential Geometry of Continua. *Int. J. Engng. Sci.* 2:219–251, 1964.

E. Kröner. Kontinuumstheorie der Versetzungen und Eigenspannungen. *Erg. Angew. Math.* 5:1–179, 1958.

E. Kröner, and A. Seeger. Nichtlineare Elastizitätstheorie der Versetzungen und Eigenspannungen. *Arch. Rat. Mech. Anal.* 3:97–119, 1959.

E. Kröner. Allgemeine Kontinuumstheorie der Versetzungen und Eigenspannungen. *Arch. Rat. Mech. Anal.* 4:237–334, 1960.

E. Kröner. Differential Geometry of Defects in Condensed Systems of Particles with only Translational Mobility. *Int. J. Engng. Sci.* 19:1507–1515, 1981.

M. Lazar, and G.A. Maugin. Nonsingular Stress and Strain Fields of Dislocations and Disclinations in First Strain Gradient Elasticity. *Int. J. Engng. Sci.* 43:1157–1184, 2005.

M. Lazar, and F.W. Hehl. Cartan's Spiral Staircase in Physics and, in Particular, in the Gauge Theory of Dislocations. *Found. Phys.* 40:1298–1325, 2010.

K.C. Le, and H. Stumpf. A Model of Elastoplastic Bodies with Continuously Distributed Dislocations. *Int. J. Plast.* 12:611–627, 1996.

T. Liebe, and P. Steinmann. Theory and Numerics of a Thermodynamically Consistent Framework for Geometrically Linear Gradient Plasticity. *Int. J. Num. Meth. Engng* 51:1437–1467, 2001.

J.E. Marsden, and T.J.R. Hughes. *Mathematical Foundations of Elasticity.* Dover, New York, 1994.

A. Menzel, and P. Steinmann. On the Continuum Formulation of Higher Gradient Plasticity for Single and Polycrystals. *J. Mech. Phys. Solids* 48:1777–1796, 2000.

A. Menzel, and P. Steinmann. On the Configurational Forces in Multiplicative Elastoplasticity. *Int. J. Solids Struct.* 44:4442–4471, 2007.

C.W. Misner, K.S. Thorne, and J.A. Wheeler. *Gravitation.* Freeman and Company, New York, 1998.

W. Noll. Materially Uniform Simple Bodies with Inhomogeneities. *Arch. Rat. Mech. Anal.* 27:1–32, 1967.

J.F. Nye. Some Geometrical Relations in Dislocated Crystals. *Acta Metallurgica* 1:153–162, 1953.

B.D. Reddy, F. Ebobisse, and A. McBride. Well-Posedness of a Model of Strain Gradient Plasticity for Plastically Irrotational Materials. *Int. J. Plast.* 24:55–73, 2008.

P. Steinmann. Views on Multiplicative Elastoplasticity and the Continuum Theory of Dislocations. *Int. J. Engng. Sci.* 34:1717–1735, 1996.

B. Svendsen. Continuum Thermodynamic Models for Crystal Plasticity Including the Effects of Geometrically-Necessary Dislocations. *J. Mech. Phys. Solids* 50:1297–1329, 2002.

J.A. Schouten. *Ricci-Calculus.* Springer, Berlin, 1954.

J.A. Schouten. *Tensor Analysis for Physicists.* Dover, New York, 1989.

R. de Wit. A View of the Relation between the Continuum Theory of Lattice Defects and Non-Euclidean Geometry in the Linear Approximation. *Int. J. Engng. Sci.* 19:1475–1506, 1981.

Cosserat Media

Holm Altenbach [†] and Victor A. Eremeyev [†‡]
[†] Lehrstuhl Technische Mechanik, Institut für Mechanik,
Fakultät für Maschinenbau, Otto-von-Guericke-Universität Magdeburg,
Germany
[‡] South Scientific Center of RASci and South Federal University,
Rostov on Don, Russia

Abstract In this chapter we discuss the three-dimensional Cosserat-type theories of continua. Originally this approach belongs to Cosserat brothers, Eugèn and François, who introduced an elastic media with kinematically independent translational and rotational degrees of freedom in their centurial book "Théorie des corps déformables" (Cosserat and Cosserat, 1909). Within the framework of the Cosserat media each point of the media can be represented as an infinitesimal rigid body. This means that in this media there exist stresses and couple stresses as responses for translational and rotational degrees of freedom. These characteristic features of the Cosserat continuum model give a possibility to describe more complex media, for example, micro-inhomogeneous materials, polycrystalline and cellular solids, foams, lattices, masonries, particle assemblies, magnetic rheological fluids, liquid crystals, etc. The aim of this chapter consists of a brief presentation of basics of Cosserat continuum mechanics including kinematics, dynamics and constitutive modeling.

1 Introduction

Mechanics of Cosserat continua is based on the idea to consider the translational and rotational degrees of freedom of material particles as independent variables and so to add into the material description not only ordinary stresses but also couple stresses. This material model now is named *Cosserat's* or *micropolar continuum*. The basic ideas of this approach were first presented by Cosserat and Cosserat (1896, 1909). Since that time there are published tenth of books and thousands of papers in this field, see e.g. Eringen (1999, 2001); Nowacki (1986); Eremeyev et al. (2012) and the recent collections edited by Maugin and Metrikine (2010); Altenbach et al. (2011); Markert (2011) where the state of the art in the field of the Cosserat theory is presented.

In the sixties of the last century there were first summarized various results on the Cosserat continuum. Stojanović (1969a,b) collects more than 500 references in the field. The first set of papers on generalized continua and Cosserat media is given in Kröner (1968); Nowacki (1970); Nowacki and Olszak (1974).

It is worth to mention that the micropolar mechanics was the subject of several CISM courses organized by Eringen, Mindlin, Nowacki, Stojanović. In particular, the second CISM course "Mechanics of Polar Continua: Theory and Applications" was organized by Prof. Stojanović (Stojanović, 1969a).

The Cosserat model is used to describe media with a complex microstructure like soils, polycrystalline and composite materials, granular and powder-like materials, masonries, cellular or porous media and foams, bones, liquid crystals, as well as electromagnetic and ferromagnetic media. The application of the Cosserat continuum model to these media is quite natural. For example, open-cell and closed-cell foams can be considered as a system of elastic beams or plates, respectively. For beams and plates the force and couple interactions are essential, so stresses and couple stresses are the primary quantities as well as in the mechanics of structures. Replacing such highly inhomogeneous material as a foam by a quasi-homogeneous medium one can expect that this medium inherits the couple stress properties and rotational degrees of freedom of the initial system of beams or plates. Applying homogenization techniques for a micro-inhomogeneous media one obtains an "effective" quasi-homogeneous Cosserat-type medium. The transition from an inhomogeneous media to a generalized quasi-homogeneous one is analogous to the transition from the equations of three-dimensional (3D) elasticity to the two-dimensional (2D) equations of the shell theory with resultant stresses and couple stresses using the through-the-thickness integration technique or other 3D–to–2D reduction procedures. In the case of an electromagnetic medium one may take into account the couples induced by the electromagnetic field that leads to the appearance of the couple stresses in the medium, see Maugin's chapter in this book. Let us us note that after the paper by Ericksen and Truesdell (1958), the Cosserat model has found applications in development of various generalized models for beams, plates, and shells, see Chapters 3 and 4 in this book.

Nowadays the Cosserat continuum has taken a significant place in Continuum Mechanics, see for example recent books by Besson et al. (2010); Clayton (2011); Maugin (1999, 2010), and Narasimhan (1993), where the Cosserat medium model is included within the context of the mechanics of materials. The contents of this chapter is almost based on the works of Eremeyev and Zubov (1994); Zubov and Eremeev (1996); Eremeyev (2005);

Eremeyev and Zubov (2007, 2009); Pietraszkiewicz and Eremeyev (2009a,b); Altenbach et al. (2010b,a); Eremeyev and Pietraszkiewicz (2012) and the monograph by Eremeyev et al. (2012). The chapter is organized as follows. In Sect. 2 we briefly discuss the state of art in the field and present some elements of rigid body dynamics and the mechanics of rods to demonstrate that Cosserat's description of continua appears naturally from the classical mechanics as well as from the mechanics of structures and fill the gap between the Continuum Mechanics and General Mechanics. In Sect. 2 we recall the kinematics of the Cosserat continuum and introduce the directors and the microrotation tensor. In Sect. 3 we consider the theory of stresses and couple stresses. Here we present the motion equations of micropolar continuum and Lagrangian and Eulerian statements of the boundary-value problems. Section 4 is devoted to the theory of constitutive equations. The general principles of constitutive modeling are considered: determinism, local action, and material frame-indifference. In addition, we present the constitutive equations of the hyperelastic micropolar continuum, introduce strain measures, material symmetry group, and some constitutive inequalities.

In what follows, we use the following notations. We apply the direct (symbolic) tensorial notations as for example in Lebedev et al. (2010); Eremeyev et al. (2012). The vectors are denoted by semibold italic font like $\boldsymbol{A}$, $\boldsymbol{a}$, the second-order tensors are denoted by semibold roman font like $\mathbf{A}$, $\mathbf{a}$, the higher-order tensors are denoted by blackboard bold font like $\mathbb{A}$, Greece indices take values 1 and 2, whereas Latin indices are arbitrary. We also employ the Einstein summation convention.

1.1 Elements of Rigid Body Dynamics

Dynamics of rigid bodies is one of the oldest parts of mechanics. Such notions of rigid body mechanics as the moment, the inertia tensor and others are also used in mechanics of the Cosserat continua. We recall some notions and equations of classical mechanics that are used afterwards in this section. For details we can refer to various textbooks, see for example Lurie (2001).

The principle notion of classical mechanics is the *mass-point* or the *material point* or the *particle*. A mass-point possesses a mass m, its position is given by a radius vector $\boldsymbol{r}(t)$ at instant time t. The vector $\boldsymbol{r}(t)$ connects some given origin O, called also pole, with the point z, see Fig. 1. In an inertial coordinate system, the motion of the material point is determined by *Newton's second law*

$$m\dot{\boldsymbol{v}} = \boldsymbol{f}, \tag{1}$$

where $\boldsymbol{v} = \dot{\boldsymbol{r}}$ is the velocity, $\boldsymbol{f}$ is the force acting on the mass-point, and the

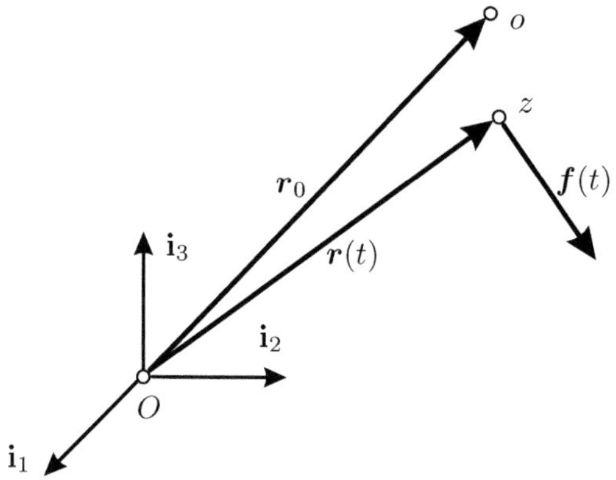

Figure 1. Mass-point motion.

overdot denotes the derivative with respect to t, for example, $\dot{\boldsymbol{r}} = \mathrm{d}\boldsymbol{r}/\,\mathrm{d}t$.

Definition 1.1. *The momentum of the mass-point (called also the linear momentum) is the quantity*

$$\mathfrak{P} = m\boldsymbol{v}.$$

Definition 1.2. *The moment of momentum of the mass-point (called also the angular momentum) with respect to a point o with radius vector* $\boldsymbol{r}_0$ *is*

$$\mathfrak{M} = (\boldsymbol{r} - \boldsymbol{r}_0) \times m\boldsymbol{v}.$$

The next object of classical mechanics is a system of n mass-points. The motion of n mass-points is determined by n vectorial equations

$$m_i \dot{\boldsymbol{v}}_i = \boldsymbol{f}_i, \qquad i = 1, 2, \ldots, n, \tag{2}$$

where $\boldsymbol{v}_i = \dot{\boldsymbol{r}}_i$, m_i and $\boldsymbol{r}_i(t)$ are the mass and the position vector of the ith point, respectively, $\boldsymbol{f}_i$ is the force acting on the ith mass-point.

Definition 1.3. *The momentum and the moment of momentum of n mass-points with respect to the point o with the radius-vector* $\boldsymbol{r}_0$ *are*

$$\mathfrak{P} = \sum_{i=1}^{n} m_i \boldsymbol{v}_i \quad \text{and} \quad \mathfrak{M} = \sum_{i=1}^{n} (\boldsymbol{r}_i - \boldsymbol{r}_0) \times m_i \boldsymbol{v}_i,$$

respectively.

For the motion of the mass-point system the following theorems hold.

Theorem 1.1. *The rate of the change of the momentum of n mass-points is equal to the total (resultant) force vector $\mathfrak{F}$, that is the sum of all the forces acting to the mass-points*

$$\frac{\mathrm{d}}{\mathrm{d}t}\mathfrak{P} = \mathfrak{F}, \qquad \text{where} \quad \mathfrak{F} \triangleq \sum_{i=1}^{n} \boldsymbol{f}_i. \tag{3}$$

Theorem 1.2. *The rate of the change of the moment of momentum with respect to pole $\boldsymbol{r}_0$ of n mass-points is equal to the total torque (resultant moment) $\mathfrak{C}$ with respect to point o of all forces acting on the mass-points*

$$\frac{\mathrm{d}}{\mathrm{d}t}\mathfrak{M} = \mathfrak{C}, \qquad \text{where} \quad \mathfrak{C} \triangleq \sum_{i=1}^{n} (\boldsymbol{r}_i - \boldsymbol{r}_0) \times \boldsymbol{f}_i. \tag{4}$$

For the system of n mass-points, the balance equations of the momentum (3) and the moment of momentum (4) are consequences of the motion equations (2).

Definition 1.4. *The kinetic energy of the n mass-points is defined as follows*

$$K = \frac{1}{2}\sum_{i=1}^{n} m_i \boldsymbol{v}_i \cdot \boldsymbol{v}_i.$$

A more complex object of classical mechanics is the *rigid body.*

Definition 1.5. *A set of material points for which the mutual distances between the points remain unchanged in motion, is called rigid body.*

The rigid body position in the tree-dimensional space $\mathbb{R}^3$ is uniquely determined by the position of one of its points and the body orientation in the space. Indeed, let $o \in \mathcal{P}$ be a point the body called the pole and $\boldsymbol{r}_0(t)$ is the position vector of o. Let us rigidly "embed" the coordinate trihedron with unit vectors

$$\boldsymbol{d}_1(t), \quad \boldsymbol{d}_2(t), \quad \boldsymbol{d}_3(t), \qquad \boldsymbol{d}_i \cdot \boldsymbol{d}_j = \delta_{ij}$$

into the body, see Fig. 2. Here δ_{ij} is the Kronecker symbol. Then the position of any point $z \in \mathcal{P}$ is given by

$$\boldsymbol{r}(t) = \boldsymbol{r}_0(t) + \boldsymbol{z}(t), \quad \boldsymbol{z} = z_i \boldsymbol{d}_i(t). \tag{5}$$

For the body, we fix an initial configuration $\varkappa$. For example, we can take the body position at instant $t = 0$ as the initial configuration. The position

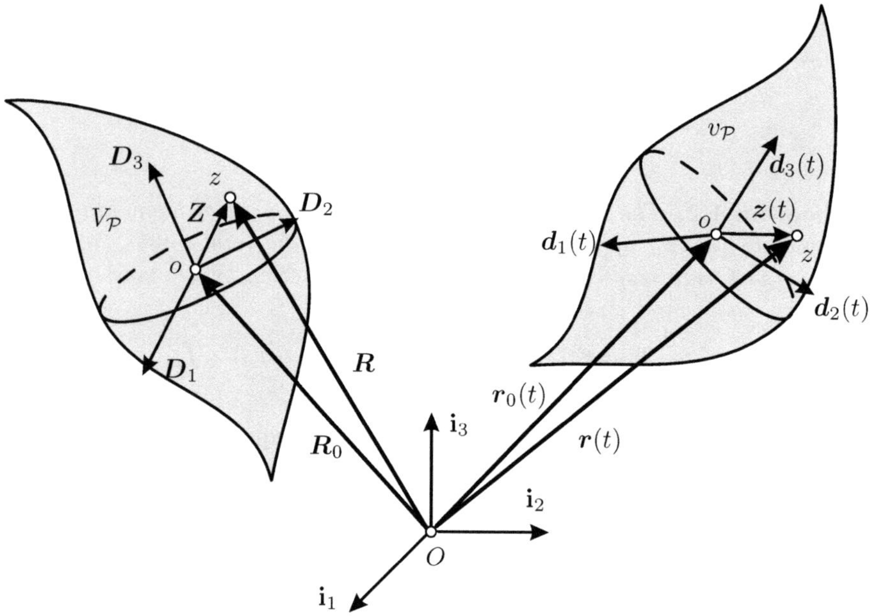

Figure 2. Rigid body motion.

of the pole o, the point z and the embedded trihedron of the coordinate axes that are $\boldsymbol{R}_0$, $\boldsymbol{R} = \boldsymbol{R}_0 + \boldsymbol{Z}$, $\boldsymbol{D}_1$, $\boldsymbol{D}_3$, $\boldsymbol{D}_3$, respectively, in the initial configuration, define the body position uniquely at any instant t. As the body is rigid, $\boldsymbol{Z} = z_i \boldsymbol{D}_i$.

To describe the body rotation, instead of vectors $\boldsymbol{d}_i$ we can introduce a proper orthogonal tensor $\mathbf{Q} = \boldsymbol{d}_i \otimes \boldsymbol{D}_i$. Then Eq. (5) takes the form

$$\boldsymbol{r}(t) = \boldsymbol{R}_0 + \boldsymbol{u}(t) + \mathbf{Q}(t) \cdot \boldsymbol{Z}. \tag{6}$$

Hence the rigid body motion is determined by two quantities, one of which is the translation vector of the point o, i.e. $\boldsymbol{u}(t) = \boldsymbol{r}_0(t) - \boldsymbol{R}_0$, and another is the rotation tensor $\mathbf{Q}(t)$. Using the representation of the proper orthogonal tensor

$$\mathbf{Q} = \frac{1}{(4+\theta^2)} \left[(4-\theta^2)\mathbf{I} + 2\boldsymbol{\theta} \otimes \boldsymbol{\theta} - 4\mathbf{I} \times \boldsymbol{\theta} \right], \quad \theta^2 = \boldsymbol{\theta} \cdot \boldsymbol{\theta} \tag{7}$$

instead of $\mathbf{Q}$ we also can use *Rodrigues's finite rotation vector* $\boldsymbol{\theta}$, cf. Lurie (2001). In classical mechanics, to determine the rigid body orientation, one can use other sets of three parameters, for example Euler's angles, ship or airplane angles, etc., see Lurie (2001). We should underline that the position of a rigid body $\mathcal{P}$ in the space is always determined by 6 scalar quantities.

Differentiating Eq. (6) we get

$$\dot{\boldsymbol{r}}(t) = \dot{\boldsymbol{u}}(t) + \dot{\mathbf{Q}}(t) \cdot \boldsymbol{Z}.$$

The tensor $\dot{\mathbf{Q}} \cdot \mathbf{Q}^T$ is skew-symmetric because $\mathbf{Q}$ is orthogonal. Any skew-symmetric tensor $\mathbf{A}$ can be represented by the vector $\boldsymbol{a}$ as follows

$$\mathbf{A} = \boldsymbol{a} \times \mathbf{I}, \tag{8}$$

where $\mathbf{I}$ is the three-dimensional unit tensor, see e.g. Lebedev et al. (2010), and $\times$ is the vector (cross) product. Hence, by (8), $\dot{\mathbf{Q}} \cdot \mathbf{Q}^T$ can be represented in the form

$$\dot{\mathbf{Q}} \cdot \mathbf{Q}^T = \boldsymbol{\omega} \times \mathbf{I}, \tag{9}$$

where $\boldsymbol{\omega}$ is the angular velocity of $\mathcal{P}$. Vector $\boldsymbol{\omega}$ can be determined as follows

$$\boldsymbol{\omega} = -\frac{1}{2}(\dot{\mathbf{Q}} \cdot \mathbf{Q}^T)_\times, \tag{10}$$

where $\mathbf{X}_\times$ denotes the vector invariant of a second-order tensor $\mathbf{X}$ by the formula

$$\mathbf{X}_\times = (X_{mn}\boldsymbol{e}^m \otimes \boldsymbol{e}^n)_\times \triangleq X_{mn}\boldsymbol{e}^m \times \boldsymbol{e}^n \tag{11}$$

for any base vectors $\boldsymbol{e}^k$. In particular, for a dyad $\boldsymbol{a} \otimes \boldsymbol{b}$ we have $(\boldsymbol{a} \otimes \boldsymbol{b})_\times = \boldsymbol{a} \times \boldsymbol{b}$.

Thus the velocity vector of a body point takes the form

$$\boldsymbol{v}(t) = \dot{\boldsymbol{u}}(t) + \boldsymbol{\omega}(t) \times \boldsymbol{Z}, \tag{12}$$

The rigid body can be considered as a system of mass-points and so we can introduce the following definitions.

Definition 1.6. *The momentum and the moment of momentum with respect to the pole o for a rigid body are the quantities*

$$\mathfrak{P} = \int_{v_\mathcal{P}} \rho \boldsymbol{v} \, \mathrm{d}v, \qquad \mathfrak{M} = \int_{v_\mathcal{P}} \rho(\boldsymbol{r} - \boldsymbol{r}_0) \times \boldsymbol{v} \, \mathrm{d}v,$$

respectively.

Here ρ is the mass density of $\mathcal{P}$. The mass m of $\mathcal{P}$ is given by the integral over the domain $v_\mathcal{P} \subset \mathbb{R}^3$ taken by the body in the space,

$$m(\mathcal{P}) = \int_{v_\mathcal{P}} \rho \, \mathrm{d}v.$$

Let us take as a pole the body mass center, that is the point whose radius vector $\boldsymbol{r}_0$ satisfies the relation

$$\int\limits_{v_{\mathcal{P}}} \rho(\boldsymbol{r} - \boldsymbol{r}_0)\, \mathrm{d}v = \mathbf{0}.$$

Then the momentum and the moment of momentum of the rigid body take the form

$$\mathfrak{P} = m\boldsymbol{v}_0, \qquad \mathfrak{M} = \int\limits_{v_{\mathcal{P}}} \rho \boldsymbol{z} \times \dot{\boldsymbol{z}}\, \mathrm{d}v = \int\limits_{v_{\mathcal{P}}} \rho \boldsymbol{z} \times (\boldsymbol{\omega} \times \boldsymbol{z})\, \mathrm{d}v = \mathbf{J} \cdot \boldsymbol{\omega}, \tag{13}$$

where $\boldsymbol{v}_0 = \dot{\boldsymbol{u}}$ and $\mathbf{J}$ is the *inertia tensor*

$$\mathbf{J} = \int\limits_{v_{\mathcal{P}}} \rho[(\boldsymbol{z} \cdot \boldsymbol{z})\mathbf{I} - \boldsymbol{z} \otimes \boldsymbol{z}]\, \mathrm{d}v. \tag{14}$$

It is obvious that $\mathbf{J}$ possesses the following property

$$\mathbf{J} = \mathbf{Q} \cdot \mathbf{J}_0 \cdot \mathbf{Q}^T, \qquad \mathbf{J}_0 = \int\limits_{V_{\mathcal{P}}} \rho[(\boldsymbol{Z} \cdot \boldsymbol{Z})\mathbf{I} - \boldsymbol{Z} \otimes \boldsymbol{Z}]\, \mathrm{d}v, \tag{15}$$

where the volume integral is taken over $V_{\mathcal{P}}$ in the initial configuration. The constant tensor $\mathbf{J}_0$ is the inertia tensor in the initial configuration. For example, for a homogeneous sphere of radius a, $\mathbf{J}$ is a spherical tensor

$$\mathbf{J} = \frac{2}{5} m a^2 \mathbf{I} = \mathbf{J}_0.$$

If the directors $\boldsymbol{d}_k$ are unit vectors along the principle axes of the inertia tensor, we see that $\mathbf{J}$ and $\mathbf{J}_0$ are diagonal

$$\mathbf{J} = J_1 \boldsymbol{d}_1 \otimes \boldsymbol{d}_1 + J_2 \boldsymbol{d}_2 \otimes \boldsymbol{d}_2 + J_3 \boldsymbol{d}_3 \otimes \boldsymbol{d}_3,$$

$$\mathbf{J}_0 = J_1 \boldsymbol{D}_1 \otimes \boldsymbol{D}_1 + J_2 \boldsymbol{D}_2 \otimes \boldsymbol{D}_2 + J_3 \boldsymbol{D}_3 \otimes \boldsymbol{D}_3, \quad \boldsymbol{D}_k = \mathbf{Q} \cdot \boldsymbol{d}_k,$$

where J_1, J_2, J_3 are principal moments of inertia with respect to the principal axes.

With regard to (9) and (15) it can be shown that the derivative of $\mathbf{J}$ satisfies the relation

$$\dot{\mathbf{J}} = \boldsymbol{\omega} \times \mathbf{J} - \mathbf{J} \times \boldsymbol{\omega}. \tag{16}$$

Taking the mass center as a pole we can rewrite the kinetic energy of the rigid body as follows

$$K = \frac{1}{2} \int\limits_{v_{\mathcal{P}}} \rho \boldsymbol{v} \cdot \boldsymbol{v}\, \mathrm{d}v = \frac{1}{2} m \boldsymbol{v}_0 \cdot \boldsymbol{v}_0 + \frac{1}{2} \boldsymbol{\omega} \cdot \mathbf{J} \cdot \boldsymbol{\omega}.$$

The following identities are valid

$$\mathfrak{P} = \frac{\partial K}{\partial \boldsymbol{v}_0}, \qquad \mathfrak{M} = \frac{\partial K}{\partial \boldsymbol{\omega}}.$$

To a rigid body we can apply the forces and torques (couples or moments). The forces relate to the translation of the body whereas the torques involve body rotation.

The rigid body motion is described by two Euler's laws of motion.

First Euler's laws of motion of a rigid body - Balance of momentum. *The time rate of the rigid body momentum is equal to the resultant vector of forces* $\mathfrak{F}$, *acting on the body*

$$\frac{\mathrm{d}}{\mathrm{d}t}\mathfrak{P} = \mathfrak{F}, \qquad \text{where} \quad \mathfrak{F} \triangleq \int\limits_{v_{\mathcal{P}}} \rho \boldsymbol{f} \, \mathrm{d}v. \tag{17}$$

Second Euler's law of motion of the rigid body - Balance of moment of momentum. *The time rate of the rigid body moment of momentum with respect to pole o is equal to the resultant moment of all forces with respect to the pole and the body moments*

$$\frac{\mathrm{d}}{\mathrm{d}t}\mathfrak{M} = \mathfrak{C}, \qquad \text{where} \quad \mathfrak{C} \triangleq \int\limits_{v_{\mathcal{P}}} \rho[(\boldsymbol{r} - \boldsymbol{r}_0) \times \boldsymbol{f} + \boldsymbol{m}] \, \mathrm{d}v. \tag{18}$$

Here $\boldsymbol{f}$ and $\boldsymbol{m}$ are the densities of the forces and the moments acting on the body, respectively.

In the case of statics, these laws reduce to the equality to zero of the resultant vector of forces and the resultant vector of moments

$$\mathfrak{F} = \mathbf{0}, \qquad \mathfrak{C} = \mathbf{0}. \tag{19}$$

Substituting (13) to (18) and taking account (16), we get the motion equations of the rigid body

$$m\dot{\boldsymbol{v}}_0 = \mathfrak{F}, \quad \mathbf{J} \cdot \dot{\boldsymbol{\omega}} + \boldsymbol{\omega} \times \mathbf{J} \cdot \boldsymbol{\omega} = \mathfrak{C}. \tag{20}$$

Naturally, equations (20) contain the n mass point motion equations (1) and (2) as a particular case.

Taking the principal axes of the inertia tensor $\mathbf{J}$ as the basis, we transform Eqs. (20) to the form

$$\begin{gathered} m\dot{v}_1 = \mathfrak{F}_1, \quad m\dot{v}_2 = \mathfrak{F}_2, \quad m\dot{v}_3 = \mathfrak{F}_3, \\ J_1\dot{\omega}_1 + (J_3 - J_2)\omega_2\omega_3 = \mathfrak{C}_1, \\ J_2\dot{\omega}_2 + (J_1 - J_3)\omega_1\omega_3 = \mathfrak{C}_2, \\ J_3\dot{\omega}_3 + (J_2 - J_1)\omega_1\omega_2 = \mathfrak{C}_3. \end{gathered}$$

Equations (20) constitute the system of nonlinear ordinary differential equations. The analytic solutions of Eqs. (20) are known in few cases for some relations between J_i. These solutions relate to works by Euler, Lagrange and Kovalevskaya. The rigid body motion can be also considered under some constraints like in the theory of rotation of a rigid body about an immovable point or axis, cf. for example Lurie (2001). It is worth of noting that rigid body dynamics are quite complicated. Many areas of modern mathematics arose from this branch of mechanics, cf. Arnold (1989).

For infinitesimal displacements $\boldsymbol{u}$ and rotations $\boldsymbol{\theta}$, we derive simplified relations:

$$\mathbf{Q} \approx \mathbf{I} - \boldsymbol{\theta} \times \mathbf{I}, \quad \boldsymbol{\omega} = \dot{\boldsymbol{\theta}}.$$

Here equations (20) reduce to a linear system of ordinary differential equations with respect to $\boldsymbol{u}$ and $\boldsymbol{\theta}$:

$$m\ddot{\boldsymbol{u}} = \mathfrak{F}, \quad \mathbf{J}_0 \cdot \ddot{\boldsymbol{\theta}} = \mathfrak{C}.$$

In Mechanics, Euler's laws are known since ancient time. For example, Eq. (19) was formulated as the lever law by Archimedes in the 3rd century BC. Euler was the first who showed that the second dynamic law, or the second equilibrium condition, is independent of the first one. From Eqs. (18) and (19) we can derive Newton's laws for the system of mass points. On the other hand as it was mentioned in Truesdell and Toupin (1960), to derive Euler's equations from Newton's law, we should introduce additional assumptions on the nature of the interaction between the mass-points. Being more general, Eqs. (18) and (19) constitute the foundation of classical mechanics as well as of continuum mechanics.

1.2 Elements of Mechanics of Elastic Rods

Transition from the Classical Mechanics to Mechanics of Structures and Continua is based on the *solidification principle*. This principle says that motion or equilibrium equations of a rigid body dynamics can be applied to *any* part of a deformable body. To illustrate this idea we consider a more simple one-dimensional (1D) case, that is mechanics of elastic rods using the direct approach.

We consider the rod as a deformable Cosserat curve whose, kinematics is described by the independent translation and rotation fields. We will employ the kinematical model of *directed curves*, presented in Antman (2005). In this approach, the thin rod-like bodies are modeled as deformable curves endowed with a triad of vectors (also called directors) attached to every point. This triad of directors rotate rigidly during deformation and gives thus information about rotations of the rod cross-sections. In other words, we consider the rod as a 1D Cosserat or micropolar continuum.

Let us briefly present the kinematical model of directed curves. We denote by the curve $\mathcal{C}$ in the initial (reference) configuration $\varkappa$ and by s the material coordinate along $\mathcal{C}$. The coordinate s is taken to be the arclength parameter along $\mathcal{C}$. The reference configuration of the rod is defined by the vector fields (see Fig. 3)

$$\boldsymbol{R} = \boldsymbol{R}(s), \qquad \boldsymbol{D}_k = \boldsymbol{D}_k(s), \quad k = 1, 2, 3, \quad s \in [0, l].$$

Here $\boldsymbol{R}$ is the position vector of points on $\mathcal{C}$, the triad $\{\boldsymbol{D}_1, \boldsymbol{D}_2, \boldsymbol{D}_3\}$ specifies the 3 unit vectors (directors) orthogonal to each other and l is the length of the rod. The vector $\boldsymbol{D}_3$ can be chosen tangent to the curve $\mathcal{C}$ (i.e. $\boldsymbol{D}_3 = \boldsymbol{R}' \equiv \boldsymbol{\tau}$), while $\boldsymbol{D}_1$ and $\boldsymbol{D}_2$ are usually taken along the principal axes of inertia of the rod cross-section. The prime designates the derivative with respect to the spatial coordinate s (i.e. $f' = \frac{\partial f}{\partial s}$).

Let χ denote the actual configuration at time t, which is determined by the position vector $\boldsymbol{r} = \boldsymbol{r}(s,t)$ of the curve $\mathcal{L}$ and the directors $\boldsymbol{d}_i = \boldsymbol{d}_i(s,t)$, $i = 1, 2, 3$. The three unit vectors $\boldsymbol{d}_i$ remain mutually orthogonal after deformation, but $\boldsymbol{d}_3$ is no longer tangent to the curve $\mathcal{L}$, which means that the model takes into account the transverse shear deformations of rods.

Differentiating $\boldsymbol{D}_k$ and $\boldsymbol{d}_k$ with respect to s we obtain the formulae

$$\boldsymbol{D}'_k = \boldsymbol{k}_0 \times \boldsymbol{D}_k, \quad \boldsymbol{d}'_k = \boldsymbol{k} \times \boldsymbol{d}_k. \tag{21}$$

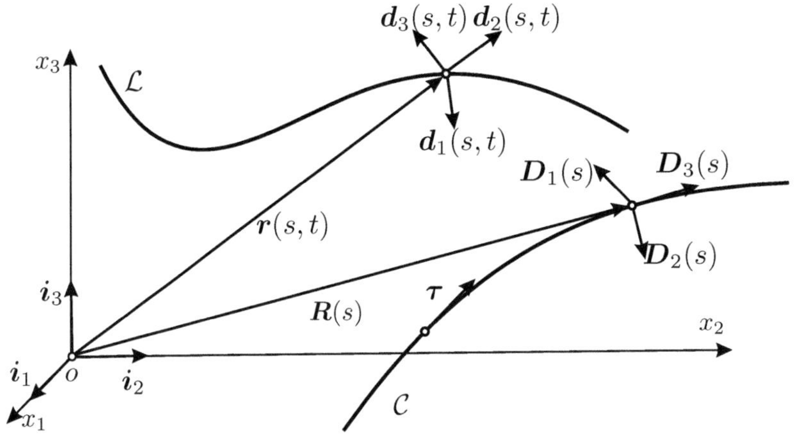

Figure 3. Actual and reference configurations of the rod.

Equations (21) are analogues of the Frenet-Serret formulae in the differential geometry, see e.g. Lebedev et al. (2010). The vectors $\boldsymbol{k}_0$ and $\boldsymbol{k}$ are analogues of the Darboux vector and can be called the *generalized Darboux vectors* in the reference and actual configurations, respectively. Note that the rotation tensor $\mathbf{Q}$ is defined by the relation

$$\mathbf{Q}(s,t) = \boldsymbol{d}_k(s,t) \otimes \boldsymbol{D}_k(s).$$

The velocity vector $\boldsymbol{v}$ is given by the standard formula $\boldsymbol{v}(s,t) = \dot{\boldsymbol{r}}(s,t)$ and the angular velocity vector $\boldsymbol{\omega}(s,t)$ – by Eq. (10). We introduce the kinetic energy density $K(s,t)$ by relation

$$K = \frac{1}{2}\rho\left(\boldsymbol{v}\cdot\boldsymbol{v} + 2\boldsymbol{v}\cdot\boldsymbol{\Theta}_1\cdot\boldsymbol{\omega} + \boldsymbol{\omega}\cdot\boldsymbol{\Theta}_2\cdot\boldsymbol{\omega}\right),$$

where ρ is the mass density in the reference configuration $\varkappa$, the second-order tensors $\boldsymbol{\Theta}_1(s,t)$ and $\boldsymbol{\Theta}_2(s,t)$ are the *tensors of inertia* (per unit mass), which characterize the distribution of the material in the cross-section. For the reference configuration, the tensors of inertia are denoted by $\boldsymbol{\Theta}_1^0(s)$ and $\boldsymbol{\Theta}_2^0(s)$. As in the case of rigid body dynamics, see Eq. $(15)_1$, we have $\boldsymbol{\Theta}_\alpha = \mathbf{Q}\cdot\boldsymbol{\Theta}_\alpha^0\cdot\mathbf{Q}^T$, $\alpha = 1, 2$.

We assume the following formulae for the momentum $\mathfrak{P}$ and the moment of momentum $\mathfrak{M}$ of any part $\mathcal{P}$ of the rod

$$\mathfrak{P}(\mathcal{P}) \triangleq \int_{s_1}^{s_2} \rho\boldsymbol{K}_1\,\mathrm{d}s, \quad \mathfrak{M}(\mathcal{P}) \triangleq \int_{s_1}^{s_2} \rho\{(\boldsymbol{r}-\boldsymbol{r}_0)\times\boldsymbol{K}_1 + \boldsymbol{K}_2\}\,\mathrm{d}s, \tag{22}$$

where $\boldsymbol{K}_1$ and $\boldsymbol{K}_2$ are given by

$$\boldsymbol{K}_1 \triangleq \frac{\partial K}{\partial \boldsymbol{v}} = \boldsymbol{v} + \boldsymbol{\Theta}_1^T\cdot\boldsymbol{\omega}, \quad \boldsymbol{K}_2 \triangleq \frac{\partial K}{\partial \boldsymbol{\omega}} = \boldsymbol{\Theta}_1\cdot\boldsymbol{v} + \boldsymbol{\Theta}_2\cdot\boldsymbol{\omega}.$$

$s_1, s_2 \in [0, l]$ are the coordinates of $\mathcal{P}$ in the reference configuration, see Fig. 4. Euler's laws of motion for the rod are one-dimensional analogues of Eq.s (17) and (18) and can be formulated as follows:

First Euler's law of motion of the rod. *The time rate of change of the momentum of an arbitrary part $\mathcal{P}$ of the rod is equal to the total force acting on $\mathcal{P}$:*

$$\frac{\mathrm{d}}{\mathrm{d}t}\mathfrak{P}(\mathcal{P}) = \mathfrak{F}, \quad \mathfrak{F} \triangleq \int_{s_1}^{s_2} \boldsymbol{f}\,\mathrm{d}s + \boldsymbol{t}\Big|_{s_1}^{s_2}. \tag{23}$$

Second Euler's law of motion of the rod. *The time rate of change of the moment of momentum of an arbitrary part $\mathcal{P}$ of the rod about a fixed*

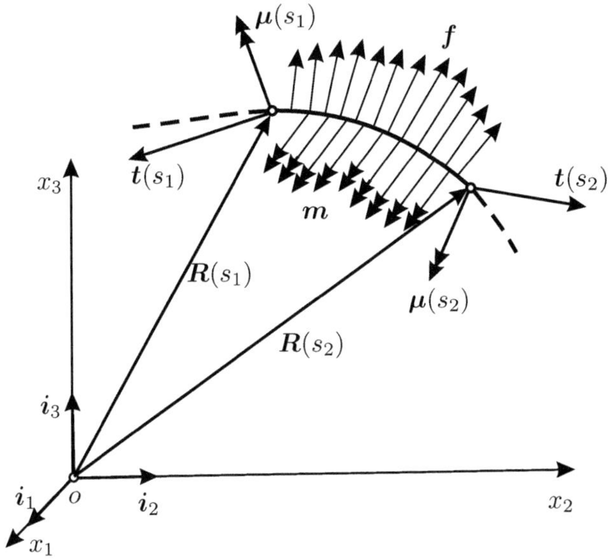

Figure 4. Part $\mathcal{P}$ of the rod and acting on $\mathcal{P}$ forces and couples.

point $\boldsymbol{r}_0$ *is equal to the total moment about* $\boldsymbol{r}_0$ *acting on* $\mathcal{P}$*:*

$$\frac{\mathrm{d}}{\mathrm{d}t}\mathfrak{M}(\mathcal{P}) = \mathfrak{C}, \quad \mathfrak{C} \triangleq \int_{s_1}^{s_2} \{(\boldsymbol{r} - \boldsymbol{r}_0) \times \boldsymbol{f} + \boldsymbol{m}\}\, \mathrm{d}s + \{(\boldsymbol{r} - \boldsymbol{r}_0) \times \boldsymbol{t} + \boldsymbol{\mu}\}\Big|_{s_1}^{s_2}. \tag{24}$$

Here $\boldsymbol{f}$ and $\boldsymbol{m}$ are the external resultant body force and moment per unit length of $\mathcal{C}$, $\boldsymbol{t}$ and $\boldsymbol{\mu}$ are the force and moment vectors, respectively. To differ the vectors that describe the moments from the vectors of forces, we will depict them by double arrows, see Fig. 4. A moment (couple) that corresponds to a double arrow acts in the clockwise direction if to see along the arrow direction.

Under suitable smoothness assumptions, we obtain the following local equations of the principles (23) and (24) that are the Lagrangian equations of motion:

$$\boldsymbol{t}'(s,t) + \boldsymbol{f} = \rho\,\frac{\mathrm{d}}{\mathrm{d}t}\,(\boldsymbol{v} + \boldsymbol{\Theta}_1 \cdot \boldsymbol{\omega}),$$
$$\boldsymbol{\mu}'(s,t) + \boldsymbol{r}' \times \boldsymbol{t}(s,t) + \boldsymbol{m} = \rho\left[\frac{\mathrm{d}}{\mathrm{d}t}\,(\boldsymbol{\Theta}_1^T \cdot \boldsymbol{v} + \boldsymbol{\Theta}_2 \cdot \boldsymbol{\omega}) + \boldsymbol{v} \times \boldsymbol{\Theta}_1 \cdot \boldsymbol{\omega}\right].$$

For an elastic rod there exists the strain energy density W. According to the equipresence principle we take W in the form $W = W(\boldsymbol{r}, \boldsymbol{r}', \mathbf{Q}, \mathbf{Q}')$. Applying to W the material frame-indifference principle we derive that W depends on two vectorial relative Lagrangian strain measures

$$W = W(\boldsymbol{E}, \boldsymbol{K}), \quad \boldsymbol{E} = \mathbf{Q}^T \cdot \boldsymbol{r}' - \boldsymbol{\tau}\,, \quad \boldsymbol{K} = -\frac{1}{2}\left(\mathbf{Q}^T \cdot \mathbf{Q}'\right)_{\times} = \mathbf{Q}^T \cdot \boldsymbol{k} - \boldsymbol{k}_0\,,$$

where $\boldsymbol{E}$ is the vector of stretching–shear and $\boldsymbol{K}$ is the vector of bending–twisting. We notice that (essentially) the same strain measures have been considered in other approaches to the rod theory Antman (2005); Eliseev (1996); Svetlitsky (2000); Simmonds (2005).

Using the virtual work principle we can derive the constitutive equations in the following form, see Eliseev (1996),

$$\boldsymbol{t} = \frac{\partial W}{\partial \boldsymbol{E}} \cdot \mathbf{Q}^T\,, \qquad \boldsymbol{\mu} = \frac{\partial W}{\partial \boldsymbol{K}} \cdot \mathbf{Q}^T\,.$$

Mechanics of elastic rods presented above gives a good example of an one-dimensional Cosserat continuum. Assuming certain constraints one can derive Timoshenko or Euler-Bernoulli models of rods Antman (2005); Rubin (2000); Eliseev (1996). On the other hand the theory of rods give also the "generalized" one-dimensional models of continuum which possess more than six independent kinematical degrees of freedom. Examples of such "generalized" models are the theory of thin-walled rods or theories taking into account warping of the rod cross-section Eliseev (1996); Hodges (2006); Librescu and Song (2006). Chapter 4 of this book contains more details on the rod theory.

2 Kinematics of Cosserat Continuum

In this section we briefly recall general kinematical relations for a Cosserat continuum. For a comprehensive approach, we refer the reader to Eringen and Kafadar (1976); Eringen (1999).

The description of motion of a particle of a Cosserat continuum is based on the assumption that every particle of the micropolar body has six degrees of freedom, see Eringen and Kafadar (1976); Eringen (1999). This is similar to the description of a rigid body in classical mechanics. Three of the degrees of freedom are translational as in classic elasticity, and other three degrees are *orientational* or *rotational*.

In the actual configuration χ at instant t, the position of a particle of micropolar continuum is given by the position vector $\boldsymbol{r}$. The particle orientation is defined by an orthonormal trihedron $\boldsymbol{d}_k$ $(k = 1, 2, 3)$ whose

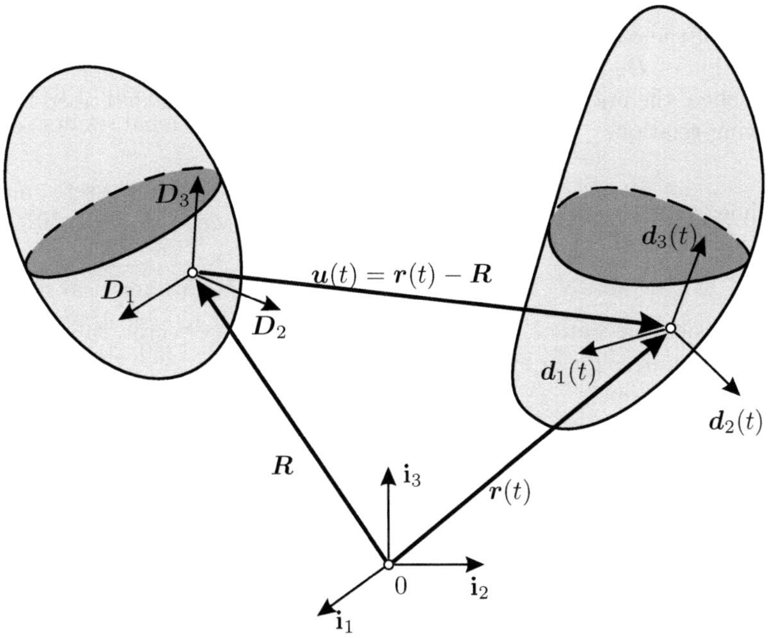

Figure 5. Deformation of a micropolar body (reference and actual configurations)

vectors are called *directors*. The two vector fields $\boldsymbol{r}$ and $\boldsymbol{d}_k$ define the translational and rotational motions of a particle.

To describe the medium relative deformation, we use some fixed position of the body that may be taken at $t = 0$ or another fixed instant; we call this position the *reference configuration* κ. Here the state of particle is defined by the position vector $\boldsymbol{R}$, whereas its orientation by directors $\boldsymbol{D}_k$ (cf. Fig. 5). Let us note that as the reference configuration can be chosen not only the real state but also any one.

The motion of a micropolar continuum can be described by the following vectorial fields

$$\boldsymbol{r} = \boldsymbol{r}(\boldsymbol{R}, t), \quad \boldsymbol{d}_k = \boldsymbol{d}_k(\boldsymbol{R}, t). \tag{25}$$

During the deformation the trihedron $\boldsymbol{d}_k$ stays orthonormal, $\boldsymbol{d}_k \cdot \boldsymbol{d}_m = \delta_{km}$. The change of the directors can be described by an orthogonal tensor that is

$$\mathbf{H} = \boldsymbol{d}_k \otimes \boldsymbol{D}_k.$$

$\mathbf{H}$ is called the *microrotation tensor*. So $\boldsymbol{r}$ describes the position of the particle of the continuum at time t, whereas $\mathbf{H}$ defines its orientation. The orientation of $\boldsymbol{D}_k$ and $\boldsymbol{d}_k$ can be selected the same, so $\mathbf{H}$ is proper orthogonal. Hence, the micropolar continuum deformation can be described by the following relations

$$\boldsymbol{r} = \boldsymbol{r}(\boldsymbol{R}, t), \quad \mathbf{H} = \mathbf{H}(\boldsymbol{R}, t). \tag{26}$$

The linear velocity is given by the relation

$$\boldsymbol{v} = \dot{\boldsymbol{r}}.$$

As in classical mechanics, see (10), the angular velocity vector, called *microgyration vector*, is given by

$$\boldsymbol{\omega} = -\frac{1}{2}\left(\mathbf{H}^T \cdot \dot{\mathbf{H}}\right)_\times . \tag{27}$$

Relation (27) means that $\boldsymbol{\omega}$ is the axial vector associated with the skew-symmetric tensor $\mathbf{H}^T \cdot \dot{\mathbf{H}}$.

3 Forces and Couples, Stress and Couple Stress Tensors in Micropolar Continua

In this section using the balance of momentum and balance of moment of momentum (Euler's laws of motion) we introduce the stress and couple stress tensors. Then we derive the motion equations of the micropolar continuum which contains the motion equations of simple (non-polar) continuum as a special case.

3.1 Forces and Couples

Forces and *couples* are the primary quantities of continuum mechanics. It is possible to introduce them in the axiomatic way as it was done by Truesdell and Noll (1965); Truesdell (1977, 1984) for simple materials.

Let us consider an arbitrary part $\mathcal{P}$ of a material body $\mathcal{B}$ at instant t, cf. Fig. 6. As usual in Continuum Mechanics we assume that there are two kinds of forces and couples acting on $\mathcal{P}$. The first ones are *body loads* acting on each material point of the body and which are independent of the rest of the body. The second ones are *contact loads* acting on the boundary of $\mathcal{P}$ and describing the interaction with the remaining part of the body or with the environment. In the case of micropolar continua one has the same situation as in general mechanics – we have to take into account forces and couples as primary variables.

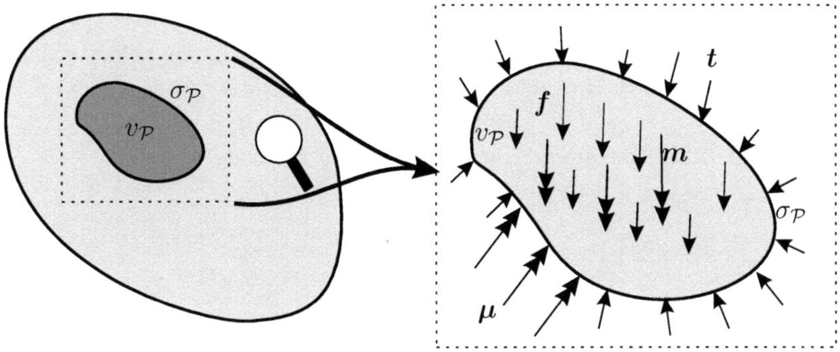

Figure 6. Forces and couples acting on a body part $\mathcal{P}$

The gravity force, centrifugal force, and ponderomotive force are examples of body forces. Body couples arise, for example, when the body is in an electromagnetic field.

Forces and couples acting on $\mathcal{P}$ can be represented as follows

$$\boldsymbol{f}(\mathcal{P}) = \boldsymbol{f}_B(\mathcal{P}) + \boldsymbol{f}_C(\mathcal{P}), \quad \boldsymbol{m}(\mathcal{P}) = \boldsymbol{m}_B(\mathcal{P}) + \boldsymbol{m}_C(\mathcal{P}),$$

where the subscript B denotes body forces and couples and C is for the surface quantities.

Let us suppose $\boldsymbol{f}_B(\mathcal{P})$, $\boldsymbol{m}_B(\mathcal{P})$ to be absolutely continuous functions of the mass of the part on which they act and $\boldsymbol{f}_C(\mathcal{P})$, $\boldsymbol{m}_C(\mathcal{P})$ to be absolutely continuous functions the surface area. This means that we can introduce their mass and surface densities

$$\boldsymbol{f}_B(\mathcal{P}) = \int\limits_{v_{\mathcal{P}}} \rho \boldsymbol{f}\,\mathrm{d}v, \quad \boldsymbol{m}_B(\mathcal{P}) = \int\limits_{v_{\mathcal{P}}} \rho \boldsymbol{m}\,\mathrm{d}v,$$

$$\boldsymbol{f}_C(\mathcal{P}) = \int\limits_{\sigma_{\mathcal{P}}} \boldsymbol{t}\,\mathrm{d}\sigma, \quad \boldsymbol{m}_C(\mathcal{P}) = \int\limits_{\sigma_{\mathcal{P}}} \boldsymbol{\mu}\,\mathrm{d}\sigma,$$

where $v_{\mathcal{P}}$ is the volume of $\mathcal{P}$ in the actual configuration, $\sigma_{\mathcal{P}} = \partial v_{\mathcal{P}}$ is the boundary of $\mathcal{P}$, $\boldsymbol{t}$, $\boldsymbol{\mu}$ are the force and couple per the area unit in the actual configuration.

Definition 3.1. *$\boldsymbol{t}$ is called the stress vector and $\boldsymbol{\mu}$ is the couple stress vector.*

3.2 Euler's Laws of Motion

The angular velocity field $\boldsymbol{\omega}$ is independent of the linear velocity field $\boldsymbol{v}$ in a micropolar body. Let us introduce two definitions.

Definition 3.2. *The momentum of part $\mathcal{P}$ of the body is*

$$\mathfrak{P}(\mathcal{P}) \triangleq \int\limits_{v_\mathcal{P}} \rho \boldsymbol{v} \, \mathrm{d}v. \tag{28}$$

Definition 3.3. *The moment of momentum of part $\mathcal{P}$ of the body is*

$$\mathfrak{M}(\mathcal{P}) \triangleq \int\limits_{v_\mathcal{P}} \{(\boldsymbol{r} - \boldsymbol{r}_0) \times \rho \boldsymbol{v} + j\boldsymbol{\omega}\} \, \mathrm{d}v. \tag{29}$$

In the definitions the following variables are introduced: ρ is the material density, $\boldsymbol{r}$ is the position vector in the actual configuration, $\boldsymbol{r}_0$ is an arbitrary position vector that does not depend on t, j is the scalar measure of rotatory inertia of "microparticles" of the material.

Let us note that in the general case, j is a tensor so we have to change it to the tensor of inertia $\mathbf{J}$ that has to be positive definite. $\mathbf{J}$ is a characteristic of the material that defines the rotatory inertia of particles of the body. In Sect. 4 we discuss the generalizations of (28) and (29) that are the constitutive relations of special kind. The mentioned above definitions generalize the definitions of the momentum and the moment of momentum of the classical mechanics to continua.

We accept the following two Eulerian dynamic laws as axioms[1].

1. First Euler's law of motion of the micropolar continuum. *The time rate of change of the momentum of an arbitrary part $\mathcal{P}$ of the body is equal to the total force acting on $\mathcal{P}$:*

$$\frac{\mathrm{d}}{\mathrm{d}t}\mathfrak{P}(\mathcal{P}) = \mathfrak{F}, \quad \mathfrak{F} \triangleq \int\limits_{v_\mathcal{P}} \rho \boldsymbol{f} \, \mathrm{d}v + \int\limits_{\sigma_\mathcal{P}} \boldsymbol{t} \, \mathrm{d}\sigma. \tag{30}$$

2. Second Euler's law of motion of the micropolar continuum. *The time rate of change of the moment of momentum of an arbitrary part $\mathcal{P}$ of the body about a fixed point $\boldsymbol{r}_0$ is equal to the total moment about $\boldsymbol{r}_0$ acting*

[1] In what follows we consider an inertial reference frame.

on $\mathcal{P}$:

$$\frac{\mathrm{d}}{\mathrm{d}t}\mathfrak{M}(\mathcal{P}) = \mathfrak{C}, \tag{31}$$

$$\mathfrak{C} \triangleq \int\limits_{v_{\mathcal{P}}} \{(\boldsymbol{r}-\boldsymbol{r}_0)\times\rho\boldsymbol{f}+\rho\boldsymbol{m}\}\,\mathrm{d}v + \int\limits_{\sigma_{\mathcal{P}}} \{(\boldsymbol{r}-\boldsymbol{r}_0)\times\boldsymbol{t}+\boldsymbol{\mu}\}\,\mathrm{d}\sigma.$$

Note that for an oriented medium, the momentum definition (28) and the formulation of the first law (30) coincide with the momentum definition for simple (non-polar) materials. For the medium with couple stresses, the definition of the moment of momentum (29) and the second law formulation (31) are different of those for simple materials; they contain additional terms related to the couple inertia tensor of a particle, the distributed body and surface couples.

Using the first dynamic law we can demonstrate that the formulation of the second law of motion does not depend on the choice of position vector $\boldsymbol{r}_0$.

3.3 Stress Tensor and Couple Stress Tensor

At instant t the stress vector and the couple stress tensor depend on the particle position $\boldsymbol{r}$ and the normal to surface $\boldsymbol{n}$ only.

Cauchy's postulate. *At a point of the body, the stress vector and the couple stress vector depend on the surface only through the unit normal to the considered surface, that is they have the same values, respectively, for all surfaces through the point which have the same normal.*

In what follows, the normal $\boldsymbol{n}$ to the body surface will be taken external with respect to the part of the body under consideration.

Cauchy's lemma. *The stress and couple stress vectors are odd functions with respect to $\boldsymbol{n}$:*

$$\boldsymbol{t}(\boldsymbol{r},\boldsymbol{n}) = -\boldsymbol{t}(\boldsymbol{r},-\boldsymbol{n}), \tag{32}$$

$$\boldsymbol{\mu}(\boldsymbol{r},\boldsymbol{n}) = -\boldsymbol{\mu}(\boldsymbol{r},-\boldsymbol{n}). \tag{33}$$

Cauchy's lemma is the re-formulation of the third Newton's axiom of reciprocal actions in micropolar continuum. Equations (32), (33) describe the interactions of the body parts that have contacting points.

Proof the Cauchy's lemma is based on the consideration of an arbitrary part $v_{\mathcal{P}}$ of the body $\mathcal{P}$ and its arbitrary subparts v_1 and v_2, $v_{\mathcal{P}} = v_1 \cup v_2$ with applications of Euler's laws of motion for domains $v_{\mathcal{P}}$, v_1, and v_2, see Eremeyev et al. (2012).

Cauchy's lemma is an essential tool for the introduction of the stress tensor and the couple stress tensor. This is given by *Cauchy's theorem*.

Theorem 3.1. *For any point of the body there exist second-order tensors* $\mathbf{T}$ *and* $\mathbf{M}$ *such that the stress vector and the couple stress vector acting at a point of the surface with normal* $\boldsymbol{n}$ *are presented by formulae*

$$\boldsymbol{t} = \mathbf{T} \cdot \boldsymbol{n}, \quad \boldsymbol{\mu} = \mathbf{M} \cdot \boldsymbol{n}.$$

Proof. First let us prove the existence of $\mathbf{T}$. Consider an arbitrary orthogonal parallelepiped Π with edges oriented along the axes of Cartesian coordinates x_1, x_2, x_3, cf. Fig. 7. The unit normal to a side of the parallelepiped coincides with one of the coordinate unit vectors $\boldsymbol{n} = \pm \boldsymbol{i}_k$ $(k = 1, 2, 3)$ up to direction. In frame $\boldsymbol{i}_k$, the stress vector is

$$\boldsymbol{t}(\boldsymbol{r}, \mathbf{H}, \boldsymbol{i}_k) = t_{sk}(\boldsymbol{r}, \mathbf{H}) \boldsymbol{i}_s. \tag{34}$$

In the rest part of the proof we will omit arguments $\boldsymbol{r}$ and $\mathbf{H}$. Here t_{sk} are the components of $\boldsymbol{t}$ in the frame $\boldsymbol{i}_k$. From the law of reciprocal actions it follows

$$\boldsymbol{t}(-\boldsymbol{i}_k) = -t_{sk} \boldsymbol{i}_s.$$

We denote by n_k the components of the normal in the frame $\boldsymbol{i}_k$. For coordinate area elements, the representation (34) is

$$\boldsymbol{t}(\boldsymbol{i}_1) = n_1 t_{s1} \boldsymbol{i}_s, \quad \boldsymbol{t}(\boldsymbol{i}_2) = n_2 t_{s2} \boldsymbol{i}_s, \quad \boldsymbol{t}(\boldsymbol{i}_3) = n_3 t_{s3} \boldsymbol{i}_s. \tag{35}$$

Let us apply the first Euler's law to the parallelepiped. We get

$$\int\limits_{v_\Pi} \rho \left(\frac{\mathrm{d}\boldsymbol{v}}{\mathrm{d}t} - \boldsymbol{f} \right) \mathrm{d}v = \int\limits_{\sigma_\Pi} \boldsymbol{t} \, \mathrm{d}\sigma.$$

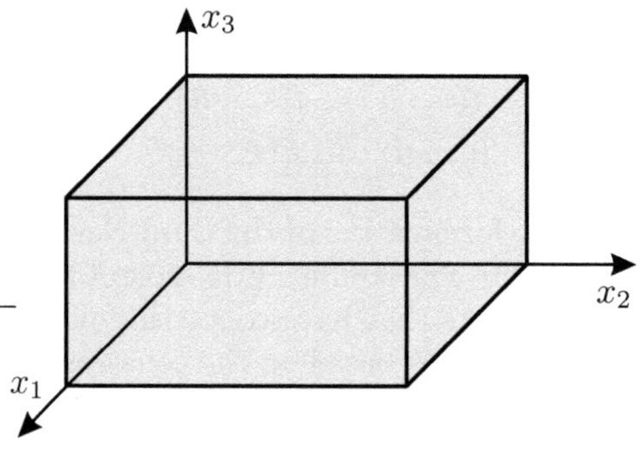

Figure 7. Parallelepiped Π

Using (35) the latter equation can be written as

$$\int_{v_\Pi} \rho\left(\frac{\mathrm{d}\boldsymbol{v}}{\mathrm{d}t} - \boldsymbol{f}\right)\mathrm{d}v = \int_{\sigma_\Pi} n_k t_{sk} \boldsymbol{i}_s \,\mathrm{d}\sigma.$$

Applying the Gauss–Ostrogradsky theorem to the surface integral we get

$$\int_{v_\Pi} \left\{\rho\left(\frac{\mathrm{d}\boldsymbol{v}}{\mathrm{d}t} - \boldsymbol{f}\right) - \frac{\partial t_{sk}}{\partial x_k}\boldsymbol{i}_s\right\}\mathrm{d}v = \mathbf{0}. \tag{36}$$

As the parallelepiped is arbitrary, from the integral equation (36) it follows the differential equation

$$\rho\left(\frac{\mathrm{d}\boldsymbol{v}}{\mathrm{d}t} - \boldsymbol{f}\right) - \frac{\partial t_{sk}}{\partial x_k}\boldsymbol{i}_s = \mathbf{0}. \tag{37}$$

Let us consider now an arbitrary tetrahedron T, Fig. 8. Applying the first Euler's law for T we get

$$\int_{v_T} \rho\left(\frac{\mathrm{d}\boldsymbol{v}}{\mathrm{d}t} - \boldsymbol{f}\right)\mathrm{d}v = \int_{\sigma_T} n_k t_{sk} \boldsymbol{i}_s \,\mathrm{d}\sigma + \int_{M_1 M_2 M_3} \boldsymbol{t}(\boldsymbol{n})\,\mathrm{d}\sigma, \tag{38}$$

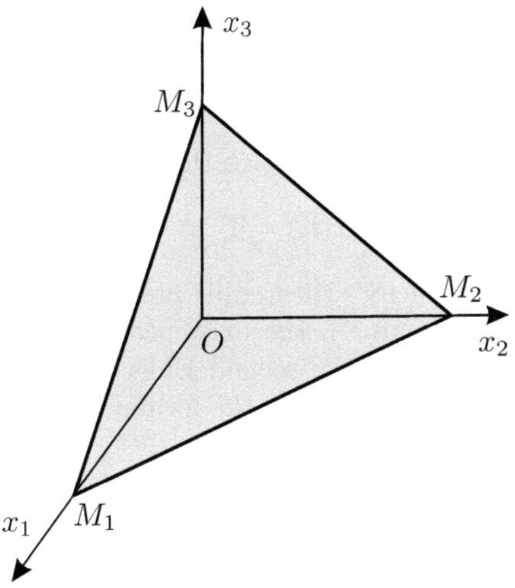

Figure 8. Tetrahedron T

where v_T is the volume, σ_T is the part of the tetrahedron sides that are parallel to the coordinate planes, $M_1M_2M_3$ is the inclined side. In Eq. (38) we have used the representation of the stress vector on the coordinate plane elements Eq. (35).

With the help of Eq. (37) we transform the left-hand side of (38)

$$\int\limits_{v_T} \frac{\partial t_{sk}}{\partial x_k} \boldsymbol{i}_s \, \mathrm{d}v = \int\limits_{\sigma_T} n_k t_{sk} \boldsymbol{i}_s \, \mathrm{d}\sigma + \int\limits_{M_1M_2M_3} \boldsymbol{t}(\boldsymbol{n}) \, \mathrm{d}\sigma. \tag{39}$$

Applying again the Gauss–Ostrogradsky formula to the volume integral in (39) we transform the equation to the form

$$\int\limits_{M_1M_2M_3} [\boldsymbol{t}(\boldsymbol{n}) - n_k t_{sk} \boldsymbol{i}_s] \, \mathrm{d}\sigma = \mathbf{0}.$$

As the tetrahedron and its side $M_1M_2M_3$ are arbitrary it follows the identity

$$\boldsymbol{t}(\boldsymbol{n}) - n_k t_{sk} \boldsymbol{i}_s = \mathbf{0}$$

that is valid for any normal $\boldsymbol{n}$. Thus, we have demonstrated that $\boldsymbol{t}$ depends on $\boldsymbol{n}$ linearly.

It is well known the following representation theorem of a linear function, see e.g. Lebedev et al. (2010).

Theorem 3.2. *A linear vector-valued function $\boldsymbol{l}(\boldsymbol{n})$ of a vectorial argument $\boldsymbol{n}$ can be represented in the form $\boldsymbol{l}(\boldsymbol{n}) = \mathbf{L} \cdot \boldsymbol{n}$, where $\mathbf{L}$ is a second-order tensor.*

By this theorem, there exists $\mathbf{T}$ such that

$$\boldsymbol{t}(\boldsymbol{n}) = \mathbf{T} \cdot \boldsymbol{n}.$$

Thus the first part of Cauchy's theorem is proved.

The second theorem part on the existence of $\mathbf{M}$ can be proven similarly. First we should apply the second Euler's law first to an arbitrary parallelepiped. As a result we obtain the formula

$$\int\limits_{v_\Pi} \left\{ j \frac{\mathrm{d}\boldsymbol{\omega}}{\mathrm{d}t} - \rho \boldsymbol{m} - t_{sk} \boldsymbol{i}_k \times \boldsymbol{i}_s - \frac{\partial m_{sk}}{\partial x_k} \boldsymbol{i}_s \right\} \mathrm{d}v = \mathbf{0},$$

from which it follows the differential equation

$$j \frac{\mathrm{d}\boldsymbol{\omega}}{\mathrm{d}t} - \rho \boldsymbol{m} = \frac{\partial m_{sk}}{\partial x_k} \boldsymbol{i}_s - t_{sk} \boldsymbol{i}_s \times \boldsymbol{i}_k. \tag{40}$$

The last addendum in (40) is the vector invariant of the second-order tensor $\mathbf{T}$.

Then we apply the second Euler's law to an arbitrary tetrahedron and derive the representation

$$\boldsymbol{\mu}(\boldsymbol{n}) = \mathbf{M} \cdot \boldsymbol{n},$$

which completes the proof of Cauchy's theorem for the Cosserat medium.

□

Definition 3.4. *Tensor* $\mathbf{T}$ *is called the stress tensor of Cauchy-type and tensor* $\mathbf{M}$ *the couple stress tensor of Cauchy-type.*

From the theorem proof we see that matrices t_{sk} and m_{sk} represent the components of the stress tensor and of the couple stress tensor in Cartesian basis $\boldsymbol{i}_k$:

$$\mathbf{T} = t_{ks}\boldsymbol{i}_k\boldsymbol{i}_s, \quad \mathbf{M} = m_{ks}\boldsymbol{i}_k\boldsymbol{i}_s.$$

Note that the stress and couple stress tensors are not symmetric, in general, that is $\mathbf{T}^T \neq \mathbf{T}$, $\mathbf{M}^T \neq \mathbf{M}$. This property distinguishes the micropolar continuum from the simple materials for which the stress tensor is always symmetric.

Non-symmetry of matrices t_{sk} and m_{sk} makes us carefully work with their indices. Let us consider an arbitrary cube in the body. The sides of the cube are oriented along the axes of Cartesian coordinate system. Second subscript of t_{sk} means the area element with normal $\boldsymbol{i}_k$ whereas the first subscript shows the direction of the stress action that is $\boldsymbol{i}_s$. For example, t_{13} is the stress acting on the cross-section of the body that is perpendicular to axis x_3, its action is in the direction of axis x_1. Note this agreement for the indices is valid for any orthonormal basis.

3.4 Principal Stresses in Micropolar Continua

The representation of the Cauchy stress tensor $\mathbf{T}$ in a general non-orthogonal basis is

$$\mathbf{T} = t_{ks}\boldsymbol{i}_s \otimes \boldsymbol{i}_k,$$

where matrix t_{sk} contains 9 components that are non-zero, in general.

Let us consider the problem of finding a basis in which matrix representation of $\mathbf{T}$ is most simple.

In the classical Cauchy continuum there is known the *spectral decomposition* of symmetric stress tensor, the matrix of representation is diagonal

$$\mathbf{T} = \sigma_1\boldsymbol{e}_1 \otimes \boldsymbol{e}_1 + \sigma_2\boldsymbol{e}_2 \otimes \boldsymbol{e}_2 + \sigma_3\boldsymbol{e}_3 \otimes \boldsymbol{e}_3, \tag{41}$$

where σ_1, σ_2, σ_3 are eigenvalues of the matrix

$$\begin{pmatrix} t_{11} & t_{12} & t_{13} \\ t_{21} & t_{22} & t_{23} \\ t_{31} & t_{32} & t_{33} \end{pmatrix},$$

that are called *principal stresses*, and $\boldsymbol{e}_1$, $\boldsymbol{e}_2$, $\boldsymbol{e}_3$, are eigenvectors of t_{ks} that are called *principal axes* of $\mathbf{T}$. As for the Cauchy continuum $\mathbf{T}$ is always symmetric, there exists an orthonormal basis $\boldsymbol{e}_1$, $\boldsymbol{e}_2$, $\boldsymbol{e}_3$ in which the matrix for $\mathbf{T}$ is diagonal. From the physical point of view the spectral decomposition (41) means that at any point of the body we can always find orthogonal surface elements on which only normal stresses, stretching or compressing, act. If the stress field is homogeneous then we can select such an elementary cube that on its facets there act normal stresses only.

In micropolar mechanics, $\mathbf{T}$ is not symmetric, in general, and so a spectral decomposition of $\mathbf{T}$ can not be established. Now an analogue of the orthogonal representation (41) of $\mathbf{T}$ is the matrix singular value decomposition that exists for all non-symmetric tensors.

Definition 3.5. *We call the singular value decomposition of a second-order tensor* $\mathbf{T}$ *the following relation*

$$\mathbf{T} = s_1 \boldsymbol{e}_1 \otimes \boldsymbol{e}'_1 + s_2 \boldsymbol{e}_2 \otimes \boldsymbol{e}'_2 + s_3 \boldsymbol{e}_3 \otimes \boldsymbol{e}'_3, \tag{42}$$

where s_k $(k = 1, 2, 3)$ *are non-negative real numbers, that are singular values of* $\mathbf{T}$, *and* $\boldsymbol{e}_k$, $\boldsymbol{e}'_j$ *are two orthonormal bases.*

Proof of the singular value decomposition can be found in many books on the theory of matrices, see for example Horn and Johnson (1985). The singular representation (42) includes two orthogonal bases $\boldsymbol{e}_k$ and $\boldsymbol{e}'_j$ that is its disadvantage.

Let us answer the question whether there exist a more simple representation of non-symmetric $\mathbf{T}$ with only one basis that can be non-orthonormal. We will use the results on the *real Jordan canonical form* of a non-symmetric real matrix, cf. Horn and Johnson (1985). It is shown the following. A non-symmetric real valued 3×3 matrix A is similar to one of the following matrices

$$\begin{pmatrix} \lambda_1 & 0 & 0 \\ 0 & \lambda_2 & 0 \\ 0 & 0 & \lambda_3 \end{pmatrix}, \tag{43}$$

$$\begin{pmatrix} \lambda_1 & 0 & 0 \\ 0 & \lambda_2 & 0 \\ 0 & 0 & \lambda_2 \end{pmatrix}, \begin{pmatrix} \lambda_1 & 0 & 0 \\ 0 & \lambda_2 & \varepsilon \\ 0 & 0 & \lambda_2 \end{pmatrix}, \tag{44}$$

$$\begin{pmatrix} \lambda_1 & 0 & 0 \\ 0 & \lambda_1 & 0 \\ 0 & 0 & \lambda_1 \end{pmatrix}, \begin{pmatrix} \lambda_1 & 0 & 0 \\ 0 & \lambda_1 & \varepsilon \\ 0 & 0 & \lambda_1 \end{pmatrix}, \begin{pmatrix} \lambda_1 & \varepsilon & 0 \\ 0 & \lambda_1 & \varepsilon \\ 0 & 0 & \lambda_1 \end{pmatrix}, \tag{45}$$

$$\begin{pmatrix} \lambda_1 & 0 & 0 \\ 0 & \alpha & \beta \\ 0 & -\beta & \alpha \end{pmatrix} \tag{46}$$

with real valued similarity matrices. Let us recall that two matrices A and B are called similar if $A = P^{-1}BP$ for some invertible matrix P. Which of the matrices should be selected, it depends on eigenvalues of A and their multiplicity. Here λ_k are real valued eigenvalues of A, ε is an arbitrarily small non-zero number, α and β are real and imaginary parts of a complex eigenvalue of A when its exists. Let us note that usually one takes $\varepsilon = 1$, cf. Horn and Johnson (1985). This corresponds to a selection of the similarity matrix. The case (43) is for existence of three different real eigenvalues $\lambda_1 \neq \lambda_2 \neq \lambda_3$ of A. The case (44) happens when $\lambda_1 \neq \lambda_2 = \lambda_3$ and they are real. If $\lambda_1 = \lambda_2 = \lambda_3$ then the representation (45) is valid. The fourth case (46) holds if one of eigenvalues is real and other two of eigenvalues are complex, $\lambda_1 \neq \lambda_2 = \overline{\lambda_3} \equiv \alpha + i\beta$, and $\overline{(\ldots)}$ denotes the complex conjugate of $(\ldots)$.

Thus, a non-symmetric $\mathbf{T}$ has a representation

$$\mathbf{T} = t^{\circ}_{mn} \boldsymbol{e}_m \otimes \boldsymbol{e}_n,$$

where t°_{mn} is one of the matrices given by formulas (43)–(46). Here vectors $\boldsymbol{e}_m$ are not orthogonal, in general.

Obviously, all these results on the representation of $\mathbf{T}$ relate to the couple stress tensor $\mathbf{M}$, that is $\mathbf{M}$ takes form $\mathbf{M} = m^{\circ}_{mn} \tilde{\boldsymbol{e}}_m \otimes \tilde{\boldsymbol{e}}_n$, where m°_{mn} is the matrix having the structure of one of (43)–(46) and $\tilde{\boldsymbol{e}}_n$ is a non-orthogonal basis.

3.5 Equations of Motion

Transforming Euler's laws of motion as was done in the proof of Cauchy's theorem 3.1, we get dynamic equations of micropolar continuum in the local form

$$\rho \frac{\mathrm{d}\boldsymbol{v}}{\mathrm{d}t} = \operatorname{div} \mathbf{T} + \rho \boldsymbol{f}, \tag{47}$$

$$j \frac{\mathrm{d}\boldsymbol{\omega}}{\mathrm{d}t} = \operatorname{div} \mathbf{M} - \mathbf{T}_{\times} + \rho \boldsymbol{m}. \tag{48}$$

Deriving Eq. (48) we have used Eq. (47).

Equations (47) and (48) in Cartesian coordinates take the form

$$\rho \frac{\mathrm{d}v_s}{\mathrm{d}t} = \frac{\partial t_{sk}}{\partial x_k} + \rho f_s, \quad j \frac{\mathrm{d}\omega_s}{\mathrm{d}t} = \frac{\partial m_{sk}}{\partial x_k} + t_{mn}\epsilon_{mns} + \rho m_s, \tag{49}$$

where $\epsilon_{mns} = -(\boldsymbol{i}_m \times \boldsymbol{i}_n) \cdot \boldsymbol{i}_s$ is the permutation symbol.

When the medium does not possesses couple's properties, that is rotation interaction of particles is negligible, then in Eq. (48) we should change to zero the following terms: the rotation inertia j, the couple stress tensor $\mathbf{M}$ and the volume couples $\boldsymbol{m}$. As a consequence of the balance of moment of momentum we obtain the following equation:

$$\mathbf{T}_\times = \mathbf{0}. \tag{50}$$

Its solution is the symmetric stress tensor, that is $\mathbf{T} = \mathbf{T}^T$. Thus when couple stresses and the distributed external couples in the balance equation of moment of momentum are absent, it follows the symmetry of the Cauchy stress tensor that is a property of the classic continuum mechanics. However, now it is impossible to consider the action of the couple loads, in particular the action of the couples distributed on the boundary or inside the body.

For classical continuum, that is when couple stresses and the external couples are absent, Eq. (50) holds automatically as a consequence of the constitutive equations. Hence, the balance equation of moment of momentum plays less important role in the Cauchy continuum.

3.6 Boundary-Value Problems

To setup a boundary-value problem for a micropolar body, we should supplement the motion equations (47) and (48) with boundary and initial conditions. *Static* boundary conditions consist of external forces and couples $\boldsymbol{t}_*$ and $\boldsymbol{\mu}_*$ given on the body boundary or on some its part $\sigma_f \subset \sigma$ (in the actual configuration)

$$\mathbf{T} \cdot \boldsymbol{n} = \boldsymbol{t}_*, \quad \mathbf{M} \cdot \boldsymbol{n} = \boldsymbol{\mu}_* \quad \text{on} \quad \sigma_f \tag{51}$$

as is shown on Fig. 9. On the rest part of the boundary we assign two *kinematic* conditions

$$\boldsymbol{r} = \boldsymbol{r}_*, \quad \mathbf{H} = \mathbf{H}_* \quad (\mathbf{H}_* \cdot \mathbf{H}_*^T = \mathbf{I}) \quad \text{on} \quad \sigma_u, \tag{52}$$

where there are given $\boldsymbol{r}_*$ and $\mathbf{H}_*$, the translations and microrotations of the body particles on $\sigma_u = \sigma / \sigma_f$. We get a particular case of kinematic boundary conditions, when the part σ_u is fixed, that is

$$\boldsymbol{r} = \boldsymbol{R}, \quad \mathbf{H} = \mathbf{I} \quad \text{on} \quad \sigma_u. \tag{53}$$

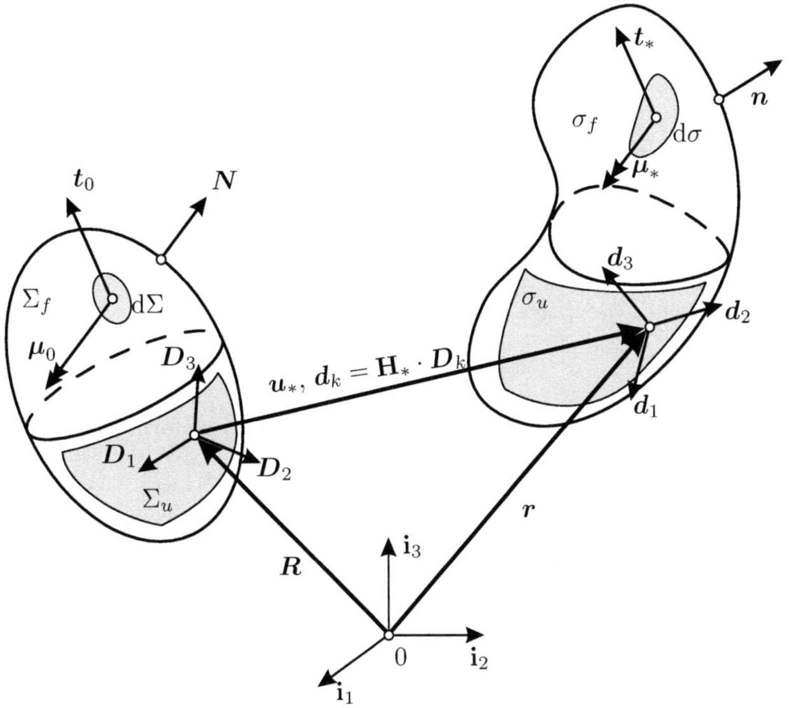

Figure 9. Boundary conditions. On the left the body is shown in the reference configuration, on the right it is in the actual configuration.

Using the vectorial representation of $\mathbf{H}$ through the finite rotation vector $\boldsymbol{\theta}$, see Eq. (7), one can rewrite the kinematical boundary conditions:

$$\boldsymbol{u} = \boldsymbol{u}_*, \quad \boldsymbol{\theta} = \boldsymbol{\theta}_* \quad \text{on} \quad \sigma_u \tag{54}$$

with given $\boldsymbol{u}_*$ and $\boldsymbol{\theta}_*$. In particular, Eqs. (53) take simple form: $\boldsymbol{u} = \mathbf{0}$, $\boldsymbol{\theta} = \mathbf{0}$ on σ_u.

There are other types of boundary conditions of the micropolar mechanics. For example, in the micropolar hydrodynamics on a free surface they use the boundary condition: $\boldsymbol{\omega} = -\frac{1}{2}\mathrm{rot}\,\boldsymbol{v}$, see Migoun and Prokhorenko (1984).

For dynamic problems we should assign the initial conditions

$$\boldsymbol{r} = \boldsymbol{r}^0, \quad \boldsymbol{v} = \boldsymbol{v}^0, \qquad \mathbf{H} = \mathbf{H}^0 \quad (\mathbf{H}^0 \cdot \mathbf{H}^{0^T} = \mathbf{I}), \quad \boldsymbol{\omega} = \boldsymbol{\omega}^0 \quad \text{at} \quad t = t_0 \tag{55}$$

with given $\boldsymbol{r}^0$, $\boldsymbol{v}^0$, $\mathbf{H}^0$, and $\boldsymbol{\omega}^0$.

Equations (47) and (48) with boundary conditions (51), (52) (or (53)) and with initial conditions (55) constitute the Eulerian boundary-value problem for the micropolar body, which is widely used in the micropolar hydrodynamics.

Using *Piola's identity*

$$\operatorname{Div}\left[(\det \mathbf{F})\mathbf{F}^{-T}\right] = \mathbf{0},$$

where $\mathbf{F} = \operatorname{Grad} \boldsymbol{r}$ is the *deformation gradient*, we transform Eqs (47) and (48) to the *Lagrangian motion equations*

$$\rho_0 \frac{\mathrm{d}\boldsymbol{v}}{\mathrm{d}t} = \operatorname{Div} \mathbf{T}_\kappa + \rho_0 \boldsymbol{f}, \quad \rho_0 \gamma \frac{\mathrm{d}\boldsymbol{\omega}}{\mathrm{d}t} = \operatorname{Div} \mathbf{M}_\kappa - \left(\mathbf{T}_\kappa \cdot \mathbf{F}^T\right)_\times + \rho_0 \boldsymbol{m} \tag{56}$$

with boundary conditions

$$\mathbf{T}_\kappa \cdot \boldsymbol{N} = \boldsymbol{t}_0, \quad \mathbf{M}_\kappa \cdot \boldsymbol{N} = \boldsymbol{\mu}_0 \quad \text{on } \Sigma_f, \qquad \boldsymbol{r} = \boldsymbol{r}_*, \quad \mathbf{H} = \mathbf{H}_* \quad \text{on} \quad \Sigma_u, \tag{57}$$

where $\rho_0 = \rho J$ is the mass density in the reference configuration, $\rho_0 \gamma = jJ$, $J = \det \mathbf{F}$,

$$\mathbf{T}_\kappa = J\, \mathbf{T} \cdot \mathbf{F}^{-T}, \quad \mathbf{M}_\kappa = J\, \mathbf{M} \cdot \mathbf{F}^{-T}$$

are the *first Piola-Kirchhoff stress tensor and the couple stress tensor*, respectively, Div is the divergence operator in the reference configuration, $\boldsymbol{t}_0 = \boldsymbol{t}_* \frac{\mathrm{d}\sigma}{\mathrm{d}\Sigma}$ and $\boldsymbol{\mu}_0 = \boldsymbol{\mu}_* \frac{\mathrm{d}\sigma}{\mathrm{d}\Sigma}$ are given external surface loads acting on the surface in the reference configuration, and $\boldsymbol{N}$ is the unit normal to the boundary surface in the reference configuration.

The equilibrium of the micropolar continuum is described by the Eulerian equilibrium equations

$$\operatorname{div} \mathbf{T} + \rho \boldsymbol{f} = \mathbf{0}, \quad \operatorname{div} \mathbf{M} - \mathbf{T}_\times + \rho \boldsymbol{m} = \mathbf{0}, \tag{58}$$

or by the Lagrangian equilibrium equations

$$\operatorname{Div} \mathbf{T}_\kappa + \rho_0 \boldsymbol{f} = \mathbf{0}, \quad \operatorname{Div} \mathbf{M}_\kappa - \left(\mathbf{T}_\kappa \cdot \mathbf{F}^T\right)_\times + \rho_0 \boldsymbol{m} = \mathbf{0}, \tag{59}$$

complemented by the boundary conditions (51) and (52) or (57), respectively.

In linear micropolar elasticity the existence of solution and some properties of solutions of boundary-value problems are established in Ieşan (1970, 1971); Nowacki (1986). Mathematical studies of micropolar hydrodynamics can be found in Łukaszewicz (1999). The study of qualitative questions of nonlinear micropolar continuum is far from completion.

4 Constitutive Equations of Cosserat Continua

For an arbitrary part of the body, Eqs (47) and (48) express the balance equations for the moment and the moment of momentum. These six scalar equations contain 18 unknown quantities that are the components of tensors $\mathbf{T}$ and $\mathbf{M}$. The dependence of $\mathbf{T}$ and $\mathbf{M}$ on medium deformations is determined by the *constitutive equations* or *constitutive relations* that depend on the material properties. They are determined experimentally. The constitutive equations must obey some principles that restrict their form, see Truesdell and Noll (1965).

In the framework of micropolar continuum mechanics we introduce the constitutive equations for $\mathbf{T}$ and $\mathbf{M}$. Besides, we can introduce another type of constitutive equations that determines the moment and the moment of momentum of more general form than those given by definitions (28) and (29). These relations are called *kinetic constitutive equations.* Note that in the general nonlinear shell theory, the dynamic equations coincide with two-dimensional dynamic equations for the micropolar continuum with more complex relations for the inertia tensors.

4.1 General Principles Restricting the Constitutive Equations

In this section we formulate the principles for constitutive equations for micropolar continuum. We restrict them to the case of the pure mechanical theory. Here we will omit the influence of temperature and do not introduce internal energy, entropy, etc. The restrictions stipulated on the form of constitutive equations by the principles must be valid for any medium, cf. Truesdell and Noll (1965).

Principle of Determinism. *At each body point, both the stress and the couple stress tensors are uniquely determined by the pre-history of the body motion.*

Principle of Local Action. *At each body point, both the stress and the couple stress tensors are uniquely determined by the motion of any neighborhood of the point that can be so small as is desired.*

Let us note that the Principle of Determinism states that the constitutive equations cannot forecast the future and so it is absolute whereas the

Principle of Local Action is not: it can be not valid for some materials. Non-local continuum models are presented for example in Eringen (2002).

Principle of Material Frame-Indifference. *Being determined by the constitutive equations, the stress and couple stress tensors must be material frame-indifferent quantities.*

To introduce the notion of frame indifference we remind the definition of equivalent motions. In classical mechanics, two motions $\boldsymbol{r}$ and $\boldsymbol{r}^*$ are called *equivalent* if they relate as follows

$$\boldsymbol{r}^* = \boldsymbol{a}(t) + \mathbf{O}(t) \cdot (\boldsymbol{r} - \boldsymbol{r}_0), \tag{60}$$

where $\mathbf{O}(t)$ is an arbitrary orthogonal tensor, $\boldsymbol{a}(t)$ is an arbitrary vector function and the constant vector $\boldsymbol{r}_0$ represents a fixed point position (a pole). We can treat equivalent motions as the one body motion considered in different reference frames.

In micropolar mechanics, body deformations are described by the position vector $\boldsymbol{r}$ and the trihedron $\boldsymbol{d}_k$ or the orthogonal tensor $\mathbf{H}$. After Kafadar and Eringen (1971); Le and Stumpf (1998), we *assume* that in the equivalent motion the directors $\boldsymbol{d}_k$ rotate similarly to $\boldsymbol{r}$:

$$\boldsymbol{d}_k^* = \mathbf{O}(t) \cdot \boldsymbol{d}_k.$$

It follows that in the equivalent motion the microrotation tensors relate by the equation:

$$\mathbf{H}^* = \mathbf{O}(t) \cdot \mathbf{H}. \tag{61}$$

We underline that two deformations of the micropolar medium are equivalent if the position vectors and microrotation tensors relate through Eqs. (60) and (61).

The principle of material frame-indifference or the principle of objectivity was originally proposed for classical continuum mechanics by Noll (1958), see Truesdell and Noll (1965). There has been an extensive discussion in the literature about the proper understanding of this principle, because its different formulations seem to reflect different physical contents, see for example recent papers by Murdoch (2000, 2003, 2005); Muschik and Restuccia (2002); Rivlin (2002, 2005, 2006); Bertram and Svendsen (2001, 2004); Svendsen and Bertram (1999), and the book by Bertram (2012).

However, for Cosserat medium, the material frame-indifference principle needs to be specified. Unlike classical mechanics, micropolar mechanics employs axial vectors and tensors that are not "real" frame-indifferent quantities. Under mirror reflection, the change of sign of axial quantities differs from the one for polar quantities. Among the introduced quantities the total moment, the couple stress vector $\boldsymbol{\mu}$, and the couple stress tensor $\mathbf{M}$ are axial ones. If in the definition of the equivalent motion we leave only proper orthogonal vectors $\mathbf{O}$ there is no difference between axial and polar vectors. However non-proper orthogonal tensors correspond to the rotation of coordinates and to the coordinate orientation change. Such transformations look admissible but axial quantities have quite certain physical meaning as well. I what follows we use the following definition.

Definition 4.1. *The scalar x, the vector $\boldsymbol{x}$ and the tensor $\mathbf{X}$ are called indifferent or objective if for two equivalent motions there hold*

$$\begin{aligned} x^* &= (\det \mathbf{O})^p x, \\ \boldsymbol{x}^* &= (\det \mathbf{O})^p \mathbf{O} \cdot \boldsymbol{x}, \\ \mathbf{X}^* &= (\det \mathbf{O})^p \mathbf{O} \cdot \mathbf{X} \cdot \mathbf{O}^T, \end{aligned} \tag{62}$$

where superscript "$*$" denotes the quantities in the equivalent motion and $p = 0$ for polar quantities and $p = 1$ for axial ones.

As an example let us consider the *mirror reflection* that is when $\mathbf{O} = -\mathbf{I}$. For polar quantities from (62) it follows the relations

$$x^* = x, \quad \boldsymbol{x}^* = -\boldsymbol{x}, \quad \mathbf{X}^* = \mathbf{X}.$$

For axial scalars, vectors and tensors, Eq. (62) results in

$$x^* = -x, \quad \boldsymbol{x}^* = \boldsymbol{x}, \quad \mathbf{X}^* = -\mathbf{X}.$$

At first glance the transformation rule (62) contradicts to definition in Truesdell and Noll (1965) or a naive understanding of independence of an objective quantity on the choice of coordinates. However, (62) is correct. The dependence on mirror reflections is a consequence of the agreement on the choice of right-hand or left-hand rules when we define such quantities as the cross product or the couple. It is similar to the agreement of what of the Earth poles is northern or what of the rotation directions we consider to be positive. Described by indifferent axial quantities, physical processes do not depend on the choice of coordinates, in particular they do not depend on the change of the frame orientation. An interesting discussion of the physical nature of axial vectors and mirror reflection can be found in Feynman et al. (1977).

4.2 Natural Lagrangian Strain Measures of Cosserat Continuum

The stretch and wryness tensors of the non-linear Cosserat continuum were originally defined by Cosserat and Cosserat (1909) through components of some fields in the common Cartesian frame. Following Pietraszkiewicz and Eremeyev (2009a,b) discussed the definitions of the Lagrangian stretch and wryness tensors in the non-linear Cosserat continuum using three different methods of introducing the strain measures into the non-linear Cosserat continuum:

a) by a direct geometrical approach,
b) by defining the strain measures as the fields work-conjugate to the respective internal stress and couple-stress tensor fields,
c) by applying the principle of material frame-indifference to the strain energy density of the polar-elastic body.

Each of the three ways allows to associate different geometrical and/or physical interpretations to the corresponding strain measures. The strain measures of Pietraszkiewicz and Eremeyev (2009a) called the *natural* ones are of the *relative type*. They have to vanish in the reference configuration.

Strain Measures by Geometrical Approach Within the geometrical approach the strain measures are defined by analyzing the difference of the fields describing position and orientation differentials of the material particles of the micropolar continuum in the 3D physical space.

Since $\boldsymbol{D}_k$ are the unit vectors, the vector $\mathrm{d}\boldsymbol{D}_k$ can be represented by the axial vector $\boldsymbol{b}$ depending linearly on $\mathrm{d}\boldsymbol{R}$, so that

$$\begin{gathered} \mathrm{d}\boldsymbol{D}_k = \boldsymbol{b} \times \boldsymbol{D}_k, \quad \boldsymbol{b} = \mathbf{B} \cdot \mathrm{d}\boldsymbol{R}, \\ \boldsymbol{b} = \frac{1}{2}\boldsymbol{D}_k \times \mathrm{d}\boldsymbol{D}_k, \quad \mathbf{B} = \frac{1}{2}\boldsymbol{D}_k \times \operatorname{Grad} \boldsymbol{D}_k. \end{gathered} \tag{63}$$

In (63), $\mathbf{B}$ is the *microstructure curvature tensor* in the reference configuration of the micropolar continuum. Two tensors $\mathbf{I}$, $\mathbf{B}$ are the basic measures of local geometry of the reference configuration κ.

Again, $\mathrm{d}\boldsymbol{d}_k$ can be represented by the axial vector $\boldsymbol{c}$ depending linearly on $\mathrm{d}\boldsymbol{r}$ by the relations

$$\mathrm{d}\boldsymbol{d}_k = \boldsymbol{c}\times\boldsymbol{d}_k, \quad \boldsymbol{c} = \mathbf{C}\cdot \mathrm{d}\boldsymbol{r}, \qquad \boldsymbol{c} = \frac{1}{2}\boldsymbol{d}_k\times \mathrm{d}\boldsymbol{d}_k, \quad \mathbf{C} = \frac{1}{2}\boldsymbol{d}_k\times\operatorname{grad}\boldsymbol{d}_k. \tag{64}$$

where $\mathbf{C}$ is the *microstructure curvature tensor* in the actual configuration of the micropolar continuum. Two tensors $\mathbf{I}$ and $\mathbf{C}$ are the basic measures of local geometry of the actual configuration χ.

In the actual configuration χ the differential of $\boldsymbol{r}$ is

$$\mathrm{d}\boldsymbol{r} = (\operatorname{grad}\boldsymbol{r})\cdot \mathrm{d}\boldsymbol{r} = (\operatorname{Grad}\boldsymbol{r})\cdot \mathrm{d}\boldsymbol{R} = \mathbf{F}\cdot \mathrm{d}\boldsymbol{R}\,, \tag{65}$$

where grad denotes the gradient operator in χ, $\operatorname{grad}\boldsymbol{r} = \mathbf{I}$, and $\mathbf{F} = \operatorname{Grad}\boldsymbol{r}$ is the classical *deformation gradient* tensor.

Since $\mathbf{H}^T \cdot \mathbf{H},_i = -(\mathbf{H}^T \cdot \mathbf{H},_i)^T$ is skew-symmetric it can be expressed through the axial vector $\boldsymbol{k}_i$,

$$\mathbf{H}^T \cdot \mathbf{H},_i = \boldsymbol{k}_i \times \mathbf{I} = \mathbf{I} \times \boldsymbol{k}_i, \qquad \boldsymbol{k}_i = -\frac{1}{2}(\mathbf{H}^T \cdot \mathbf{H},_i)_\times, \tag{66}$$

where $(\ldots),_i$, $i = 1, 2, 3$, denotes the derivative with respect to the Cartesian coordinate x_i in the reference configuration. This allows one to introduce the second-order tensor

$$\begin{gathered} \mathbf{K} = \boldsymbol{k}_j \otimes \boldsymbol{i}_j = -\frac{1}{2}(\mathbf{H}^T \cdot \mathbf{H},_j)_\times \otimes \boldsymbol{i}_j = -\frac{1}{2}\boldsymbol{\mathcal{E}} : \left(\mathbf{H}^T \cdot \operatorname{Grad}\mathbf{H}\right), \\ \mathbf{H}^T \cdot \operatorname{Grad}\mathbf{H} = \mathbf{I} \times \mathbf{K}, \end{gathered} \tag{67}$$

where the double dot product : of two third-order tensors $\mathbb{A}$, $\mathbb{B}$ represented in the base $\boldsymbol{i}_k$ is defined as $\mathbb{A} : \mathbb{B} = A_{amn}B_{mnb}\boldsymbol{i}_a \otimes \boldsymbol{i}_b$, and $\boldsymbol{\mathcal{E}} = -\mathbf{I} \times \mathbf{I}$ is the *permutation tensor* that is also called *Levi–Civita's tensor*. The tensor $\mathbf{K}$ characterizes uniquely the third-order tensor $\mathbf{H}^T \cdot \operatorname{Grad}\mathbf{H}$ skew-symmetric with regard to first two tensor places. The tensor $\mathbf{K}$ is frequently called the *wryness tensor* in the literature, cf. Kafadar and Eringen (1971).

Using the chain rule $\operatorname{grad}\boldsymbol{d}_k = (\operatorname{Grad}\boldsymbol{d}_k) \cdot \mathbf{F}^{-1}$ with (66) and (67), the tensor $\mathbf{C}$ can now be represented by

$$\mathbf{C} = \mathbf{H} \cdot (\mathbf{K} + \mathbf{B}) \cdot \mathbf{F}^{-1}.$$

The relative changes of lengths and orientations of the micropolar continuum during deformation are governed by differences of differentials (63), (64), and (65) brought to the comparable orientation by the tensor $\mathbf{H}$,

$$\mathrm{d}\boldsymbol{r} - \mathbf{H} \cdot \mathrm{d}\boldsymbol{R} = \mathbf{X} \cdot \mathrm{d}\boldsymbol{R} = \mathbf{e} \cdot \mathrm{d}\boldsymbol{r}, \tag{68}$$

$$\mathbf{C} \cdot \mathrm{d}\boldsymbol{r} - \mathbf{H} \cdot \mathbf{B} \cdot \mathrm{d}\boldsymbol{R} = \boldsymbol{\Phi} \cdot \mathrm{d}\boldsymbol{R} = \mathbf{k} \cdot \mathrm{d}\boldsymbol{r}, \tag{69}$$

$$\mathbf{X} = \mathbf{F} - \mathbf{H}, \quad \mathbf{e} = \mathbf{I} - \mathbf{H} \cdot \mathbf{F}^{-1} = \mathbf{X} \cdot \mathbf{F}^{-1}, \tag{70}$$

$$\boldsymbol{\Phi} = \mathbf{C} \cdot \mathbf{F} - \mathbf{H} \cdot \mathbf{B}, \quad \mathbf{k} = \mathbf{C} - \mathbf{H} \cdot \mathbf{B} \cdot \mathbf{F}^{-1} = \boldsymbol{\Phi} \cdot \mathbf{F}^{-1}. \tag{71}$$

Scalar products of (68) and (69) by itself leads to the quadratic forms

$$\mathrm{d}\boldsymbol{R} \cdot \mathbf{X}^T \cdot \mathbf{X} \cdot d\boldsymbol{R} = \mathrm{d}\boldsymbol{r} \cdot \mathbf{e}^T \cdot \mathbf{e} \cdot \mathrm{d}\boldsymbol{r}, \quad \mathrm{d}\boldsymbol{R} \cdot \boldsymbol{\Phi}^T \cdot \boldsymbol{\Phi} \cdot \mathrm{d}\boldsymbol{R} = \mathrm{d}\boldsymbol{r} \cdot \mathbf{k}^T \cdot \mathbf{k} \cdot \mathrm{d}\boldsymbol{r}. \tag{72}$$

However, the relative changes of lengths and orientations can also be calculated by the alternative back-rotated expressions

$$\mathbf{H}^T \cdot \mathrm{d}\boldsymbol{r} - d\boldsymbol{R} = \mathbf{E} \cdot \mathrm{d}\boldsymbol{R} = \mathbf{x} \cdot \mathrm{d}\boldsymbol{r}, \tag{73}$$

$$\mathbf{H}^T \cdot \mathbf{C} \cdot \mathrm{d}\boldsymbol{r} - \mathbf{B} \cdot \mathrm{d}\boldsymbol{R} = \mathbf{K} \cdot \mathrm{d}\boldsymbol{R} = \boldsymbol{\varphi} \cdot \mathrm{d}\boldsymbol{r}, \tag{74}$$

$$\mathbf{E} = \mathbf{H}^T \cdot \mathbf{F} - \mathbf{I} = \mathbf{H}^T \cdot \mathbf{X}, \tag{75}$$

$$\mathbf{x} = \mathbf{H}^T - \mathbf{F}^{-1} = \mathbf{E} \cdot \mathbf{F}^{-1} = \mathbf{H}^T \cdot \mathbf{e} = \mathbf{H}^T \cdot \mathbf{X} \cdot \mathbf{F}^{-1}, \tag{76}$$

$$\mathbf{K} = \mathbf{H}^T \cdot \mathbf{C} \cdot \mathbf{F} - \mathbf{B} = \mathbf{H}^T \cdot \boldsymbol{\Phi}, \tag{77}$$

$$\boldsymbol{\varphi} = \mathbf{H}^T \cdot \mathbf{C} - \mathbf{B} \cdot \mathbf{F}^{-1} = \mathbf{K} \cdot \mathbf{F}^{-1} = \mathbf{H}^T \cdot \mathbf{k} = \mathbf{H}^T \cdot \boldsymbol{\Phi} \cdot \mathbf{F}^{-1}. \tag{78}$$

From (67), $(70)_2$, $(71)_2$, (75), and (77) we obtain the following relations:

$$\mathbf{e} = \mathbf{H} \cdot \mathbf{E} \cdot \mathbf{F}^{-1}, \quad \mathbf{k} = \mathbf{H} \cdot \mathbf{K} \cdot \mathbf{F}^{-1} = -\frac{1}{2}\mathbf{H} \cdot \mathcal{E} : (\mathbf{H}^T \cdot \operatorname{grad} \mathbf{H}). \tag{79}$$

Scalar products of each of (73) by itself give the alternative quadratic forms

$$\mathrm{d}\boldsymbol{R} \cdot \mathbf{E}^T \cdot \mathbf{E} \cdot \mathrm{d}\boldsymbol{R} = \mathrm{d}\boldsymbol{r} \cdot \mathbf{x}^T \cdot \mathbf{x} \cdot \mathrm{d}\boldsymbol{r}, \quad \mathrm{d}\boldsymbol{R} \cdot \mathbf{K}^T \cdot \mathbf{K} \cdot \mathrm{d}\boldsymbol{R} = \mathrm{d}\boldsymbol{r} \cdot \boldsymbol{\varphi}^T \cdot \boldsymbol{\varphi} \cdot \mathrm{d}\boldsymbol{r}. \tag{80}$$

From (72) and (80) it follows that each of the tensors $\mathbf{X}$, $\mathbf{E}$, or $\mathbf{x}$, $\mathbf{e}$ and $\boldsymbol{\Phi}$, $\mathbf{K}$ or $\boldsymbol{\varphi}$, $\mathbf{k}$ is the corresponding measure of deformation, stretch or orientation change of the non-linear micropolar continuum in the Lagrangian or Eulerian description, respectively.

The quadratic forms (72) and (80) do not change if $\mathbf{X}$, $\mathbf{E}$, $\boldsymbol{\Phi}$, $\mathbf{K}$ and their counterparts are replaced by $\mathbf{R} \cdot \mathbf{X}$, $\mathbf{R} \cdot \mathbf{E}$, $\mathbf{R} \cdot \boldsymbol{\Phi}$, $\mathbf{R} \cdot \mathbf{K}$, etc., respectively, where $\mathbf{R}$ is a orthogonal tensor. Hence, any so transformed tensor can also be regarded as the possible strain measure of the non-linear micropolar continuum. In particular when such a transformation with $\mathbf{R} = \mathbf{H}^T$ is applied to the measures $\mathbf{X}$, $\mathbf{e}$, $\boldsymbol{\Phi}$, $\mathbf{k}$ entering the quadratic form (72) the measures become $\mathbf{E}$, $\mathbf{x}$, $\mathbf{K}$, $\boldsymbol{\varphi}$, i.e. those entering the quadratic form (80).

It follows from (70), (71), and (76) that $\mathbf{X}$, $\boldsymbol{\Phi}$ (and $\mathbf{x}$, $\boldsymbol{\varphi}$) are two-point tensors with the left leg associated with the actual configuration and the right leg with the reference one (and reverse for $\mathbf{x}$, $\boldsymbol{\varphi}$). Such measures may also be called the deformation measures. The tensors $\mathbf{e}$, $\mathbf{k}$ are the *relative Eulerian strain measures*, while the tensors $\mathbf{E}$, $\mathbf{K}$ are the *relative Lagrangian strain measures*, or the *Lagrangian stretch and wryness tensors*, respectively. The latter are given by the formulae

$$\mathbf{E} = \mathbf{H}^T \cdot \mathbf{F} - \mathbf{I}, \quad \mathbf{K} = \mathbf{H}^T \cdot \mathbf{C} \cdot \mathbf{F} - \mathbf{B} = -\frac{1}{2}\mathcal{E} : \left(\mathbf{H}^T \cdot \operatorname{Grad} \mathbf{H}\right). \tag{81}$$

Let us note some interesting features of the relative Lagrangian strain measures:

1. They are given in the common coordinate-free notation; their various component representations can easily be generated, if necessary.
2. Definitions of the measures are valid for finite translations and rotations as well as for unrestricted stretches and changes of the microstructure orientation of the Cosserat body.
3. The measures are expressed in terms of the rotation tensor $\mathbf{H}$; for any specific parametrization of the rotation group $SO(3)$ by various finite rotation vectors, Euler angles, quaternions, etc. appropriate expressions for the measures can easily be found, if necessary.
4. The measures vanish in the rigid-body deformation $\boldsymbol{r} = \mathbf{O} \cdot \boldsymbol{R} + \boldsymbol{a}$, $\boldsymbol{d}_k = \mathbf{O} \cdot \boldsymbol{D}_k$ with a constant vector $\boldsymbol{a}$ and a constant proper orthogonal tensor $\mathbf{O}$ defined for the whole body.
5. In the absence of deformation from the reference configuration, that is when $\mathbf{F} = \mathbf{H} = \mathbf{I}$, the measures identically vanish.
6. The measures are not symmetric, in general: $\mathbf{E}^T \neq \mathbf{E}$, $\mathbf{K}^T \neq \mathbf{K}$.

In purely geometrical approach there is no need for discussion whether these measures might be defined as transposed ones or with opposite signs. Elements of geometrical approach in Cartesian components were used already by Cosserat and Cosserat (1909) and more recently by Merlini (1997) who took explicitly into account the microstructure curvature tensors describing spatial changes of orientations of the material particles in the reference and actual configurations. These tensors are independently introduced also in Zubov and Eremeev (1996); Yeremeyev and Zubov (1999) within the theory of viscoelastic micropolar fluids, and in Chróścielewski et al. (2004) within the general theory of shells. The microstructure curvature tensors are extensively used in discussions on the local material symmetry group of elastic shells and polar-elastic media by Eremeyev and Pietraszkiewicz (2006, 2012).

Principle of Virtual Work and Work-Conjugate Strain Measures
Reissner (1973, 1975) noted that the internal structure of two local equilibrium equations of the Cosserat elastic body requires two specific strain measures expressed in terms of independent translation and rotation vectors as the only field variables. The *principle of virtual work* of the non-linear Cosserat continuum takes the form

$$\begin{aligned}\int_V \left[\mathbf{T}_\kappa \bullet (\operatorname{Grad}\mathbf{v} - \boldsymbol{\omega}\times\mathbf{F}) + \mathbf{M}_\kappa \bullet \operatorname{Grad}\boldsymbol{\omega}\right]\,\mathrm{d}V \\ = \int_V \rho_0\,(\boldsymbol{f}\cdot\boldsymbol{v} + \boldsymbol{m}\cdot\boldsymbol{\omega})\,dV + \int_{\Sigma_f} (\boldsymbol{t}_0\cdot\boldsymbol{v} + \boldsymbol{\mu}_0\cdot\boldsymbol{\omega})\,\mathrm{d}\Sigma.\end{aligned} \tag{82}$$

We can show that $\delta\mathbf{E} = \mathbf{H}^T \cdot (\operatorname{Grad} \boldsymbol{v} - \boldsymbol{\omega} \times \mathbf{F})$ and $\delta\mathbf{K} = \mathbf{H}^T \cdot \operatorname{Grad} \boldsymbol{\omega}$, and the *internal virtual work density* under the first volume integral of (82) can now be given by the expressions

$$\sigma = \mathbf{T}_\kappa \bullet (\mathbf{H} \cdot \delta\mathbf{E}) + \mathbf{M}_\kappa \bullet (\mathbf{H} \cdot \delta\mathbf{K}) = \mathbf{P}_1 \bullet \delta\mathbf{E} + \mathbf{P}_2 \bullet \delta\mathbf{K}, \tag{83}$$

where $\mathbf{P}_1 = \mathbf{H}^T \cdot \mathbf{T}_\kappa$, $\mathbf{P}_2 = \mathbf{H}^T \cdot \mathbf{M}_\kappa$ are the non-symmetric stress and couple-stress tensors whose natural components are referred entirely to the reference configuration. We call $\mathbf{P}_1$ and $\mathbf{P}_2$ the *second Piola-Kirchhoff stress and couple stress tensors*, respectively. The stress measures $\mathbf{P}_1$, $\mathbf{P}_2$ require the relative Lagrangian strain measures $\mathbf{E}$, $\mathbf{K}$ as their work-conjugate counterparts.

Invariance of the Strain Energy Density In the polar-elastic body the constitutive relations are defined through the strain energy density W_κ per unit volume of the reference configuration κ. In general, the density W_κ can be assumed in the following form:

$$W_\kappa = W_\kappa(\boldsymbol{r}, \mathbf{F}, \mathbf{H}, \operatorname{Grad} \mathbf{H}; \boldsymbol{R}, \mathbf{B})\ . \tag{84}$$

But W_κ in (84) should satisfy the principle of material frame-indifference or invariance under the superposed rigid-body deformations. Applying (60) and (61) we obtain that the principle of material frame-indifference requires the values of W_κ to be the same for both deformations $\boldsymbol{r}$, $\mathbf{H}$ and $\boldsymbol{r}^*$, $\mathbf{H}^*$,

$$W_\kappa(\boldsymbol{r},\, \mathbf{F},\, \mathbf{H},\, \operatorname{Grad} \mathbf{H};\, \boldsymbol{R}, \mathbf{B}) = W_\kappa(\mathbf{O}\cdot\boldsymbol{r}+\boldsymbol{a}, \mathbf{O}\cdot\mathbf{F}, \mathbf{O}\cdot\mathbf{H}, \mathbf{O}\cdot\operatorname{Grad} \mathbf{H};\, \boldsymbol{R}, \mathbf{B}). \tag{85}$$

From Eq. (85) it follows the representation

$$W_\kappa = W_\kappa\,(\mathbf{E}, \mathbf{K};\, \boldsymbol{R}, \mathbf{B})\ . \tag{86}$$

This again confirms that the relative Lagrangian strain measures $\mathbf{E}$, $\mathbf{K}$ are required to be the independent fields in the polar-elastic strain energy density in order it to be invariant under the superposed rigid-body deformation. This way of introducing the Lagrangian strain measures is most common in the literature and various such procedures are used, for example, in Kafadar and Eringen (1971); Stojanović (1974); Eremeyev and Zubov (1994); Zubov and Eremeev (1996); Zubov (1997); Nikitin and Zubov (1998).

Since for the micropolar elastic continuum $\sigma = \delta W_\kappa$ from (83) it follows the constitutive equations for the stress measures

$$\mathbf{P}_1 = \frac{\partial W_\kappa}{\partial \mathbf{E}}, \quad \mathbf{P}_2 = \frac{\partial W_\kappa}{\partial \mathbf{K}}, \qquad \mathbf{T}_\kappa = \mathbf{H} \cdot \frac{\partial W_\kappa}{\partial \mathbf{E}}, \quad \mathbf{M}_\kappa = \mathbf{H} \cdot \frac{\partial W_\kappa}{\partial \mathbf{K}}. \tag{87}$$

The geometrical approach, the structure of equilibrium conditions and the invariance of the polar-elastic strain energy density all require the tensors $\mathbf{E}$, $\mathbf{K}$ as the most appropriate Lagrangian strain measures for the non-linear Cosserat continuum. We call the measures the *natural* stretch and wryness tensors, respectively.

Pietraszkiewicz and Eremeyev (2009a) were shown that the stretch and wryness tensors introduced in many papers do not agree with each other and with the Lagrangian strain measures defined in (81). Most definitions differ only by transpose of the measures or by opposite signs. Some measures do not vanish in the absence of deformation. Such differences are not essential for the theory, although one should be aware of them.

4.3 Vectorial Parameterizations of Strain Measures

While three components of $\boldsymbol{u}$ in (81) are all independent, nine components of $\mathbf{H}$ in (81) are subjected to six constraints following from the orthogonality conditions $\mathbf{H}^{-1} = \mathbf{H}^T$, $\det \mathbf{H} = +1$, so that only three rotational parameters of $\mathbf{H}$ are independent. In many applications it is more convenient to use the strain measures expressed in terms of six disconfiguration parameters all of which are independent.

Many techniques how to parameterize the rotation group $SO(3)$ are developed, which can roughly be classified as vectorial and non-vectorial ones. Various finite rotation vectors as well as the Cayley-Gibbs and exponential map parameters are examples of the vectorial parameterization, for they all have three independent scalar parameters as Cartesian components of a generalized vector in the 3D vector space. The non-vectorial parameterizations are expressed either in terms of three scalar parameters that cannot be treated as vector components, such as Euler-type angles for example, or through more scalar parameters subject to additional constraints, such as unit quaternions, Cayley-Klein parameters, or direction cosines. Each of these expressions may appear to be more convenient than others when solving specific problems of the non-linear Cosserat continuum.

The microrotation tensor $\mathbf{H}$ can be expressed by the Gibbs formula

$$\mathbf{Q} = (\mathbf{I} - \boldsymbol{e} \otimes \boldsymbol{e}) \cos\phi + \boldsymbol{e} \otimes \boldsymbol{e} - \boldsymbol{e} \times \mathbf{I} \sin\phi, \tag{88}$$

where φ is the rotation angle about the axis with the unit vector $\boldsymbol{e}$. In the vectorial parameterization of $\mathbf{H}$ one introduces a scalar function $p(\phi)$ generating three components of the *finite rotation vector* $\boldsymbol{p}$ defined as $\boldsymbol{p} = p(\phi)\boldsymbol{e}$, see for example Bauchau and Trainelli (2003). The generating function $p(\phi)$

has to be an odd function of ϕ with the limit behavior

$$\lim_{\phi \to 0} \frac{p(\phi)}{\phi} = \kappa,$$

where κ is a positive real normalization factor (usually 1 or $\frac{1}{2}$), and $p(0) = 0$. Then the tensor $\mathbf{H}$ can be represented as

$$\mathbf{H} = \cos\phi \mathbf{I} + \frac{1-\cos\phi}{p^2} \boldsymbol{p} \otimes \boldsymbol{p} - \frac{\sin\phi}{p} \boldsymbol{p} \times \mathbf{I}\,. \tag{89}$$

Taking the gradient of (89) and substituting it into (81), after appropriate transformations the natural Lagrangian stretch $\mathbf{E}$ and wryness $\mathbf{K}$ tensors can be represented in terms of the finite rotation vector $\boldsymbol{p}$ and the translation vector $\boldsymbol{u}$ by the relations

$$\mathbf{E} = \left(\cos\phi \mathbf{I} + \frac{1-\cos\phi}{p^2} \boldsymbol{p} \otimes \boldsymbol{p} + \frac{\sin\phi}{p} \boldsymbol{p} \times \mathbf{I}\right)(\mathbf{I} + \operatorname{Grad} \boldsymbol{u}) - \mathbf{I}\,, \tag{90}$$

$$\mathbf{K} = \left[\frac{-\sin\phi}{p} \mathbf{I} + \frac{1}{p^2}\left(\frac{1}{p'} + \frac{\sin\phi}{p}\right) \boldsymbol{p} \otimes \boldsymbol{p} - \frac{1-\cos\phi}{p^2} \boldsymbol{p} \times \mathbf{I}\right] \operatorname{Grad} \boldsymbol{p}\,. \tag{91}$$

Among definitions of $\boldsymbol{p}$ used in the literature let us mention the finite rotation vectors defined as

$$\boldsymbol{\theta} = 2\tan\frac{\phi}{2}\,\boldsymbol{e}\,, \quad \boldsymbol{\phi} = \phi\,\boldsymbol{e}\,, \quad \bar{\boldsymbol{\omega}} = \sin\phi\,\boldsymbol{e}\,, \quad \boldsymbol{\rho} = \tan\frac{\phi}{2}\,\boldsymbol{e}\,, \tag{92}$$

$$\boldsymbol{\sigma} = 2\sin\frac{\phi}{2}\,\boldsymbol{e}\,, \quad \boldsymbol{\mu} = 4\tan\frac{\phi}{4}\,\boldsymbol{e}\,, \quad \boldsymbol{\beta} = 4\sin\frac{\phi}{4}\,\boldsymbol{e}\,, \tag{93}$$

where the generating functions are

$$\theta = 2\tan\frac{\phi}{2},\ \phi,\ \varpi = \sin\phi, \rho = \tan\frac{\phi}{2}, \sigma = 2\sin\frac{\phi}{2},\ \mu = 4\tan\frac{\phi}{4},\ \ \beta = 4\sin\frac{\phi}{4},$$

respectively. The explicit formulae for $\mathbf{E}$ and $\mathbf{K}$ expressed in terms of the corresponding finite rotation vectors (92) and (93) are summarized in Pietraszkiewicz and Eremeyev (2009b).

4.4 Kinetic Constitutive Equations

Deriving expressions for momentum and moment of momentum of an arbitrary body part, that is (28) and (28), we have used the simplest relations for linear and angular velocities $\boldsymbol{v}$ and $\boldsymbol{\omega}$. However, even for a rigid body we see that the momentum and moment of momentum can have a complicated form. Let us determine $\mathfrak{P}$ and $\mathfrak{M}$ as follows

$$\mathfrak{P}(\mathcal{P}) \triangleq \int\limits_{v_\mathcal{P}} \rho \boldsymbol{K}_1 \, \mathrm{d}v, \quad \mathfrak{M}(\mathcal{P}) \triangleq \int\limits_{v_\mathcal{P}} \rho \left\{ (\boldsymbol{r} - \boldsymbol{r}_0) \times \boldsymbol{K}_1 + \boldsymbol{K}_2 \right\} \mathrm{d}v, \tag{94}$$

where

$$\begin{aligned} \boldsymbol{K}_1 &= \boldsymbol{K}_1(\boldsymbol{v}, \boldsymbol{\omega}; \boldsymbol{r}, \mathbf{H}, \operatorname{Grad} \boldsymbol{r}, \operatorname{Grad} \mathbf{H}), \\ \boldsymbol{K}_2 &= \boldsymbol{K}_2(\boldsymbol{v}, \boldsymbol{\omega}; \boldsymbol{r}, \mathbf{H}, \operatorname{Grad} \boldsymbol{r}, \operatorname{Grad} \mathbf{H}). \end{aligned} \tag{95}$$

These are the expressions for the density of momentum and moment of momentum of micropolar continuum. Such constitutive equations are called *kinetic*. Note that up to now we have dealt with the constitutive equation for frame-indifferent quantities like the strain energy density W. That allowed us to use the material frame-indifference principle to reduce the number of arguments in W and to get indifferent constitutive equations. Now $\boldsymbol{K}_1$ and $\boldsymbol{K}_2$ are not frame-indifferent quantities that is a consequence of the fact that $\boldsymbol{v}$ and $\boldsymbol{\omega}$ are not frame-indifferent. So we cannot use the material frame-indifference principle to simplify equations (95).

A rational way to simplify (95) is to use the analogy with the rigid body. Let us assume that $\boldsymbol{K}_1$ and $\boldsymbol{K}_2$ are linear functions of $\boldsymbol{v}$ and $\boldsymbol{\omega}$:

$$\boldsymbol{K}_1 = \mathbf{J}_1 \cdot \boldsymbol{v} + \mathbf{J}_2 \cdot \boldsymbol{\omega}, \quad \boldsymbol{K}_2 = \mathbf{J}_3 \cdot \boldsymbol{v} + \mathbf{J}_4 \cdot \boldsymbol{\omega},$$

where $\mathbf{J}_k$, $k = 1, 2, 3, 4$, are *tensors of inertia*. In general, $\mathbf{J}_k$ are tensorial functions of arguments $\boldsymbol{r}$, $\mathbf{H}$, $\operatorname{Grad} \boldsymbol{r}$, and $\operatorname{Grad} \mathbf{H}$.

For micropolar medium, it is natural to postulate the existence of the kinetic energy density $K = K(\boldsymbol{v}, \boldsymbol{\omega}; \boldsymbol{r}, \mathbf{H}, \operatorname{Grad} \boldsymbol{r}, \operatorname{Grad} \mathbf{H})$ that is the potential for $\boldsymbol{K}_1$ and $\boldsymbol{K}_2$. Now we get

$$\begin{aligned} K &= \frac{1}{2}(\boldsymbol{v} \cdot \mathbf{J}_1 \cdot \boldsymbol{v} + \boldsymbol{v} \cdot \mathbf{J}_2 \cdot \boldsymbol{\omega} + \boldsymbol{\omega} \cdot \mathbf{J}_3 \cdot \boldsymbol{v} + \boldsymbol{\omega} \cdot \mathbf{J}_4 \cdot \boldsymbol{\omega}), \\ \boldsymbol{K}_1 &= \frac{\partial K}{\partial \boldsymbol{v}}, \quad \boldsymbol{K}_2 = \frac{\partial K}{\partial \boldsymbol{\omega}}. \end{aligned} \tag{96}$$

Equations (96) imply the following properties of the inertia tensors

$$\mathbf{J}_1 = \mathbf{J}_1^T, \qquad \mathbf{J}_2^T = \mathbf{J}_3, \qquad \mathbf{J}_4 = \mathbf{J}_4^T.$$

Let us require K to be positive definite. This implies that $\mathbf{J}_1$ and $\mathbf{J}_4$ are positive definite tensors.

We get a particular case of equations (96) taking $\mathbf{J}_k$ in the form

$$\mathbf{J}_1 = \mathbf{I}, \qquad \mathbf{J}_2^T = \mathbf{J}_3 = \mathbf{H} \cdot \mathbf{J}_3^0 \cdot \mathbf{H}^T, \qquad \mathbf{J}_4 = \mathbf{H} \cdot \mathbf{J}_4^0 \cdot \mathbf{H}^T, \tag{97}$$

where $\mathbf{J}_3^0$ and $\mathbf{J}_4^0$ are constant tensors. We will call $\mathbf{J}_3$ and $\mathbf{J}_4$ microinertia tensors in the actual configuration and $\mathbf{J}_3^0$ and $\mathbf{J}_4^0$ microinertia tensors in the reference configuration. Equations (97) are completely analogous to equations $(15)_1$ for the inertia tensor of rigid body.

For further simplification of kinetic equations we can take

$$\mathbf{J}_1 = \mathbf{I}, \quad \mathbf{J}_2 = \mathbf{J}_3 = \mathbf{0}, \quad \mathbf{J}_4 = \rho^{-1} j \, \mathbf{I},$$

that are widely used in the literature, see Dyszlewicz (2004); Eringen (1999); Nowacki (1986). They correspond to the description of a micropolar particle as a solid homogeneous sphere of finite radius.

4.5 Material Symmetry Group

In general, the form of elastic strain energy density W_κ of the micropolar body depends upon the choice of the reference configuration κ. Particularly important are sets of reference configurations which leave unchanged the form of the energy density. Transformations of the reference configuration under which the energy density remains unchanged are called here *invariant transformations*. In other words invariant transformations are density-preserving deformations and all microrotations of the reference configuration of the micropolar continuum that cannot be experimentally detected. All such invariant transformations constitute the *material symmetry group*. Knowledge of the material symmetry group allows one to precisely define the fluid, the solid, the liquid crystal or the subfluid as well as to introduce notions of isotropic, hemitropic or orthotropic polar-elastic continua among others. A similar approach is used in classical continuum mechanics and in non-linear elasticity in Truesdell (1964, 1977); Truesdell and Noll (1965); Wang and Truesdell (1973); Rivlin (1980), as well as in non-linear theories of shells by Eremeyev and Pietraszkiewicz (2006). The material symmetry group of the non-linear micropolar continuum was first characterized by Eringen and Kafadar (1976) and modified by Eremeyev and Pietraszkiewicz (2012) by taking into account dependence of W_κ on the microstructure curvature tensor $\mathbf{B}$ as well as different transformation properties of the polar and axial tensors.

Let us introduce another reference configuration κ_* of the micropolar body $\mathcal{B}$, in which the position of point x is given by the vector $\mathbf{R}_*$ relative to the same origin o and its orientation is fixed by three orthonormal directors $\boldsymbol{D}_{*k}$. Let $\mathbf{P}$, $\det \mathbf{P} \neq 0$, be the deformation gradient transforming $\mathrm{d}\boldsymbol{R}$ into $\mathrm{d}\boldsymbol{R}_*$, and $\mathbf{R}$ be the orthogonal tensor transforming $\boldsymbol{D}_k$ into $\boldsymbol{D}_{*k}$, so that

$$\mathrm{d}\boldsymbol{R}_* = \mathbf{P} \cdot \mathrm{d}\boldsymbol{R}, \quad \boldsymbol{D}_{*k} = \mathbf{R} \cdot \boldsymbol{D}_k. \tag{98}$$

In what follows all fields associated with deformation relative to the reference configuration κ_* will be marked by the lower index $*$. From (98) it follows that

$$\mathbf{F} = \mathbf{F}_* \cdot \mathbf{P}, \quad \mathbf{H} = \mathbf{H}_* \cdot \mathbf{R}.$$

The transformations of the strain measures and the microstructure tensors are given by

$$\mathbf{E}_* = \mathbf{R} \cdot \mathbf{E} \cdot \mathbf{P}^{-1} + \mathbf{R} \cdot \mathbf{P}^{-1} - \mathbf{I}, \quad \mathbf{K}_* = (\det \mathbf{R})\mathbf{R} \cdot \mathbf{K} \cdot \mathbf{P}^{-1} + \mathbf{L}, \tag{99}$$

$$\mathbf{B}_* = (\det \mathbf{R})\mathbf{R} \cdot \mathbf{B} \cdot \mathbf{P}^{-1} - \mathbf{L}, \tag{100}$$

where $\mathbf{L} = \mathbf{R} \cdot \mathbf{Z} \cdot \mathbf{P}^{-1}$, $\mathbf{Z} = -\frac{1}{2}\boldsymbol{\mathcal{E}} : (\mathbf{R} \cdot \mathrm{Grad}\, \mathbf{R}^T)$.

The assumption that the constitutive relation is insensitive to the change of the reference configuration κ into κ_* means that the explicit form of the strain energy densities W_κ and W_* should coincide, that is

$$W_\kappa(\mathbf{E}, \mathbf{K}; \boldsymbol{R}, \mathbf{B}) = W_\kappa(\mathbf{E}_*, \mathbf{K}_*; \boldsymbol{R}, \mathbf{B}_*). \tag{101}$$

In other words, this means that one may use the same function for the strain energy density independently on the choice of κ or κ_*, but with different expressions for stretch and wryness tensors as well as for the microstructure curvature tensor. In what follows we not always explicitly indicate that all the functions depend also on the position vector $\boldsymbol{R}$ and W is taken relative to the reference configuration κ.

The relations (99), (100), and (101) hold locally, i.e. it should be satisfied at any x and $\mathbf{B}$, and the tensors $\mathbf{P}$, $\mathbf{R}$, $\mathbf{L}$ are treated as independent here. As a result, the local invariance of W under change of the reference configuration is described by the triple of tensors $(\mathbf{P}, \mathbf{R}, \mathbf{L})$. Using (99) and (100) we obtain the following invariance requirement for W and introduce the following definition:

Definition 4.2. *By the material symmetry group $\mathcal{G}_\kappa$ at x and $\mathbf{B}$ of the micropolar elastic continuum we call all sets of ordered triples of tensors*

$$\mathbb{X} = (\mathbf{P} \in Unim, \mathbf{R} \in Orth, \mathbf{L} \in Lin), \tag{102}$$

satisfying the relation

$$\begin{aligned} &W(\mathbf{E}, \mathbf{K}; \mathbf{B}) \\ &\quad = W\big[\mathbf{R} \cdot \mathbf{E} \cdot \mathbf{P}^{-1} + \mathbf{R} \cdot \mathbf{P}^{-1} - \mathbf{I},\ (\det \mathbf{R})\mathbf{R} \cdot \mathbf{K} \cdot \mathbf{P}^{-1} + \mathbf{L}; \\ &\qquad (\det \mathbf{R})\mathbf{R} \cdot \mathbf{B} \cdot \mathbf{P}^{-1} - \mathbf{L}\big] \end{aligned} \tag{103}$$

for any tensors $\mathbf{E}$, $\mathbf{K}$, $\mathbf{B}$ *in the domain of definition of the function* W.

The set $\mathcal{G}_\kappa$ is the group relative to the group operation $\circ$ defined by

$$(\mathbf{P}_1, \mathbf{R}_1, \mathbf{L}_1) \circ (\mathbf{P}_2, \mathbf{R}_2, \mathbf{L}_2) = \left[\mathbf{P}_1 \cdot \mathbf{P}_2, \mathbf{R}_1 \cdot \mathbf{R}_2, \mathbf{L}_1 + (\det \mathbf{R}_1)\mathbf{R}_1 \cdot \mathbf{L}_2 \cdot \mathbf{P}_1^{-1}\right].$$

The unit element of $\mathcal{G}_\kappa$ is $\mathbb{I} = (\mathbf{I}, \mathbf{I}, \mathbf{0})$ and the inverse element to $\mathbb{X} \in \mathcal{G}_\kappa$ is given by

$$\mathbb{X}^{-1} \equiv (\mathbf{P}, \mathbf{R}, \mathbf{L})^{-1} = \left[\mathbf{P}^{-1}, \mathbf{R}^T, -(\det \mathbf{R})\mathbf{R}^T \cdot \mathbf{L} \cdot \mathbf{P}\right].$$

Since the material symmetry group depends not only on the particle $x \in \mathcal{B}$ but also upon the choice of the reference configuration, let us analyze how the symmetry groups corresponding to different reference configurations are related. Let κ_1 and κ_2 be two different reference configurations, and $\mathcal{G}_1$ and $\mathcal{G}_2$ be the material symmetry groups relative to these reference configurations, respectively. Let now $\mathbf{P}$ be the non-singular deformation gradient, $\det \mathbf{P} \neq 0$, $\mathbf{R}$ be the orthogonal tensor associated with transformation $\kappa_1 \to \kappa_2$, as well as $\mathbf{P}^{-1}$ and $\mathbf{R}^T$ be the inverse deformation gradient and the inverse orthogonal tensor associated with the inverse transformation $\kappa_2 \to \kappa_1$, respectively.

The definition 4.2 allows one to establish an analogue of Noll's rule given for the classical simple material in Truesdell and Noll (1965). In what follows quantities described in the configurations κ_1 and κ_2 are marked by the respective lower indices 1 and 2. Let the element $\mathbb{X}_1 \equiv (\mathbf{P}_1, \mathbf{R}_1, \mathbf{L}_1) \in \mathcal{G}_1$. Then the element $\mathbb{X}_2 = \mathbb{P} \circ \mathbb{X}_1 \circ \mathbb{P}^{-1} \in \mathcal{G}_2$, where $\mathbb{P} \equiv (\mathbf{P}, \mathbf{R}, \mathbf{L})$. Thus the material symmetry group under change of the reference configuration transforms according to the analogue of the *Noll rule*

$$\mathcal{G}_2 = \mathbb{P} \circ \mathcal{G}_1 \circ \mathbb{P}^{-1}. \tag{104}$$

As in the case of non-polar elastic materials, the property of *isotropy* of the micropolar elastic material is expressed in terms of the orthogonal group.

Definition 4.3. *The micropolar elastic continuum is called isotropic at x and $\mathbf{B}$ if there exists a reference configuration κ, called undistorted, such that the material symmetry group relative to κ contains the group $\mathcal{S}_\kappa$,*

$$\mathcal{S}_\kappa \subset \mathcal{G}_\kappa, \qquad \mathcal{S}_\kappa \equiv \{(\mathbf{P} = \mathbf{O}, \mathbf{O}, \mathbf{0}) : \quad \mathbf{O} \in Orth\} \tag{105}$$

From the physical point of view this definition means that uniform rotations and mirror reflections of the undistorted reference configuration κ cannot be recognized by any experiment.

To define the *micropolar elastic fluid* we apply the requirement that its strain energy density should be insensitive to any change of the reference configuration.

Definition 4.4. *The micropolar elastic continuum is called polar micropolar elastic fluid at x and* $\mathbf{B}$ *if there exists a reference configuration κ, named undistorted, such that the material symmetry group relative to κ is given by*

$$\mathcal{G}_\kappa = \mathcal{U}_\kappa \equiv \{(\mathbf{P} \in Unim, \mathbf{R} \in Orth, \mathbf{L} \in Lin)\}. \tag{106}$$

Hence, the strain energy density of the polar-elastic fluid satisfies the relation

$$\begin{aligned} &W_\kappa(\mathbf{E}, \mathbf{K}; \mathbf{B}) \\ &\quad = W_\kappa \big[\mathbf{R}\cdot\mathbf{E}\cdot\mathbf{P}^{-1} + \mathbf{R}\cdot\mathbf{P}^{-1} - \mathbf{I},\ (\det\mathbf{R})\mathbf{R}\cdot\mathbf{K}\cdot\mathbf{P}^{-1} + \mathbf{L}; \\ &\qquad\qquad (\det\mathbf{R})\mathbf{R}\cdot\mathbf{B}\cdot\mathbf{P}^{-1} - \mathbf{L}\big], \\ &\forall\mathbf{P}\in Unim, \quad \forall\mathbf{R}\in Orth, \quad \forall\mathbf{L}\in Lin. \end{aligned}$$

From the Noll rule (104) it is easy to verify that for the polar-elastic fluid any reference configuration becomes undistorted, similarly as it is for the non-polar elastic fluid, because the symmetry group becomes here to be maximal. Obviously, the micropolar fluid is also isotropic.

The constitutive equation of the polar-elastic fluid are given by, see Eremeyev and Pietraszkiewicz (2012); Zubov and Eremeev (1996); Yeremeyev and Zubov (1999),

$$W = W^\times(\det\mathbf{F}, \mathbf{C}) = \overline{W}(\rho, \mathbf{C}), \tag{107}$$

where $\mathbf{C}$ is the microstructure curvature tensor of the deformed configuration χ defined in (64). $W^\times$ can be considered as a function of only six invariants $j_n(\mathbf{C})$, $n = 1, \ldots, 6$, $W = W^\times(\det\mathbf{F}, j_1, j_2, \ldots, j_6)$. The constitutive equations for $\mathbf{T}$ and $\mathbf{M}$ corresponding to (107) are

$$\mathbf{T} = -p\mathbf{I} - \mathbf{C}^T\cdot\mathbf{M}, \quad \mathbf{M} = J\frac{\partial\overline{W}}{\partial\mathbf{C}}, \quad p = \rho J\frac{\partial\overline{W}}{\partial\rho}. \tag{108}$$

A simple example of the strain energy density is the following quadratic function:

$$\overline{W}(\rho, \mathbf{C}) = \alpha_0(\rho) + \alpha_1 j_1^2 + \alpha_2 j_2 + \alpha_3 j_4, \tag{109}$$

where α_i, $i = 1, 2, 3$, are the material constants.

To define the *micropolar elastic solid* we apply the requirement that its material symmetry group consists of the orthogonal tensors.

Definition 4.5. *The micropolar elastic continuum is called micropolar elastic solid at x and* $\mathbf{B}$ *if there exists a reference configuration κ, called undistorted, such that the material symmetry group relative to κ is given by*

$$\mathcal{G}_\kappa = \mathcal{R}_\kappa \equiv \{(\mathbf{P} = \mathbf{O}, \mathbf{O}, \mathbf{0}) : \quad \mathbf{O} \in \mathcal{O}_\kappa \subset Orth\}. \tag{110}$$

The group $\mathcal{R}_\kappa$ is fully described by a subgroup $\mathcal{O}_\kappa$ of the orthogonal group $Orth$. The invariance requirement of W leads here to finding the subgroup $\mathcal{O}_\kappa$ such that

$$W(\mathbf{E}, \mathbf{K}; \mathbf{B}) = W\left[\mathbf{O} \cdot \mathbf{E} \cdot \mathbf{O}^T, (\det \mathbf{O})\mathbf{O} \cdot \mathbf{K} \cdot \mathbf{O}^T; (\det \mathbf{O})\mathbf{O} \cdot \mathbf{B} \cdot \mathbf{O}^T\right], \qquad (111)$$
$$\forall \mathbf{O} \in \mathcal{O}_\kappa.$$

The strain energy density of a polar-elastic continuum may also admit other material symmetry groups, in general. For example, it is possible to construct the material symmetry groups of W in analogy to the symmetry groups used to model liquid crystals in continuum mechanics of simple materials, see Truesdell and Noll (1965); Wang and Truesdell (1973).

Definition 4.6. *The micropolar elastic continuum is called micropolar elastic liquid crystal at x and $\mathbf{B}$ if the material symmetry group $\mathcal{G}_\kappa$ does not coincide with $\mathcal{U}_\kappa$, but there exist elements $\mathbb{X} \in \mathcal{G}_\kappa$, which are not the members of any group constructed using only the orthogonal tensors.*

In other words the micropolar polar elastic liquid crystal is the material which is neither fluid nor solid.

4.6 Non-Linear Micropolar Isotropic Solids

Let us consider the micropolar isotropic elastic solids.

Definition 4.7. *The micropolar elastic solid is called isotropic at point x and $\mathbf{B}$ if there exists a reference configuration κ, called undistorted, such that the material symmetry group relative to κ takes the form*

$$\mathcal{G}_\kappa = \mathcal{S}_\kappa \equiv \{(\mathbf{P} = \mathbf{O}, \mathbf{O}, \mathbf{0}) : \quad \mathbf{O} \in Orth\}. \qquad (112)$$

This definition means that the strain energy density of the polar-elastic isotropic solid satisfies the relation

$$W(\mathbf{E}, \mathbf{K}; \mathbf{B}) = W\left[\mathbf{O} \cdot \mathbf{E} \cdot \mathbf{O}^T, (\det \mathbf{O})\mathbf{O} \cdot \mathbf{K} \cdot \mathbf{O}^T; (\det \mathbf{O})\mathbf{O} \cdot \mathbf{B} \cdot \mathbf{O}^T\right],$$
$$\forall \mathbf{O} \in Orth.$$

Scalar-valued isotropic functions of second-order tensors can be expressed by the so-called representation theorems in terms of joint invariants of tensorial arguments, see Spencer (1965, 1971); Zheng (1994). The integrity basis for the proper orthogonal group is given by Spencer, see Table 1 in Spencer (1965) or Table II in Spencer (1971). The number of members of the integrity basis of $\mathbf{E}$, $\mathbf{K}$, $\mathbf{B}$ is much larger than the number of components of these tensors. However, there are some polynomial dependencies (syzygies) between elements of the integrity basis of three non-symmetric tensors. For the proper orthogonal group there is no difference in transformations of axial and polar tensors. It is not the case if one considers transformations using the full orthogonal group. Since $\mathbf{K}$ and $\mathbf{B}$ are axial tensors, not all invariants listed in Spencer (1965, 1971) are *absolute invariants* under orthogonal transformations, because some of them change sign under non-proper orthogonal transformations. Following Spencer (1971), we call such invariants *relative invariants*. Examples of relative invariants are $\operatorname{tr}\mathbf{K}$, $\operatorname{tr}\mathbf{K}^3$, $\operatorname{tr}\mathbf{E}\cdot\mathbf{K}$, $\operatorname{tr}\mathbf{E}\cdot\mathbf{B}$, etc. This gives us the following property of W:

$$W(\mathbf{E},\mathbf{K};\mathbf{B}) = W(\mathbf{E},-\mathbf{K};-\mathbf{B}). \tag{113}$$

The full list of 119 absolute and relative invariants for the orthogonal group is presented in Eremeyev and Pietraszkiewicz (2012).

If we neglect the explicit dependence of W on $\mathbf{B}$, or assume that $\mathbf{B}=\mathbf{0}$, then $W = W(\mathbf{E},\mathbf{K})$. The integrity basis of two non-symmentric tensors under the orthogonal group contains 39 members. Following Zheng (1994), Ramezani et al. (2009) listed these invariants for the non-linear polar-elastic solids and proposed the corresponding constitutive equations. Let us note, however, that not all 39 elements of this integrity basis are functionnally independent. Kafadar and Eringen (1971) constructed the functional basis for two non-symmetric tensors taking into account these functional dependencies. According to Kafadar and Eringen (1971), as the isotropic scalar-valued function of two non-symmetric tensors $\mathbf{E}$ and $\mathbf{K}$, W is expressible in terms of 15 invariants,

$$W = W(I_1, I_2, \ldots, I_{15}), \tag{114}$$

where I_k are given by

$$\begin{array}{llllll}
I_1 = \operatorname{tr}\mathbf{E}, & I_2 = \operatorname{tr}\mathbf{E}^2, & I_3 = \operatorname{tr}\mathbf{E}^3, \\
I_4 = \operatorname{tr}\mathbf{E}\cdot\mathbf{E}^T, & I_5 = \operatorname{tr}\mathbf{E}^2\cdot\mathbf{E}^T, & I_6 = \operatorname{tr}\mathbf{E}^2\cdot(\mathbf{E}^T)^2, \\
I_7 = \operatorname{tr}\mathbf{E}\cdot\mathbf{K}, & I_8 = \operatorname{tr}\mathbf{E}^2\cdot\mathbf{K}, & I_9 = \operatorname{tr}\mathbf{E}\cdot\mathbf{K}^2, \\
I_{10} = \operatorname{tr}\mathbf{K}, & I_{11} = \operatorname{tr}\mathbf{K}^2, & I_{12} = \operatorname{tr}\mathbf{K}^3, \\
I_{13} = \operatorname{tr}\mathbf{K}\cdot\mathbf{K}^T, & I_{14} = \operatorname{tr}\mathbf{K}^2\cdot\mathbf{K}^T, & I_{15} = \operatorname{tr}\mathbf{K}^2\cdot(\mathbf{K}^T)^2.
\end{array}$$

Taking into account that $W = W(\mathbf{E},\mathbf{K})$ is an even function with respect to $\mathbf{K}$, W becomes also a even function with respect to some invariants

$$\begin{aligned} &W(I_1, I_2, I_3, I_4, I_5, I_6, I_7, I_8, I_9, I_{10}, I_{11}, I_{12}, I_{13}, I_{14}, I_{15}) \\ &\quad = W(I_1, I_2, I_3, I_4, I_5, I_6, -I_7, -I_8, I_9, -I_{10}, I_{11}, -I_{12}, I_{13}, -I_{14}, I_{15}). \end{aligned} \tag{115}$$

Expanding W a the Taylor series with respect to $\mathbf{E}$ and $\mathbf{K}$, and keeping up to quadratic terms, we obtain with (115) the approximate polynomial representation of (114)

$$W = w_0 + a_1 I_1 + b_1 I_1^2 + b_2 I_2 + b_3 I_4 + b_4 I_{10}^2 + b_5 I_{11} + b_6 I_{13}, \tag{116}$$

where w_0, a_1, a_2, $b_1, \ldots, b_6$ are material parameters.

Let us consider the representation of W which takes the form of a sum of two scalar functions each depending on one strain measure

$$W = W_1(\mathbf{E}) + W_2(\mathbf{K}). \tag{117}$$

For example, this form has W in (116). The form (117) is also used by Ramezani et al. (2009) in order to generalize the classical neo-Hookean and Mooney-Rivlin models to the micropolar elastic solids. Using the representation theorem for the isotropic scalar-valued function of one non-symmetric tensor, we obtain the following representation of W:

$$W = \widetilde{W}_1(I_1(\mathbf{E}), \ldots, I_6(\mathbf{E})) + \widetilde{W}_2(I_1(\mathbf{K}), \ldots, I_6(\mathbf{K})),$$

where $\widetilde{W}_2$ has the property

$$\widetilde{W}_2(I_1, I_2, I_3, I_4, I_5, I_6) = \widetilde{W}_2(-I_1, I_2, -I_3, I_4, -I_5, I_6).$$

If in (112) we use only the proper orthogonal tensors then the resulting constitutive equations correspond to the hemitropic polar-elastic continuum.

4.7 Physically Linear Micropolar Solids

Let us consider the strain energy density as a quadratic function of $\mathbf{E}$ and $\mathbf{K}$

$$W = \frac{1}{2}\mathbf{E} \bullet \mathbb{C} \bullet \mathbf{E} + \mathbf{E} \bullet \mathbb{B} \bullet \mathbf{K} + \frac{1}{2}\mathbf{K} \bullet \mathbb{D} \bullet \mathbf{K}, \tag{118}$$

where $\mathbb{C}$, $\mathbb{B}$, and $\mathbb{D}$ are the fourth-order tensors of elastic moduli of the micropolar elastic solid. The components of tensors $\mathbb{C}$ and $\mathbb{D}$ have the symmetry properties

$$C_{ijmn} = C_{mnij}, \quad D_{ijmn} = D_{mnij}.$$

With (118) the corresponding stress measures $\mathbf{P}_1$ and $\mathbf{P}_2$ take the form

$$\mathbf{P}_1 = \mathbb{C} \bullet \mathbf{E} + \mathbb{B} \bullet \mathbf{K}, \quad \mathbf{P}_2 = \mathbf{E} \bullet \mathbb{B} + \mathbb{D} \bullet \mathbf{K}.$$

The model based on (118) can be called the *physically linear micropolar elastic solid*. Since $\mathbf{E}$ and $\mathbf{K}$ are non-symmetric tensors, W contains 171 independent material parameters, in general.

In the case of isotropic solids one can use the representation of an isotropic fourth-order tensor (Lebedev et al., 2010). The corresponding tensors $\mathbb{C}$ and $\mathbb{D}$ take the form

$$\mathbb{C} = \lambda \mathbf{I} \otimes \mathbf{I} + \mu \mathbf{i}_a \otimes \mathbf{I} \otimes \mathbf{i}_a + (\mu + \kappa)\mathbf{i}_a \otimes \mathbf{i}_b \otimes \mathbf{i}_a \otimes \mathbf{i}_b,$$
$$\mathbb{D} = \beta_1 \mathbf{I} \otimes \mathbf{I} + \beta_2 \mathbf{i}_a \otimes \mathbf{I} \otimes \mathbf{i}_a + \beta_3 \mathbf{i}_a \otimes \mathbf{i}_b \otimes \mathbf{i}_a \otimes \mathbf{i}_b,$$

where λ, μ, κ, β_i, $i = 1, 2, 3$, are the independent elastic moduli, while $\mathbb{B} = \mathbb{O}$. Thus, the strain energy density of the physically linear polar-elastic isotropic solid contains only six scalar elastic moduli. W takes form

$$2W = \lambda \mathrm{tr}^2 \mathbf{E} + \mu I_2 \mathrm{tr}\, \mathbf{E}^2 + (\mu + \kappa) \mathrm{tr}\, (\mathbf{E} \cdot \mathbf{E}^T) + \beta_1 \mathrm{tr}^2 \mathbf{K} + \beta_2 \mathrm{tr}\, \mathbf{K}^2 + \beta_3 \mathrm{tr}\, (\mathbf{K} \cdot \mathbf{K}^T). \tag{119}$$

This constitutive relation corresponds to the linear isotropic elastic Cosserat continuum described by (121). Within the linear micropolar elasticity the explicit structure of tensors $\mathbb{C}$ and $\mathbb{D}$ is presented by Zheng and Spencer (1993) for 14 symmetry groups, see also Xiao (1998).

4.8 Linear Micropolar Isotropic Solids

Let us consider the case when the translations and microrotations are very small. Now we assume that $\boldsymbol{u}$, $\boldsymbol{\theta}$, and their spatial derivatives are infinitesimally small:

$$\|\boldsymbol{u}\| \ll 1, \quad \|\mathrm{Grad}\, \boldsymbol{u}\| \ll 1, \quad \|\boldsymbol{\theta}\| \ll 1, \quad \|\mathrm{Grad}\, \boldsymbol{\theta}\| \ll 1,$$

where $\boldsymbol{u} \triangleq \boldsymbol{r} - \boldsymbol{R}$ is the *infinitesimal translation vector*, and $\boldsymbol{\theta} = \varphi \boldsymbol{e}$ is the *infinitesimal rotation vector* now.

There hold

$$\mathbf{H} \approx \mathbf{I} + \mathbf{I} \times \boldsymbol{\theta}, \quad \mathbf{K} \approx \boldsymbol{\varkappa} \triangleq \mathrm{Grad}\, \boldsymbol{\theta}, \quad \mathbf{E} \approx \boldsymbol{\varepsilon} \triangleq \mathrm{Grad}\, \boldsymbol{u} - \mathbf{I} \times \boldsymbol{\theta}. \tag{120}$$

Definition 4.8. *The tensor* $\boldsymbol{\varepsilon} \triangleq \mathrm{Grad}\, \boldsymbol{u} - \mathbf{I} \times \boldsymbol{\theta}$ is called the linear stretch tensor and $\boldsymbol{\varkappa} \triangleq \mathrm{Grad}\, \boldsymbol{\theta}$ – the linear wryness tensor.

We should note the introduced linear stretch tensor is different of the strain tensor of linear elasticity $\mathbf{e} \triangleq \frac{1}{2}\left[\mathrm{Grad}\, \boldsymbol{u} + (\mathrm{Grad}\, \boldsymbol{u})^T\right]$. The difference for the two theories exists for other quantities, for example, for the tensors of the time rate of strains.

Note in the present linear approximation of the theory, there is no difference between the gradient operators grad and Grad, that is $\mathrm{Grad}\, \boldsymbol{u} \approx$

grad $\boldsymbol{u}$, Grad $\boldsymbol{\theta} \approx$ grad $\boldsymbol{\theta}$ as well as there is no difference between the stress tensors $\mathbf{T} \approx \mathbf{T}_\kappa \approx \mathbf{P}_1$ and $\mathbf{M} \approx \mathbf{M}_\kappa \approx \mathbf{P}_2$.

We can rewrite the constitutive equations for isotropic material in the form

$$\mathbf{T} = \frac{\partial W}{\partial \boldsymbol{\varepsilon}}, \quad \mathbf{M} = \frac{\partial W}{\partial \boldsymbol{\varkappa}},$$

where W is the quadratic form of $\boldsymbol{\varepsilon}$, $\boldsymbol{\varkappa}$

$$\begin{aligned} W =& \frac{\lambda}{2} \mathrm{tr}^2 \boldsymbol{\varepsilon} + \frac{\mu + \kappa}{2} \mathrm{tr}\,(\boldsymbol{\varepsilon} \cdot \boldsymbol{\varepsilon}^T) + \frac{\mu}{2} \mathrm{tr}\,(\boldsymbol{\varepsilon} \cdot \boldsymbol{\varepsilon}) \\ &+ \frac{\beta_1}{2} \mathrm{tr}^2 \boldsymbol{\varkappa} + \frac{\beta_2}{2} \mathrm{tr}\,(\boldsymbol{\varkappa} \cdot \boldsymbol{\varkappa}) + \frac{\beta_3}{2} \mathrm{tr}\,(\boldsymbol{\varkappa} \cdot \boldsymbol{\varkappa}^T). \end{aligned} \tag{121}$$

Thus $\mathbf{T}$ and $\mathbf{M}$ are given by linear relations

$$\mathbf{T} = \lambda \mathbf{I}\,\mathrm{tr}\,\boldsymbol{\varepsilon} + (\mu + \kappa)\boldsymbol{\varepsilon} + \mu \boldsymbol{\varepsilon}^T, \quad \mathbf{M} = \beta_1 \mathbf{I}\,\mathrm{tr}\,\boldsymbol{\varkappa} + \beta_2 \boldsymbol{\varkappa}^T + \beta_3 \boldsymbol{\varkappa}. \tag{122}$$

In the framework of the linear theory of Cosserat continuum, the rotation velocity is given through the material derivative of the microrotation vector: $\boldsymbol{\omega} = \dot{\boldsymbol{\theta}}$.

Quadratic form (121) is assumed to be non-negative. This holds if the following inequalities are valid:

$$3\lambda + 2\mu + \kappa \geq 0, \quad 2\mu + \kappa \geq 0, \quad \kappa \geq 0, \tag{123}$$

$$3\beta_1 + \beta_2 + \beta_3 \geq 0, \quad \beta_2 + \beta_3 \geq 0, \quad \beta_3 - \beta_2 \geq 0. \tag{124}$$

The linear theory of micropolar elasticity is discussed in many works, see for example Nowacki (1986); Eringen (1999); Dyszlewicz (2004) and their reference lists.

In mechanics, it is also known a model of *Cosserat's pseudocontinuum*, also called the *medium with constrained rotations*. By this model, the macrorotation of a body particle that is determined by displacement $\boldsymbol{u}$, coincides with the microrotations defined by $\boldsymbol{\theta}$. In linear elasticity, the rotation of a body particle is given by formula $-1/2\mathrm{Rot}\,\boldsymbol{u}$, cf. Lurie (2005). The identity hypothesis for macro- and microrotations brings us to the relation

$$\boldsymbol{\theta} = -\frac{1}{2} \mathrm{Rot}\,\boldsymbol{u}. \tag{125}$$

Substituting (125) to the expression for $\boldsymbol{\varepsilon}$, we get

$$\boldsymbol{\varepsilon} = \frac{1}{2} \left(\mathrm{Grad}\,\boldsymbol{u} + (\mathrm{Grad}\,\boldsymbol{u})^T \right) = \mathbf{e}.$$

This means that the stretch tensor coincides with the strain tensor of classical linear elasticity. Substituting (125) to the expression for $\boldsymbol{\varkappa}$ we can show that $\operatorname{tr}\boldsymbol{\varkappa} = 0$. From (122) it follows that for Cosserat's pseudocontinuum the stress and couple tensors possess the properties

$$\mathbf{T} = \mathbf{T}^T, \quad \operatorname{tr}\mathbf{M} = 0.$$

Relation (125) also implies that $\boldsymbol{\varkappa}$ can be expressed through the second spatial derivatives of $\boldsymbol{u}$. Thus Cosserat's pseudocontinuum model can be considered as a particular case of the gradient elasticity theory.

4.9 Constraints

In mechanics, there are various models for the media with constraints. Restrictions on medium deformations are called *constraints*. In nonlinear elasticity, an example of such constraints is the incompressibility of material, it is widely used to describe the deformation of rubber-like materials, see Ogden (1984). In nonlinear elasticity, under the incompressibility constraint, there are established some analytical solutions, so-called universal solutions, see Truesdell and Noll (1965). In this section we consider mechanics of micropolar materials with constraints.

A scalar constraint takes the form

$$l(\boldsymbol{r}, \mathbf{H}, \operatorname{Grad}\boldsymbol{r}, \operatorname{Grad}\mathbf{H}) = 0. \tag{126}$$

As (126) reflects physical properties of the material, it should be material frame-indifferent. Repeating the transformations used to derive W in Sect. 4.5, we get that the frame-indifferent constraint reduces to the following

$$l(\mathbf{E}, \mathbf{K}) = 0. \tag{127}$$

Similarly we can show that a vectorial frame-indifferent constraint is of the form

$$\mathbf{H} \cdot \boldsymbol{l}(\mathbf{E}, \mathbf{K}) = \mathbf{0}. \tag{128}$$

The constitutive equations for materials with constraints are deduced with use of Lagrange's multipliers, for example see Lurie (1990). Stress tensors $\mathbf{P}_1$ and $\mathbf{P}_2$ are given by formulas

$$\mathbf{P}_1 = \frac{\partial W^*}{\partial \mathbf{E}}, \quad \mathbf{P}_2 = \frac{\partial W^*}{\partial \mathbf{K}},$$

where

$$W^*(\mathbf{E}, \mathbf{K}) = W(\mathbf{E}, \mathbf{K}) + \lambda l(\mathbf{E}, \mathbf{K}) + \boldsymbol{\lambda} \cdot \mathbf{H} \cdot \boldsymbol{l}(\mathbf{E}, \mathbf{K}),$$

λ and $\boldsymbol{\lambda}$ are the scalar and vectorial Lagrange multipliers related with constrains (127) and (128), respectively. The both Lagrange's multipliers are frame-indifferent quantities, they arise as a result of constraint reactions. So we have

$$\mathbf{P}_1 = \frac{\partial W}{\partial \mathbf{E}} + \lambda \frac{\partial l}{\partial \mathbf{E}} + \boldsymbol{\lambda} \cdot \mathbf{H} \cdot \frac{\partial \boldsymbol{l}}{\partial \mathbf{E}}, \quad \mathbf{P}_2 = \frac{\partial W}{\partial \mathbf{K}} + \lambda \frac{\partial l}{\partial \mathbf{K}} + \boldsymbol{\lambda} \cdot \mathbf{H} \cdot \frac{\partial \boldsymbol{l}}{\partial \mathbf{K}}.$$

Note that instead of $\boldsymbol{\lambda}$ we can use a non-indifferent vector $\tilde{\boldsymbol{\lambda}} = \boldsymbol{\lambda} \cdot \mathbf{H}$.

An example of scalar constraints is the *incompressibility* condition,

$$\det \mathbf{F} = 1.$$

In terms of strain measures, it takes the form

$$l \equiv \det(\mathbf{E} + \mathbf{I}) - 1 = 0. \tag{129}$$

For a micropolar body, the incompressibility condition was used in Eremeyev and Zubov (1994); Ramezani et al. (2009); Zubov (1997).

The model of Cosserat's pseudocontinuum gives an example of vectorial constraint

$$\boldsymbol{l} \equiv \mathbf{E}_\times = \mathbf{0}. \tag{130}$$

Condition (130) was applied in Eremeyev and Zubov (1994); Zubov (1997). Using $(120)_3$ for infinitesimal deformations, we can see that (130) takes the form of (125) of Cosserat's linear pseudocontinuum model.

Note that for finite deformations, constraint (130) affects the stress tensor definition but does not restrict the form of the couple stress tensor. However in Cosserat's linear pseudocontinuum theory, constraint (125) involves that $\operatorname{tr}\mathbf{M} = 0$. As its consequence in this theory we cannot assign three independent couple boundary conditions.

For constraints (129) and (130), $\mathbf{P}_1$ takes the form

$$\mathbf{P}_1 = \frac{\partial W}{\partial \mathbf{E}} + \lambda \mathbf{H}^T \cdot \mathbf{F}^{-T} - \tilde{\boldsymbol{\lambda}} \times \mathbf{I}.$$

To derive this, we should use the differentiation formulae

$$(\det \mathbf{X})_{,\mathbf{X}} = (\det \mathbf{X})\mathbf{X}^{-T}, \quad (\mathbf{X}_\times)_{,\mathbf{X}} = \boldsymbol{\mathcal{E}}.$$

As an example of constitutive equations with constraints we present a generalization of Varga's incompressible material model (Varga, 1996) to micropolar elasticity with rotation constraints:

$$W = \mu \operatorname{tr}\mathbf{E} + \frac{\beta_1}{2}\operatorname{tr}^2\mathbf{K} + \frac{\beta_2}{2}\operatorname{tr}(\mathbf{K}\cdot\mathbf{K}^T) + \frac{\beta_3}{2}\operatorname{tr}(\mathbf{K}\cdot\mathbf{K}). \tag{131}$$

For model (131), tensors $\mathbf{P}_1$ and $\mathbf{P}_2$ are

$$\mathbf{P}_1 = \mu\mathbf{I} + \lambda\mathbf{H}^T \cdot \mathbf{F}^{-T} - \tilde{\boldsymbol{\lambda}} \times \mathbf{I}, \quad \mathbf{P}_2 = \beta_1\mathbf{I}\,\mathrm{tr}\,\mathbf{K} + \beta_2\mathbf{K} + \beta_3\mathbf{K}^T.$$

In a similar way, well known models of incompressible elastic materials by Mooney–Rivlin, Ogden, etc., can be extended to Cosserat's nonlinear pseudocontinuum.

It is worth of noting that the rigid body motion can be considered as a particular case of the motion of a micropolar body with the following constraints: $\mathbf{E} = \mathbf{K} = \mathbf{0}$. In this case, stress tensor $\mathbf{P}_1$ and couple tensor $\mathbf{P}_2$ are Lagrange "multipliers" that correspond to the constraints and do not depend on the deformation. The Euler laws of motion take the form of (20).

4.10 Constitutive Inequalities

In nonlinear elasticity there are known so-called constitutive restrictions. They are the strong ellipticity condition, Hadamard inequality, generalized Coleman-Noll condition (GCN-condition), and some others, see Truesdell and Noll (1965); Truesdell (1977, 1984). Each of them plays a role in nonlinear elasticity. They express mathematically precise and physically intuitive restrictions for constitutive equations of elastic bodies. In particular, the GCN condition proposed by Coleman and Noll asserts that "the transformation from deformation gradient to first Piola-Kirchhoff stress tensor shall be monotone with respect to pairs of deformations differing from one another by a pure stretch" (see, Truesdell and Noll (1965)).

The simplest example of the constitutive restrictions gives the linear micropolar elasticity. Suppose that the strain energy density of the linear elastic micropolar solids $W(\boldsymbol{\varepsilon}, \boldsymbol{\varkappa})$ is positive definite

$$W(\boldsymbol{\varepsilon}, \boldsymbol{\varkappa}) > 0, \quad \forall\, \boldsymbol{\varepsilon}, \boldsymbol{\varkappa} \neq \mathbf{0}$$

For an isotropic micropolar elastic solid, W takes the form (121). Positivity of W with respect to $\boldsymbol{\varepsilon}$ and $\boldsymbol{\varkappa}$ is equivalent to the strict inequalities (123) and (124). If the inequalities fail (123) and (124) this leads to a number of pathological mathematical and physical consequences. For example, boundary value problems of linear micropolar elasticity can have more than one solutions or can have no solution for some loads.

For finite strains, the positive definiteness of the strain energy density $W(\mathbf{E}, \mathbf{K})$ with respect to the strain measures is not enough for the desired properties of boundary value problems. As in the case of non-linear elasticity, one should introduce some additional restrictions.

The Coleman-Noll constitutive inequality is one of well-known in nonlinear elasticity, see Truesdell and Noll (1965); Truesdell (1977, 1984). Its differential form, a so-called Generalized Coleman-Noll (GCN) condition, expresses the property that the work of the stress rate cannot be negative or the property that the first Piola-Kirchhoff stress tensor shall be monotone with respect to pairs of deformations differing from one another by a pure stretch.

Generalizing the classical GCN-condition we will use the following definition.

Definition 4.9. *The generalized Coleman-Noll condition of the Cosserat continuum is given by the inequality*

$$\dot{\mathbf{P}}_1 \bullet \dot{\mathbf{E}} + \dot{\mathbf{P}}_2 \bullet \dot{\mathbf{K}} > 0, \quad \forall\, \dot{\mathbf{E}} \neq \mathbf{0},\ \dot{\mathbf{K}} \neq \mathbf{0}. \tag{132}$$

The strain rates are given by

$$\dot{\mathbf{E}} = \mathbf{H}^T \cdot (\operatorname{Grad} \boldsymbol{v} - \boldsymbol{\omega} \times \mathbf{F}) = \mathbf{H}^T \cdot \boldsymbol{\varepsilon} \cdot \mathbf{F}, \quad \dot{\mathbf{K}} = \mathbf{H}^T \cdot \operatorname{Grad} \boldsymbol{\omega} = \mathbf{H}^T \cdot \boldsymbol{\varkappa} \cdot \mathbf{F}, \tag{133}$$

where $\boldsymbol{\varepsilon}$ and $\boldsymbol{\varkappa}$ are now the strain rates given by

$$\boldsymbol{\varepsilon} \triangleq \operatorname{grad} \boldsymbol{v} - \boldsymbol{\omega} \times \mathbf{I}, \quad \boldsymbol{\varkappa} \triangleq \operatorname{grad} \boldsymbol{\omega}. \tag{134}$$

Taking into account (87) and (133), (134), we rewrite (132) in the equivalent form

$$\left. \frac{\mathrm{d}^2}{\mathrm{d}\tau^2} W(\mathbf{E} + \tau\boldsymbol{\varepsilon}, \mathbf{K} + \tau\boldsymbol{\varkappa}) \right|_{\tau=0} > 0 \quad \forall\, \boldsymbol{\varepsilon} \neq \mathbf{0}, \quad \boldsymbol{\varkappa} \neq \mathbf{0}. \tag{135}$$

The conditions (135) or (132) are analogues of the *Coleman-Noll inequality* in the 3D elasticity. Note that the inequality (135) satisfies the principle of material frame-indifference. This means that (135) can be chosen as a constitutive inequality for micropolar elastic continuum.

The *strong ellipticity* condition is also one of well known constitutive restrictions in nonlinear elasticity, cf. Truesdell and Noll (1965). Mathematically, it expresses a precise and physically intuitive restriction for the constitutive equations of elastic materials. In general, the strong ellipticity of the equations is a property of the material in some area of deformation. For some deformations it can be fulfilled whereas for others it does not. For a number of real media that are modeled by Cosserat continuum, arising of discontinuous solution is physically possible. For example, such media are soils, granular and porous media. For some other materials such discontinuities are physically impossible. Thus, the strong ellipticity condition allows us "to sort" admissible types of the constitutive equations as well as to determine "dangerous" deformations. Besides, the condition of strong ellipticity is a criterion to find "dangerous" domains in a body.

To formulate the strong ellipticity condition of the equilibrium equations (59) we use the general results of the partial differential equations theory (PDE) presented in Lions and Magenes (1968); Fichera (1972); Hörmander (1976).

Definition 4.10. *The strong ellipticity condition of the equilibrium equations (59) for the micropolar elastic solids is the inequality*

$$\boldsymbol{\xi}\cdot\mathbb{Q}(\boldsymbol{N})\cdot\boldsymbol{\xi} > 0, \quad \forall\, \boldsymbol{N} \neq \mathbf{0}, \quad \forall\, \boldsymbol{\xi} \in \mathrm{I\!R}^6, \quad \boldsymbol{\xi} \neq \mathbf{0}, \tag{136}$$

where $\boldsymbol{\xi} = (\boldsymbol{a}, \boldsymbol{b}) \in \mathrm{I\!R}^6$ *and matrix* $\mathbb{Q}(\boldsymbol{N})$ *is*

$$\mathbb{Q}(\boldsymbol{N}) \triangleq \begin{bmatrix} \dfrac{\partial^2 W}{\partial \mathbf{E} \partial \mathbf{E}}\{\boldsymbol{N}\} & \dfrac{\partial^2 W}{\partial \mathbf{E} \partial \mathbf{K}}\{\boldsymbol{N}\} \\ \dfrac{\partial^2 W}{\partial \mathbf{K} \partial \mathbf{E}}\{\boldsymbol{N}\} & \dfrac{\partial^2 W}{\partial \mathbf{K} \partial \mathbf{K}}\{\boldsymbol{N}\} \end{bmatrix},$$

where for arbitrary forth-order tensor $\mathbb{G}$ *and vector* $\boldsymbol{N}$ *that are represented in a Cartesian basis* $\boldsymbol{i}_k$ $(k = 1, 2, 3)$*, we have used the notation*

$$\mathbb{G}\{\boldsymbol{N}\} \equiv G_{klmn} N_l N_n \boldsymbol{i}_k \otimes \boldsymbol{i}_m.$$

For example, if $\mathbb{G} = \boldsymbol{c} \otimes \boldsymbol{d} \otimes \boldsymbol{e} \otimes \boldsymbol{f}$ then $\mathbb{G}\{\boldsymbol{N}\} = (\boldsymbol{d} \cdot \boldsymbol{N})(\boldsymbol{f} \cdot \boldsymbol{N})\boldsymbol{c} \otimes \boldsymbol{e}$.

A weak form of inequality (136) is an analogue of the *Hadamard inequality*, that is

$$\boldsymbol{\xi}\cdot\mathbb{Q}(\boldsymbol{N})\cdot\boldsymbol{\xi} \geq 0, \quad \forall\, \boldsymbol{N} \neq \mathbf{0}, \quad \forall\, \boldsymbol{\xi} \in \mathrm{I\!R}^6, \quad \boldsymbol{\xi} \neq \mathbf{0}, \tag{137}$$

$\mathbb{Q}$ plays a role of the *acoustic tensor* in the micropolar elasticity.

The strong ellipticity condition and the Hadamard inequality are also examples of constitutive restrictions of the constitutive equations of the micropolar elastic solids under finite deformations. As for the theory of non-polar materials, a failure in inequality (137) can lead to the existence of non-smooth solutions to equilibrium equations (59).

As the Coleman–Noll inequality the strong ellipticity condition can be written in the equivalent form

$$\left.\frac{\mathrm{d}^2}{\mathrm{d}\tau^2} W\left(\mathbf{E} + \tau \boldsymbol{a} \otimes \boldsymbol{N}, \mathbf{K} + \tau \boldsymbol{b} \otimes \boldsymbol{N}\right)\right|_{\tau=0} > 0 \quad \forall\, \boldsymbol{N},\, \boldsymbol{a},\, \boldsymbol{b} \neq \mathbf{0}. \tag{138}$$

Comparing (135) and (138) we prove the following theorem:

Theorem 4.1. *For the non-liner micropolar elastic solid the Coleman–Noll inequality implies the strong ellipticity condition.*

Proof. Indeed, inequality (135) holds for any tensors $\boldsymbol{\varepsilon}$ and $\boldsymbol{\varkappa}$. Unlike to 3D non-linear elasticity here $\boldsymbol{\varepsilon}$ and $\boldsymbol{\varkappa}$ may be nonsymmetric tensors. Substituting relations $\boldsymbol{\varepsilon} = \boldsymbol{a} \otimes \boldsymbol{N}$ and $\boldsymbol{\varkappa} = \boldsymbol{b} \otimes \boldsymbol{N}$ into (135), we immediately obtain (138).

□

Thus, for the micropolar elastic continuum the strong ellipticity condition is a particular case of the Coleman–Noll inequality. This is an essential difference between the micropolar elasticity and the non-linear elasticity of non-polar elastic materials: in the latter these two properties are independent in the sense that neither of them implies the other.

Let us consider the constitutive relation (117). Then the Hadamard inequality (136) is equivalent to two independent inequalities

$$\boldsymbol{a}\cdot\frac{\partial^2 W_1}{\partial \mathbf{E}\partial \mathbf{E}}\{\boldsymbol{N}\}\cdot\boldsymbol{a} > 0, \quad \boldsymbol{b}\cdot\frac{\partial^2 W_2}{\partial \mathbf{K}\partial \mathbf{K}}\{\boldsymbol{N}\}\cdot\boldsymbol{b} > 0, \quad \forall\ \boldsymbol{N},\ \boldsymbol{a},\ \boldsymbol{b} \neq \mathbf{0}.$$

As an example, let us consider consequences of (136) for the physically linear isotropic micropolar solid (119). In this case we have

$$\begin{aligned} \frac{\partial^2 W_1}{\partial \mathbf{E}\partial \mathbf{E}}\{\boldsymbol{N}\} &= (\mu+\kappa)\,\mathbf{I} + (\lambda+\mu)\boldsymbol{N}\otimes\boldsymbol{N}, \\ \frac{\partial^2 W_2}{\partial \mathbf{K}\partial \mathbf{K}}\{\boldsymbol{N}\} &= \beta_3\mathbf{I} + (\beta_1+\beta_2)\boldsymbol{N}\otimes\boldsymbol{N}. \end{aligned} \tag{139}$$

Using (139) we show that the inequality (136) is equivalent to the conditions

$$2\mu+\kappa+\lambda > 0, \quad \mu+\kappa > 0, \qquad \beta_1+\beta_2+\beta_3 > 0, \quad \beta_3 > 0. \tag{140}$$

Thus for the strain energy density (119), the strong ellipticity condition reduces to simple inequalities (140). They are expressed in terms of the elastic moduli and do not depend on strains. For simple non-linear solids the analogue of (119) is the semi-linear material, see Lurie (1990, 2005). For the latter, the strong elipticity condition depends on strains as well as the elastic moduli. When the ellipticity condition breaks down, singular solutions arise – these may correspond to infinite rotations, for example, Lurie (1990). Hence it may be simpler to check the strong ellipticity condition for a micropolar material than for a simple nonlinearly elastic material.

In the case of small strains, the strain energy density (119) takes the form of (121). Following from the condition of positive semidefiniteness of W, the inequalities (123) and (124) imply (138) but not conversely. Obviously, the strong ellipticity conditions are less restrictive than the requirement of positive semidefiniteness of the energy under infinitesimal deformations.

Eremeyev and Zubov (1994) proved that the infinitesimal stability of micropolar body implies the strong ellipticity condition. The proof follows from the analysis of the second variation of the total energy functional of the micropolar elastic body. Unlike to the 3D non-linear elasticity, for the micropolar elasticity the fulfillment of the strong ellipticity condition does not imply the infinitesimal stability of a homogeneous deformation of a micropolar body with Dirichlet boundary conditions.

Let us introduce another type of ellipticity that is the ordinary ellipticity.

Definition 4.11. *The ellipticity condition of the equilibrium equations of micropolar elastic continuum named also the condition of the Petrovsky ellipticity or of the ordinary ellipticity is the following inequality:*

$$\det \mathbb{Q}(\boldsymbol{N}) \neq 0, \quad \forall\, \boldsymbol{N}, \tag{141}$$

The definition follows from the general definition of ellipticity in the partial differential equations theory of Agranovich (1997); Hörmander (1976); Nirenberg (2001). Obviously, the inequalities (141) are weaker than the strong ellipticity condition (136).

For the constitutive relation (117), condition (141) splits into two conditions

$$\det \frac{\partial^2 W_1}{\partial \mathbf{E} \partial \mathbf{E}}\{\boldsymbol{N}\} \neq 0, \quad \det \frac{\partial^2 W_2}{\partial \mathbf{K} \partial \mathbf{K}}\{\boldsymbol{N}\} \neq 0. \tag{142}$$

As an example, we consider conditions (142) for the constitutive relations of a physically linear micropolar solid (119). Using Eqs (139), we reduce (142) to the inequalities

$$\mu + \kappa \neq 0, \quad 2\mu + \lambda + \kappa \neq 0, \qquad \beta_3 \neq 0, \quad \beta_1 + \beta_2 + \beta_3 \neq 0.$$

Let us note that for the micropolar elastic solids with constraints we have mixed order systems of PDE. In this case one may use a more general definition of ellipticity such as the ellipticity in the sense of Douglis–Nirenberg, see Agranovich (1997).

4.11 Micropolar Fluid

Although the basic content of the chapter is devoted to the micropolar elastic solids in this section we briefly recall the constitutive equations of viscous micropolar fluids. The model was proposed by Aero et al. (1965) and Eringen (1966), see Migoun and Prokhorenko (1984); Łukaszewicz (1999); Eringen (2001). Some generalizations of the model are presented in Eringen (1997a,b, 2000); Yeremeyev and Zubov (1999); Zubov and Eremeev (1996). For the sake of simplicity we restrict ourselves by the equations of the *linear viscous incompressible micropolar fluid.* The constitutive equations for the stress and couple stress tensors are given by

$$\mathbf{T} = -p\,\mathbf{I} + \mu_1 \boldsymbol{\varepsilon} + \mu_2 \boldsymbol{\varepsilon}^T, \quad \mathbf{M} = \nu_1 (\operatorname{tr} \boldsymbol{\varkappa})\,\mathbf{I} + \nu_2 \boldsymbol{\varkappa} + \nu_3 \boldsymbol{\varkappa}^T, \tag{143}$$

where p is the pressure, the strain rates $\boldsymbol{\varepsilon}$ and $\boldsymbol{\varkappa}$ are defined by Eqs. (134), and μ_1, μ_2, ν_1, ν_2, ν_3 are the viscosities. The incompressibility condition of the fluid is

$$\operatorname{tr} \boldsymbol{\varepsilon} \equiv \operatorname{div} \boldsymbol{v} = 0. \tag{144}$$

Equations (143) satisfy the material frame-indifference principle and represent the general possible linear dependence of $\mathbf{T}$ and $\mathbf{M}$ on $\boldsymbol{\varepsilon}$ and $\boldsymbol{\varkappa}$.

Assuming that the dissipation should be positive we obtain the following inequality

$$\mathbf{T} \bullet \boldsymbol{\varepsilon} + \mathbf{M} \bullet \boldsymbol{\varkappa} \geq 0, \quad \forall\, \boldsymbol{\varepsilon}, \boldsymbol{\varkappa} \neq \mathbf{0}. \tag{145}$$

This implies the constitutive inequalities for the viscosity parameters

$$\mu_1 + \mu_2 \geq 0, \quad \mu_1 - \mu_2 \geq 0, \quad \nu_1 \geq 0, \quad \nu_2 + \nu_3 \geq 0, \quad \nu_2 - \nu_3 \geq 0.$$

4.12 Some Sources of Cosserat's Constitutive Equations

For any mechanical theory, the key point is the formulation of constitutive equations. The theory should propose the general form of constitutive equations and the ways how to get their numerical realization for certain materials. For Cosserat's continuum, experimental finding of material parameters is a harder problem than determining of elastic moduli in elasticity. This can be seen from the fact that an linear elastic isotropic material is determined by two elastic moduli whereas the description of an isotropic linear elastic micropolar material, Eqs. (121) and (122), needs six elastic parameters. We will touch the question of construction of the constitutive equations for Cosserat's continuum in brief.

Experimental Data Elastic moduli for micropolar materials can be determined from the experiments with specimens on tension, volume compression, torsion, flexure, and some dynamic experiments. For small deformations of homogeneous materials like steels, we cannot expect micropolar properties. They are watched in experiments with such micro-nonhomogeneous materials as foams, bones, rocks, granular media, etc. Besides micropolar properties arise at a neighborhood of stress concentrators like crack tips or notches. It seems that micropolar properties should be taken into account in nanoscale effects, see for example Ivanova et al. (2003). It should be recognized that the experimental data on micropolar solids are rather scarce. We can note the works on foams by Lakes (1986, 1991, 1995) and on bones by Park and Lakes (1986); Yang and Lakes (1982). In these, to define elastic micropolar modulus, it is used a so-called *size-effect* that is typical for Cosserat's continuum. For example, the size-effect becomes apparent in torsion tests on finding the shear modulus with specimens of few sizes. Lakes established the values of elastic modulus for some foams. We also should note experimental works by Gauthier and Jahsman (1975, 1981), who studied a composite material with aluminum shot uniformly distributed throughout an epoxy matrix. Cosserat's constants also are determined for polycarbonate honeycombs in Mora and Waas (2000). The values of micropolar moduli can be also found in Eringen (1999); Erofeev (2003). Using the concept of bounded stiffness and infinitesimal conformal invariance, Neff and Jeong (2009); Jeong and Neff (2010); Neff et al. (2010) presented the mathematical analysis of Cosserat's constitutive equations that reduces the number of independent elastic moduli of a micropolar medium.

Homogenization Various homogenization methods applied to lattices, granular medias, soils, cellular structures, etc. are a more significant source of known micropolar modules. The literature on this is more broad, we note the papers by Besdo (2010); Bigoni and Drugan (2007); Diebels (1999); Ehlers et al. (2003); Forest (1998); Forest and Sab (1998); Forest and Sievert (2006); Suiker and de Borst (2005); Larsson and Diebels (2007); Larsson and Zhang (2007); Pasternak and Mühlhaus (2005); van der Sluis et al. (1999). The idea of homogenization is to replace a composite material or assembly of particles by an effective generalized continuum model. For example, consider a system of interacting rigid bodies for which the force and couple interaction are essential. Changing the system to an effective continuum medium we expect the medium to inherit couple properties and rotational freedom degrees of rigid bodies. Changing open-cell foam that can be considered as a system of elastic beams to an effective continuum medium, we get a similar modeling situation. Thus the homogenization technique for a micro-inhomogeneous medium replaces direct experiments and can subject us with the values of micropolar modules. Note that the homogenization can give us more general models of continuum medium like micromorphic continuum, see Chapter 5, and, for example, the papers by Forest and Sievert (2006); Forest and Duy (2011); Kouznetsova et al. (2002); Neff and Forest (2007); Cielecka et al. (2000).

Numerics At first glance, numerical methods are an unexpected source of constitutive equations for Cosserat's continuum. Here Cosserat' s constitutive equations are used to accelerate numerical solving of classical problem of elasticity or elasto-plasticity. The choice of the elastic moduli is determined by the need to accelerate convergence to a solution or to regularize the solution. Note that the introduction of the rotations as independent kinematic variables for more accurate description of 3D continuum relates to Reissner's idea in the plate theory. In this approach, micropolar modules are a kind of numerical tuning values, they have no certain physical meaning. In elasticity and elasto-plasticity, the approach is used by Neff et al. (2007); Neff and Chełmiński (2007). Among other numerical works that used the idea of Cosserat's continuum we should note Rubin (1985, 2000); Nadler and Rubin (2003); Jabareen and Rubin (2007, 2008). There were developed special types of enhanced finite elements that use the representation of a finite element as *Cosserat's point.*

Micropolar Fluids For micropolar fluids, the situation of constitutive equations is different. Starting with pioneering papers by Aero et al. (1965) and Eringen (1966) the micropolar continuum is applied to model magnetic liquids, polymer suspensions, liquid crystals, and other types of fluids with microstructure, see also the books by Migoun and Prokhorenko (1984); Łukaszewicz (1999); Eringen (2001); Eremeyev and Zubov (2009). In particular, the use of magnetic fluids named also ferrofluids caused the development of micropolar hydromechanics where a magnetic field induces voluminous couples (Rosensweig, 1987). To describe the couples, liquid corpuscules should possess rotational degrees of freedom. Comparing with micropolar elasticity, micropolar hydrodynamics is a more extensive part of mechanics with well established experimentally constitutive equations.

Acknowledgement. The second author was supported by the DFG grant No. AL 341/33-1 and by the RFBR grant No. 12-01-00038.

Bibliography

E. L. Aero, A. N. Bulygin, and E. V. Kuvshinskii. Asymmetric hydromechanics. *Journal of Applied Mathematics and Mechanics*, 29(2):333–346, 1965.

M. Agranovich. Elliptic boundary problems. In M. Agranovich, Y. Egorov, and M. Shubin, editors, *Partial Differential Equations IX: Elliptic Boundary Problems. Encyclopaedia of Mathematical Sciences*, volume 79, pages 1–144. Springer, Berlin, 1997.

H. Altenbach, V. A. Eremeyev, L. P. Lebedev, and L. A. Rendón. Acceleration waves and ellipticity in thermoelastic micropolar media. *Archive of Applied Mechanics*, 80(3):217–227, 2010a.

H. Altenbach, G. A. Maugin, and V. Erofeev, editors. *Mechanics of Generalized Continua*, volume 7 of *Advanced Structured Materials*. Springer, Berlin, 2011.

J. Altenbach, H. Altenbach, and V. A. Eremeyev. On generalized Cosserat-type theories of plates and shells. A short review and bibliography. *Archive of Applied Mechanics*, 80(1):73–92, 2010b.

S. S. Antman. *Nonlinear Problems of Elasticity*. Springer Science Media, New York, 2nd edition, 2005.

V. I. Arnold. *Mathematical Methods of Classical Mechanics*. Springer, New York, 2nd edition, 1989.

O. A. Bauchau and L. Trainelli. The vectorial parameterization of rotation. *Nonlinear Dynamics*, 32(1):71–92, 2003.

A. Bertram. *Elasticity and Plasticity of Large Deformations: An Introduction*. Springer, Berlin, 3rd edition, 2012.

A. Bertram and B. Svendsen. On material objectivity and reduced constitutive equations. *Archive of Mechanics*, 53(6):653–675, 2001.

A. Bertram and B. Svendsen. Reply to Rivlin's Material symmetry revisited or Much Ado About Nothing. *GAMM Mitteilungen*, 27(1):88–93, 2004.

D. Besdo. Towards a Cosserat-theory describing motion of an originally rectangular structure of blocks. *Archive of Applied Mechanics*, 80(1): 25–45, 2010.

J. Besson, G. Cailletaud, J.-L. Chaboche, S. Forest, and M. Blétry. *Non-Linear Mechanics of Materials*, volume 167 of *Solid Mechanics and Its Applications*. Springer, Dordrecht, 2010.

D. Bigoni and W. J. Drugan. Analytical derivation of Cosserat moduli via homogenization of heterogeneous elastic materials. *Transactions of ASME. Journal of Applied Mechanics*, 74(4):741–753, 2007.

J. Chróścielewski, J. Makowski, and W. Pietraszkiewicz. *Statics and Dynamics of Multyfolded Shells. Nonlinear Theory and Finite Elelement Method (in Polish)*. Wydawnictwo IPPT PAN, Warszawa, 2004.

I. Cielecka, M. Woźniak, and C. Woźniak. Elastodynamic behaviour of honeycomb cellular media. *Journal of Elasticity*, 60(1):1–17, 2000.

J. D. Clayton. *Nonlinear Mechanics of Crystals*, volume 177 of *Solid Mechanics and Its Applications*. Springer, Dordrecht, 2011.

E. Cosserat and F. Cosserat. *Théorie des corps déformables*. Herman et Fils, Paris, 1909.

E. Cosserat and F. Cosserat. Sur la théorie de l'elasticité. *Ann. Toulouse*, 10:1–116, 1896.

S. Diebels. A micropolar theory of porous media: constitutive modelling. *Transport in Porous Media*, 34(1–3):193–208, 1999.

J. Dyszlewicz. *Micropolar Theory of Elasticity*. Springer, Berlin, 2004.

W. Ehlers, E. Ramm, S. Diebels, and G. D. A. d'Addetta. From particle ensembles to Cosserat continua: Homogenization of contact forces towards stresses and couple stresses. *International Journal of Solids and Structures*, 40(24):6681–6702, 2003.

V. V. Eliseev. *Mechanics of Elastic Bodies (in Russian)*. Politekhn. Univ. Publ., St. Petersburg, 1996.

V. A. Eremeyev. Acceleration waves in micropolar elastic media. *Doklady Physics*, 50(4):204–206, 2005.

V. A. Eremeyev and W. Pietraszkiewicz. Local symmetry group in the general theory of elastic shells. *Journal of Elasticity*, 85(2):125–152, 2006.

V. A. Eremeyev and W. Pietraszkiewicz. Material symmetry group of the non-linear polar-elastic continuum. *International Journal of Solids and Structures*, doi: 10.1016/j.ijsolstr.2012.04.007, 2012.

V. A. Eremeyev and L. M. Zubov. On the stability of elastic bodies with couple stresses. *Mechanics of Solids*, 29(3):172–181, 1994.

V. A. Eremeyev and L. M. Zubov. On constitutive inequalities in nonlinear theory of elastic shells. *ZAMM*, 87(2):94–101, 2007.

V. A. Eremeyev and L. M. Zubov. *Principles of Viscoelastic Micropolar Fluid Mechanics (in Russian)*. SSC of RASci Publishers, Rostov on Don, 2009.

V. A. Eremeyev, L. P. Lebedev, and H. Altenbach. *Foundations of Microplar Mechanics*. Springer, Berlin, 2012.

J. L. Ericksen and C. Truesdell. Exact theory of stress and strain in rods and shells. *Archive for Rational Mechanics and Analysis*, 1(1):295–323, 1958.

A. C. Eringen. *Microcontinuum Field Theory*, volume II. *Fluent Media*. Springer, New York, 2001.

A. C. Eringen. A unified continuum theory for electrodynamics of polymeric liquid crystals. *International Journal of Engineering Science*, 38(9–10): 959–987, 2000.

A. C. Eringen. *Nonlocal continuum field theories*. Springer, New York, 2002.

A. C. Eringen. Theory of micropolar fluids. *Journal of Mathematics and Mechanics*, 16(1):1–18, 1966.

A. C. Eringen. A unified continuum theory of liquid crystals. *ARI - An International Journal for Physical and Engineering Sciences*, 73–84(2): 369–374, 1997a.

A. C. Eringen. A unified continuum theory of electrodynamics of liquid crystals. *International Journal of Engineering Science*, 35(12–13):1137–1157, 1997b.

A. C. Eringen. *Microcontinuum Field Theory*, volume I. *Foundations and Solids*. Springer, New York, 1999.

A. C. Eringen and C. B. Kafadar. Polar field theories. In A. C. Eringen, editor, *Continuum Physics*, volume IV, pages 1–75. Academic Press, New York, 1976.

V. I. Erofeev. *Wave Processes in Solids with Microstructure.* World Scientific, Singapore, 2003.

R. P. Feynman, R. B. Leighton, and M. Sands. *The Feynman Lectures on Physics*, volume 1. Addison-Wesley, Reading, Massachusetts, 6th edition, 1977.

G. Fichera. Existence theorems in elasticity. In S. Flügge, editor, *Handbuch der Physik*, volume VIa/2, pages 347–389. Springer, Berlin, 1972.

S. Forest. Mechanics of generalized continua: construction by homogenizaton. *Journal de Physique IV France*, 8(PR4):Pr4–39–Pr4–48, 1998.

S. Forest and Kh. T. Duy. Generalized continua and non-homogeneous boundary conditions in homogenisation methods. *ZAMM*, 91(2):90–109, 2011.

S. Forest and K. Sab. Cosserat overal modelling of heterogeneous materials. *Mechanics Research Communications*, 25(4):449–454, 1998.

S. Forest and R. Sievert. Nonlinear microstrain theories. *International Journal of Solids and Structures*, 43(24):7224–7245, 2006.

R. D. Gauthier and W. E. Jahsman. Quest for micropolar elastic-constants. *Transactions of the ASME. Journal of Applied Mechanics*, 42(2):369–374, 1975.

R. D. Gauthier and W. E. Jahsman. Quest for micropolar elastic-constants. 2. *Archives of Mechanics*, 33(5):717–737, 1981.

D. H. Hodges. *Nonlinear Composite Beam Theory*, volume 213 of *Progress in Astronautics and Aeronautics.* American Institute of Aeronautics and Astronautics, Inc., Reston, 2006.

L. Hörmander. *Linear Partial Differential Equations*, volume 116 of *A Series of Comprehensive Studies in Mathematics.* Springer, Berlin, 4th edition, 1976.

R. A. Horn and Ch. R. Johnson. *Matrix Analysis.* Cambridge University Press, Cambridge, 1985.

D. Ieşan. Existence theorems in the theory of micropolar elasticity. *International Journal of Engineering Science*, 8:777–791, 1970.

D. Ieşan. Existence theorems in micropolar elastostatics. *International Journal of Engineering Science*, 9:59–78, 1971.

E. A. Ivanova, A. M. Krivtsov, N. F. Morozov, and A. D. Firsova. Description of crystal packing of particles with torque interaction. *Mechanics of Solids*, 38(4):76–88, 2003.

M. Jabareen and M. Rubin. An improved 3-D brick Cosserat point element for irregular shaped elements. *Computational Mechanics*, 40(6): 979–1004, 2007.

M. Jabareen and M. Rubin. Modified torsion coefficients for a 3-D brick Cosserat point element. *Computational Mechanics*, 41(4):517–525, 2008.

J. Jeong and P. Neff. Existence, uniqueness and stability in linear Cosserat elasticity for weakest curvature conditions. *Mathematics and Mechanics of Solids*, 15(1):78–95, 2010.

C. B. Kafadar and A. C. Eringen. Micropolar media – I. The classical theory. *International Journal of Engineering Science*, 9(3):271–305, 1971.

V. Kouznetsova, M. G. D. Geers, and W. A. M. Brekelmans. Multi-scale constitutive modelling of heterogeneous materials with a gradient-enhanced computational homogenization scheme. *International Journal for Numerical Methods in Engineering*, 54(8):1235–1260, 2002.

E. Kröner, editor. *Mechanics of Generalized Continua. Proceedings of the IUTAM-Symposium on the Generalized Cosserat Continuum and the Continuum Theory of Dislocations with Applications, Freudenstadt and Stuttgart (Germany), 1967*, Berlin, 1968. Springer.

R. S. Lakes. Experimental methods for study of Cosserat elastic solids and other generalized continua. In H. Mühlhaus, editor, *Continuum Models for Materials with Micro-Structure*, pages 1–22. Wiley, N. Y., 1995.

R. S. Lakes. Experimental microelasticity of two porous solids. *International Journal of Solids and Structures*, 22(1):55–63, 1986.

R. S. Lakes. Experimental micro mechanics methods for conventional and negative Poisson's ratio cellular solids as Cosserat continua. *Transactions of the ASME. Journal of Engineering Materials and Technology*, 113(1): 148–155, 1991.

R. Larsson and S. Diebels. A second-order homogenization procedure for multi-scale analysis based on micropolar kinematics. *International Journal for Numerical Methods in Engineering*, 69(12):2485–2512, 2007.

R. Larsson and Y. Zhang. Homogenization of microsystem interconnects based on micropolar theory and discontinuous kinematics. *Journal of the Mechanics and Physics of Solids*, 55(4):819–841, 2007.

K. C. Le and H. Stumpf. Strain measures, integrability condition and frame indifference in the theory of oriented media. *International Journal of Solids and Structures*, 35(9–10):783–798, 1998.

L. P. Lebedev, M. J. Cloud, and V. A. Eremeyev. *Tensor Analysis with Applications in Mechanics.* World Scientific, New Jersey, 2010.

L. Librescu and O. Song. *Thin-Walled Composite Beams: Theory and Applications*, volume 131 of *Solid Mechanics and Its Applications.* Springer, Dordrecht, 2006.

J.-L. Lions and E. Magenes. *Problèmes aux limites non homogènes et applications.* Dunod, Paris, 1968.

G. Łukaszewicz. *Micropolar Fluids: Theory and Applications.* Birkhäuser, Boston, 1999.

A. I. Lurie. *Theory of Elasticity.* Springer, Berlin, 2005.

A. I. Lurie. *Nonlinear Theory of Elasticity.* North-Holland, Amsterdam, 1990.

A. I. Lurie. *Analytical Mechanics.* Springer, Berlin, 2001.

B. Markert, editor. *Advances in Extended and Multifield Theories for Continua*, volume 59 of *Lecture Notes in Applied and Computational Mechanics.* Springer, Berlin, 2011.

G. A. Maugin. *Configurational Forces: Thermomechanics, Physics, Mathematics, and Numerics.* CRC Press, Boca Raton, 2010.

G. A. Maugin. *Nonlinear Waves on Elastic Crystals.* Oxford University Press, Oxford, 1999.

G. A. Maugin and A. V. Metrikine, editors. *Mechanics of Generalized Continua: One Hundred Years After the Cosserats.* Springer, New York, 2010.

T. Merlini. A variational formulation for finite elasticity with independent rotation and biot-axial fields. *Computational Mechanics*, 19(3):153–168, 1997.

N. P. Migoun and P. P. Prokhorenko. *Hydrodynamics and Heattransfer in Gradient Flows of Microstructured Fluids (in Russian).* Nauka i Technika, Minsk, 1984.

R. Mora and A. M. Waas. Measurement of the Cosserat constant of circular-cell polycarbonate honeycomb. *Philosophical Magazine A. Physics of Condensed Matter Structure Defects and Mechanical Properties*, 80(7): 1699–1713, 2000.

A. I. Murdoch. On objectivity and material symmetry for simple elastic solids. *Journal of Elasticity*, 60(3):233–242, 2000.

A. I. Murdoch. Objectivity in classical continuum physics: a rationale for discarding the 'principle of invariance under superposed rigid body motions' in favour of purely objective considerations. *Continuum Mechanics and Thermodynamics*, 15(3):309–320, 2003.

A. I. Murdoch. On criticism of the nature of objectivity in classical continuum physics. *Continuum Mechanics and Thermodynamics*, 17(2):135–148, 2005.

W. Muschik and L. Restuccia. Changing the observer and moving materials in continuum physics: objectivity and frame-indifference. *Technische Mechanik*, 22(2):152–160, 2002.

B. Nadler and M.B. Rubin. A new 3-D finite element for nonlinear elasticity using the theory of a Cosserat point. *International Journal of Solids and Structures*, 40(17):4585–4614, 2003.

M. N. L. Narasimhan. *Principles of Continuum Mechanics.* Wiley, New York, 1993.

P. Neff and K. Chełmiński. Well-posedness of dynamic Cosserat plasticity. *Applied Mathematics and Optimization*, 56(1):19–35, 2007.

P. Neff and S. Forest. A geometrically exact micromorphic model for elastic metallic foams accounting for affine microstructure. Modelling, existence of minimizers, identification of moduli and computational results. *Journal of Elasticity*, 87(2–3):239–276, 2007.

P. Neff and J. Jeong. A new paradigm: the linear isotropic Cosserat model with conformally invariant curvature energy. *ZAMM*, 89(2):107–122, 2009.

P. Neff, K. Chełmiński, W. Müller, and Ch. Wieners. A numerical solution method for an infinitesimal elasto-plastic Cosserat model. *Mathematical Models & Methods in Applied Sciences*, 17(8):1211–1239, 2007.

P. Neff, J. Jeong, and A. Fischle. Stable identification of linear isotropic Cosserat parameters: bounded stiffness in bending and torsion implies conformal invariance of curvature. *Acta Mechanica*, 211(3–4):237–249, 2010.

E. Nikitin and L. M. Zubov. Conservation laws and conjugate solutions in the elasticity of simple materials and materials with couple stress. *Journal of Elasticity*, 51(1):1–22, 1998.

L. Nirenberg. *Topics in Nonlinear Functional Analysis.* American Mathematical Society, New York, 2001.

W. Noll. A mathematical theory of the mechanical behavior of continuous media. *Archive for Rational Mechanics and Analysis*, 2(1):197–226, 1958.

W. Nowacki, editor. *Theory of Micropolar Elasticity*, volume 25 of *CISM Courses and Lectures.* Springer, Wien, 1970.

W. Nowacki. *Theory of Asymmetric Elasticity.* Pergamon-Press, Oxford et al., 1986.

W. Nowacki and W. Olszak, editors. *Micropolar Elasticity*, volume 151 of *CISM Courses and Lectures.* Springer, Wien, 1974.

R. W. Ogden. *Non-linear Elastic Deformations.* Ellis Horwood, Chichester, 1984.

H. C. Park and R. S. Lakes. Cosserat micromechanics of human bone: strain redistribution by a hydration-sensitive constituent. *Journal of Biomechanics*, 19(5):385–397, 1986.

E. Pasternak and H.-B. Mühlhaus. Generalised homogenisation procedures for granular materials. *Journal of Engineering Mathematics*, 52(1):199–229, 2005.

W. Pietraszkiewicz and V. A. Eremeyev. On natural strain measures of the non-linear micropolar continuum. *International Journal of Solids and Structures*, 46(3–4):774–787, 2009a.

W. Pietraszkiewicz and V. A. Eremeyev. On vectorially parameterized natural strain measures of the non-linear Cosserat continuum. *International Journal of Solids and Structures*, 46(11–12):2477–2480, 2009b.

S. Ramezani, R. Naghdabadi, and S. Sohrabpour. Constitutive equations for micropolar hyper-elastic materials. *International Journal of Solids and Structures*, 46(14–15):2765–2773, 2009.

E. Reissner. On kinematics and statics of finite-strain force and moment stress elasticity. *Studies in Applied Mathematics*, 52:93–101, 1973.

E. Reissner. Note on the equations of finite-strain force and moment stress elasticity. *Studies in Applied Mathematics*, 54:1–8, 1975.

R. S. Rivlin. Material symmetry and constitutive equations. *Ingenieur-Archiv*, 49(5-6):325–336, 1980.

R. S. Rivlin. Material symmetry revisited. *GAMM Mitteilungen*, 25(1/2): 109–126, 2002.

R. S. Rivlin. Frame indifference and relative frame indifference. *Mathematics and Mechanics of Solids*, 10(2):145–154, 2005.

R. S. Rivlin. Some thoughts on frame indifference. *Mathematics and Mechanics of Solids*, 11(2):113–122, 2006.

R. E. Rosensweig. Magnetic fluids. *Annual Review of Fluid Mechanics*, 19: 437–461, 1987.

M. B. Rubin. *Cosserat Theories: Shells, Rods and Points.* Kluwer, Dordrecht, 2000.

M. B. Rubin. On the theory of a Cosserat point and its application to the numerical solution of continuum problems. *Transactions of ASME. Journal of Applied Mechanics*, 52(2):368–372, 1985.

J. G. Simmonds. A simple nonlinear thermodynamic theory of arbitrary elastic beams. *Journal of Elasticity*, 81(1):51–62, 2005.

A. J. M. Spencer. Isotropic integrity bases for vectors and second-order tensors. Part II. *Archive for Rational Mechanics and Analysis*, 18(1): 51–82, 1965.

A. J. M. Spencer. Theory of invariants. In A. C. Eringen, editor, *Continuum Physics*, volume 1, pages 239–353. Academic Press, New-York, 1971.

R. Stojanović. *Mechanics of Polar Continua: Theory and Applications*, volume 2 of *CISM Courses and Lectures*. Springer, Wien, 1969a.

R. Stojanović. *Recent Developments in the Theory of Polar Continua*, volume 27 of *CISM Courses and Lectures No. 27*. Springer, Wien, 1969b.

R. Stojanović. Nonlinear micropolar elasticity. In W. Nowacki and W. Olszak, editors, *Micropolar Elasticity*, volume 151, pages 73–103. Springer, Wien, 1974.

A. S. J. Suiker and R. de Borst. Enhanced continua and discrete lattices for modelling granular assemblies. *Philosophical Transactions of the Royal Society A* , 363(1836):2543–2580, 2005.

B. Svendsen and A. Bertram. On frame-indifference and form-invariance in constitutive theory. *Acta Mechanica*, 132(1–4):195–207, 1999.

V. A. Svetlitsky. *Statics of Rods*. Springer, Berlin, 2000.

C. Truesdell. Die Entwicklung des Drallsatzes. *ZAMM*, 44(4/5):149–158, 1964.

C. Truesdell. *A First Course in Rational Continuum Mechanics*. Academic Press, New York, 1977.

C. Truesdell. *Rational Thermodynamics*. Springer, New York, 2nd edition, 1984.

C. Truesdell and W. Noll. The nonlinear field theories of mechanics. In S. Flügge, editor, *Handbuch der Physik*, volume III/3, pages 1–602. Springer, Berlin, 1965.

C. Truesdell and R. Toupin. The classical field theories. In S. Flügge, editor, *Handbuch der Physik*, volume III/1, pages 226–793. Springer, Berlin, 1960.

O. van der Sluis, P. H. J. Vosbeek, P. J. G. Schreurs, and H. E. H. Meijer. Homogenization of heterogeneous polymers. *International Journal of Solids and Structures*, 36(21):3193–3214, 1999.

A. O. Varga. *Stress-Strain Behavior of Elastic Materials: Selected Problems of Large Deformations*. Interscience, New York, 1996.

C. C. Wang and C. Truesdell. *Introduction to Rational Elasticity*. Noordhoof Int. Publishing, Leyden, 1973.

H. Xiao. On symmetries and anisotropies of classical and micropolar linear elasticities: a new method based upon a complex vector basis and some systematic results. *Journal of Elasticity*, 49(2):129–162, 1998.

J. F. C. Yang and R. S. Lakes. Experimental study of micropolar and couple stress elasticity in compact bone in bending. *Journal of Biomechanics*, 15(2):91–98, 1982.

V. A. Yeremeyev and L. M. Zubov. The theory of elastic and viscoelastic micropolar liquids. *Journal of Applied Mathematics and Mechanics*, 63 (5):755–767, 1999.

Q. S. Zheng. Theory of representations for tensor functions – a unified invariant approach to constitutive equations. *Applied Mechanics Reviews*, 47(11):545–587, 1994.

Q. S. Zheng and A. J. M. Spencer. On the canonical representations for Kronecker powers of orthogonal tensors with application to material symmetry problems. *International Journal of Engineering Science*, 31(4): 617–635, 1993.

L. M. Zubov. *Nonlinear Theory of Dislocations and Disclinations in Elastic Bodies*. Springer, Berlin, 1997.

L. M. Zubov and V. A. Eremeev. Equations for a viscoelastic micropolar fluid. *Doklady Physics*, 41(12):598–601, 1996.

Cosserat-Type Shells

Holm Altenbach † and Victor A. Eremeyev †‡

† Lehrstuhl Technische Mechanik, Institut für Mechanik, Fakultät für Maschinenbau, Otto-von-Guericke-Universität Magdeburg, Germany

‡ South Scientific Center of RASci and South Federal University, Rostov on Don, Russia

Abstract In this chapter we discuss the Cosserat-type theories of plates and shells. We call Cosserat-type shell theories various theories of shells based on the consideration of a shell base surface as a deformable directed surface, that is the surface with attached deformable or non-deformable (rigid) vectors (directors), or based on the derivation of two-dimensional (2D) shell equations from the three-dimensional (3D) micropolar (Cosserat) continuum equations.

Originally the first approach of such a kind belongs to Cosserat brothers who considered a shell as a deformable surface with attached three unit orthogonal directors. In the literature are known theories of shells which kinematics is described by introduction of the translation vector and additionally p deformable directors or one deformable director or three unit orthogonal to each other directors. Additional vector fields of directors describe the rotational (in some special cases additional) degrees of freedom of the shell. The most popular theories use the one deformable director or three unit directors. In both cases the so-called direct approach is applied.

Another approach is based on the 3D-to-2D reduction procedure applied to the 3D motion or equilibrium equations of the micropolar shell-like body. In the literature the various reduction methods are known using for example asymptotic methods, the-through-the thickness integration procedure, expansion in series, etc.

The aim of the chapter is to present the various Cosserat-type theories of plates of shells and discuss the peculiarities and differences between these approaches.

1 Introduction

The mechanics of the Cosserat continuum is based on the introduction of translations and rotations as kinematically independent quantities. The two- and one-dimensional analogues of the Cosserat continuum were presented by Cosserat and Cosserat (1909). On the other hand in the theory of plates and shells the independence of rotations was recognized by Reissner (1944, 1945, 1947, 1985). Since the paper by Ericksen and Truesdell (1958) the Cosserat model has found applications in construction of various generalized models for beams, rods, plates, and shells. Within the framework of the direct approach applied by Ericksen and Truesdell (1958), the shell is modeled as a deformable surface at each point of which a set of deformable directors is attached. Hence, the deformation of a shell is described by the position vector $\boldsymbol{r}$ and p directors $\boldsymbol{d}_i$, $i = 1, \ldots p$. This approach is developed in the original papers by Green et al. (1965); Green and Naghdi (1967a,b, 1968, 1970, 1971, 1974, 1979); Ericksen (1971, 1972a, 1973a, 1977); Naghdi and Trapp (1972); deSilva and Tsai (1973); Naghdi (1975); Naghdi and Srinivasa (1993); Naghdi and Rubin (1995). This variant of a shell theory is also named Cosserat shell theory, or the theory of Cosserat surfaces, or the Naghdi shell theory. The direct approach in the Cosserat surface theories is critisized, see, for example, Simmonds (1997). But the theory is developed successfully and there are various applications, see Itou and Atsumi (1970); Gürgöze (1971); Kurlandz (1972, 1973); Chowdhur and Glockner (1973); Glockner and Malcolm (1974); Farshad and Tabarrok (1976); Gurtin and Murdoch (1975); Turner and Nicol (1980); Kayuk and Zhukovskii (1981); Cohen and Thomas (1984, 1986); Ramsey (1986, 1987); Cohen and Wang (1989); Antman (1990); Krishnaswamy et al. (1998); Bhattacharya and James (1999); Rubin (2004); Rubin and Benveniste (2004); Neff (2004, 2007); Bîrsan (2006b,a, 2007, 2008, 2009d,b,a,c, 2010, 2011); Antman and Lacarbonara (2009); Antman and Bourne (2010); Plotnikov and Toland (2011), which partly cannot be analyzed with the help of the classical (Love) shell theory. In particular, the theory of symmetry of the constitutive equations was developed by Ericksen (1972b, 1973b); Murdoch and Cohen (1979, 1981). Finite element formulations of the Cosserat shell theory are presented in Sansour and Bednarczyk (1995); Chinosi et al. (1998); Yang et al. (2000); Jog (2004); Kreja (2007). Let us mention only the books by Naghdi (1972); Rubin (2000); Antman (2005) and Bîrsan (2009d), where the theory of Cosserat shells is summarized.

The theory of Cosserat shells contains as a special case the linear theory of Cosserat plates. This theory is mostly formulated with the help of the introduction of one deformable director, see Green and Naghdi (1967a,b). A variant with few directors is discussed, for example, in Provan and Koeller (1970). Applications of the Cosserat surface theory to sandwich plates are given in Malcolm and Glockner (1972a,b); Glockner and Malcolm (1974). In the case of the theory of Cosserat plates with one director the unknown functions are the vector of translations of the surface, representing the plate, and the vector describing the change of length and rotation of the director. Thus one assumes that such theory contains six degrees of freedom and as a consequence one has to establish six boundary conditions. In the case of Cosserat shells it is also possible to describe the thickness changes. So one can conclude that in this case for each material point six degrees of freedom are assumed: three translational degrees of freedom, two rotational degrees of freedom describing the rotations of the director and one degree of freedom which is related to the thickness changes.

Independently Eringen (1967) has formulated a linear theory of micropolar plates, see also the monograph by Eringen (1999). The two-dimensional equations of this theory are deduced with the help of independent integration over the thickness of both the first and the second Euler laws of motion of the linear elastic micropolar continuum. The theories of the zeroth and the first order are presented applying a special linear approximation of the displacement and the microrotation fields. Eringen's theory is based on eight unknowns: three averaged displacements, two averaged macrorotations of the cross-sections and three averaged microrotations. This means that one has to introduce eight boundary conditions. The static boundary conditions in Eringen's plate theory cannot be presented as forces and moments at the boundaries like in the Kirchhoff-type theories, see Timoshenko and Woinowsky-Krieger (1985). From the point of view of the direct approach Eringen's micropolar plate is a deformable surface with eight degrees of freedom. Eringen's approach is widely discussed, for example, by Ariman (1968); Nowacki and Nowacki (1969); Boschi (1973); Korman et al. (1974); Constanda (1977); Schiavone (1989, 1991); Schiavone and Constanda (1989); Wang (1990); Wang and Zhou (1991); Tomar (2005); Kumar and Deswal (2006); Kumar and Partap (2006, 2007); Pompei and Rigano (2006) and in the monograph by Eringen (1999).

The theories of plates and shells and the theories based on the reduction of the three-dimensional (3D) equations of the micropolar continuum are presented in several publications. Gevorkyan (1967); Reissner (1970, 1977); Jemielita (1995); Ambartsumian (1996, 1999) applied various averaging procedures in the thickness direction together with the approximation of the displacements and rotations or the force and moment stresses in the thickness direction. As a result, one gets different numbers of unknowns and the number of two-dimensional equilibrium equations differs. For example, Reissner (1977) presented a generalized linear theory of shells containing nine equilibrium equations. Additionally, Reissner (1972) worked out the two-dimensional theory of a sandwich plate with a core having the properties of a Cosserat continuum. The variants of the micropolar plate theory based on the asymptotic methods are developed in Nazarov (1993); Erbay (2000); Aganović et al. (2006); Sargsyan (2005, 2008, 2009, 2011); Tambača and Velčić (2009); Velčić and Tambača (2010); Zubov (2009); Sargsyan and Sargsyan (2011). The non-linear theory of elastic shells derived from the pseudo-Cosserat continuum is considered in Badur and Pietraszkiewicz (1986). The linear theory of micropolar plates is presented in Altenbach and Eremeyev (2009a) where a discussion on the reduction procedure is given. The Γ-convergence based approach to the Cosserat-type theory of plates and shells is discussed by Neff (2004, 2005); Neff and Chełmiński (2007).

In both the Cosserat and the Eringen micropolar plate theories one has additional kinematic variables – the rotations. It should be noted that in the theories of plates and shells the rotations are introduced as independent kinematic variables before the Cosserat theory was established. The term “angle of rotation” is also introduced in Kirchhoff’s theory – but the rotations are expressed by the displacement field. After the pioneering work by Kirchhoff (1850) thousands of publications are presented, which try to give the foundations and the methods of deduction of the equations of the Kirchhoff–Love theory, but also of improvements, see, for example, Fox et al. (1993); Podio-Guidugli (1989); Nicotra et al. (1999); Ciarlet (1997, 2000); Friesecke et al. (2002a,b,c, 2003); Kienzler (2002); Kienzler et al. (2004). Considering sandwich structures with a soft core Reissner (1947, 1985) worked out a theory by taking into account the transverse shear which was ignored by Kirchhoff. Similar governing equations (only some effects are not included) were derived by Mindlin (1951) introducing additional degrees of freedom for the points of the midplane. The order of the system of the partial differential static equilibrium equations in the case of Reissner-type theories is equal to ten. That means, that the number of boundary equations is equal to five. In the theories of Reissner and Mindlin only two angles of rotations are independent of the displacements, and the transverse shear can be taken into account. The third angle of rotation (rotation about the normal to the surface, so-called drilling rotation) is not considered as an independent variable. In Reissner’s theory the static boundary conditions are equivalent to the introduction of distributed forces and moments on the contour, the last one has no components in the normal direction. The Reissner plate as well as the Kirchhoff plate are not able to react on the distributed moments on the surfaces or boundaries which are directed along the surface normal (so-called drilling moments). That means that Reissner’s plate is modeled by a material surface with points having five degrees of freedom (three translation and two rotations) while Kirchhoff’s plate is a material surface each point of which has only three degrees of freedom (three translations)[1]. Now we have thousands of papers and monographs on Reissner's and Mindlin's approach, see for example the reviews Grigolyuk and Selezov (1973); Nardinocchi and Podio-Guidugli (1994).

In the last decades the so-called higher order theories are very popular. Starting with the pioneering contributions by Levinson (1980) and Reddy (1984), new theories are established systematically. If one discusses higher order theories in the point of view of the direct approach one assumes deformable surfaces with additional degrees of freedom. For example, the third order theory presented in Wang et al. (2000) can be regarded as a theory with seven degrees of freedom including rotations of the plate cross-sections. Kienzler (2002) discussed the consistent higher order theories of plates using series expansions. Let us mention also the contributions by Simmonds and Danielson (1972); Pietraszkiewicz (1979a); Başar (1987); Fox and Simo (1992); Hodges et al. (2004); Brank et al. (1997); Hodges (2006); Kreja (2007); Wiśniewski (2010); Merlini and Morandini (2011a,b), where the rotations in shells are considered, while an extensive discussion of the application of the rotations in Continuum Mechanics is given in Pietraszkiewicz and Badur (1983).

The direct approach in the theory of shells based on Cosserat's ideas is applied also in Zhilin (1976). In contrast to Ericksen and Truesdell (1958), the shell is regarded as a deformable surface with material points at which three directors are attached. The directors are orthogonal unit vectors. The deformations of the shell are presented by a position vector and a properly orthogonal tensor. This variant of the shell and plate theories based on the direct approach is investigated, for example, in Shkutin (1980, 1985); Pal'mov (1982); Palmow and Altenbach (1982); Eremeyev (2005); Eremeyev and Zubov (2007, 2008); Zhilin (2006); Zubov (1997, 1999, 2001, 2011). It must be noted that this variant is very similar to the one presented within the general non-linear theory of shells discussed in the monographs by Libai and Simmonds (1998), and Chróścielewski et al. (2004), see also Pietraszkiewicz (2001) and the contributions by Libai and Simmonds (1983); Simmonds (1984, 1997); Makowski and Stumpf (1990); Chróścielewski (1996); Chróścielewski et al. (1997, 2002); Makowski et al. (1999); Makowski and Pietraszkiewicz (2002); Lubowiecka and Chróścielewski (2002); Eremeyev and Pietraszkiewicz (2004, 2006, 2009); Pietraszkiewicz et al. (2005); Konopińska and Pietraszkiewicz (2006); Pietraszkiewicz et al. (2007); Chróścielewski and Witkowski (2010, 2011); Chróścielewski et al. (2011); Altenbach and Eremeyev (2011); Pietraszkiewicz (2011). The two-dimensional motion equations given in Libai and Simmonds (1983, 1998);

[1]The original Kirchhoff's plate theory has only one degree of freedom (the deflection).

Pietraszkiewicz (2001); Chróścielewski et al. (2004) are derived by the exact integration over the thickness of the equations of motion of a shell-like body. The deformation measures, which are the same as those introduced within the framework of the direct approach, can be defined as work-conjugate fields to the stress and the stress couple tensors.

Altenbach and Zhilin (1988) transformed the general theory with six degrees of freedom to a theory of shells with five degrees of freedom (similar to Reissner's theory) introducing some constraints for the deformations. This variant of the theory is discussed in Altenbach (1988); Altenbach and Zhilin (2004); Zhilin (2006); Altenbach and Eremeyev (2008); Bîrsan and Altenbach (2010, 2011). The method presented in Altenbach and Zhilin (1988) is applied by Grekova and Zhilin (2001) to the three-dimensional case. It should be noted that the main problem in application of the direct approach is both the establishment and the identification of the constitutive equations. They should be formulated for the two-dimensional measures of stresses and measures of deformations. This means that some effective stiffness properties should be introduced. For anisotropic elastic plates the identification procedure for the effective stiffness properties is discussed by Altenbach and Zhilin (1988); Zhilin (2006); Altenbach and Eremeyev (2008, 2011) and for the viscoelastic case in Altenbach (1988); Altenbach and Eremeyev (2009b). The last one approach has some similarities with the presented one by Rothert (1975, 1977).

It is worth mentioning here the bibliographical papers on he shell theory by Noor (1990, 2004); Pietraszkiewicz (1992) and by Jemielita (see Jemilita (2001)), who has discussed papers since 1767.

The aim of this chapter is a discussion of the Cosserat-type theories of plates and shells. The chapter is organized as follows. In Sect. 2 we present the basic equations of the theory of the directed surfaces. We restricted ourselves to a variant of the theory with one deformed director. In Sect. 3 we consider a 6-parametric nonlinear theory of shells using the direct approach. In Sect. 4 we present the equations of the linear theory of micropolar plates based on the reduction of three-dimensional equations of the micropolar medium to two-dimensional equations used in plates and shells theories. Finally, we discuss in brief the difference in the presented approaches.

Further we use the direct tensorial notations, see for example Lebedev et al. (2010); Lurie (2005). Vectors are denoted by semibold italic font like $\boldsymbol{a}$, $\boldsymbol{A}$. Tensors are denoted by semibold normal font like $\mathbf{a}$, $\mathbf{A}$. Functionals are denoted by calligraphic letters like $\mathcal{A}$. Greek indices take values 1 and 2, while Latin indices are arbitrary.

2 Cosserat Surface

Following Naghdi (1972); Rubin (2000) and using the direct approach we consider the basic relations of the nonlinear theory of Cosserat shells. The advantage of the latter approach is discussed in many papers, see for example Ericksen and Truesdell (1958); Green and Naghdi (1974) or Cohen and Wang (1989).

2.1 Kinematics

We introduce the Cosserat surface as a deformable surface with an attached deformable director. The kinematics of the Cosserat surface may be described as follows. Let Σ be the shell surface in the reference configuration (undeformed state) represented by the Gaussian coordinates q^α ($\alpha = 1, 2$), and $\boldsymbol{R}(q^1, q^2)$ is the position-vector describing the material points on the surface Σ. The surface σ of the shell after deformation is also represented by the coordinates q^α, the position of the material points on σ is given by $\boldsymbol{r}(q^1, q^2)$, see Fig. 1. Here $\boldsymbol{N}$ and $\boldsymbol{n}$ are the vectors of the unit normals to the shell base surfaces Σ and σ, while $\boldsymbol{\nu}$ and $\widetilde{\boldsymbol{\nu}}$ are the unit normal vectors to the shell boundary contours $\omega = \partial\sigma$ and $\Omega = \partial\Sigma$, respectively, $\boldsymbol{\nu} \cdot \boldsymbol{N} = 0$, $\widetilde{\boldsymbol{\nu}} \cdot \boldsymbol{n} = 0$. $\boldsymbol{r}_\beta$ and $\boldsymbol{r}^\alpha$ are the co- and contravariant vector bases on σ and $\boldsymbol{R}_\beta$, $\boldsymbol{R}^\alpha$ are the co- and contravariant vector bases on Σ

$$\boldsymbol{r}^\alpha \cdot \boldsymbol{r}_\beta = \delta^\alpha_\beta, \quad \boldsymbol{r}^\alpha \cdot \boldsymbol{n} = 0, \quad \boldsymbol{r}_\beta = \frac{\partial \boldsymbol{r}}{\partial q^\beta},$$

$$\boldsymbol{R}^\alpha \cdot \boldsymbol{R}_\beta = \delta^\alpha_\beta, \quad \boldsymbol{R}^\alpha \cdot \boldsymbol{N} = 0, \quad \boldsymbol{R}_\beta = \frac{\partial \boldsymbol{R}}{\partial q^\beta} \quad (\alpha, \beta = 1, 2),$$

where δ^α_β is the Kronecker symbol.

The Cosserat surface is a material surface, on which is given a director-vector field. By this field the changes of the orientation and the length of the material fibres are presented. In this sense the material fibres of the shell behave like three-dimensional rigid bodies. In the actual configuration we denote this field by $\boldsymbol{d}$, while in the reference configuration by $\boldsymbol{D}$ (Fig. 1). We assume that $\boldsymbol{D} \cdot \boldsymbol{N} \neq 0$ and $\boldsymbol{d} \cdot \boldsymbol{n} \neq 0$. This means that $\boldsymbol{d}$ and $\boldsymbol{D}$ cannot be tangential vectors to the shell surface.

Thus, the deformation of the Cosserat surface is given by two vector fields

$$\boldsymbol{r} = \boldsymbol{R}(q^1, q^2), \quad \boldsymbol{d} = \boldsymbol{d}(q^1, q^2). \tag{1}$$

Let us note that the directors $\boldsymbol{d}$ and $\boldsymbol{D}$ are not unit normal vectors to the surfaces σ and Σ, in general. According to (1) any material point of the Cosserat surface has six degrees of freedom. Considering various constraints

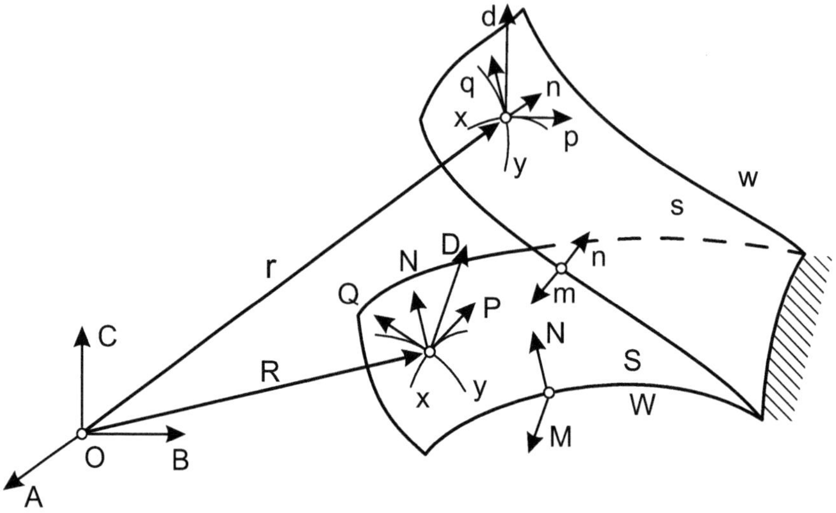

Figure 1. Deformation of a Cosserat surface (reference and actual configurations)

on $\boldsymbol{d}$, one can derive special theories of shells. For example, assuming the constraint $\boldsymbol{d} = \boldsymbol{n}$ we obtain the Kirchhoff-Love theory of shells. A Reissner-type shell theory can be obtained on the assumptions $\boldsymbol{d} \neq \boldsymbol{n}$, $\boldsymbol{d} \cdot \boldsymbol{d} = 1$, see Rubin (2000) for details. Within the framework of the model the thickness changes can be modeled.

2.2 Strain Energy Density of an Elastic Cosserat Surface

Let us assume an elastic behavior of the Cosserat shell. In this case there exists the strain energy density W which is a function of $\boldsymbol{r}$, $\boldsymbol{d}$ and their first spatial gradients

$$W = W(\boldsymbol{r}, \nabla \boldsymbol{r}, \boldsymbol{d}, \nabla \boldsymbol{d}), \quad \nabla (\ldots) \stackrel{\triangle}{=} \boldsymbol{R}^\alpha \otimes \frac{\partial(\ldots)}{\partial q^\alpha}. \tag{2}$$

In what follows we denote $\nabla \boldsymbol{r}$ by $\mathbf{F}$. For the transformation of W we apply the principle of material frame-indifference formulated in Truesdell (1964) or Truesdell and Noll (1965). The strain energy density should be invariant under the superposed rigid-body motion. In other words, W does not change under the transformations

$$\boldsymbol{r} \to \boldsymbol{a} + \boldsymbol{r} \cdot \mathbf{O}, \quad \boldsymbol{d} \to \boldsymbol{d} \cdot \mathbf{O}, \tag{3}$$

where $\boldsymbol{a}$ is an arbitrary constant vector, and $\mathbf{O}$ is an arbitrary constant orthogonal tensor. From (3) it follows that W does not depend on $\boldsymbol{r}$ and satisfies the relation

$$W\left[\mathbf{F}\cdot\mathbf{O}, \boldsymbol{d}\cdot\mathbf{O}, (\nabla\,\boldsymbol{d})\cdot\mathbf{O}\right] = W\left(\mathbf{F}, \boldsymbol{d}, \nabla\,\boldsymbol{d}\right) \tag{4}$$

for the arbitrary orthogonal tensor $\mathbf{O}$.

To construct the strain energy function satisfying (4) let us find the polar decomposition of the tensor $\mathbf{F}$, see for example Eremeyev and Zubov (2008). Since $\boldsymbol{N}\cdot\mathbf{F} = \mathbf{0}$, $\mathbf{F}$ is a singular tensor, while $\mathbf{F} + \boldsymbol{N}\otimes\boldsymbol{n}$ is a non-singular one. We have the polar decomposition in the form,

$$\mathbf{F} + \boldsymbol{N}\otimes\boldsymbol{n} = (\mathbf{U} + \boldsymbol{N}\otimes\boldsymbol{N})\cdot\mathbf{R},$$

where $\mathbf{U}$ is a positive definite symmetric tensor on the surface (two-dimensional tensor), which can be computed by $\mathbf{U} = \mathbf{F}\cdot\mathbf{F}^T$, $\mathbf{R}$ is a rotation tensor of the surface such that $\boldsymbol{N}\cdot\mathbf{R} = \boldsymbol{n}$. Finally, we have

$$\mathbf{F} = \mathbf{U}\cdot\mathbf{R}.$$

Assuming in Eq. (4) $\mathbf{O} = \mathbf{R}^T$, we obtain the constitutive equation satisfying the principle of the material frame-indifference

$$W = W\left(\mathbf{U}, \boldsymbol{d}\cdot\mathbf{R}^T, (\nabla\,\boldsymbol{d})\cdot\mathbf{R}^T\right). \tag{5}$$

2.3 Principle of Virtual Work and the Equilibrium Conditions

The equilibrium equation of the material surface we derive with the help of the variational calculus. The starting point is the principle of virtual work

$$\delta\iint\limits_{\Sigma} W\,\mathrm{d}\Sigma - \delta'\mathcal{A} = 0. \tag{6}$$

Here $\delta'\mathcal{A}$ is the elementary work of the external loadings, δ denotes the variation symbol. For the sake of simplicity we use Eq. (2) for the variation of the strain energy density W. Then we obtain

$$\delta W = \mathbf{T}\bullet\delta\mathbf{F} + \mathbf{M}\bullet\nabla\,\delta\boldsymbol{d} + \frac{\partial W}{\partial\boldsymbol{d}}\cdot\delta\boldsymbol{d} \tag{7}$$

with

$$\mathbf{T} \triangleq \frac{\partial W}{\partial\mathbf{F}}, \quad \mathbf{M} \triangleq \frac{\partial W}{\partial\nabla\,\boldsymbol{d}} \tag{8}$$

as the surface stress and the couple stress tensors of first Piola-Kirchoff-type (two-point stress measures), $\bullet$ is the scalar product in the space of second-order tensors, that is $\mathbf{X}\bullet\mathbf{Y} = \mathbf{X}\cdot\cdot\mathbf{Y}^T = \mathrm{tr}\,(\mathbf{X}\cdot\mathbf{Y})$ for any second-order tensors $\mathbf{X}$ and $\mathbf{Y}$, see Lebedev et al. (2010).

Taking into account Eqs. (7) and (8) and the surface divergence theorem, we obtain

$$\delta \iint\limits_{\Sigma} W \, \mathrm{d}\Sigma = \oint\limits_{\Omega} \widetilde{\boldsymbol{\nu}} \cdot (\mathbf{T} \cdot \delta \boldsymbol{r} + \mathbf{M} \cdot \delta \boldsymbol{d}) \, \mathrm{d}s$$

$$- \iint\limits_{\Sigma} (\nabla \cdot \mathbf{T}) \cdot \delta \boldsymbol{r} \, \mathrm{d}\Sigma - \iint\limits_{\Sigma} \left[\nabla \cdot \mathbf{M} - \frac{\partial W}{\partial \boldsymbol{d}} \right] \cdot \delta \boldsymbol{d} \, \mathrm{d}\Sigma, \qquad (9)$$

Considering the variational equation (6) and Eq. (9) the elementary work of the external loadings $\delta' \mathcal{A}$ acting on the Cosserat surface can be taken in the following form:

$$\delta' \mathcal{A} = \iint\limits_{\Sigma} (\boldsymbol{f} \cdot \delta \boldsymbol{r} + \boldsymbol{\ell} \cdot \delta \boldsymbol{d}) \, \mathrm{d}\Sigma + \int\limits_{\Omega_2} \boldsymbol{\varphi} \cdot \delta \boldsymbol{r} \, \mathrm{d}s + \int\limits_{\Omega_4} \boldsymbol{\gamma} \cdot \delta \boldsymbol{d} \, \mathrm{d}s. \qquad (10)$$

In Eq. (10) $\boldsymbol{f}$, $\boldsymbol{\ell}$, $\boldsymbol{\varphi}$ and $\boldsymbol{\gamma}$ are given functions on the contour parts Ω_2 and Ω_4, respectively. Here we use the decomposition $\Omega = \Omega_1 \cup \Omega_2 = \Omega_3 \cup \Omega_4$. On Ω_1 and Ω_3 the position-vector $\boldsymbol{r}$ and the director $\boldsymbol{d}$ are given, respectively.

From the variational equation (6) considering Eq. (10) the formulation of the boundary-value problem follows for the equilibrium of a non-linear elastic Cosserat shell in the reference configuration

$$\nabla \cdot \mathbf{T} + \boldsymbol{f} = \mathbf{0}, \quad \nabla \cdot \mathbf{M} - \frac{\partial W}{\partial \boldsymbol{d}} + \boldsymbol{\ell} = \mathbf{0}, \qquad (11)$$

$$\Omega_1 \; : \; \boldsymbol{r} = \boldsymbol{\rho}(s), \qquad (12)$$

$$\Omega_2 \; : \; \widetilde{\boldsymbol{\nu}} \cdot \mathbf{T} = \boldsymbol{\varphi}(s), \qquad (13)$$

$$\Omega_3 \; : \; \boldsymbol{d} = \boldsymbol{d}^0(s), \qquad (14)$$

$$\Omega_4 \; : \; \widetilde{\boldsymbol{\nu}} \cdot \mathbf{M} = \boldsymbol{\gamma}(s). \qquad (15)$$

$\boldsymbol{\rho}(s)$, $\boldsymbol{d}^0(s)$ are given functions in the kinematic boundary conditions (12) and (14). $\boldsymbol{\rho}(s)$ defines the position of the part of the shell boundary Ω_1 in the space, while $\boldsymbol{d}^0(s)$ defines the director on Ω_3. The functions $\boldsymbol{f}$ and $\boldsymbol{\varphi}$ in Eqs. $(11)_1$ and (13) represent the distributed on the shell surface and boundary part Ω_2 forces, respectively. The functions $\boldsymbol{\ell}$ and $\boldsymbol{\gamma}$ describe the surface and boundary loads which are both combinations of bending couples and force dipoles.

3 Micropolar Shells

In this section we use again the direct approach to the formulation of the basic equations of micropolar shell theory. A micropolar shell is a two-dimensional analogue to the Cosserat continuum, i.e. a micropolar shell is a material surface, each particle of which has six degrees of freedom of the rigid body. Further we will use the notations given in Eremeyev (2005); Eremeyev and Zubov (2007, 2008).

3.1 Kinematics

Let Σ be a base surface of the micropolar shell in the reference configuration, q^α ($\alpha = 1, 2$) be Gaussian coordinates on Σ, and $\boldsymbol{R}(q^1, q^2)$ be a position-vector of Σ, see Fig. 2. As an example one can use an undeformed state as the reference configuration. In the actual (or deformed) configuration the shell base surface is denoted by σ, and the position of its material points (infinitesimal point-bodies) is given by the vector $\boldsymbol{r}(q^1, q^2)$. The orientation of the point-bodies is described by the so-called *microrotation tensor* $\mathbf{Q}(q^1, q^2)$, which is a proper orthogonal tensor. If we introduce three orthonormal vectors $\boldsymbol{D}_k$ ($k = 1, 2, 3$), which describe the orientation in the reference configuration, and three orthonormal vectors $\boldsymbol{d}_k$, which determine the orientation in the actual configuration, then the tensor $\mathbf{Q}$ is given by $\mathbf{Q} = \boldsymbol{D}_k \otimes \boldsymbol{d}_k$. Thus, the micropolar shell is described by two kinematically independent fields

$$\boldsymbol{r} = \boldsymbol{r}(q^1, q^2), \quad \mathbf{Q} = \mathbf{Q}(q^1, q^2). \tag{16}$$

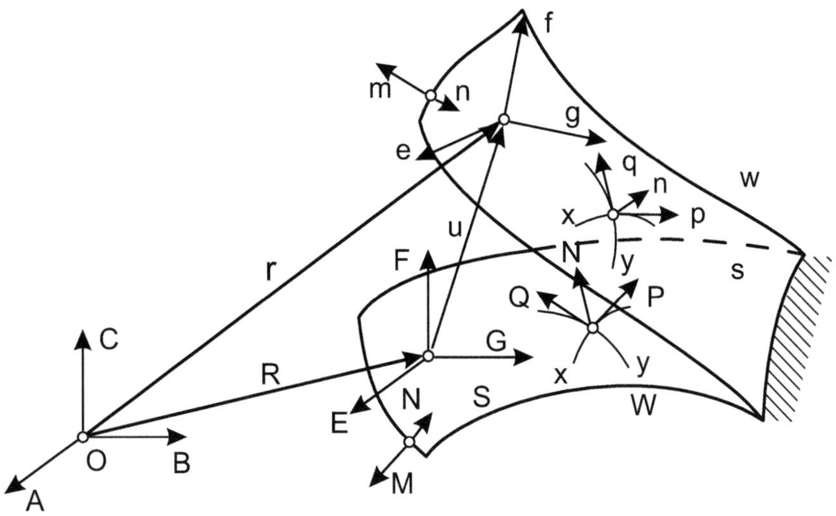

Figure 2. Kinematics of a micropolar shell, reference base surface Σ and actual base surface σ

A proper orthogonal tensor describes the rotation about an arbitrary axis. It can be represented by Gibbs' formula

$$\mathbf{Q} = (\mathbf{I} - \boldsymbol{e} \otimes \boldsymbol{e}) \cos \chi + \boldsymbol{e} \otimes \boldsymbol{e} - \boldsymbol{e} \times \mathbf{I} \sin \chi, \tag{17}$$

where χ and $\boldsymbol{e}$ are the angle of rotation about the axis with the unit vector $\boldsymbol{e}$, and $\mathbf{I}$ is the 3D unit tensor, respectively. Introducing the finite rotation vector

$$\boldsymbol{\theta} = 2\boldsymbol{e} \tan \frac{\chi}{2}$$

and using the formulae

$$\cos \chi = \frac{1 - \tan^2 \chi/2}{1 + \tan^2 \chi/2}, \quad \sin \chi = \frac{2 \tan \chi/2}{1 + \tan^2 \chi/2}$$

we obtain the representation of $\mathbf{Q}$ in a form which does not contain trigonometric functions

$$\mathbf{Q} = \frac{1}{(4 + \theta^2)} \left[(4 - \theta^2)\mathbf{I} + 2\boldsymbol{\theta} \otimes \boldsymbol{\theta} - 4\mathbf{I} \times \boldsymbol{\theta} \right], \quad \theta^2 = \boldsymbol{\theta} \cdot \boldsymbol{\theta}. \tag{18}$$

In the rigid body kinematics the vector $\boldsymbol{\theta}$ is called *Rodrigues' finite rotation vector*, cf. Lurie (2001). In the theory of micropolar shells we call it the *microrotation vector*. From Eq. (18) for a given proper orthogonal tensor $\mathbf{Q}$ we find uniquely the vector $\boldsymbol{\theta}$

$$\boldsymbol{\theta} = 2(1 + \operatorname{tr} \mathbf{Q})^{-1} \mathbf{Q}_{\times}. \tag{19}$$

Here $\mathbf{T}_{\times}$ is the vectorial invariant of a second-order tensor $\mathbf{T}$ defined by

$$\mathbf{T}_{\times} = (T^{mn} \boldsymbol{e}_m \otimes \boldsymbol{e}_n)_{\times} = T^{mn} \boldsymbol{e}_m \times \boldsymbol{e}_n$$

for any base $\boldsymbol{e}_m$, $\times$ is the vector product, see e.g. Lebedev et al. (2010). Other vectorial parameterizations of an orthogonal tensor are summarized in Pietraszkiewicz and Eremeyev (2009b).

For an elastic micropolar shell there exists a strain energy density W. By using the principle of local action formulated in Truesdell and Noll (1965); Truesdell (1991) the constitutive equation for the function W can be written as follows, see Eremeyev (2005); Eremeyev and Zubov (2007, 2008),

$$W = W(\boldsymbol{r}, \nabla \boldsymbol{r}, \mathbf{Q}, \nabla \mathbf{Q}).$$

Using the principle of material frame-indifference we conclude that W depends on two *Cosserat strain measures* $\mathbf{E}$ and $\mathbf{K}$

$$W = W(\mathbf{E}, \mathbf{K}),$$

$$\mathbf{E} = \mathbf{F}\cdot\mathbf{Q}^T - \mathbf{A}, \quad \mathbf{K} = \frac{1}{2}\boldsymbol{R}^\alpha \otimes \left(\frac{\partial \mathbf{Q}}{\partial q^\alpha}\cdot\mathbf{Q}^T\right)_\times, \tag{20}$$

where $\mathbf{F} = \nabla\, \boldsymbol{r}$ is the surface deformation gradient, $\mathbf{A} = \mathbf{I} - \boldsymbol{N} \otimes \boldsymbol{N}$.

Using the finite rotation vector we express $\mathbf{K}$ as follows

$$\mathbf{K} = \frac{4}{4+\theta^2}\nabla\,\boldsymbol{\theta}\cdot\left(\mathbf{I} + \frac{1}{2}\mathbf{I}\times\boldsymbol{\theta}\right). \tag{21}$$

The strain measures $\mathbf{E}$ and $\mathbf{K}$ are the two-dimensional analogues of the strain measures introduced in 3D Cosserat continuum mechanics and discussed by Pietraszkiewicz and Eremeyev (2009a,b). We call $\mathbf{E}$ and $\mathbf{K}$ the surface stretch and wryness tensors, respectively.

3.2 Principle of Virtual Work and Boundary-Value Problems

The Lagrangian equilibrium equations of a micropolar shell can be derived from *the principle of virtual work*

$$\delta \iint\limits_\Sigma W\, d\Sigma = \delta' A, \tag{22}$$

where

$$\delta' A = \iint\limits_\Sigma (\boldsymbol{f}\cdot\delta\boldsymbol{r} + \boldsymbol{c}\cdot\delta'\boldsymbol{\psi})\, d\Sigma + \int\limits_{\Omega_2} \boldsymbol{\varphi}\cdot\delta\boldsymbol{r}\, ds + \int\limits_{\Omega_4} \boldsymbol{\eta}\cdot\delta'\boldsymbol{\psi}\, ds, \tag{23}$$

$$\mathbf{I}\times\delta'\boldsymbol{\psi} = -\mathbf{Q}^T\cdot\delta\mathbf{Q}.$$

In Eq. (22) δ is the symbol of variation, $\delta'\boldsymbol{\psi}$ is the virtual rotation vector, $\boldsymbol{f}$ is the surface force density distributed on Σ, $\boldsymbol{c}$ is the surface couple density distributed on Σ, $\boldsymbol{\varphi}$ and $\boldsymbol{\eta}$ are linear densities of forces and couples distributed along corresponding parts of the shell boundary Ω, respectively.

Using the formulae introduced by Eremeyev and Zubov (2008)

$$\delta W = \frac{\partial W}{\partial \mathbf{E}} \bullet \delta\mathbf{E} + \frac{\partial W}{\partial \mathbf{K}} \bullet \delta\mathbf{K},$$

$$\delta\mathbf{E} = (\nabla\,\delta\boldsymbol{r})\cdot\mathbf{Q}^T + \mathbf{F}\cdot\delta\mathbf{Q}^T, \quad \delta\mathbf{K} = (\nabla\,\delta'\boldsymbol{\psi})\cdot\mathbf{Q}^T,$$

$$\delta'\boldsymbol{\psi} = \frac{4}{4+\theta^2}\left(\delta\boldsymbol{\theta} + \frac{1}{2}\boldsymbol{\theta}\times\delta\boldsymbol{\theta}\right),$$

from Eq. (22) we obtain the *Lagrangian shell equations*

$$\nabla\cdot\mathbf{D} + \boldsymbol{f} = \mathbf{0}, \quad \nabla\cdot\mathbf{G} + \left[\mathbf{F}^T\cdot\mathbf{D}\right]_\times + \boldsymbol{c} = \mathbf{0}, \tag{24}$$

$$\mathbf{D} = \mathbf{P}_1 \cdot \mathbf{Q}, \quad \mathbf{G} = \mathbf{P}_2 \cdot \mathbf{Q}, \qquad \mathbf{P}_1 = \frac{\partial W}{\partial \mathbf{E}}, \quad \mathbf{P}_2 = \frac{\partial W}{\partial \mathbf{K}}, \tag{25}$$

$$\begin{aligned} \Omega_1 : &\quad \boldsymbol{r} = \boldsymbol{\rho}(s), \\ \Omega_2 : &\quad \boldsymbol{\nu} \cdot \mathbf{D} = \boldsymbol{\varphi}(s), \\ \Omega_3 : &\quad \mathbf{Q} = \mathbf{h}(s), \quad \mathbf{h} \cdot \mathbf{h}^T = \mathbf{I}, \\ \Omega_4 : &\quad \boldsymbol{\nu} \cdot \mathbf{G} = \boldsymbol{\eta}(s). \end{aligned} \tag{26}$$

Here $\boldsymbol{\rho}(s)$, $\mathbf{h}(s)$ are given vector and tensor functions, and $\boldsymbol{\nu}$ is the external unit normal to the boundary curve Ω ($\boldsymbol{\nu} \cdot \boldsymbol{N} = 0$). Equations (24) are the local balance equations of the linear momentum and angular momentum of any shell part. The tensors $\mathbf{D}$ and $\mathbf{G}$ are *the surface stress and couple stress tensors* of the first Piola-Kirchhoff-type, and the corresponding stress measures $\mathbf{P}_1$ and $\mathbf{P}_2$ in Eqs. (24) are the stress tensors of the second Piola-Kirchhoff-type, respectively. The strain measures $\mathbf{E}$ and $\mathbf{K}$ are work-conjugated to the stress measures $\mathbf{P}_1$ and $\mathbf{P}_2$. The boundary Ω of Σ is divided into two parts $\Omega = \Omega_1 \cup \Omega_2 = \Omega_3 \cup \Omega_4$. The following relations are valid

$$\boldsymbol{N} \cdot \mathbf{D} = \boldsymbol{N} \cdot \mathbf{G} = \boldsymbol{N} \cdot \mathbf{P}_1 = \boldsymbol{N} \cdot \mathbf{P}_2 = \mathbf{0}. \tag{27}$$

The *equilibrium equations* (24) may be transformed to the *Eulerian form*

$$\nabla_\Sigma \cdot \mathbf{T} + J^{-1} \boldsymbol{f} = \mathbf{0}, \quad \nabla_\Sigma \cdot \mathbf{M} + \mathbf{T}_\times + J^{-1} \boldsymbol{c} = \mathbf{0}, \tag{28}$$

where

$$\begin{gathered} \nabla_\Sigma (\ldots) \triangleq \boldsymbol{r}^\alpha \frac{\partial}{\partial q^\alpha} (\ldots), \\ \mathbf{T} = J^{-1} \mathbf{F}^T \cdot \mathbf{D}, \quad \mathbf{M} = J^{-1} \mathbf{F}^T \cdot \mathbf{G}, \\ J = \sqrt{\frac{1}{2} \left\{ \left[\mathrm{tr}\, (\mathbf{F} \cdot \mathbf{F}^T) \right]^2 - \mathrm{tr} \left[(\mathbf{F} \cdot \mathbf{F}^T)^2 \right] \right\}}. \end{gathered} \tag{29}$$

Here $\mathbf{T}$ and $\mathbf{M}$ are the Cauchy-type surface stress and couple stress tensors, ∇_Σ is the surface nabla operator on Σ associated with ∇ by the formula

$$\nabla = \mathbf{F} \cdot \nabla_\Sigma .$$

The *equations of motion* of the micropolar shell are given by the relations (see, for example, Zhilin (1976); Libai and Simmonds (1983, 1998); Altenbach and Zhilin (1988); Chróścielewski et al. (2004); Eremeyev and Zubov (2008))

$$\begin{gathered} \nabla \cdot \mathbf{D} + \boldsymbol{f} = \rho \frac{\mathrm{d} \boldsymbol{K}_1}{\mathrm{d}t}, \\ \nabla \cdot \mathbf{G} + \left[\mathbf{F}^T \cdot \mathbf{D} \right]_\times + \boldsymbol{c} = \rho \left(\frac{\mathrm{d} \boldsymbol{K}_2}{\mathrm{d}t} + \boldsymbol{v} \times \boldsymbol{\Theta}_1^T \cdot \boldsymbol{\omega} \right), \end{gathered} \tag{30}$$

where

$$\boldsymbol{K}_1 \triangleq \frac{\partial K}{\partial \boldsymbol{v}} = \boldsymbol{v} + \boldsymbol{\Theta}_1^T \cdot \boldsymbol{\omega}, \quad \boldsymbol{K}_2 \triangleq \frac{\partial K}{\partial \boldsymbol{\omega}} = \boldsymbol{\Theta}_1 \cdot \boldsymbol{v} + \boldsymbol{\Theta}_2 \cdot \boldsymbol{\omega},$$

$$K(\boldsymbol{v}, \boldsymbol{\omega}) = \frac{1}{2}\boldsymbol{v} \cdot \boldsymbol{v} + \boldsymbol{\omega} \cdot \boldsymbol{\Theta}_1 \cdot \boldsymbol{v} + \frac{1}{2}\boldsymbol{\omega} \cdot \boldsymbol{\Theta}_2 \cdot \boldsymbol{\omega},$$

$$\boldsymbol{v} = \frac{\mathrm{d}\boldsymbol{r}}{dt}, \qquad \boldsymbol{\omega} = \frac{1}{2}\left(\mathbf{Q}^T \cdot \frac{\mathrm{d}\mathbf{Q}}{dt}\right)_\times$$

$\boldsymbol{v}$ and $\boldsymbol{\omega}$ are the linear and angular velocities, respectively, ρ is the surface mass density in the reference configuration, ρK is the surface density of the kinetic energy, and $\rho\boldsymbol{\Theta}_1$, $\rho\boldsymbol{\Theta}_2$ are *the rotatory inertia tensors* ($\boldsymbol{\Theta}_2^T = \boldsymbol{\Theta}_2$). By physical meaning, $\boldsymbol{\Theta}_1$ and $\boldsymbol{\Theta}_2$ have the following properties

$$\boldsymbol{\Theta}_1 = \mathbf{Q}^T \cdot \boldsymbol{\Theta}_1^\circ \cdot \mathbf{Q}, \quad \boldsymbol{\Theta}_2 = \mathbf{Q}^T \cdot \boldsymbol{\Theta}_2^\circ \cdot \mathbf{Q}, \quad \frac{\mathrm{d}\boldsymbol{\Theta}_1^\circ}{dt} = \frac{\mathrm{d}\boldsymbol{\Theta}_2^\circ}{dt} = \mathbf{0}. \tag{31}$$

The tensors $\boldsymbol{\Theta}_1^\circ$ and $\boldsymbol{\Theta}_2^\circ$ are called the inertia tensors in the reference configuration.

For the dynamic problem (30), the initial conditions are given by

$$\boldsymbol{r}\big|_{t=0} = \boldsymbol{r}^\circ, \quad \boldsymbol{v}\big|_{t=0} = \boldsymbol{v}^\circ, \quad \mathbf{Q}\big|_{t=0} = \mathbf{Q}^\circ, \quad \boldsymbol{\omega}\big|_{t=0} = \boldsymbol{\omega}^\circ,$$

where $\boldsymbol{r}^\circ$, $\boldsymbol{v}^\circ$, $\mathbf{Q}^\circ$, $\boldsymbol{\omega}^\circ$ are prescribed initial values.

Under some conditions the equilibrium problem of a micropolar shell can be transformed to the system of equations with respect to the strain measures

$$\nabla \cdot \mathbf{P}_1 - \left(\mathbf{P}_1^T \cdot \mathbf{K}\right)_\times + \boldsymbol{f}^* = \mathbf{0}; \tag{32}$$

$$\nabla \cdot \mathbf{P}_2 - \left(\mathbf{P}_2^T \cdot \mathbf{K} + \mathbf{P}_1^T \cdot \mathbf{E}\right)_\times + \boldsymbol{c}^* = \mathbf{0}, \tag{33}$$

$$\Omega_2 \,:\, \boldsymbol{\nu} \cdot \mathbf{P}_1 = \boldsymbol{\varphi}^*, \qquad \Omega_4 \,:\, \boldsymbol{\nu} \cdot \mathbf{P}_2 = \boldsymbol{\eta}^*, \tag{34}$$

$$\boldsymbol{f}^* \triangleq \boldsymbol{f} \cdot \mathbf{Q}^T, \;\; \boldsymbol{c}^* \triangleq \boldsymbol{c} \cdot \mathbf{Q}^T, \quad \boldsymbol{\varphi}^* \triangleq \boldsymbol{\varphi} \cdot \mathbf{Q}^T, \;\; \boldsymbol{\eta}^* \triangleq \boldsymbol{\eta} \cdot \mathbf{Q}^T.$$

Let the vectors $\boldsymbol{f}^*$, $\boldsymbol{c}^*$, $\boldsymbol{\varphi}^*$, $\boldsymbol{\eta}^*$ be given as functions of the coordinates q^1, q^2. From the physical point of view it means that the shell is loaded by tracking forces and couples. Then Eqs. (32)–(34) depend on $\mathbf{E}$, $\mathbf{K}$ as the only independent fields.

3.3 On the Constitutive Equations

For an elastic shell the constitutive equations consist in dependence of the surface strain energy density on two strain measures. An example of constitutive equations is the model of a *physically linear isotropic shell* presented in Chróścielewski et al. (2004); Eremeyev and Pietraszkiewicz (2006); Eremeyev and Zubov (2008), the energy of which is given by the quadratic form

$$\begin{aligned} 2W &= \alpha_1 \mathrm{tr}^2 \mathbf{E}_{\|} + \alpha_2 \mathrm{tr}\, \mathbf{E}_{\|}^2 + \alpha_3 \mathrm{tr}\left(\mathbf{E}_{\|}\cdot\mathbf{E}_{\|}^T\right) + \alpha_4 \boldsymbol{n}\cdot\mathbf{E}^T\cdot\mathbf{E}\cdot\boldsymbol{n} \\ &+ \beta_1 \mathrm{tr}^2 \mathbf{K}_{\|} + \beta_2 \mathrm{tr}\, \mathbf{K}_{\|}^2 + \beta_3 \mathrm{tr}\left(\mathbf{K}_{\|}\cdot\mathbf{K}_{\|}^T\right) + \beta_4 \boldsymbol{n}\cdot\mathbf{K}^T\cdot\mathbf{K}\cdot\boldsymbol{n}, \\ \mathbf{E}_{\|} &\triangleq \mathbf{E}\cdot\mathbf{A}, \quad \mathbf{K}_{\|} \triangleq \mathbf{K}\cdot\mathbf{A}. \end{aligned} \tag{35}$$

In Eq. (35) the terms that are bilinear in $\mathbf{E}$ and $\mathbf{K}$ are not considered. It is a consequence of the fact that the wryness tensor $\mathbf{K}$ is a pseudo-tensor that changes the sign of the value when we apply the mirror reflection of the space. Note that the constitutive equations contain 8 parameters α_k, β_k $(k = 1, 2, 3, 4)$.

Chróścielewski et al. (2004) used the following relations for the elastic moduli appearing in Eqs. (35)

$$\begin{aligned} &\alpha_1 = C\nu, \quad \alpha_2 = 0, \quad \alpha_3 = C(1-\nu), \quad \alpha_4 = \alpha_s C(1-\nu), \\ &\beta_1 = D\nu, \quad \beta_2 = 0, \quad \beta_3 = D(1-\nu), \quad \beta_4 = \alpha_t D(1-\nu), \\ &C = \frac{Eh}{1-\nu^2}, \quad D = \frac{Eh^3}{12(1-\nu^2)}, \end{aligned} \tag{36}$$

where E and ν are the Young modulus and the Poisson ratio of the bulk material, respectively, α_s and α_t are dimensionless shear correction factors, while h is the shell thickness. α_s is the shear correction factor introduced in the plate theory by Reissner (1944) ($\alpha_s = 5/6$) or by Mindlin (1951) ($\alpha_s = \pi^2/12$). The parameter α_t plays the same role for the couple stresses. The value $\alpha_t = 0.7$ was proposed by Pietraszkiewicz (1979b,a), see also Chróścielewski et al. (2010). In Chróścielewski et al. (2004, 2010); Chróścielewski and Witkowski (2010) the influence of α_s and α_t on the solution is investigated numerically for several boundary-value problems.

3.4 Compatibility Conditions

Let us consider how to determine the position-vector $\boldsymbol{r}(q^1, q^2)$ of σ from the surface stretch tensor $\mathbf{E}$ and micro-rotation tensor $\mathbf{Q}$, which are assumed to be given as continuously differentiable functions on Σ. By using the equation

$$\mathbf{F} = (\mathbf{E} + \mathbf{A}) \cdot \mathbf{Q} \tag{37}$$

the problem is reduced to

$$\nabla \boldsymbol{r} = \mathbf{F}. \tag{38}$$

The necessary and sufficient condition for solvability of Eq. (38) is given by the relation

$$\nabla\cdot(\mathbf{e}\cdot\mathbf{F}) = \mathbf{0}, \quad \mathbf{e} \stackrel{\triangle}{=} -\mathbf{I}\times\boldsymbol{N}, \tag{39}$$

which we call the compatibility condition for the distortion tensor $\mathbf{F}$. Here $\mathbf{e}$ is the skew-symmetric discriminant tensor on the surface Σ. For a simply-connected region Σ, if the condition (39) is satisfied, the vector field $\boldsymbol{r}$ may be deduced from Eq. (38) only up to an additive vector.

Let us consider a more complex problem of determination of both the translations and rotations of the micropolar shell from the given fields of $\mathbf{E}$ and $\mathbf{K}$. At first, let us deduce the field $\mathbf{Q}(q^1, q^2)$ by using the system of equations following from definition (20) of $\mathbf{K}$

$$\frac{\partial\mathbf{Q}}{\partial q^\alpha} = -\boldsymbol{K}_\alpha\times\mathbf{Q}, \quad \boldsymbol{K}_\alpha \stackrel{\triangle}{=} \boldsymbol{R}_\alpha\cdot\mathbf{K}. \tag{40}$$

The integrability conditions for the system (40) are given by the relation

$$\frac{\partial\boldsymbol{K}_\alpha}{\partial q^\beta} - \frac{\partial\boldsymbol{K}_\beta}{\partial q^\alpha} = \boldsymbol{K}_\alpha\times\boldsymbol{K}_\beta \quad (\alpha,\beta=1,2). \tag{41}$$

Equations (41) were obtained by Pietraszkiewicz (1979a, 1989); Libai and Simmonds (1983) as the conditions of existence of the rotation field of the shell. They may be written in the following coordinate-free form

$$\nabla\cdot(\mathbf{e}\cdot\mathbf{K}) + \mathbf{K}^\perp\cdot\boldsymbol{n} = \mathbf{0}, \tag{42}$$

$$\mathbf{K}^\perp \stackrel{\triangle}{=} \frac{1}{2}(\boldsymbol{K}_\alpha\times\boldsymbol{K}_\beta)\otimes\left(\boldsymbol{R}^\alpha\times\boldsymbol{R}^\beta\right) = \mathbf{K}^2 - \mathbf{K}\mathrm{tr}\,\mathbf{K} + \frac{1}{2}\left(\mathrm{tr}^{\,2}\mathbf{K} - \mathrm{tr}\,\mathbf{K}^2\right)\mathbf{I}.$$

Using Eqs. (37) and (20) we transform the compatibility condition (39) into the coordinate-free form

$$\nabla\cdot(\mathbf{e}\cdot\mathbf{E}) + \left[(\mathbf{E}+\mathbf{A})^T\cdot\mathbf{e}\cdot\mathbf{K}\right]_\times = \mathbf{0}. \tag{43}$$

Two coordinate-free vector equations (42), (43) are the compatibility conditions for the nonlinear micropolar shell. These conditions and the system of equations (32)–(34) constitute the complete boundary-value problem of statics of micropolar shells expressed entirely in terms of the surface strain measures $\mathbf{E}$ and $\mathbf{K}$.

3.5 Variational Statements

The presented above static and dynamic problems of the micropolar shell theory have corresponding variational statements. Two of them for statics and one for dynamics are presented below.

Lagrange-Type Principle Let us assume that the external forces and couples are conservative. In the Lagrange-type variational principle

$$\delta \mathcal{E}_1 = 0$$

we use the total energy functional

$$\mathcal{E}_1[\boldsymbol{r}, \mathbf{Q}] = \iint\limits_{\Sigma} W \, \mathrm{d}\Sigma - \mathcal{A}[\boldsymbol{r}, \mathbf{Q}], \tag{44}$$

where $\mathcal{A}$ is the potential of the external loads.

Here the translations and the rotations have to satisfy the kinematic boundary conditions $(26)_1$ and $(26)_3$ on Ω_1 and Ω_3, respectively. The stationarity of $\mathcal{E}_1$ is equivalent to the equilibrium equations (24), (25) and the static boundary conditions $(26)_2$ and $(26)_4$ on Ω_2 and Ω_4.

Hu-Washizu-Type Principle For this principle the functional is given by

$$\begin{aligned}
\mathcal{E}_2[\boldsymbol{r}, \mathbf{Q}, \mathbf{E}, \mathbf{K}, \mathbf{D}, \mathbf{P}_2] &= \iint\limits_{\Sigma} \Bigg\{ W(\mathbf{E}, \mathbf{K}) - \mathbf{D} \bullet (\mathbf{E} \cdot \mathbf{Q} - \nabla \boldsymbol{r}) \\
&- \mathbf{P}_2 \bullet \left[\mathbf{K} - \frac{1}{2} \boldsymbol{r}^{\alpha} \otimes \left(\frac{\partial \mathbf{Q}}{\partial q^{\alpha}} \cdot \mathbf{Q}^T \right)_{\times} \right] \Bigg\} \mathrm{d}\Sigma \\
&- \int\limits_{\Omega_1} \boldsymbol{\nu} \cdot \mathbf{D} \cdot (\boldsymbol{r} - \boldsymbol{\rho}) \, \mathrm{d}s - \mathcal{A}[\boldsymbol{r}, \mathbf{Q}].
\end{aligned}$$

From the condition $\delta \mathcal{E}_2 = 0$ the equilibrium equations (24) and (25), the constitutive equations, and the relations (20) can be deduced. For this principle the natural boundary conditions are given by the relations $(26)_1$, $(26)_2$ and $(26)_4$, respectively.

Several other variational statements are given in Eremeyev and Zubov (2008). Mixed-type variational functionals are constructed in Chróścielewski et al. (2004). They are used for the development of a family of finite elements with six degrees of freedom in each node. A number of nonlinear simulations of complex multifold shell structures were performed on the base of these elements.

Hamilton-Type Principle The *kinetic energy of micropolar shells* can be expressed as

$$\mathcal{K} = \iint\limits_{\Sigma} \rho K(\boldsymbol{v}, \boldsymbol{\omega}) \, \mathrm{d}\Sigma. \tag{45}$$

It is obvious that we should assume the kinetic energy to be a positive definite function that imposes some restriction on the form of the inertia tensors.

The Hamilton principle is a variational principle of dynamics. In real motion, the functional

$$\mathcal{E}_3[\boldsymbol{R}, \mathbf{H}] = \int\limits_{t_0}^{t_1} (\mathcal{K} - \mathcal{E}_1)\, \mathrm{d}t \tag{46}$$

takes a stationary value on the set of all possible shell motions that at the range t_0, t_1 take given values of the real motion values and satisfy the kinematic boundary values. In other words, its first variation on a real motion is zero. From condition $\delta\mathcal{E}_3 = 0$, Eqs. (30) can be established.

3.6 Linear Theory of Micropolar Shells

Let us suppose that the strains are small. Then we can simplify the equations of the shell theory significantly. In the geometrically linear case we do not distinguish Eulerian and Lagrangian descriptions. The difference of surfaces σ and Σ is infinitesimal. It is not necessary to introduce two operators ∇ and ∇_Σ as well as earlier different types of stress and couple stress tensors. Let us introduce the *vector of infinitesimal translation* $\boldsymbol{u}$ and the *vector of infinitesimal rotation* $\boldsymbol{\vartheta}$ such that there hold

$$\boldsymbol{r} = \boldsymbol{R} + \boldsymbol{u}, \quad \mathbf{Q} \approx \mathbf{I} - \mathbf{I} \times \boldsymbol{\vartheta}, \tag{47}$$

$$\|\boldsymbol{u}\| \ll 1, \quad \|\boldsymbol{\vartheta}\| \ll 1, \quad \|\nabla \boldsymbol{u}\| \ll 1, \quad \|\nabla \boldsymbol{\vartheta}\| \ll 1. \tag{48}$$

In Eqs. (47) the last formula follows from the representation of a proper orthogonal tensor through the finite rotation vector (18) if $|\boldsymbol{\theta}| \ll 1$. In this case $\boldsymbol{\theta} \approx \boldsymbol{\vartheta}$.

Up to the linear addendum, the strain measures $\mathbf{E}$ and $\mathbf{K}$ can be expressed in terms of the *linear stretch tensor* and *linear wryness tensor* $\boldsymbol{\epsilon}$ and $\boldsymbol{\kappa}$

$$\mathbf{E} \approx \boldsymbol{\epsilon}, \quad \mathbf{K} \approx \boldsymbol{\kappa}, \quad \boldsymbol{\epsilon} \triangleq \nabla \boldsymbol{u} + \mathbf{A} \times \boldsymbol{\vartheta}, \quad \boldsymbol{\kappa} \triangleq \nabla \boldsymbol{\vartheta}. \tag{49}$$

The tensors $\boldsymbol{\epsilon}$ and $\boldsymbol{\kappa}$ are used in the linear theory of micropolar shells, cf. Zhilin (1976) and Zubov (1997). Assuming Eqs. (48) in the linear shell theory the stress tensors $\mathbf{D}$, $\mathbf{P}_1$, $\mathbf{T}$ and the couple tensors $\mathbf{G}$, $\mathbf{P}_2$, $\mathbf{M}$ coincide. In what follows we will denote the stress tensor by $\mathbf{T}$ and the couple stress tensor by $\mathbf{M}$.

The constitutive equations of an elastic shell can be represented through the surface strain energy density $W = W(\boldsymbol{\epsilon}, \boldsymbol{\kappa})$ as follows

$$\mathbf{T} = \frac{\partial W}{\partial \boldsymbol{\epsilon}}, \quad \mathbf{M} = \frac{\partial W}{\partial \boldsymbol{\kappa}}. \tag{50}$$

In the linear theory the equilibrium equations take the form

$$\nabla\cdot \mathbf{T} + \boldsymbol{f} = \mathbf{0}, \quad \nabla\cdot \mathbf{M} + \mathbf{T}_{\times} + \boldsymbol{c} = \mathbf{0}, \tag{51}$$

whereas the boundary conditions are transformed to

$$\begin{aligned} \Omega_1 &: \quad \boldsymbol{u} = \boldsymbol{u}_0(s), \\ \Omega_2 &: \quad \boldsymbol{\nu} \cdot \mathbf{T} = \boldsymbol{\varphi}(s), \\ \Omega_3 &: \quad \boldsymbol{\vartheta} = \boldsymbol{\vartheta}_0(s), \\ \Omega_4 &: \quad \boldsymbol{\nu} \cdot \mathbf{M} = \boldsymbol{\eta}(s), \end{aligned} \tag{52}$$

where $\boldsymbol{u}_0(s)$ and $\boldsymbol{\vartheta}_0(s)$ are given functions of the arc length that respectively define the displacements and rotations on a part of the shell contour.

If the strains are small, an example of the constitutive equation is the following quadratic form

$$\begin{aligned} 2W &= \alpha_1 \mathrm{tr}^2 \boldsymbol{\epsilon}_{\|} + \alpha_2 \mathrm{tr}\, \boldsymbol{\epsilon}_{\|}^2 + \alpha_3 \mathrm{tr} \left(\boldsymbol{\epsilon}_{\|} \cdot \boldsymbol{\epsilon}_{\|}^T\right) + \alpha_4 \boldsymbol{N} \cdot \boldsymbol{\epsilon}^T \cdot \boldsymbol{\epsilon} \cdot \boldsymbol{N} \\ &+ \beta_1 \mathrm{tr}^2 \boldsymbol{\kappa}_{\|} + \beta_2 \mathrm{tr}\, \boldsymbol{\kappa}_{\|}^2 + \beta_3 \mathrm{tr} \left(\boldsymbol{\kappa}_{\|} \cdot \boldsymbol{\kappa}_{\|}^T\right) + \beta_4 \boldsymbol{N} \cdot \boldsymbol{\kappa}^T \cdot \boldsymbol{\kappa} \cdot \boldsymbol{N}. \end{aligned} \tag{53}$$

This form describes *physically linear isotropic shells*. Here α_k and β_k are elastic moduli ($k = 1, 2, 3, 4$) and

$$\boldsymbol{\epsilon}_{\|} \stackrel{\triangle}{=} \boldsymbol{\epsilon} \cdot \mathbf{A}, \qquad \boldsymbol{\kappa}_{\|} \stackrel{\triangle}{=} \boldsymbol{\kappa} \cdot \mathbf{A}.$$

Considering Eqs. (50) and (53), the stress tensor and the couple stress tensor are expressed by the formulas

$$\mathbf{T} = \alpha_1 \mathbf{A} \mathrm{tr}\, \boldsymbol{\epsilon}_{\|} + \alpha_2 \boldsymbol{\epsilon}_{\|}^T + \alpha_3 \boldsymbol{\epsilon}_{\|} + \alpha_4 \boldsymbol{\epsilon} \cdot \boldsymbol{N} \otimes \boldsymbol{N}, \tag{54}$$

$$\mathbf{M} = \beta_1 \mathbf{A} \mathrm{tr}\, \boldsymbol{\kappa}_{\|} + \beta_2 \boldsymbol{\kappa}_{\|}^T + \beta_3 \boldsymbol{\kappa}_{\|} + \beta_4 \boldsymbol{\kappa} \cdot \boldsymbol{N} \otimes \boldsymbol{N}. \tag{55}$$

Supplemented with Eqs. (51) and (52), the linear constitutive equations (54), (55) close the linear boundary-value problem with respect to the fields of translations and rotations. It describes the equilibrium of the micropolar shell when strains are infinitesimal.

When the strains are small, the Lagrange-type variational principle (44) is transformed to the following form

$$\mathcal{E}_1[\boldsymbol{u}, \boldsymbol{\vartheta}] = \iint_{\Sigma} W(\boldsymbol{\epsilon}, \boldsymbol{\kappa})\, \mathrm{d}\Sigma - \mathcal{A}[\boldsymbol{u}, \boldsymbol{\vartheta}], \tag{56}$$

where the potential of the external loads $\mathcal{A}[\boldsymbol{u}, \boldsymbol{\vartheta}]$ is defined by the equation

$$\mathcal{A}[\boldsymbol{u}, \boldsymbol{\vartheta}] \triangleq \iint_{\Sigma} (\boldsymbol{f} \cdot \boldsymbol{u} + \boldsymbol{c} \cdot \boldsymbol{\vartheta}) \, \mathrm{d}\Sigma + \int_{\Omega_2} \boldsymbol{\varphi} \cdot \boldsymbol{u} \, \mathrm{d}s + \int_{\Omega_4} \boldsymbol{\eta} \cdot \boldsymbol{\vartheta} \, \mathrm{d}s.$$

Let functional (56) be given on a of twice differentiable fields of displacements and rotations of the surface Σ that satisfy the boundary conditions $(52)_1$ and $(52)_3$ on Ω_1 and Ω_3, respectively. It is easy to check that the condition of the functional to have a stationary value is equivalent to the equilibrium equations (51) and the boundary conditions $(52)_2$ and $(52)_4$ on Ω_2 and Ω_4, respectively. Let us note that when the strains are small and the form $W(\boldsymbol{\epsilon}, \boldsymbol{\kappa})$ is positive definite, the Lagrange-type variational principle is a minimal principle, this means functional (56) takes a minimal value on the equilibrium solution. Existence and uniqueness of weak solutions to the boundary value problems of statics and dynamics of micropolar linear shells were established by Eremeyev and Lebedev (2011).

In the linear theory it is valid a variational principle for free oscillations. By linearity, eigen-solutions are proportional to $e^{i\Omega t}$

$$\boldsymbol{u} = \boldsymbol{u}^\circ e^{i\Omega t}, \quad \boldsymbol{\vartheta} = \boldsymbol{\vartheta}^\circ e^{i\Omega t}.$$

Now the variational *Rayleigh-type principle* can be formulated: the forms of the eigen-oscillations of the shell are stationary points of the strain energy functional

$$\mathcal{E}_4[\boldsymbol{u}^\circ, \boldsymbol{\vartheta}^\circ] = \iint_{\Sigma} W(\boldsymbol{\epsilon}^\circ, \boldsymbol{\kappa}^\circ) \, \mathrm{d}\Sigma, \tag{57}$$

where

$$\boldsymbol{\epsilon}^\circ = \nabla \boldsymbol{u}^\circ + \mathbf{A} \times \boldsymbol{\vartheta}^\circ, \qquad \boldsymbol{\kappa}^\circ = \nabla \boldsymbol{\vartheta}^\circ,$$

on a set of functions that satisfy the following conditions

$$\Omega_1 : \quad \boldsymbol{u}^\circ = \mathbf{0}, \quad \Omega_3 : \quad \boldsymbol{\vartheta}^\circ = \mathbf{0} \tag{58}$$

and the restriction

$$\iint_{\Sigma} \rho K \left(\boldsymbol{u}^\circ, \boldsymbol{\vartheta}^\circ\right) \, \mathrm{d}\Sigma = 1. \tag{59}$$

Functions $\boldsymbol{u}^\circ$, $\boldsymbol{\vartheta}^\circ$ represent the amplitudes of oscillations of the translations and small rotations.

The Rayleigh-type variational principle is equivalent to the stationary principle for the *Rayleigh-type quotient*

$$\mathcal{R}[\boldsymbol{u}^\circ, \boldsymbol{\vartheta}^\circ] = \frac{\displaystyle\iint_\Sigma W(\boldsymbol{\epsilon}^\circ, \boldsymbol{\kappa}^\circ)\,\mathrm{d}\Sigma}{\displaystyle\iint_\Sigma \rho K\left(\boldsymbol{u}^\circ, \boldsymbol{\vartheta}^\circ\right)\,\mathrm{d}\Sigma}, \tag{60}$$

that is defined on kinematically admissible functions $\boldsymbol{u}^\circ$, $\boldsymbol{\vartheta}^\circ$. Note that the least squared eigenfrequency for the shell corresponds to the minimal value of $\mathcal{R}$

$$\Omega^2_{\mathrm{min}} = \inf \mathcal{R}[\boldsymbol{u}^\circ, \boldsymbol{\vartheta}^\circ]$$

on $\boldsymbol{u}^\circ$, $\boldsymbol{\vartheta}^\circ$ satisfying (58). Using the Courant minimax principle, the Rayleigh-type quotient (60) allows us to estimate the values of higher eigenfrequencies, see Courant and Hilbert (1991). For this we should consider $\mathcal{R}$ on the set of functions that are orthogonal to the previous modes of eigen-oscillations in some sense, see Eremeyev and Lebedev (2011) for details.

3.7 Constitutive Restrictions for Micropolar Shells

In the field of nonlinear elasticity there are well known so-called constitutive restrictions. They are the strong ellipticity condition, the Hadamard inequality, the Generalized Coleman-Noll condition (GCN-condition), and some others, see Truesdell and Noll (1965); Truesdell (1984, 1991). Each of them plays some role in non-linear elasticity. Here we formulate similar constitutive restrictions in the general nonlinear theory of micropolar shells.

The linear micropolar elastic shell theory gives a simple example of a constitutive restriction. In this case the surface strain energy density is a quadratic form of both the linear stretch tensor and the linear wryness tensor. Assuming that

$$W(\boldsymbol{\epsilon}, \boldsymbol{\kappa}) > 0, \quad \forall\, \boldsymbol{\epsilon}, \boldsymbol{\kappa} \neq \mathbf{0} \tag{61}$$

we obtain the following set of inequalities

$$\begin{array}{llll} 2\alpha_1 + \alpha_2 + \alpha_3 > 0, & \alpha_2 + \alpha_3 > 0, & \alpha_3 - \alpha_2 > 0, & \alpha_4 > 0, \\ 2\beta_1 + \beta_2 + \beta_3 > 0, & \beta_2 + \beta_3 > 0, & \beta_3 - \beta_2 > 0, & \beta_4 > 0. \end{array} \tag{62}$$

Hence the inequality (61) and following from this the inequalities for the elastic moduli of an isotropic shell (62) are the simplest example of additional inequalities in the shell theory. When (61) is violated, it leads to a number of pathological consequences such as non-uniqueness of the solution of boundary value problems in the linear shell theory that implies that for some loads no solution exist. In the case of finite strains the positive definiteness of the surface strain energy density $W(\mathbf{E}, \mathbf{K})$ cannot guarantee the required properties of constitutive equations.

The Coleman-Noll inequality is one of the well-known constitutive inequalities in the nonlinear elasticity, see Truesdell and Noll (1965); Truesdell (1984, 1991). In the nonlinear elasticity the differential form of the Coleman-Noll inequality (so-called GCN-condition) expresses the property that the increment of the elastic energy density under arbitrary infinitesimal non-zero pure strains for any arbitrary reference configuration should be positive. The generalization of the GCN-condition in the case of micropolar shells is obtained by Eremeyev and Zubov (2007). It is given by

$$\left.\frac{\mathrm{d}^2}{\mathrm{d}\tau^2} W\left(\mathbf{E} + \tau\boldsymbol{\epsilon}, \mathbf{K} + \tau\boldsymbol{\kappa}\right)\right|_{\tau=0} > 0 \quad \forall\, \boldsymbol{\epsilon} \neq \mathbf{0}, \quad \boldsymbol{\kappa} \neq \mathbf{0}. \tag{63}$$

Let us note that the inequality (63) satisfies the principle of material frame-indifference and can serve as a constitutive inequality for elastic shells.

Other important constitutive inequalities in the nonlinear elasticity are the strong ellipticity condition and its weak form known as the Hadamard inequality. Following the theory of systems of partial differential equations (PDE) presented in Lions and Magenes (1968); Fichera (1972); Hörmander (1976); Agranovich (1997), the strong ellipticity condition of the equilibrium equations (24) can be formulated as follows

$$\left.\frac{\mathrm{d}^2}{\mathrm{d}\tau^2} W\left(\mathbf{E} + \tau\boldsymbol{\nu}\otimes\boldsymbol{a}, \mathbf{K} + \tau\boldsymbol{\nu}\otimes\boldsymbol{b}\right)\right|_{\tau=0} > 0 \quad \forall\, \boldsymbol{\nu},\, \boldsymbol{a},\, \boldsymbol{b} \neq \mathbf{0},\, \boldsymbol{\nu}\cdot\boldsymbol{N} = 0. \tag{64}$$

A weak form of the inequality (64) is an analogue of the *Hadamard inequality* for the shell. These inequalities are examples of possible restrictions of the constitutive equations of elastic shells at finite deformations. As in the case of simple materials, a violation of the inequality (64) means the possibility of existence of non-smooth solutions of the equilibrium equations (24).

Comparing the condition of strong ellipticity (64) and the Coleman-Noll inequality (63) one can see that the latter implies the former. Indeed, the inequality (63) holds for any tensors $\boldsymbol{\epsilon}$ and $\boldsymbol{\kappa}$. Note that the tensors $\boldsymbol{\epsilon}$ and $\boldsymbol{\kappa}$ may be non-symmetric, in general. If we substitute in inequality (63) the relations $\boldsymbol{\epsilon} = \boldsymbol{\nu}\otimes\boldsymbol{a}$ and $\boldsymbol{\kappa} = \boldsymbol{\nu}\otimes\boldsymbol{b}$ then we immediately obtain the inequality (64). Thus, the strong ellipticity condition is the special case of the Coleman–Noll inequality.

For the constitutive equations of an isotropic micropolar shell (35) the inequality (64) reduces to the system of inequalities

$$\begin{aligned} &\alpha_3 > 0, \quad \alpha_1 + \alpha_2 + \alpha_3 > 0, \quad \alpha_4 > 0, \\ &\beta_3 > 0, \quad \beta_1 + \beta_2 + \beta_3 > 0, \quad \beta_4 > 0. \end{aligned} \tag{65}$$

For a linear isotropic shell, the inequalities (65) provide strong ellipticity of equilibrium equations (51), they are more weak in comparison with the condition of positive definiteness of (62).

In the theory of partial differential equations there is another definition of ellipticity, called the *ellipticity in the sense of Petrovsky*, see Agranovich (1997). Now we introduce this definition within the framework of the micropolar shell theory. We assume singular time-independent curves of second order. Suppose on the shell surface Σ there exists a curve γ on which a jump in the second derivatives of the position-vector $\boldsymbol{r}$ or microrotation tensor $\mathbf{Q}$ happens. We will call such a jump the *weak discontinuity*. For example, the curvature of σ is determined through second derivatives of $\boldsymbol{r}$ so such discontinuity can be exhibited in the form of cusps of the shell surface.

After transformations of the equilibrium equations (24) we derive the ellipticity condition in the form

$$\det \mathbb{A}(\boldsymbol{\nu}) \neq 0, \tag{66}$$

where the matrix $\mathbb{A}$ is given by the relation

$$\mathbb{A} = \begin{bmatrix} \dfrac{\partial^2 W}{\partial \mathbf{E} \partial \mathbf{E}} \circledast \boldsymbol{\nu} & \dfrac{\partial^2 W}{\partial \mathbf{E} \partial \mathbf{K}} \circledast \boldsymbol{\nu} \\ \\ \dfrac{\partial^2 W}{\partial \mathbf{K} \partial \mathbf{E}} \circledast \boldsymbol{\nu} & \dfrac{\partial^2 W}{\partial \mathbf{K} \partial \mathbf{L}} \circledast \boldsymbol{\nu} \end{bmatrix}, \tag{67}$$

and the operation $\circledast$ is defined by the rule

$$\mathbb{G} \circledast \boldsymbol{\nu} \equiv G_{klmn} \nu_l \nu_n \boldsymbol{i}_k \otimes \boldsymbol{i}_m.$$

for any arbitrary forth-order tensor $\mathbb{G}$ and vector $\boldsymbol{\nu}$ represented in a Cartesian basis $\boldsymbol{i}_k$ $(k = 1, 2, 3)$. $\mathbb{A}$ plays the role of an acoustic tensor in the micropolar shell theory.

As an example we consider the conditions (66) for the constitutive equations of a physically linear shell (35). Here the conditions (66) reduce to the inequalities

$$\begin{aligned} &\alpha_3 \neq 0, \quad \alpha_1 + \alpha_2 + \alpha_3 \neq 0, \quad \alpha_4 \neq 0, \\ &\beta_3 \neq 0, \quad \beta_1 + \beta_2 + \beta_3 \neq 0, \quad \beta_4 \neq 0. \end{aligned}$$

Condition (66) is also called the *ordinary ellipticity condition*, see Knowles and Sternberg (1976, 1978); Zee and Sternberg (1983). It is weaker than the strong ellipticity condition (64).

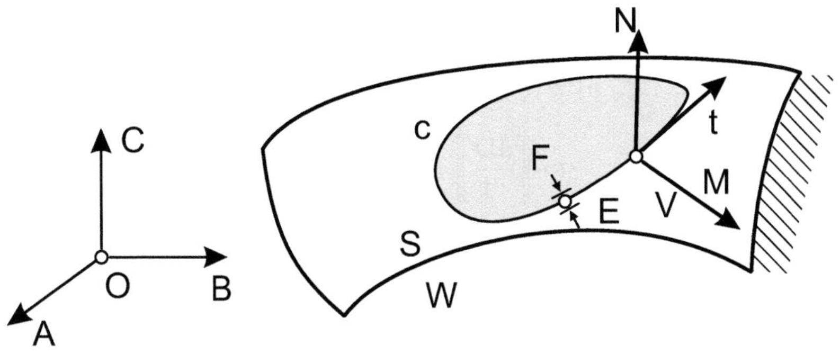

Figure 3. Singular curve

3.8 Strong Ellipticity Condition and Acceleration Waves

Following Eremeyev (2005); Eremeyev and Zubov (2007, 2008); Altenbach et al. (2011), we show that the inequality (64) coincides with the conditions for propagation of acceleration waves in a micropolar shell. We consider a weak discontinuity motion of a shell, that is when some kinematic and dynamic quantities on a certain smooth curve $C(t)$ may be discontinuous. We assume that the limit values of these quantities exist on C and that they are generally different on the opposite sides of C. The jump of an arbitrary quantity $\boldsymbol{\Psi}$ on C is denoted by $[\![\boldsymbol{\Psi}]\!] = \boldsymbol{\Psi}^+ - \boldsymbol{\Psi}^-$ (Fig. 3).

The acceleration wave (weak-discontinuity wave or second-order singular curve) in a shell is a moving singular curve C on which the second derivatives (with respect to the spatial coordinates and time) of the radius-vector $\boldsymbol{r}$ and the microrotation tensor $\mathbf{Q}$ are discontinuous, while the quantities themselves and their first derivatives are continuous. This means

$$[\![\mathbf{F}]\!] = \mathbf{0}, \quad [\![\nabla\,\mathbf{Q}]\!] = \mathbf{0}, \quad [\![\boldsymbol{v}]\!] = \mathbf{0}, \quad [\![\boldsymbol{\omega}]\!] = \mathbf{0} \tag{68}$$

are valid on C. According to Eqs. (20), the stretch tensor $\mathbf{E}$ and the wryness tensor $\mathbf{K}$ are continuous near C, and, with respect to constitutive equations (25), there are no jumps in tensors $\mathbf{D}$ and $\mathbf{G}$. The application of the Maxwell theorem to continuous fields of velocities $\boldsymbol{v}$ and $\boldsymbol{\omega}$, surface stress tensor $\mathbf{D}$, and the surface couple stress tensor $\mathbf{G}$ yields a system of equations that relate the jumps of their derivatives with respect to the spatial coordinates and time, see Truesdell (1984, 1991),

$$\left[\!\!\left[\frac{\mathrm{d}\boldsymbol{v}}{\mathrm{d}t}\right]\!\!\right] = -V\boldsymbol{a}, \quad [\![\nabla\, \boldsymbol{v}]\!] = \boldsymbol{\nu}\otimes\boldsymbol{a}, \quad \left[\!\!\left[\frac{\mathrm{d}\boldsymbol{\omega}}{\mathrm{d}t}\right]\!\!\right] = -V\boldsymbol{b}, \quad [\![\nabla\, \boldsymbol{\omega}]\!] = \boldsymbol{\nu}\otimes\boldsymbol{b}, \tag{69}$$
$$V\,[\![\nabla\cdot \mathbf{D}]\!] = -\boldsymbol{\nu}\cdot\left[\!\!\left[\frac{\mathrm{d}\mathbf{D}}{\mathrm{d}t}\right]\!\!\right], \quad V\,[\![\nabla\cdot \mathbf{G}]\!] = -\boldsymbol{\nu}\cdot\left[\!\!\left[\frac{\mathrm{d}\mathbf{G}}{\mathrm{d}t}\right]\!\!\right].$$

Here $\boldsymbol{a}$ and $\boldsymbol{b}$ are the vector amplitudes for the jumps of the linear and angular accelerations, $\boldsymbol{\nu}$ is the unit normal vector to C such that $\boldsymbol{n}\cdot\boldsymbol{\nu} = 0$, $\boldsymbol{\tau}$ is the unit tangent vector to C such that $\boldsymbol{n}\cdot\boldsymbol{\tau} = \boldsymbol{\nu}\cdot\boldsymbol{\tau} = 0$, and V is the velocity of the surface C in the direction $\boldsymbol{\nu}$. If external forces and couples are continuous, the relations

$$[\![\nabla\cdot \mathbf{D}]\!] = \rho\left[\!\!\left[\frac{\mathrm{d}\boldsymbol{K}_1}{\mathrm{d}t}\right]\!\!\right], \quad [\![\nabla\cdot \mathbf{G}]\!] = \rho\left[\!\!\left[\frac{\mathrm{d}\boldsymbol{K}_2}{\mathrm{d}t}\right]\!\!\right]$$

follow immediately from the equations of motion (30).

Differentiating constitutive Eqs. (25) and using equations (68) and (69), we express latter relations only in terms of the vector amplitudes $\boldsymbol{a}$ and $\boldsymbol{b}$

$$\boldsymbol{\nu}\cdot\frac{\partial^2 W}{\partial\mathbf{E}\partial\mathbf{E}}\bullet\left(\boldsymbol{\nu}\otimes\boldsymbol{a}\cdot\mathbf{Q}^T\right) + \boldsymbol{\nu}\cdot\frac{\partial^2 W}{\partial\mathbf{E}\partial\mathbf{K}}\bullet\left(\boldsymbol{\nu}\otimes\boldsymbol{b}\cdot\mathbf{Q}^T\right)$$
$$= \rho V^2\left[\boldsymbol{a}\cdot\mathbf{Q}^T + \left(\mathbf{Q}\cdot\boldsymbol{\Theta}_1^T\cdot\mathbf{Q}^T\right)\cdot\left(\boldsymbol{b}\cdot\mathbf{Q}^T\right)\right],$$
$$\boldsymbol{\nu}\cdot\frac{\partial^2 W}{\partial\mathbf{K}\partial\mathbf{E}}\bullet\left(\boldsymbol{\nu}\otimes\boldsymbol{a}\cdot\mathbf{Q}^T\right) + \boldsymbol{\nu}\cdot\frac{\partial^2 W}{\partial\mathbf{K}\partial\mathbf{K}}\bullet\left(\boldsymbol{\nu}\otimes\boldsymbol{b}\cdot\mathbf{Q}^T\right)$$
$$= \rho V^2\left[\left(\mathbf{Q}\cdot\boldsymbol{\Theta}_1\cdot\mathbf{Q}^T\right)\cdot\left(\boldsymbol{a}\cdot\mathbf{Q}^T\right) + \left(\mathbf{Q}\cdot\boldsymbol{\Theta}_2\cdot\mathbf{Q}^T\right)\cdot\left(\boldsymbol{b}\cdot\mathbf{Q}^T\right)\right].$$

These relations can also be written in a more compact form

$$\mathbb{A}(\boldsymbol{\nu})\cdot\boldsymbol{\xi} = \rho V^2\mathbb{B}\cdot\boldsymbol{\xi}, \tag{70}$$

where $\mathbb{A}$ is defined by (67), $\boldsymbol{\xi} \triangleq (\mathbf{Q}\cdot\boldsymbol{a}, \mathbf{Q}\cdot\boldsymbol{b})$, and the matrix $\mathbb{B}$ is given by

$$\mathbb{B} = \begin{bmatrix} \mathbf{I} & \mathbf{Q}\cdot\boldsymbol{\Theta}_1^T\cdot\mathbf{Q}^T \\ \mathbf{Q}\cdot\boldsymbol{\Theta}_1\cdot\mathbf{Q}^T & \mathbf{Q}\cdot\boldsymbol{\Theta}_2\cdot\mathbf{Q}^T \end{bmatrix}.$$

Thus, the problem of acceleration wave propagation in the shell has been reduced to the spectral problem given by the algebraic equations (70). Owing to the existence of the potential-energy function W, $\mathbb{A}(\boldsymbol{\nu})$ is symmetric. Matrix $\mathbb{B}$ is also symmetric and positive definite. This property enables to formulate an analogue of the Fresnel–Hadamard–Duhem theorem for the elastic shell.

Theorem 3.1. *The squares of the velocities of a second order singular curve (acceleration wave) in the elastic shell are real for arbitrary propagation directions specified by the vector $\boldsymbol{\nu}$.*

Note that the positive definiteness of $\mathbb{A}(\boldsymbol{\nu})$, which is necessary and sufficient for the wave velocity V to be real, coincides with the strong ellipticity inequality (64).

Theorem 3.2. *The condition for existence of a acceleration wave for all directions of propagations in a micropolar thermoelastic shell is equivalent to the condition of strong ellipticity of the equilibrium equations of the shell.*

As an example we present the solution of problem (70) for a physically linear shell. Following Pietraszkiewicz (2011) we suppose that $\boldsymbol{\Theta}_1$ is zero and $\boldsymbol{\Theta}_2$ is the spherical part of the tensor, that is $\boldsymbol{\Theta}_2 = \jmath\mathbf{I}$, where $\jmath$ is the rotatory inertia measure. Let us assume that the inequalities (65) are valid. Then solutions of equation (70) are

$$
\begin{aligned}
U_1 &= \sqrt{\frac{\alpha_3}{\rho}}, && \boldsymbol{\xi}_1 = (\boldsymbol{\tau}, \mathbf{0}),\\
U_2 &= \sqrt{\frac{\alpha_1+\alpha_2+\alpha_3}{\rho}}, && \boldsymbol{\xi}_2 = (\boldsymbol{\nu}, \mathbf{0}),\\
U_3 &= \sqrt{\frac{\alpha_4}{\rho}}, && \boldsymbol{\xi}_3 = (\boldsymbol{n}, \mathbf{0}),\\
U_4 &= \sqrt{\frac{\beta_3}{\rho\jmath}}, && \boldsymbol{\xi}_4 = (\mathbf{0}, \boldsymbol{\tau}),\\
U_5 &= \sqrt{\frac{\beta_1+\beta_2+\beta_3}{\rho\jmath}}, && \boldsymbol{\xi}_5 = (\mathbf{0}, \boldsymbol{\nu}),\\
U_6 &= \sqrt{\frac{\beta_4}{\rho\jmath}}, && \boldsymbol{\xi}_6 = (\mathbf{0}, \boldsymbol{n}).
\end{aligned}
\tag{71}
$$

Solutions (71) describe transversal and longitudinal waves of acceleration and microrotation accelerations.

3.9 Principle Peculiarities of the Micropolar Shell Theory

Let us summarize principal peculiarities of the shell theory under consideration:

1. The shell equilibrium equation constitute a nonlinear system of partial differential equations. In general, the system is elliptic but in some circumstances the ellipticity condition can be violated.

2. General theorems of existence of equilibrium or dynamic solutions do not occur. Moreover, there are examples in which under certain loads no equilibrium solutions exist. As for other nonlinear systems, a solution of the equilibrium problem can be non-unique, in general.
3. The Lagrange-type variational principle is not minimal, it is only a stationary variational principle. The only exception is for the linear theory.
4. For the linear theory of micropolar shells the theorems of existence and uniqueness of a solution can be proven, see Eremeyev and Lebedev (2011).

Further developments of this version of shell theory can be performed in the following directions:

1. Development of a mathematical theory that should be based on the methods of partial differential equations theory, functional analysis and calculus of variations.
2. Numerical algorithms for solution of the reduced systems of nonlinear equations. For example, it can be done within the framework of the Finite Element Method, see for example the numerical results in Chróścielewski et al. (2004, 2010, 2011); Chróścielewski and Witkowski (2010, 2011).
3. Analysis of the restrictions of the non-linear constitutive equations.
4. Extension of the two-dimensional constitutive equations to shells made of various materials. In particular, the extension can include visco-elasticity, thermal effects, etc. In particular, the thermodynamics of thermo-elastic and thermo-visco-elastic shells with phase transitions is developed in Eremeyev and Pietraszkiewicz (2004, 2009, 2010, 2011) and Pietraszkiewicz (2011).

4 Theories of Shells and Plates by Reduction of the Three-dimensional Micropolar Continuum

We have mentioned in Sect. 1 that there are approaches based on the reduction of the three-dimensional Cossserat continuum equations to the two-dimensional equations. These two-dimensional theories inherit some micropolar properties from the three-dimensional continuum. Here we discuss these reduction techniques.

4.1 Basic Equations of Three-dimensional Linear Cosserat Continuum

The small strains of the micropolar media are usually described by using the vector of translation $\boldsymbol{u}$ and the vector of microrotation $\boldsymbol{\vartheta}$, see Nowacki

(1986); Eringen (1999). From the physical point of view, $\boldsymbol{u}$ describes the displacement of a particle of a micropolar body while $\boldsymbol{\vartheta}$ corresponds to the particle rotation.

The equilibrium conditions of any part of a micropolar body occupying the arbitrary volume $V_* \subset V$ consist of the following relations, see Eringen (1999),

$$\int\limits_{V_*} \rho \boldsymbol{F}\, \mathrm{d}V + \int\limits_{S_*} \boldsymbol{t}\, \mathrm{d}A = \boldsymbol{0}, \quad \int\limits_{V_*} \rho(\boldsymbol{r} \times \boldsymbol{F} + \boldsymbol{L})\, \mathrm{d}V + \int\limits_{S_*} (\boldsymbol{r} \times \boldsymbol{t} + \boldsymbol{m})\, \mathrm{d}A = \boldsymbol{0}, \quad (72)$$

where $\boldsymbol{F}$ and $\boldsymbol{L}$ are the mass forces and couples vectors, respectively, ρ is the density, $\boldsymbol{r}$ the position-vector, $S_* = \partial V_*$, $\boldsymbol{t}$ and $\boldsymbol{m}$ are the stress and couple stress vectors, respectively. Hence, for any part of the micropolar body Eq. $(72)_1$ states that the vector of total force is zero, while Eq. $(72)_2$ states that the vector of total moment is zero. With the relations $\boldsymbol{n} \cdot \boldsymbol{\sigma} = \boldsymbol{t}$, $\boldsymbol{n} \cdot \boldsymbol{\mu} = \boldsymbol{m}$, the local equilibrium equations of a micropolar continuum are

$$\nabla_{\mathbf{x}} \cdot \boldsymbol{\sigma} + \rho \boldsymbol{F} = \boldsymbol{0}, \quad \nabla_{\mathbf{x}} \cdot \boldsymbol{\mu} + \boldsymbol{\sigma}_{\times} + \rho \boldsymbol{L} = \boldsymbol{0}, \quad (73)$$

where $\boldsymbol{\sigma}$ and $\boldsymbol{\mu}$ are the stress and couple stress tensors, respectively, $\nabla_{\mathbf{x}}$ is the three-dimensional nabla operator. Equation $(73)_1$ is the local form of the balance of momentum while Eq. $(73)_2$ is the balance of moment of momentum.

The static boundary conditions have the following form

$$\boldsymbol{n} \cdot \boldsymbol{\sigma} = \boldsymbol{t}^0, \quad \boldsymbol{n} \cdot \boldsymbol{\mu} = \boldsymbol{m}^0 \quad \text{at } S_f. \quad (74)$$

Here $\boldsymbol{t}^0$ and $\boldsymbol{m}^0$ are the external surface forces and couples acting on the corresponding part of the surface S_f of the micropolar body,

$$S = S_u \cup S_f \equiv \partial V.$$

The kinematic boundary conditions consist of the following relations

$$\boldsymbol{u} = \boldsymbol{u}^0, \quad \boldsymbol{\vartheta} = \boldsymbol{\vartheta}^0 \quad \text{at } S_u, \quad (75)$$

where $\boldsymbol{u}^0$ and $\boldsymbol{\vartheta}^0$ are given functions at S_u. Other types of the boundary conditions may also be formulated.

The linear stain measures, i.e. the linear stretch tensor $\boldsymbol{\varepsilon}$ and the linear wryness tensor $\boldsymbol{\chi}$, are given by the relations

$$\boldsymbol{\varepsilon} = \nabla_{\mathbf{x}} \boldsymbol{u} + \boldsymbol{\vartheta} \times \mathbf{I}, \quad \boldsymbol{\chi} = \nabla_{\mathbf{x}} \boldsymbol{\vartheta}. \quad (76)$$

Eringen (1999) used $(\nabla_{\mathbf{x}} \boldsymbol{\vartheta})^T$ as linear wryness tensor. Here we use the definition $(76)_2$ for the consistency with the definition of $\boldsymbol{\varepsilon}$.

For an isotropic solid the constitutive equations are

$$\boldsymbol{\sigma} = \lambda \mathbf{I} \operatorname{tr} \boldsymbol{\varepsilon} + \mu \boldsymbol{\varepsilon}^T + (\mu + \kappa)\boldsymbol{\varepsilon}, \quad \boldsymbol{\mu} = \alpha \mathbf{I} \operatorname{tr} \boldsymbol{\chi} + \beta \boldsymbol{\chi}^T + \gamma \boldsymbol{\chi}, \tag{77}$$

where λ, μ, κ, α, β, γ are the elastic moduli which satisfy the following inequalities, see Eringen (1999),

$$2\mu + \kappa \geq 0, \quad \kappa \geq 0, \quad 3\lambda + 2\mu + \kappa \geq 0,$$
$$\beta + \gamma \geq 0, \quad \gamma - \beta \geq 0, \quad 3\alpha + \beta + \gamma \geq 0.$$

4.2 Transition to the Two-Dimensional Equilibrium Equations: Eringen's Approach

Eringen (1967, 1999) proposed the 3D-to-2D reduction technique on the base of the through-the-thickness integration of local equilibrium equations. Eringen's transition to the two-dimensional equations is based on the linear in z approximation of the translation and rotation together with independent integration of the equilibrium equations (73) through the thickness. Let the plate-like body occupy the volume

$$V = \{(x, y, z) \in \mathbb{R}^3 : (x, y) \in \mathcal{M} \subset \mathbb{R}^2, z \in [-h/2, h/2]\},$$

see Fig. 4. Here h is the plate thickness and $\boldsymbol{N} = \boldsymbol{i}_3$. For the sake of simplicity we assume that $h = \text{const}$. For the translations and rotations of the plate-like body Eringen introduced the following approximation

$$\boldsymbol{u}(x, y, z) = \boldsymbol{v}(x, y) - z\boldsymbol{\varphi}(x, y), \quad \boldsymbol{\vartheta}(x, y, z) = \boldsymbol{\phi}(x, y), \quad \boldsymbol{\varphi} \cdot \boldsymbol{i}_3 = 0 \tag{78}$$

with three independent vector fields $\boldsymbol{v}(x, y)$, $\boldsymbol{\varphi}(x, y)$, and $\boldsymbol{\phi}(x, y)$. Hence, in Eringen's theory of plates one has 8 kinematically independent scalar fields: v_1, v_2, v_3, φ_1, φ_2, ϕ_1, ϕ_2, ϕ_3.

To illustrate the transformations of (73) we assume homogeneous boundary conditions at $z = \pm h/2$, $(x, y) \in \mathcal{M}$

$$\boldsymbol{n}^{\pm} \cdot \boldsymbol{\sigma} = \boldsymbol{0}, \quad \boldsymbol{n}^{\pm} \cdot \boldsymbol{\mu} = \boldsymbol{0}, \tag{79}$$

where $\boldsymbol{n}^{\pm} = \pm \boldsymbol{i}_3$. Then the integration of $(73)_1$ over the thickness leads to the equation

$$\nabla \cdot \mathbf{T} + \boldsymbol{f} = \boldsymbol{0}, \tag{80}$$

where

$$\mathbf{T} = \langle \mathbf{A} \cdot \boldsymbol{\sigma} \rangle, \quad \boldsymbol{f} = \langle \rho \boldsymbol{F} \rangle, \quad \langle (\ldots) \rangle = \int_{-h/2}^{h/2} (\ldots) \, \mathrm{d}z.$$

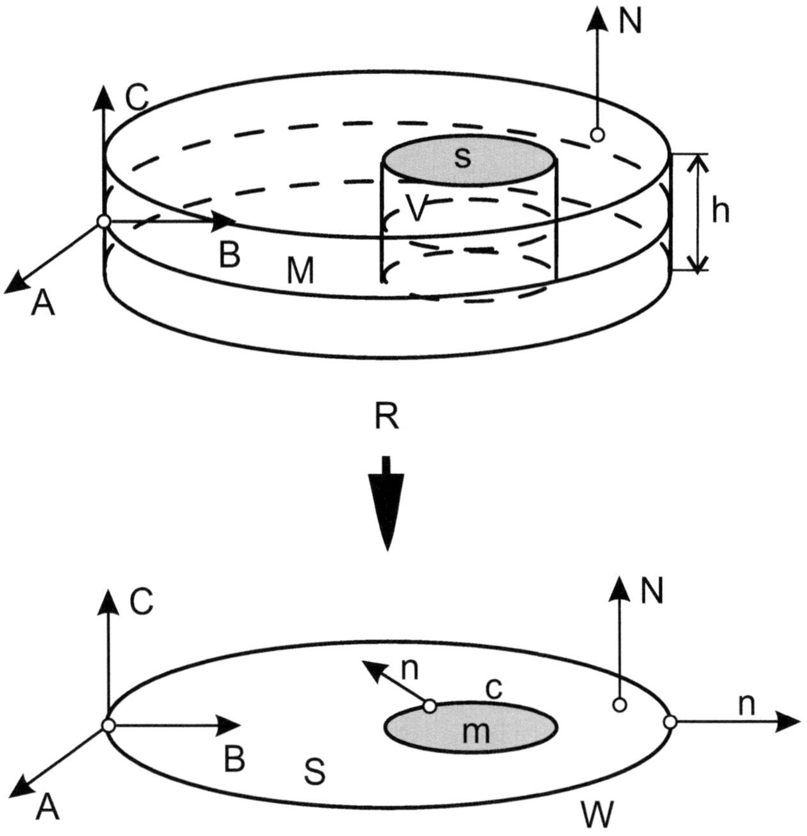

Figure 4. Plate-like body and its 2D analogue

Integration of $(73)_2$ gives us the relation

$$\nabla \cdot \mathbf{M}^{\mu} + \mathbf{T}_{\times} + \langle \sigma_{3\alpha} \rangle \boldsymbol{i}_3 \times \boldsymbol{i}_{\alpha} + \boldsymbol{c}_{\mu} = \mathbf{0}, \tag{81}$$

where

$$\mathbf{M}^{\mu} = \langle \mathbf{A} \cdot \boldsymbol{\mu} \rangle, \quad \boldsymbol{c}_{\mu} = \langle \rho \boldsymbol{L} \rangle.$$

Equations (80) and (81) constitute the balance equations of the zeroth-order theory of micropolar plates.

Additionally, cross-multiplying $(73)_1$ by $z\boldsymbol{i}_3$ and integrating over thickness we obtain the equation of the first-order theory

$$\nabla \cdot \mathbf{M}^{\sigma} - \langle \sigma_{3\alpha} \rangle \boldsymbol{i}_3 \times \boldsymbol{i}_{\alpha} + \boldsymbol{c}_{\sigma} = \mathbf{0}, \tag{82}$$

where

$$\mathbf{M}^\sigma = -\langle \mathbf{A} \cdot z\boldsymbol{\sigma} \times \boldsymbol{i}_3 \rangle, \quad \boldsymbol{c}_\sigma = \boldsymbol{i}_3 \times \langle \rho z \boldsymbol{F} \rangle.$$

In this theory the stress resultant $\mathbf{T}$, the force-stress resultant $\mathbf{M}^\sigma$, and the moment-stress resultant $\mathbf{M}^\mu$ are defined. Hence, in the case of static boundary conditions one needs to assign the values of $\boldsymbol{\nu}\cdot\mathbf{T}$, $\boldsymbol{\nu}\cdot\mathbf{M}^\mu$, $\boldsymbol{\nu}\cdot\mathbf{M}^\sigma$ at the plate boundary contour. Using Eqs. (77) and (78) one may obtain the constitutive equations for $\mathbf{T}$, $\mathbf{M}^\sigma$, and $\mathbf{M}^\mu$, which are presented in Eringen (1999) in the component form.

4.3 Transition to the Two-Dimensional Equilibrium Equations: Other Reduction Procedures

Let us mention that Eringen's approach is not unique. For example, Gevorkyan (1967) obtained the constitutive equations for the shell using the linear pseudo-Cosserat continuum. The drilling moments in his theory are generated by the couple stress tensor $\boldsymbol{\mu}$ only. Reissner (1977) introduced the stress resultants $\mathbf{T}$, the force-stress resultant $\mathbf{M}^\sigma$, and the moment-stress resultant $\mathbf{M}^\mu$ taking into account the transverse shear forces and the drilling moment. He has used the linear approximation in z direction for the stresses acting in the three-dimensional plate-like body.

The derivation of the micropolar plate theory proposed by Altenbach and Eremeyev (2009a) leads to the 6-parametric theory. The integration procedure is performed as follows. Let us consider that our plate-like body occupies a volume with one dimension which is significantly smaller in comparison with the other two. The coordinate z denotes this special direction and h is the plate thickness z which takes the values $-h/2 \le z \le h/2$. The boundary conditions of the upper (+) and lower (−) plate surfaces are given by Eqs. (79).

The main idea of the reduction procedure is the application of the 3D equilibrium conditions (72) to any volume V_* of the plate-like body and the transformation of the results to the 2D case as in Eqs. (51). Following Altenbach and Eremeyev (2009a) we transform Eqs. (72) to the relations

$$\begin{aligned}
&\int\limits_{\mathcal{M}_*} \boldsymbol{f}\,\mathrm{d}\Sigma + \int\limits_{\mathcal{C}_*} \boldsymbol{\nu}\cdot\langle\boldsymbol{\sigma}\rangle\,\mathrm{d}s = \mathbf{0},\\
&\int\limits_{\mathcal{M}_*} [\boldsymbol{x}\times\boldsymbol{f}+\boldsymbol{c}]\,\mathrm{d}\Sigma\\
&\qquad + \int\limits_{\mathcal{C}_*} [\boldsymbol{\nu}\cdot\langle\boldsymbol{\mu}\rangle - \boldsymbol{\nu}\cdot\langle z\boldsymbol{\sigma}\times\boldsymbol{i}_3\rangle - \boldsymbol{\nu}\cdot\langle\boldsymbol{\sigma}\rangle\times\boldsymbol{x}]\,\mathrm{d}s = \mathbf{0},
\end{aligned} \tag{83}$$

where $\boldsymbol{x} = \boldsymbol{r}\big|_{z=0}$, $\mathcal{C}_* = \partial\mathcal{M}_*$ and

$$\boldsymbol{f} = \langle \rho \boldsymbol{F} \rangle, \quad \boldsymbol{c} = \langle \rho \boldsymbol{L} \rangle + \boldsymbol{i}_3 \times \langle \rho z \boldsymbol{F} \rangle.$$

Equations (83) lead to the local equilibrium equations in the form of (51) where the stress resultant and stress couple tensors are determined by the following relations

$$\mathbf{T} = \langle \mathbf{A} \cdot \boldsymbol{\sigma} \rangle, \quad \mathbf{M} = \langle \mathbf{A} \cdot \boldsymbol{\mu} \rangle - \langle \mathbf{A} \cdot z\boldsymbol{\sigma} \times \boldsymbol{i}_3 \rangle. \tag{84}$$

From Eq. $(84)_2$ it follows that the components $M_{\alpha 3}$ depend only upon the couple stress tensor $\boldsymbol{\mu}$. Indeed, $\mathbf{M} \cdot \boldsymbol{i}_3 = \langle \boldsymbol{A} \cdot \boldsymbol{\mu} \cdot \boldsymbol{i}_3 \rangle$. It is obvious that the couple stress tensor $\mathbf{M}$ is the sum of Eringen's tensors $\mathbf{M}^\sigma$ and $\mathbf{M}^\mu$:

$$\mathbf{M} = \mathbf{M}^\sigma + \mathbf{M}^\mu.$$

Moreover, Eq. $(51)_2$ is the sum of Eqs. (81) and (82).

To establish the relations with the vectors $\boldsymbol{u}$ and $\boldsymbol{\vartheta}$ used in the 3D theory and their analogues in the 2D theory, we use the following approximation of $\boldsymbol{u}$ and $\boldsymbol{\vartheta}$, see Altenbach and Eremeyev (2009a) for details,

$$\boldsymbol{u}(x,y,z) = \boldsymbol{v}(x,y) - z\boldsymbol{\phi}(x,y), \quad \boldsymbol{\vartheta} = \boldsymbol{\phi}(x,y) \times \boldsymbol{i}_3 + \vartheta_3(x,y)\boldsymbol{i}_3, \quad \boldsymbol{\phi} \cdot \boldsymbol{i}_3 = 0. \tag{85}$$

The approximation (85) is more restrictive than Eqs. (78) proposed by Eringen because it contains only six scalar fields v_1, v_2, v_3, ϕ_1, ϕ_2, ϑ_3. With Eqs. (85) the couple stress tensor $\boldsymbol{\mu}$ does not depend on z, while the stress tensor $\boldsymbol{\sigma}$ depends on z linearly as in Eringen (1999). The similar procedure of 3D-to-2D reduction is applied by Zubov (2009) in the case of finite deformations of micropolar solids.

5 Conclusions and Discussion

In this paper the basic relations of the Cosserat-type theories of plates and shells are discussed. Let us note that the above presented theories are different, in general, because they have different kinematics and can describe different stress fields. We discuss here this difference in brief.

The structure of the elementary work of a micropolar shell (23) and Cosserat surface (10) are similar, but the mechanical sense of $\boldsymbol{\ell}$, $\boldsymbol{\gamma}$ and $\boldsymbol{l}$, $\boldsymbol{\eta}$ is different. The definition of the director $\boldsymbol{d}$ does not fix the orientation of an absolute rigid body in the space since any arbitrary rotations about $\boldsymbol{d}$ can be considered. This means that the elementary work of the external loads acting on the Cosserat shells (10) does not take into account the drilling moments about $\boldsymbol{d}$ since they do not perform the work on rotations about

$\boldsymbol{d}$. In other words the rotations about $\boldsymbol{d}$ are workless. On the other hand let us consider the variation of the director $\boldsymbol{d}$ only due to length changes $\delta\boldsymbol{d} = (\delta\boldsymbol{d}) \cdot \boldsymbol{d}\boldsymbol{d}$. It is obvious that the length changes are not related to any rotation. This means that they are not related to any moment. These variations describe strains (microstrains) of the material points constituting the shell. The corresponding loading characteristics are hyper-stresses (force dipoles). In this case in Eqs. $(11)_2$ and (15) the components of the vector functions $\boldsymbol{\ell}$ and $\boldsymbol{\gamma}$ have different nature. Let us present these functions as a sum of two terms parallel and normal to the director: $\boldsymbol{\ell} = \boldsymbol{\ell} \cdot \boldsymbol{d}\boldsymbol{d} + \boldsymbol{\ell} \times \boldsymbol{d}$ and $\boldsymbol{\gamma} = \boldsymbol{\gamma} \cdot \boldsymbol{d}\boldsymbol{d} + \boldsymbol{\gamma} \times \boldsymbol{d}$. The first terms correspond to the force dipoles acting in the $\boldsymbol{d}$-direction, the second terms are the bending moments. Thus $\boldsymbol{\ell}$ and $\boldsymbol{\gamma}$ do not contain the drilling moment.

Within the framework of the Cosserat-surface shell model one can discuss the material surface composed of deformable particles on which forces and moments and some hyper-stresses act. At the same time the micropolar shell can be represented by a surface composed of rigid microparticles in the form of an arbitrary ellipsoid. The interaction between these particles are given by forces and moments only. Now the Cosserat-surface shell model can be presented by a surface composed of microparticles having the shape of straight rod changing its length during the deformation, but they do not reflect the rotation about there axis. Summarizing one can state that the Cosserat-surface shell model does not follow from the micropolar shell and vice versa. It should be noted that the micropolar model seems to be more complete since it is only to prescribe forces and moments. Let us note that the difference between the Cosserat-surface shell model and the micropolar or 6-parametric shell model is analogous to the difference between the Ericksen liquid crystals and the Eringen micropolar fluids, see Ericksen (1998) and Eringen (1966, 2001).

It is obvious that Eringen's plate theory does not coincide with the linear variant of Cosserat surface as well as with the linear variant of micropolar shell theories discussed above. The vector $\boldsymbol{\varphi}$ is an analogue of the rotation vector used in the Reissner-type theories while $\boldsymbol{\phi}$ is an analogue of the microrotation vector used in the linear theory of micropolar continuum. In contrast to the Cosserat surface theory Eringen's micropolar plates theory takes into account the drilling moment but cannot take into account the force dipoles. On the other hand, this theory differs from the 6-parametric shell theory because the latter has only 6 degrees of freedom. In the first order micropolar plate theory by Eringen the two-dimensional couple stress tensor is split into two independent parts, i.e. the force-stress and moment-stress resultants.

The external loads acting on the shell surface for the above considered approaches are schematically presented in Fig. 5.

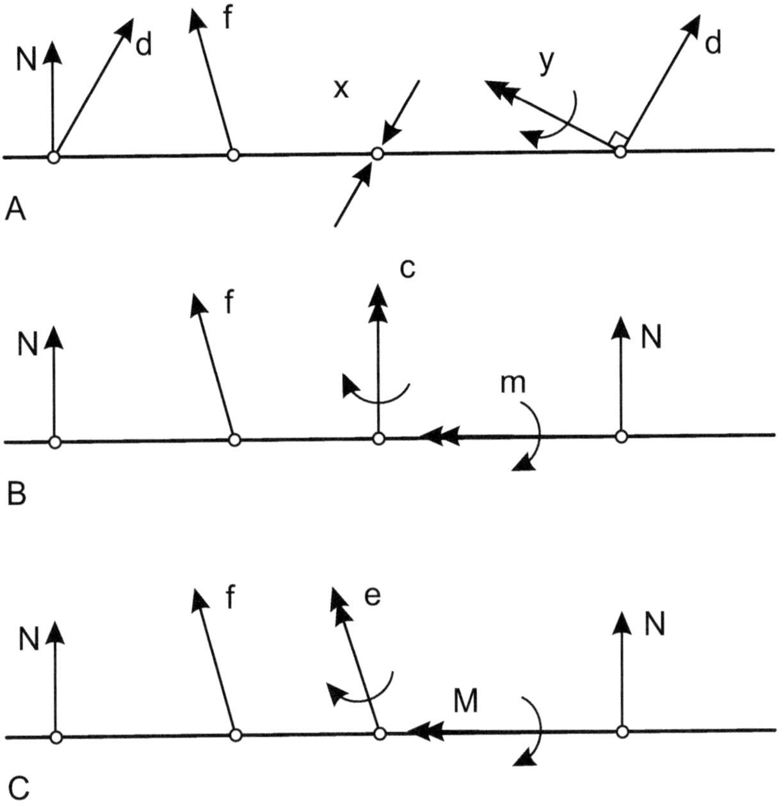

Figure 5. External surface loads: a) Cosserat's plate; b) micropolar plate; c) Eringen's plate.

Acknowledgement: The second author was supported by the DFG grant No. AL 341/33-1.

Bibliography

I. Aganović, J. Tambača, and Z. Tutek. Derivation and justification of the models of rods and plates from linearized three-dimensional micropolar elasticity. *J Elasticity*, 84:131–152, 2006.

M. Agranovich. Elliptic boundary problems. In M. Agranovich, Y. Egorov, and M. Shubin, editors, *Partial Differential Equations IX, Encyclopaedia of Mathematical Sciences 79*, pages 1–144. Springer, Berlin, 1997.

H. Altenbach. Eine direkt formulierte lineare Theorie für viskoelastische Platten und Schalen. *Ing Arch*, 58:215–228, 1988.

H. Altenbach and V. A. Eremeyev. On the shell theory on the nanoscale with surface stresses. *Int J Eng Sci*, 49(12):1294–1301, 2011.

H. Altenbach and V. A. Eremeyev. Direct approach based analysis of plates composed of functionally graded materials. *Arch Appl Mech*, 78(10): 775–794, 2008.

H. Altenbach and V. A. Eremeyev. On the linear theory of micropolar plates. *ZAMM*, 89(4):242–256, 2009a.

H. Altenbach and V. A. Eremeyev. On the bending of viscoelastic plates made of polymer foams. *Acta Mech*, 204(3–4):137–154, 2009b.

H. Altenbach and P. A. Zhilin. A general theory of elastic simple shells (in Russian). *Uspekhi Mekhaniki (Advances in Mechanics)*, 11(4):107–148, 1988.

H. Altenbach and P. A. Zhilin. The theory of simple elastic shells. In R. Kienzler, H. Altenbach, and I. Ott, editors, *Critical Review of the Theories of Plates and Shells*, volume 16 of *Lect. Notes Appl. Comp. Mech.*, pages 1–12. Springer, Berlin, 2004.

H. Altenbach, V. A. Eremeyev, and L. P. Lebedev. Micropolar shells as two-dimensional generalized continua models. In H. Altenbach, G. A. Maugin, and V. Erofeev, editors, *Mechanics of Generalized Continua*, volume 7 of *Advanced Structured Materials*, pages 23–55. Springer, Berlin, 2011.

S. A. Ambartsumian. The theory of transverse bending of plates with asymmetric elasticity. *Mech Compos Mater*, 32(1):30–38, 1996.

S. A. Ambartsumian. *The micropolar theory of plates and shells (in Russian)*. NAN Armenii, Yerevan, 1999.

S. S. Antman. *Nonlinear Problems of Elasticity*. Springer Science Media, New York, 2nd edition, 2005.

S. S. Antman. Global properties of buckled states of plates that can suffer thickness changes. *Arch Ration Mech An*, 110(2):103–117, 1990.

S. S. Antman and D. Bourne. Rotational symmetry vs. axisymmetry in shell theory. *Int J Eng Sci*, 48(11):991–1005, 2010.

S. S. Antman and W. Lacarbonara. Forced radial motions of nonlinearly viscoelastic shells. *J Elasticity*, 96(2):155–190, 2009.

T. Ariman. On circular micropolar plates. *Ing Arch*, 37(3):156–160, 1968.

Ya. Başar. A consistent theory of geometrically non-linear shells with an independent rotation vector. *Int J Solids Struct*, 23(10):1401–1415, 1987.

J. Badur and W. Pietraszkiewicz. On geometrically non-linear theory of elastic shells derived from pseudo-Cosserat continuum with constrained micro-rotations. In W. Pietraszkiewicz, editor, *Finite Rotations in Structural Mechanics*, pages 19–32. Springer, Wien, 1986.

K. Bhattacharya and R. D. James. A theory of thin films of martensitic materials with applications to microactuators. *J Mech Phys Solids*, 47 (3):531–576, 1999.

M. Bîrsan. On the theory of elastic shells made from a material with voids. *Int J Solids Struct*, 43(10):3106–3123, 2006a.

M. Bîrsan. On a thermodynamic theory of porous Cosserat elastic shells. *J Therm Stresses*, 29(9):879–899, 2006b.

M. Bîrsan. On Saint-Venant's principle in the theory of Cosserat elastic shells. *Int J Eng Sci*, 45(2–8):187–198, 2007.

M. Bîrsan. Inequalities of Korn's type and existence results in the theory of Cosserat elastic shells. *J Elasticity*, 90(3):227–239, 2008.

M. Bîrsan. Thermal stresses in cylindrical Cosserat elastic shells. *Eur J Mech A-Solid*, 28(1):94–101, 2009a.

M. Bîrsan. On Saint-Venant's problem for anisotropic, inhomogeneous, cylindrical Cosserat elastic shells. *Int J Eng Sci*, 47(1):21–38, 2009b.

M. Bîrsan. On the problems of Almansi and Michell for anisotropic Cosserat elastic shells. *Archive of Mechanics*, 61(3–4):195–227, 2009c.

M. Bîrsan. Thermal stresses in anisotropic cylindrical elastic shells. *Math Methods Appl Sci*, 33(6):799–810, 2010.

M. Bîrsan. On a problem of Truesdell for anisotropic elastic shells. *Analele Ştiinţifice Ale Universităţii "Al.I. Cuza" Din Iaşi (S.N.) Matematică,*, LVII(1):91–110, 2011.

M. Bîrsan. *Linear Cosserat Elastic Shells: Mathemathical Theory and Applications.* Matrix Rom, Bucureşti, 2009d.

M. Bîrsan and H. Altenbach. A mathematical study of the linear theory for orthotropic elastic simple shells. *Math Methods Appl Sci*, 33(12): 1399–1413, 2010.

M. Bîrsan and H. Altenbach. On the dynamical theory of thermoelastic simple shells. *ZAMM*, 91(6):443–457, 2011.

E. Boschi. Lamb and Love wave-propagation in an infinite micropolar elastic plate. *Annali di Geofisica*, 26(2-3):341–355, 1973.

B. Brank, D. Peric, and F. B. Damjanic. On large deformations of thin elasto-plastic shells: Implementation of a finite rotation model for quadrilateral shell element. *Int J Numer Meth Eng*, 40(4):689–726, 1997.

C. Chinosi, L. Della Croce, and T. Scapolla. Hierarchic finite elements for thin Naghdi shell model. *Int J Solids Struct*, 35(16):1863–1880, 1998.

K. L. Chowdhur and P. G. Glockner. Bending of an annular elastic Cosserat plate. *B Acad Pol Sci Tech*, 21(3):211–218, 1973.

J. Chróścielewski. Rodzina elementów skończonych klasy C^0 w nieliniowej sześcioparametrowej teorii powłok. *Zesz. Nauk. Politechniki Gdańskiej*, LIII(540):1–291, 1996.

J. Chróścielewski and W. Witkowski. On some constitutive equations for micropolar plates. *ZAMM*, 90(1):53–64, 2010.

J. Chróścielewski and W. Witkowski. FEM analysis of Cosserat plates and shells based on some constitutive relations. *ZAMM*, 91(5):400–412, 2011.

J. Chróścielewski, J. Makowski, and H. Stumpf. Finite element analysis of smooth, folded and multi-shell structures. *Comput Methods Appl Mech Eng*, 141:1–46, 1997.

J. Chróścielewski, J. Makowski, and W. Pietraszkiewicz. Non-linear dynamics of flexible shell structures. *Comp Assisted Mech Eng Sci*, 9:341–357, 2002.

J. Chróścielewski, J. Makowski, and W. Pietraszkiewicz. *Statics and Dynamics of Multyfold Shells. Nonlinear Theory and Finite Element Method*. Wydawnictwo IPPT PAN, Warszawa, 2004.

J. Chróścielewski, W. Pietraszkiewicz, and W. Witkowski. On shear correction factors in the non-linear theory of elastic shells. *Int J Solids Struct*, 47(25–26):3537–3545, 2010.

J. Chróścielewski, I. Kreja, A. Sabik, and W. Witkowski. Modeling of composite shells in 6-parameter nonlinear theory with drilling degree of freedom. *Mech Adv Mater Struc*, 18(6):403–419, 2011.

P. G. Ciarlet. *Mathematical Elasticity*, volume II. Theory of Plates. Elsevier, Amsterdam, 1997.

P. G. Ciarlet. *Mathematical Elasticity*, volume III. Theory of Shells. Elsevier, Amsterdam, 2000.

H. Cohen and R. S. D. Thomas. Transient waves in inhomogeneous isotropic elastic plates. *Acta Mech*, 53(3-4):141–161, 1984.

H. Cohen and R. S. D. Thomas. Transient waves in inhomogeneous anisotropic elastic plates. *Acta Mech*, 58(1-2):41–57, 1986.

H. Cohen and C.-C. Wang. A mathematical analysis of the simplest direct models for rods and shells. *Arch Ration Mech An*, 108(1):35–81, 1989.

C. Constanda. Complex variable treatment of bending of micropolar plates. *Int J Eng Sci*, 15(11):661–669, 1977.

E. Cosserat and F. Cosserat. *Théorie des corps déformables*. Herman et Fils, Paris, 1909.

R. Courant and D. Hilbert. *Methods of Mathematical Physics*, volume 1. Wiley, New York, 1991.

C. N. deSilva and P. J. Tsai. A general theory of directed surfaces. *Acta Mech*, 18(1-2):89–101, 1973.

H. A. Erbay. An asymptotic theory of thin micropolar plates. *Int J Eng Sci*, 38(13):1497–1516, 2000.

V. A. Eremeyev. Nonlinear micropolar shells: theory and applications. In W. Pietraszkiewicz and C. Szymczak, editors, *Shell Structures: Theory and Applications*, pages 11–18. Taylor & Francis, London, 2005.

V. A. Eremeyev and L. P. Lebedev. Existence theorems in the linear theory of micropolar shells. *ZAMM*, 91(6):468–476, 2011.

V. A. Eremeyev and W. Pietraszkiewicz. The non-linear theory of elastic shells with phase transitions. *J Elasticity*, 74(1):67–86, 2004.

V. A. Eremeyev and W. Pietraszkiewicz. Local symmetry group in the general theory of elastic shells. *J Elasticity*, 85(2):125–152, 2006.

V. A. Eremeyev and W. Pietraszkiewicz. Phase transitions in thermoelastic and thermoviscoelastic shells. *Arch Mech*, 61(1):41–67, 2009.

V. A. Eremeyev and W. Pietraszkiewicz. Thermomechanics of shells undergoing phase transition. *J Mech Phys Solids*, 59(7):1395–1412, 2011.

V. A. Eremeyev and W. Pietraszkiewicz. On tension of a two-phase elastic tube. In W. Pietraszkiewicz and I. Kreja, editors, *Shell Structures. Theory and Applications. Vol. 2.*, pages 63–66. CRC Press, Boca Raton, 2010.

V. A. Eremeyev and L. M. Zubov. On constitutive inequalities in nonlinear theory of elastic shells. *ZAMM*, 87(2):94–101, 2007.

V. A. Eremeyev and L. M. Zubov. *Mechanics of Elastic Shells (in Russian)*. Nauka, Moscow, 2008.

J. L. Ericksen. Wave propagation in thin elastic shells. *J Elasticity*, 43(3): 167–178, 1971.

J. L. Ericksen. The simplest problems for elastic Cosserat surfaces. *J Elasticity*, 2(2):101–107, 1972a.

J. L. Ericksen. Plane infinitesimal waves in homogeneous elastic plates. *J Elasticity*, 3(3):161–167, 1973a.

J. L. Ericksen. Simpler problems for elastic Cosserat surfaces. *J Elasticity*, 7(1):1–11, 1977.

J. L. Ericksen. *Introduction to the Thermodynamics of Solids*. Springer, New York, 2nd edition, 1998.

J. L. Ericksen. Symmetry transformations for thin elastic shells. *Arch Ration Mech An*, 47:1–14, 1972b.

J. L. Ericksen. Apparent symmetry of certain thin elastic shells. *J. Mećanique*, 12:12–18, 1973b.

J. L. Ericksen and C. Truesdell. Exact theory of stress and strain in rods and shells. *Arch Ration Mech An*, 1(1):295–323, 1958.

A. C. Eringen. *Microcontinuum Field Theory. II. Fluent Media*. Springer, New York, 2001.

A. C. Eringen. Theory of micropolar fluids. *Journal of Mathematics and Mechanics*, 16(1):1–18, 1966.

A. C. Eringen. Theory of micropolar plates. *ZAMP*, 18(1):12–30, 1967.
A. C. Eringen. *Microcontinuum Field Theory. I. Foundations and Solids.* Springer, New York, 1999.
M. Farshad and B. Tabarrok. Dualities in analysis of Cosserat plate. *Mech Res Comm*, 3(5):399–406, 1976.
G. Fichera. Existence theorems in elasticity. In S. Flügge, editor, *Handbuch der Physik*, volume VIa/2, pages 347–389. Springer, Berlin, 1972.
D. D. Fox and J. C. Simo. A drill rotation formulation for geometrically exact shells. *Comput Methods Appl Mech Eng*, 98(3):329—343, 1992.
D. D. Fox, A. Raoult, and J. C. Simo. A justification of nonlinear properly invariant plate theories. *Arch Ration Mech An*, 124(2):157–199, 1993.
G. Friesecke, R. D. James, M. G. Mora, and S. Müller. The Föppl-von Kármán plate theory as a low energy Γ-limit of nonlinear elasticity. *C R Acad Sci Paris Ser I*, 335:201–206, 2002a.
G. Friesecke, R. D. James, M. G. Mora, and S. Müller. A theorem on geometric rigidity and the derivation of nonlinear plate theory from three-dimensional elasticity. *Commun Pure Appl Math*, LV:1461–1506, 2002b.
G. Friesecke, R. D. James, M. G. Mora, and S. Müller. Rigorous derivation of nonlinear plate theory and geometric rigidity. *C R Acad Sci Paris Ser I*, 334:173—178, 2002c.
G. Friesecke, R. D. James, M. G. Mora, and S. Müller. Derivation of nonlinear bending theory for shells from three-dimensional nonlinear elasticity by Γ-convergence. *C R Acad Sci Paris Ser I*, 336 :697–702, 2003.
G. A. Gevorkyan. The basic equations of flexible plates for a medium of Cosserat. *Int Appl Mech*, 3(11):41–45, 1967.
P. G. Glockner and D. J. Malcolm. Cosserat surface – model for idealized sandwich shells. *ZAMM*, 54(4):T78, 1974.
A. E. Green and P. M. Naghdi. Micropolar and director theories of plates. *Q J Mech Appl Math*, 20:183–199, 1967a.
A. E. Green and P. M. Naghdi. The linear elastic Cosserat surface and shell theory. *Int J Solids Struct*, 4(6):585–592, 1968.
A. E. Green and P. M. Naghdi. Non-isothermal theory of rods, plates and shells. *Int J Solids Struct*, 6:209–244, 1970.
A. E. Green and P. M. Naghdi. On superposed small deformations on a large deformation of an elastic Cosserat surface. *J Elasticity*, 1(1):1–17, 1971.
A. E. Green and P. M. Naghdi. Derivation of shell theories by direct approach. *Trans ASME. J Appl Mech*, 41(1):173–176, 1974.
A. E. Green and P. M. Naghdi. On thermal effects in the theory of shells. *Proc R Soc London A*, 365A:161–190, 1979.
A. E. Green and P. M. Naghdi. Linear theory of an elastic Cosserat plate. *Proc Camb Philos S-M*, 63(2):537– 550, 1967b.

A. E. Green, P. M. Naghdi, and W. L. Wainwright. A general theory of a Cosserat surface. *Arch Ration Mech An*, 20(4):287–308, 1965.

E. Grekova and P. Zhilin. Basic equations of Kelvin's medium and analogy with ferromagnets. *J Elasticity*, 64:29–70, 2001.

E. I. Grigolyuk and I. T. Selezov. Nonclassical theories of vibration of beams, plates and shells (in Russian). In *Itogi nauki i tekhniki*, volume 5 of *Mekhanika tverdogo deformiruemogo tela*. VINITI, Moskva, 1973.

İ. T. Gürgöze. Thermostatics of an elastic Cosserat plate containing a circular hole. *Proc Camb Philos S-M*, 70(JUL):169–174, 1971.

M. E. Gurtin and A. I. Murdoch. A continuum theory of elastic material surfaces. *Arch Ration Mech An*, 57:291–323, 1975.

D. H. Hodges. *Nonlinear Composite Beam Theory*, volume 213 of *Progress in Astronautics and Aeronautics*. American Institute of Aeronautics and Astronautics, Inc., Reston, 2006.

D. H. Hodges, A. R. Atilgan, and D. A. Danielson. A geometrically nonlinear theory of elastic plates. *Trans ASME. J Appl Mech*, 60(1):109–116, 2004.

L. Hörmander. *Linear Partial Differential Equations*. Springer, Berlin, 4th edition, 1976.

S. Itou and A. Atsumi. Effect of couple-stresses on elastic Cosserat plate with loads travelling at uniform velocity along bounding surfaces. *Arch Mech Stos*, 22(4):375–384, 1970.

G. Jemielita. Rotational representation of a plate made of Grioli-Toupin material. *Trans ASME. J Appl Mech*, 62(2):414–418, 1995.

G. Jemilita. Bibliografia. In Cz. Woźniak, editor, *Mechanika sprężystych płyt i powłok*, pages 675–766. PWN, Warszawa, 2001.

C. S. Jog. Higher-order shell elements based on a Cosserat model, and their use in the topology design of structures. *Comput Methods Appl Mech Eng*, 193(23–26):2191–2220, 2004.

Ya. F. Kayuk and A. P. Zhukovskii. Theory of plates and shells based on the concept of Cosserat surfaces. *Int Appl Mech*, 17(10):922–926, 1981.

R. Kienzler. On consistent plate theories. *Arch Appl Mech*, 72:229 - 247, 2002.

R. Kienzler, H. Altenbach, and I. Ott, editors. *Critical review of the theories of plates and shells, new applications*, volume 16 of *Lect. Notes Appl. Comp. Mech.*, Berlin, 2004. Springer.

G. R. Kirchhoff. Über das Gleichgewicht und die Bewegung einer elastischen Scheibe. *Crelles Journal für die reine und angewandte Mathematik*, 40: 51–88, 1850.

J. K. Knowles and E. Sternberg. On the failure of ellipticity of the equations for finite elastostatic plane strain. *Arch Ration Mech An*, 63:321–336, 1976.

J. K. Knowles and E. Sternberg. On the failure of ellipticity and the emergence of discontinuous deformation gradients in plane finite elastostatics. *J Elasticity*, 8:329–379, 1978.

V. Konopińska and W. Pietraszkiewicz. Exact resultant equilibrium conditions in the non-linear theory of branched and self-intersecting shells. *Int J Solids Struct*, 44(1):352–369, 2006.

T. Korman, F. T. Morghem, and M. H. Baluch. Bending of micropolar plates. *Nucl Eng Des*, 26(3):432–439, 1974.

I. Kreja. *Geometrically non-linear analysis of layered composite plates and shells.* Gdańsk University of Technology, Gdańsk, 2007.

S. Krishnaswamy, Z. H. Jin, and R. C. Batra. Stress concentration in an elastic Cosserat plate undergoing extensional deformations. *Trans ASME. J Appl Mech*, 65(1):66–70, 1998.

R. Kumar and S. Deswal. Some problems of wave propagation in a micropolar elastic medium with voids. *J Vib Control*, 12(8):849–879, 2006.

R. Kumar and G. Partap. Axisymmetric free vibrations of infinite micropolar thermoelastic plate. *Appl Math Mech*, 28(3):369–383, 2007.

R. Kumar and G. Partap. Rayleigh Lamb waves in micropolar isotropic elastic plate. *Appl Math Mech*, 27(8):1049–1059, 2006.

Z. T. Kurlandz. Derivation of linear shell theory from theory of Cosserat medium. *B Acad Pol Sci Tech*, 20(10):789–794, 1972.

Z. T. Kurlandz. Anisotropic linear Cosserat surface and linear shell theory. *Arch Mech*, 25(4):613–620, 1973.

L. P. Lebedev, M. J. Cloud, and V. A. Eremeyev. *Tensor Analysis with Applications in Mechanics.* World Scientific, New Jersey, 2010.

M. Levinson. An accurate, simple theory of the statics and dynamics of elastic plates. *Mech Res Comm*, 7(6):343–350, 1980.

A. Libai and J. G. Simmonds. Nonlinear elastic shell theory. *Adv Appl Mech*, 23:271–371, 1983.

A. Libai and J. G. Simmonds. *The Nonlinear Theory of Elastic Shells.* Cambridge University Press, Cambridge, 2nd edition, 1998.

J.-L. Lions and E. Magenes. *Problèmes aux limites non homogènes et applications.* Dunod, Paris, 1968.

I. Lubowiecka and J. Chróścielewski. On dynamics of flexible branched shell structures undergoing large overall motion using finite elements. *Comput Struct*, 80:891–898, 2002.

A. I. Lurie. *Theory of Elasticity.* Foundations of Engineering Mechanics. Springer, Berlin, 2005.

A. I. Lurie. *Analytical Mechanics.* Foundations of Engineering Mechanics. Springer, Berlin, 2001.

J. Makowski and W. Pietraszkiewicz. *Thermomechanics of shells with singular curves.* Zesz. Nauk. No 528/1487/2002, Gdańsk, 2002.

J. Makowski and H. Stumpf. Buckling equations for elastic shells with rotational degrees of freedom undergoing finite strain deformation. *Int J Solids Struct*, 26:353–368, 1990.

J. Makowski, W. Pietraszkiewicz, and H. Stumpf. Jump conditions in the nonlinear theory of thin irregular shells. *J Elasticity*, 54:1–26, 1999.

D. J. Malcolm and P. G. Glockner. Cosserat surface and sandwich shell equilibrium. *Trans ASCE. J. Eng Mech Div*, 98(EM5):1075–1086, 1972a.

D. J. Malcolm and P. G. Glockner. Nonlinear sandwich shell and Cosserat surface theory. *Trans ASCE. J Eng Mech Div*, 98(EM5):1183–1203, 1972b.

T. Merlini and M. Morandini. Computational shell mechanics by helicoidal modeling I: Theory. *J Mech Mater Struc*, 6(5):659–692, 2011a.

T. Merlini and M. Morandini. Computational shell mechanics by helicoidal modeling II: Shell element. *J Mech Mater Struct*, 6(5):693–728, 2011b.

R. D. Mindlin. Influence of rotatory inertia and shear on flexural motions of isotropic elastic plates. *Trans ASME. J Appl Mech*, 18:31–38, 1951.

A. I. Murdoch and H. Cohen. Symmetry considerations for material surfaces. *Arch Ration Mech An*, 72(1):61–98, 1979.

A. I. Murdoch and H. Cohen. Symmetry considerations for material surfaces. Addendum. *Arch Ration Mech An*, 76(4):393–400, 1981.

P. M. Naghdi. The theory of plates and shells. In S. Flügge, editor, *Handbuch der Physik*, volume VIa/2, pages 425–640. Springer, Heidelberg, 1972.

P. M. Naghdi. On the formulation of contact problems of shells and plates. *J Elasticity*, 5(3-4):379–398, 1975.

P. M. Naghdi and M. B. Rubin. Restrictions on nonlinear constitutive equations for elastic shells. *J Elasticity*, 39:133–163, 1995.

P. M. Naghdi and A. R. Srinivasa. A dynamical theory of structured solids. 1. Basic developments. *Phil Trans Royal Soc London A. Math Phys Eng Sci*, 345(1677):425–458, 1993.

P. M. Naghdi and J. A. Trapp. A uniqueness theorem in the theory of Cosserat surface. *J Elasticity*, 2(1):9–20, 1972.

R. Nardinocchi and P. Podio-Guidugli. The equations of Reissner-Mindlin plates obtained by the method of internal constraints. *Meccanica*, 29: 143–157, 1994.

S. A. Nazarov. Asymptotic behavior of the solution of an elliptic boundary value problem in a thin domain. *J Math Sci*, 64(6):1351–1362, 1993.

P. Neff. A geometrically exact Cosserat shell-model including size effects, avoiding degeneracy in the thin shell limit. Part I: Formal dimensional reduction for elastic plates and existence of minimizers for positive Cosserat couple modulus. *Continuum Mech Therm*, 16(6):577–628, 2004.

P. Neff. The Γ-limit of a finite strain Cosserat model for asymptotically thin domains and a consequence for the Cosserat couple modulus. *PAMM*, 5 (1):629–630, 2005.

P. Neff. A geometrically exact planar Cosserat shell-model with microstructure: Existence of minimizers for zero Cosserat couple modulus. *Math Models & Methods Appl Sci*, 17(3):363–392, 2007.

P. Neff and K. Chełmiński. A geometrically exact Cosserat shell-model for defective elastic crystals. Justification via Γ-convergence. *Interfaces Free Bound*, 9:455–492, 2007.

V. Nicotra, P. Podio-Guidugli, and A. Tiero. Exact equilibrium solutions for linearly elastic plate-like bodies. *J Elasticity*, 56:231–245, 1999.

A. K. Noor. Bibliography of monographs and surveys on shells. *Appl Mech Rev*, 43:223–234, 1990.

A. K. Noor. List of books, monographs, and survey papers on shells. In A. K. Noor, T. Belytschko, and J. C. Simo, editors, *Analytical and Computational Models of Shells, CED*, volume 3, pages vii–xxxiv. ASME, New York, 2004.

W. Nowacki. *Theory of Asymmetric Elasticity.* Pergamon-Press, Oxford et al., 1986.

W. Nowacki and W. K. Nowacki. Propagation of monochromatic waves in an infinite micropolar elastic plate. *B Acad Pol Sci Tech*, 17(1):45–53, 1969.

V. A. Pal'mov. Contribution to Cosserat's theory of plates (in Russian). *Trudy LPI*, 386:3–8, 1982.

W. A. Palmow and H. Altenbach. Über eine Cosseratsche Theorie für elastische Platten. *Techn Mech*, 3(3):5–9, 1982.

W. Pietraszkiewicz. Refined resultant thermomechanics of shells. *Int J Eng Sci*, 49(10):1112–1124, 2011.

W. Pietraszkiewicz. Teorie nieliniowe powłok. In Cz. Woźniak, editor, *Mechanika sprężystych płyt i powłok*, pages 424–497. PWN, Warszawa, 2001.

W. Pietraszkiewicz. *Finite Rotations and Langrangian Description in the Non-linear Theory of Shells.* Polish Sci. Publ, Warszawa-Poznań, 1979a.

W. Pietraszkiewicz. Consistent second approximation to the elastic strain energy of a shell. *ZAMM*, 59:206–208, 1979b.

W. Pietraszkiewicz. Geometrically nonlinear theories of thin elastic shells. *Uspekhi Mechaniki (Advances in Mechanics)*, 12(1):51–130, 1989.

W. Pietraszkiewicz. Addendum to: Bibliography of monographs and surveys on shells. *Appl Mech Rev*, 45:249–250, 1992.

W. Pietraszkiewicz and J. Badur. Finite rotations in the description of continuum deformation. *Int J Eng Sci*, 21:1097–1115, 1983.

W. Pietraszkiewicz and V. A. Eremeyev. On natural strain measures of the non-linear micropolar continuum. *Int J Solids Struct*, 46(3–4):774–787, 2009a.

W. Pietraszkiewicz and V. A. Eremeyev. On vectorially parameterized natural strain measures of the non-linear Cosserat continuum. *Int J Solids Struct*, 46(11–12):2477–2480, 2009b.

W. Pietraszkiewicz, J. Chróścielewski, and J. Makowski. On dynamically and kinematically exact theory of shells. In W. Pietraszkiewicz and C. Szymczak, editors, *Shell Structures: Theory and Applications*, pages 163–167. Taylor & Francis, London, 2005.

W. Pietraszkiewicz, V. A. Eremeyev, and V. Konopińska. Extended non-linear relations of elastic shells undergoing phase transitions. *ZAMM*, 87 (2):150–159, 2007.

P. I. Plotnikov and J. F. Toland. Modelling nonlinear hydroelastic waves. *Phil Trans Royal Society A*, 369(1947):2942–2956, 2011.

P. Podio-Guidugli. An exact derivation of the thin plate equation. *J Elasticity*, 22:121–133, 1989.

A. Pompei and M. A. Rigano. On the bending of micropolar viscoelastic plates. *Int J Eng Sci*, 44(18-19):1324–1333, 2006.

J. W. Provan and R. C. Koeller. On the theory of elastic plates. *Int J Solids Struct*, 6(7):933–950, 1970.

H. Ramsey. Comparison of buckling deformations in compressed elastic Cosserat plates with three-dimensional plates. *Int J Solids Struct*, 22(3): 257–266, 1986.

H. Ramsey. Axisymmetrical buckling of a cylindrical elastic Cosserat shell under axial-compression. *Quart J Mech Appl Math*, 40(3):415–429, 1987.

J. N. Reddy. A simple higher-order theory for laminated composite plates. *Trans ASME. J Appl Mech*, 51:745–752, 1984.

E. Reissner. On the theory of bending of elastic plates. *J Math Phys*, 23: 184–194, 1944.

E. Reissner. The effect of transverse shear deformation on the bending of elastic plates. *Trans ASME. J Appl Mech*, 12(11):A69 – A77, 1945.

E. Reissner. On bending of elastic plates. *Quart Appl Math*, 5:55–68, 1947.

E. Reissner. Reflection on the theory of elastic plates. *Appl Mech Rev*, 38 (11):1453–1464, 1985.

E. Reissner. A note on pure bending and flexure in plane stress including the effect of moment stresses. *Arch Mech*, 28(6):633–642, 1970.

E. Reissner. On sandwich-type plates with cores capable of supporting moment stresses. *Acta Mech*, 14(1):43–51, 1972.

E. Reissner. A note on generating generalized two-dimensional plate and shell theories. *ZAMP*, 28:633–642, 1977.

H. Rothert. Lineare konstitutive Gleihungen der viskoelastischen Cosseratfläche. *ZAMM*, 55:647–656, 1975.

H. Rothert. Lösungsmöglichkeiten des Umkehrproblems viskoelastischer Cosseratfächen. *ZAMM*, 57:429–437, 1977.

M. B. Rubin. Restrictions on linear constitutive equations for a rigid heat conducting Cosserat shell. *Int J Solids Struct*, 41(24-25):7009–7033, 2004.

M. B. Rubin. *Cosserat Theories: Shells, Rods and Points.* Kluwer, Dordrecht, 2000.

M. B. Rubin and Y. Benveniste. A Cosserat shell model for interphases in elastic media. *J Mech Phys Solids*, 52(5):1023–1052, 2004.

C. Sansour and H. Bednarczyk. The Cosserat surface as a shell-model, theory and finite-element formulation. *Comput Methods Appl Mech Eng*, 120(1–2):1–32, 1995.

S. Sargsyan. The general dynamic theory of micropolar elastic thin shells. *Doklady Physics*, 56(1):39–42, 2011.

S. H. Sargsyan. Thermoelasticity of thin shells on the basis of asymmetrical theory of elasticity. *J Therm Stresses*, 32(8):791–818, 2009.

S. H. Sargsyan and A. H. Sargsyan. General dynamic theory of micropolar elastic thin plates with free rotation and special features of their natural oscillations. *Acoust Phys*, 57(4):473–481, 2011.

S. O. Sargsyan. On some interior and boundary effects in thin plates based on the asymmetric theory of elasticity. In R. Kienzler, H. Altenbach, and I. Ott, editors, *Theories of Plates and Shells: Critical Review and New Applications*, pages 201–210. Springer, Berlin, 2005.

S. O. Sargsyan. Boundary-value problems of the asymmetric theory of elasticity for thin plates. *J Appl Math Mech*, 72(1):77–86, 2008.

P. Schiavone. On existence theorems in the theory of extensional motions of thin micropolar plates. *Int J Eng Sci*, 27(9):1129–1133, 1989.

P. Schiavone. Uniqueness in dynamic problems of thin micropolar plates. *Appl Math Let*, 4(2):81–83, 1991.

P. Schiavone and C. Constanda. Existence theorems in the theory of bending of micropolar plates. *Int J Eng Sci*, 27(4):463–468, 1989.

L. I. Shkutin. Nonlinear models of deformable momental continua (in Russian). *J Appl Mech Techn Phys*, 6:111–117, 1980.

L. I. Shkutin. *Mechanics of Deformations of Flexible Bodies (in Russian).* Nauka, Novosibirsk, 1985.

J. G. Simmonds. The thermodynamical theory of shells: Descent from 3-dimensions without thickness expansions. In E. L. Axelrad and F. A. Emmerling, editors, *Flexible shells, theory and applications*, pages 1–11. Springer, Berlin, 1984.

J. G. Simmonds. Some comments on the status of shell theory at the end of the 20th century: Complaints and correctives. In *AIAA/ASME/ASCE/AHS/ASC Structures, Structural Dynamics, and Materials Conference and Exhibit, 38th, and AIAA/ASME/AHS Adaptive Structures Forum, Kissimmee, FL, Apr. 7-10, 1997, Collection of Technical Papers, Pt. 3 (A97-24112 05-39)*, pages 1912–1921. AIAA, 1997.

J. G. Simmonds and D. A Danielson. Nonlinear shell theory with finite rotation and stress-function vectors. *Trans ASME. J Appl Mech*, 39(4): 1085–1090, 1972.

J. Tambača and I. Velčić. Evolution model for linearized micropolar plates by the Fourier method. *J Elasticity*, 96(2):129–154, 2009.

S. P. Timoshenko and S. Woinowsky-Krieger. *Theory of Plates and Shells.* McGraw Hill, New York, 1985.

S. K. Tomar. Wave propagation in a micropolar elastic plate with voids. *J Vib Control*, 11(6):849–863, 2005.

C. Truesdell. Die Entwicklung des Drallsatzes. *ZAMM*, 44(4/5):149–158, 1964.

C. Truesdell. *A First Course in Rational Continuum Mechanics.* Academic Press, New York, 2nd edition, 1991.

C. Truesdell. *Rational Thermodynamics.* Springer, New York, 2nd edition, 1984.

C. Truesdell and W. Noll. The nonlinear field theories of mechanics. In S. Flügge, editor, *Handbuch der Physik, Vol. III/3*, pages 1–602. Springer, Berlin, 1965.

J. R. Turner and D. A. C. Nicol. The Signorini problem for a Cosserat plate. *J Inst Math and its Applications*, 25(2):133–145, 1980.

I. Velčić and J. Tambača. Relaxation theorem and lower-dimensional models in micropolar elasticity. *Math Mech Solids*, 15(8):812–853, 2010.

C. M. Wang, J. N. Reddy, and K. H. Lee. *Shear Deformable Beams and Shells.* Elsevier, Amsterdam, 2000.

F.-Y. Wang. On the solutions of Eringen's micropolar plate equations and of other approximate equations. *Int J Eng Sci*, 28(9):919–925, 1990.

F.-Y. Wang and Y. Zhou. On the vibration modes of three-dimensional micropolar elastic plates. *J Sound Vib*, 146(1):1–16, 1991.

K. Wiśniewski. *Finite Rotation Shells: Basic Equations and Finite Elements for Reissner Kinematics.* Lecture Notes on Numerical Methods in Engineering and Sciences. Springer, Berlin, 2010.

H. T. Y. Yang, S. Saigal, A. Masud, and R. K. Kapania. A survey of recent shell finite elements. *Int J Numer Meth Eng*, 47(1-3):101–127, 2000.

L. Zee and E. Sternberg. Ordinary and strong ellipticity in the equilibrium theory of incompressible hyperelastic solids. *Arch Ration Mech An*, 83 (1):53–90, 1983.

P. A. Zhilin. Mechanics of deformable directed surfaces. *Int J Solids Struct*, 12(9–10):635 – 648, 1976.

P. A. Zhilin. *Applied Mechanics. Foundations of the Theory of Shells (in Russian)*. St. Petersburg State Polytechnical University, Saint Petersburg, 2006.

L. M. Zubov. Micropolar-shell equilibrium equations. *Doklady Physics*, 54 (6):290–293, 2009.

L. M. Zubov. On universal deformations of nonlinear isotropic elastic shells. In H. Altenbach and V. A. Eremeyev, editors, *Shell-like Structures*, volume 15 of *Advanced Structured Materials*, pages 279–294. Springer, Berlin, 2011.

L. M. Zubov. Nonlinear theory of isolated and continuosly distributed dislocations in elastic shells. *Arch Civ Eng*, XLV(2):385–396, 1999.

L. M. Zubov. Semi-inverse solutions in nonlinear theory of elastic shells. *Arch Mech*, 53(4-5):599–610, 2001.

L. M. Zubov. *Nonlinear Theory of Dislocations and Disclinations in Elastic Bodies*. Springer, Berlin, 1997.

Cosserat-Type Rods

Holm Altenbach [†], Mircea Bîrsan [*♭], and Victor A. Eremeyev [†‡]
[†] Lehrstuhl Technische Mechanik, Institut für Mechanik,
Fakultät für Maschinenbau, Otto-von-Guericke-Universität Magdeburg,
Germany
[*] Lehrstuhl Nichtlineare Analysis und Modellierung, Fachbereich Mathematik,
Universität Duisburg-Essen, Germany
[♭] Department of Mathematics, University "A.I. Cuza" of Iaşi,
Romania
[‡] South Scientific Center of RASci and South Federal University, Rostov on Don,
Russia

Abstract In this chapter we discuss a Cosserat-type theory of rods. Cosserat-type rod theories are based on the consideration of a rod base curve as a deformable directed curve, that is a curve with attached deformable or non-deformable (rigid) vectors (directors), or based on the derivation of one-dimensional (1D) rod equations from the three-dimensional (3D) micropolar (Cosserat) continuum equations. In the literature are known theories of rods kinematics of which described by introduction of the translation vector and additionally p deformable directors or one deformable director or three unit orthogonal each other directors. The additional vector fields of directors describe the rotational (in some special cases additional) degrees of freedom of the rod. The aim of the chapter is to present a Cosserat-type theory of rods and to show various applications.

1 Introduction

The theory of rods is one of the oldest branches of mechanics. The first significant studies on the behavior of thin rods have been elaborated in the seventeenth century by Galilei and Bernoulli, then in the eighteenth century by Euler and D'Alembert, and in the nineteenth century by Bresse, Clebsch, Kirchhoff and Rayleigh, among others. In the twentieth century the classical models were significantly improved, for example by, Timoshenko and Levinson. Due to these contributions in the theory of rods, several new mathematical concepts and mechanical challenges have been formulated, which played an important role in the history of development of natural sciences.

Nowadays, the modern studies on the mechanical behavior of beams and rods have received considerable attention. The growing interest in this field is due to the intensive use of rod-like structures in mechanical and civil engineering. The emergence of new technologies and advanced materials in connection with rod manufacturing leads to the necessity of elaborating adequate models and to extend the existing theories.

In general, the theories of beams and rods allow for the approximate analysis of the stress-strain state of three-dimensional bodies for which two dimensions are much smaller in comparison with the third one. In this sense, all these theories are based on the thinness hypothesis. To obtain a set of one-dimensional approximate equations, one of the following main directions can be pursued: the application of kinematical and/or stress hypotheses, the use of mathematical techniques like series expansions and asymptotic analysis, and the so-called *direct approach* based on the deformable curve model. As examples in the first direction we can mention the beam theories of Euler and of Timoshenko, see e.g. Timoshenko (1921); Svetlitsky (2000); Hodges (2006) for details. Mathematical techniques for the study of rods and beams include the use of formal asymptotic expansions (Trabucho and Viaño, 1996; Tiba and Vodak, 2005; Berdichevsky, 2009), the Γ-convergence analysis (Freddi et al., 2007) and other variational methods (Meunier, 2008; Sprekels and Tiba, 2009).

The direct approach has been introduced for the first time by Euler. He started with the stress resultants and formulated two independent equations of motion (in particular, static equilibrium). Later the Cosserat brothers (Cosserat and Cosserat, 1909) elaborated a general theory. As a model for rods, they have considered a deformable curve in which every material point is connected to a triad of orthonormal vectors (also called *directors*) to characterize its orientation. Later, this idea has been modified and developed by Green and Naghdi (Green et al., 1974; Green and Naghdi, 1979) who created the so-called theory of Cosserat curves, in which every material point is attached to a pair of deformable directors. We mention that the Cosserat theory for rods was developed in parallel with the Cosserat theory for shells. These two models have been presented and analyzed in details in the books of Antman (1995) and Rubin (2000).

Another, but also direct approach for shells and rods has been elaborated by Zhilin (1976, 2006a,b, 2007), who followed the original idea of the Cosserats and considered deformable continua (surfaces or curves) endowed with a triad of rigidly rotating orthonormal vectors connected to each point. In his approach, Zhilin has supplemented the kinematical model suggested by Cosserat with appropriate constitutive equations, thus making the model applicable to solve practical problems (Altenbach et al., 2006). We also mention that the latter approach has the attribute of simplicity, in comparison with the theories of Cosserat surfaces or curves given by Green and Naghdi. The approach to rods proposed by Zhilin is also called the theory of *directed curves*.

In principle, the main advantage of any direct approach is that it does not require hypotheses about the through-the-thickness distributions of displacement and stress fields or the mathematical manipulations with three-dimensional equations. Furthermore, it can be employed to study very thin structures, where the use of a three-dimensional theory is not convenient. In the direct approach, the basic laws of mechanics and thermodynamics are applied directly to a dimensionally reduced continuum (i.e. surface or curve) and thus one can obtain quite accurate equations. However, the formulation of constitutive equations for such models presents some difficulties. Indeed, it is known that the elasticity tensors depend on the geometry of the rod or shell. In order to establish their structure, it is necessary to apply the generalized theory of tensor symmetry (Zhilin, 2006b). Then, a crucial aspect in this model is the identification of the effective properties (stiffness, etc.) of the structure. In the case of shells and plates, the determination of effective stiffness has been realized for various types of materials in Altenbach and Zhilin (1988); Altenbach (2000); Altenbach and Eremeyev (2008, 2009). For the direct approach to rods, the effective properties have been determined only in the case of isotropic and homogeneous materials in Zhilin (2006a, 2007).

Below we present the identification of effective stiffness for rods made of an orthotropic material with voids. We extend the theory of directed curves to include porosity and thermal effects. To describe the porosity, we employ the Nunziato-Cowin theory for elastic materials with voids (Nunziato and Cowin, 1979; Cowin and Nunziato, 1983). An additional degree of freedom is introduced in this theory, namely the volume fraction field, which characterizes the continuous distribution of voids in the body. The Nunziato-Cowin theory is intended for the study of porous solids and granular materials, and it can be regarded as a special case of the theory of materials with microstructure (Capriz, 1989; Ciarletta and Ieşan, 1993).

2 Kinematical Model of Directed Curves

In what follows, we consider thin rods modeled as directed curves, i.e. deformable curves endowed with a triad of directors $\boldsymbol{d}_i$ attached to every point (Zhilin, 2006a, 2007). The motion of the directors describes the rotations of rod's cross-sections during deformation.

Let us denote by $\mathcal{C}_0$ the curve for the reference (initial) configuration of

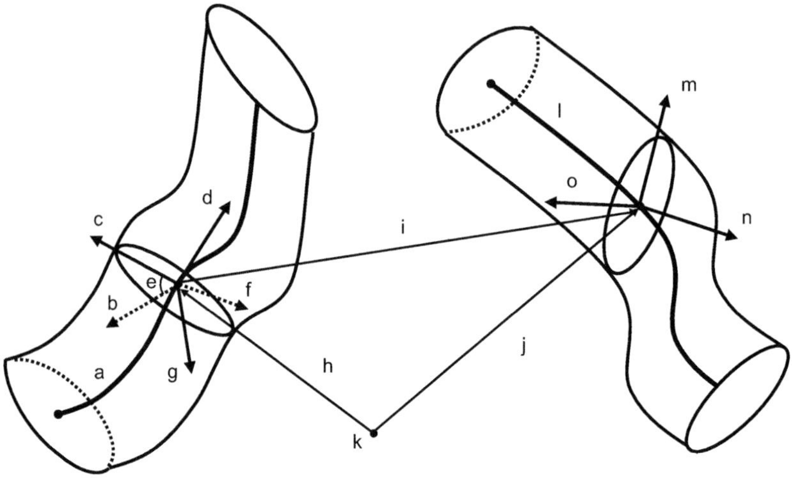

Figure 1. Deformation of a thin rod. Reference configuration and deformed configuration at time t.

the rod and by s the material coordinate along this curve which is chosen to be the arclength parameter of $\mathcal{C}_0$. The directed curve is specified by the position vector $\boldsymbol{r}(s)$ and the triad of directors $\boldsymbol{d}_i(s)$, $i = 1, 2, 3$ (see Fig. 1). We designate by $(\ldots)' = \frac{\mathrm{d}}{\mathrm{d}s}$ the derivative with respect to s. The initial directors $\boldsymbol{d}_1, \boldsymbol{d}_2, \boldsymbol{d}_3$ are unit vectors orthogonal to each other, and $\boldsymbol{d}_3$ coincides with the unit tangent vector

$$\boldsymbol{d}_3 = \boldsymbol{t} \equiv \boldsymbol{r}'(s), \qquad \boldsymbol{d}_i \cdot \boldsymbol{d}_j = \delta_{ij} \qquad \text{(the Kronecker symbol).}$$

We can see that the plane $(\boldsymbol{d}_1, \boldsymbol{d}_2)$ is the cross-section plane. In general, the directors $\boldsymbol{d}_1$ and $\boldsymbol{d}_2$ are chosen along the principal axes of inertia of the cross-section, and the curve $\mathcal{C}_0$ is the line of centroids.

Consider also the orthonormal triad $\{\boldsymbol{t}(s), \boldsymbol{n}(s), \boldsymbol{b}(s)\}$ naturally attached to the curve, consisting of the tangent vector $\boldsymbol{t}$, the principal normal $\boldsymbol{n}$ and the binormal $\boldsymbol{b}$. We denote by $\sigma(s)$ the angle between the vectors $\boldsymbol{d}_1(s)$ and $\boldsymbol{n}(s)$, which is called the *angle of natural twisting* of the rod, and we have

$$\boldsymbol{d}_1 = \boldsymbol{n}\cos\sigma + \boldsymbol{b}\sin\sigma, \qquad \boldsymbol{d}_2 = -\boldsymbol{n}\sin\sigma + \boldsymbol{b}\cos\sigma. \tag{1}$$

From the geometry of curves, we know that the following relations hold

$$\begin{gathered} \boldsymbol{t}' = \boldsymbol{\tau} \times \boldsymbol{t}, \qquad \boldsymbol{n}' = \boldsymbol{\tau} \times \boldsymbol{n}, \qquad \boldsymbol{b}' = \boldsymbol{\tau} \times \boldsymbol{b}, \\ \boldsymbol{\tau} = \frac{1}{R_t}\boldsymbol{t} + \frac{1}{R_c}\boldsymbol{b}, \qquad \frac{1}{R_c} = |\boldsymbol{r}''(s)|, \qquad \frac{1}{R_t} = \frac{|(\boldsymbol{r}', \boldsymbol{r}'', \boldsymbol{r}''')|}{|\boldsymbol{r}''|^2}, \end{gathered} \tag{2}$$

where $\boldsymbol{\tau}(s)$ is the Darboux vector, R_c the radius of curvature and R_t the radius of twisting for the curve $\mathcal{C}_0$. Then, in view of (1), we can show that

$$\boldsymbol{d}_i' = \boldsymbol{q} \times \boldsymbol{d}_i, \quad i = 1, 2, 3, \qquad \text{where} \quad \boldsymbol{q} = \sigma' \boldsymbol{t} + \boldsymbol{\tau}.$$

The deformed configuration $\mathcal{C}$ of the rod at time t is described by

$$\boldsymbol{R} = \boldsymbol{R}(s,t), \qquad \boldsymbol{D}_i = \boldsymbol{D}_i(s,t), \qquad i = 1, 2, 3, \quad s \in [0, l], \tag{3}$$

where the three directors after deformation $\boldsymbol{D}_i$ satisfy the relations $\boldsymbol{D}_i \cdot \boldsymbol{D}_j = \delta_{ij}$, but $\boldsymbol{D}_3$ is no longer tangent to the curve $\mathcal{C}$. This means that the initial cross-sections are not necessarily normal to the middle curve after deformation. We will assume that the rod's cross-sections do not change their shape during deformation, but only rotate with respect to the the middle curve.

The *displacement vector* $\boldsymbol{u}$ and the *rotation tensor* $\boldsymbol{P}$ are defined by

$$\boldsymbol{u}(s,t) = \boldsymbol{R}(s,t) - \boldsymbol{r}(s), \qquad \boldsymbol{P}(s,t) = \boldsymbol{D}_k(s,t) \otimes \boldsymbol{d}_k(s). \tag{4}$$

Denoting the time derivative by a superposed dot, the velocity vector $\boldsymbol{V}$ and the angular velocity vector $\boldsymbol{\omega}$ are given by

$$\boldsymbol{V}(s,t) = \dot{\boldsymbol{R}}(s,t), \qquad \dot{\boldsymbol{P}}(s,t) = \boldsymbol{\omega}(s,t) \times \boldsymbol{P}(s,t). \tag{5}$$

Hence, $\boldsymbol{\omega}$ is the axial vector of the antisymmetric tensor $\dot{\boldsymbol{P}} \cdot \boldsymbol{P}^{\mathrm{T}}$ and it can be expressed as $\boldsymbol{\omega} = -\frac{1}{2}\left[\dot{\boldsymbol{P}} \cdot \boldsymbol{P}^{\mathrm{T}}\right]_{\times}$, where $[\ldots]_{\times}$ designates the vector invariant (or Gibbsian cross). We employ the usual summation convention over repeated indices and the direct tensor notation in the sense of Lurie (2005), see e.g. Naumenko and Altenbach (2007); Lebedev et al. (2010).

The absolute temperature field in the points of the curve $\mathcal{C}$ will be denoted by $\theta(s,t)$, with $\theta > 0$. In the reference configuration $\mathcal{C}_0$ we assume that the temperature θ_0 is constant. To describe the porosity of the rod, we follow the Nunziato-Cowin theory for elastic material with voids (Nunziato and Cowin, 1979; Cowin and Nunziato, 1983) and introduce the so-called *volume fraction field* $\nu = \nu(s,t)$, which is an additional independent variable $(0 < \nu \leq 1)$. In this model the mass density of the porous rod $\rho(s,t)$ can be decomposed as the product of the mass density of the matrix material γ and the volume fraction field: $\rho(s,t) = \nu(s,t)\gamma(s,t)$. Thus, the porosity function ν will express the continuous distribution of pores along the rod's middle curve. In the reference configuration $\mathcal{C}_0$ we have the relation $\rho_0(s) = \nu_0(s)\gamma_0(s)$.

3 Governing Equations of the Non-Linear Theory

To describe the deformation of thermoelastic porous rods, one needs to formulate appropriate balance equations. These balance laws are postulated directly for the one-dimensional continuum and they express, respectively, the following principles accepted in Continuum Mechanics: the balances of mass, linear momentum, angular momentum, energy, and entropy, see, for example, (Altenbach and Altenbach, 1994). In this case one has to modify these balances slightly - instead of the balance of mass, the conservation of mass is assumed. In addition, we use the principle of equilibrated force which was introduced first in Goodman and Cowin (1972); Nunziato and Cowin (1979). The last principle can be seen as the special case of a balance equation which arises in the microstructural theories of elastic materials developed by Mindlin (1964) and Toupin (1964), where only the dilatation of the micromedium is considered; see also Capriz and Podio–Guidugli (1981); Capriz (1989); Jenkins (1975); Cowin and Goodman (1976); Cowin and Leslie (1980) for the discussions on various approaches leading to this type of balance equation.

The principle of mass conservation can be written as

$$\int_{s_1}^{s_2} \rho_0(s)\mathrm{d}s = \int_{s_1}^{s_2} \rho(s,t)\mathrm{d}S(s,t), \quad \forall s_1, s_2 \in [0,l], \tag{6}$$

where $\mathrm{d}S$ is the length element on the deformed curve $\mathcal{C}$ and we have $\mathrm{d}S(s,t) = |\boldsymbol{R}'(s,t)|\mathrm{d}s$. The relation (6) can be put in the local form

$$\rho(s,t) = \frac{\rho_0(s)}{1+\epsilon(s,t)} \qquad \text{or} \qquad \nu(s,t)\gamma(s,t) = \frac{\nu_0(s)\gamma_0(s)}{1+\epsilon(s,t)}, \tag{7}$$

where $\epsilon(s,t) \equiv \dfrac{\mathrm{d}S(s,t) - \mathrm{d}s}{\mathrm{d}s} = |\boldsymbol{R}'(s,t)| - 1$ is the relative dilatation of the rod. From Eq. (7) we can determine the mass density of the matrix elastic material $\gamma(s,t)$ in terms of $\boldsymbol{R}(s,t)$ and $\nu(s,t)$.

The kinetic energy $\mathcal{K}$, the linear momentum $\boldsymbol{\mathcal{K}}_1$ and the moment of momentum $\boldsymbol{\mathcal{K}}_2$ per unit mass of the rod are defined by

$$\begin{aligned}
\mathcal{K}(s,t) &= \frac{1}{2}\boldsymbol{V}\cdot\boldsymbol{V} + \boldsymbol{V}\cdot\boldsymbol{\Theta}_1(s,t)\cdot\boldsymbol{\omega} + \frac{1}{2}\boldsymbol{\omega}\cdot\boldsymbol{\Theta}_2(s,t)\cdot\boldsymbol{\omega} + \frac{1}{2}\varkappa\dot{\nu}^2,\\
\boldsymbol{\mathcal{K}}_1(s,t) &= \frac{\partial\mathcal{K}}{\partial\boldsymbol{V}} = \boldsymbol{V} + \boldsymbol{\Theta}_1\cdot\boldsymbol{\omega},\\
\boldsymbol{\mathcal{K}}_2(s,t) &= \frac{\partial\mathcal{K}}{\partial\boldsymbol{\omega}} + \boldsymbol{R}\times\frac{\partial\mathcal{K}}{\partial\boldsymbol{V}} = \boldsymbol{\Theta}_1^{\mathrm{T}}\cdot\boldsymbol{V} + \boldsymbol{\Theta}_2\cdot\boldsymbol{\omega} + \boldsymbol{R}\times\boldsymbol{\mathcal{K}}_1.
\end{aligned} \tag{8}$$

Here $\varkappa = \varkappa(s) > 0$ is an inertia coefficient associated to the porosity variable, and the second order tensors $\boldsymbol{\Theta}_1(s,t), \boldsymbol{\Theta}_2(s,t)$ are the tensors of inertia per unit mass.

Remark 3.1. In order to present the expressions of the inertia tensors, let us consider the three-dimensional rod and the position vector $\boldsymbol{r}^*(s,x,y)$ of a generic point in the reference configuration

$$\boldsymbol{r}^*(s,x,y) = \boldsymbol{r}(s) + \boldsymbol{a}(s,x,y), \quad \boldsymbol{a}(s,x,y) = x\boldsymbol{d}_1(s) + y\boldsymbol{d}_2(s), \quad (x,y) \in \Sigma,$$

where Σ is the cross-section domain and (x,y) are the material coordinates referred to the base vectors $\{\boldsymbol{d}_1, \boldsymbol{d}_2\}$. Since we assume that the cross-sections do not deform, the same material point will have the following position vector at the time t

$$\boldsymbol{R}^*(s,x,y,t) = \boldsymbol{R}(s,t) + x\boldsymbol{D}_1(s,t) + y\boldsymbol{D}_2(s,t), \qquad (x,y) \in \Sigma.$$

From these relations we see that the cross-sections of the rod can rotate with respect to the middle curve, but the deformation of the cross-sections is not taken into account. The inertia tensors in the reference configuration $\boldsymbol{\Theta}^0_\alpha(s)$, $\alpha = 1,2$, have the representation (Zhilin, 2007)

$$\rho_0\boldsymbol{\Theta}^0_1 = -\int_\Sigma (\mathbf{1}\times\boldsymbol{a})\rho^*\tilde{\mu}\mathrm{d}x\mathrm{d}y, \quad \rho_0\boldsymbol{\Theta}^0_2 = -\int_\Sigma [(\boldsymbol{a}\cdot\boldsymbol{a})\mathbf{1} - \boldsymbol{a}\otimes\boldsymbol{a}]\rho^*\tilde{\mu}\mathrm{d}x\mathrm{d}y, \quad (9)$$

where $\rho_0(s)$ is the mass density of the deformable curve $\mathcal{C}_0$, $\mathbf{1}$ is the unit tensor of second order, ρ^* is the mass density in the three-dimensional rod and the scalar $\tilde{\mu}(s,x,y) \equiv 1 + (\boldsymbol{a}\cdot\boldsymbol{n}/R_c)$. The inertia tensors in the current configuration $\mathcal{C}$ are expressed by

$$\boldsymbol{\Theta}_\alpha(s,t) = \boldsymbol{P}(s,t)\cdot\boldsymbol{\Theta}^0_\alpha(s)\cdot\boldsymbol{P}^{\mathrm{T}}(s,t), \qquad \alpha = 1,2. \qquad (10)$$

From relations (9) and (10) we can prove that $\boldsymbol{\Theta}_1$ is antisymmetric, $\boldsymbol{\Theta}_2$ is symmetric and we can prove that the tensor $\boldsymbol{\Theta}_2 - \boldsymbol{\Theta}_1^{\mathrm{T}}\cdot\boldsymbol{\Theta}_1$ is symmetric and positive definite (see, for example, Bîrsan and Bîrsan, 2011).

The principles of linear momentum and the moment of momentum are expressed by the following relations: for any part of the deformable curve, delimited by the coordinates $s_1, s_2 \in [0,l]$ (where l is the length of the curve $\mathcal{C}_0$), hold

$$\begin{aligned}
\int_{s_1}^{s_2} \rho_0\dot{\boldsymbol{\mathcal{K}}}_1(s,t)\mathrm{d}s &= \int_{s_1}^{s_2} \rho_0\boldsymbol{\mathcal{F}}(s,t)\mathrm{d}s + \boldsymbol{N}_{(\boldsymbol{t})}(s_2,t) + \boldsymbol{N}_{(-\boldsymbol{t})}(s_1,t),\\
\int_{s_1}^{s_2} \rho_0\dot{\boldsymbol{\mathcal{K}}}_2(s,t)\mathrm{d}s &= \int_{s_1}^{s_2} \rho_0\big[\boldsymbol{R}(s,t)\times\boldsymbol{\mathcal{F}}(s,t) + \boldsymbol{\mathcal{L}}(s,t)\big]\mathrm{d}s \qquad (11)\\
&\quad + \boldsymbol{R}(s_2,t)\times\boldsymbol{N}_{(\boldsymbol{t})}(s_2,t) + \boldsymbol{R}(s_1,t)\times\boldsymbol{N}_{(-\boldsymbol{t})}(s_1,t)\\
&\quad + \boldsymbol{M}_{(\boldsymbol{t})}(s_2,t) + \boldsymbol{M}_{(-\boldsymbol{t})}(s_1,t),
\end{aligned}$$

where $\boldsymbol{\mathcal{F}}$ and $\boldsymbol{\mathcal{L}}$ are the external body force and moment per unit mass, while $\boldsymbol{N}_{(\boldsymbol{t})}$ and $\boldsymbol{M}_{(\boldsymbol{t})}$ represent the external forces and moments acting on the boundary points $s = s_1, s_2$ of an arbitrary portion of the rod.

The porosity fields are governed by the principle of equilibrated force, which in the case of rods has the form

$$\begin{aligned}\int_{s_1}^{s_2} \rho_0 \frac{\mathrm{d}}{\mathrm{d}t}\big(\varkappa(s)\dot{\nu}(s,t)\big)\mathrm{d}s = &\int_{s_1}^{s_2} \big(\rho_0(s)p(s,t) - g(s,t)\big)\mathrm{d}s \\ &+ h_{(\boldsymbol{t})}(s_2,t) + h_{(-\boldsymbol{t})}(s_1,t), \forall s_1, s_2 \in [0,l],\end{aligned} \tag{12}$$

where p is the assigned equilibrated body force, g is the internal equilibrated body force and $h_{(\boldsymbol{t})}$ represents the equilibrated stress. This principle has been formulated by Goodman and Cowin (1972) for granular materials and by Nunziato and Cowin (1979) for elastic materials with voids. It can be regarded as a special case of a balance equation which arises in the microstructural theories of continua established by Mindlin (1964) and Toupin (1964), when only the dilatation of the micromedium is taken into account. This equation was derived from a variational argument by Cowin and Goodman (1976) and given mechanical interpretation by Jenkins (1975). There exist various approaches leading to this type of balance equations, which are discussed in Cowin and Leslie (1980); Capriz and Podio–Guidugli (1981); Capriz (1989).

Applying the principles (11) and (12) to parts of the rods with the infinitesimal length, we deduce that we can introduce the fields $\boldsymbol{N}$, $\boldsymbol{M}$ and h by

$$\boldsymbol{N} \equiv \boldsymbol{N}_{(\boldsymbol{t})} = -\boldsymbol{N}_{(-\boldsymbol{t})}, \boldsymbol{M} \equiv \boldsymbol{M}_{(\boldsymbol{t})} = -\boldsymbol{M}_{(-\boldsymbol{t})}, h \equiv h_{(\boldsymbol{t})} = -h_{(-\boldsymbol{t})}. \tag{13}$$

Concerning the interpretation given to the porosity variables, the function ν expresses the compression-dilation of the continuously distributed pores, while the equilibrated stress h are identified in Cowin and Nunziato (1983) with singular stress systems known in classical elasticity as double force systems without moments (Love, 1944).

Denoting by $\mathcal{U}(s,t)$ the internal energy per unit mass of the rod, the balance of energy is expressed by

$$\int_{s_1}^{s_2} \rho_0(\dot{\mathcal{K}}+\dot{\mathcal{U}})\mathrm{d}s = \int_{s_1}^{s_2} \rho_0\big(\boldsymbol{\mathcal{F}}\cdot\boldsymbol{V}+\boldsymbol{\mathcal{L}}\cdot\boldsymbol{\omega}+p\dot{\nu}+S\big)\mathrm{d}s+\big(\boldsymbol{N}\cdot\boldsymbol{V}+\boldsymbol{M}\cdot\boldsymbol{\omega}+h\dot{\nu}+q\big)\Big|_{s_1}^{s_2} \tag{14}$$

for all $s_1, s_2 \in [0,l]$, where we employ the notation $f\Big|_{s_1}^{s_2} = f(s_2) - f(s_1)$ for any f. Here S is the external rate of the heat supply per unit mass and $q \equiv q_{(\boldsymbol{t})} = -q_{(-\boldsymbol{t})}$ is the heat flux along the rod.

The force vector $\boldsymbol{N}$ and the moment vector $\boldsymbol{M}$ are defined by the relations (11) and $(13)_{1,2}$ and they are dual mechanical quantities to the corresponding kinematical fields such that $\boldsymbol{N}\cdot\boldsymbol{V}+\boldsymbol{M}\cdot\boldsymbol{\omega}$ represents mechanical power. Also, the equilibrated stress h defined by relations (12) and $(13)_3$ is dual to the porosity variable ν, and the product $h\dot{\nu}$ is included in the expression of the mechanical power for porous rods, see equation (14).

Let $\eta(s,t)$ designate the specific entropy function. Then, the Clausius-Duhem inequality for the entropy of the rod is written as

$$\int_{s_1}^{s_2}\rho_0\dot{\eta}\mathrm{d}s \geq \int_{s_1}^{s_2}\rho_0\frac{S}{\theta}\mathrm{d}s+\left(\frac{q}{\theta}\right)\Big|_{s_1}^{s_2},\quad \forall s_1,s_2\in[0,l]. \tag{15}$$

In view of the integral forms of the principles (11)-(15), we obtain the following local equations:

- equations of motion:

$$\begin{aligned}&\boldsymbol{N}'(s,t)+\rho_0\boldsymbol{\mathcal{F}}=\rho_0\frac{\mathrm{d}}{\mathrm{d}t}(\boldsymbol{V}+\boldsymbol{\Theta}_1\cdot\boldsymbol{\omega}),\\&\boldsymbol{M}'(s,t)+\boldsymbol{R}'\times\boldsymbol{N}(s,t)+\rho_0\boldsymbol{\mathcal{L}}\\&=\rho_0\big[\boldsymbol{V}\times\boldsymbol{\Theta}_1\cdot\boldsymbol{\omega}+\frac{\mathrm{d}}{\mathrm{d}t}(\boldsymbol{V}\cdot\boldsymbol{\Theta}_1+\boldsymbol{\Theta}_2\cdot\boldsymbol{\omega})\big];\end{aligned} \tag{16}$$

- equation of equilibrated force:

$$h'(s,t)-g(s,t)+\rho_0 p=\rho_0\frac{\mathrm{d}}{\mathrm{d}t}(\varkappa\dot{\nu}); \tag{17}$$

- energy balance equation:

$$\rho_0\dot{\mathcal{U}}=\mathcal{P}+\rho_0 S+q', \tag{18}$$

where

$$\mathcal{P}=\boldsymbol{N}\cdot(\boldsymbol{V}'+\boldsymbol{R}'\times\boldsymbol{\omega})+\boldsymbol{M}\cdot\boldsymbol{\omega}'+g\dot{\nu}+h\dot{\nu}'. \tag{19}$$

- entropy inequality:

$$\rho_0\theta\dot{\eta}\geq\rho_0 S+q'-\frac{\theta'}{\theta}q, \tag{20}$$

which must be valid for every point s and time t.

Equations (16) are similar to the one-dimensional equations for rods found in many previous works (especially for the left-hand sides), but Eq. (17) represents the equation of equilibrated force which governs the porosity fields in our model. The assigned body loads $\boldsymbol{\mathcal{F}}$, $\boldsymbol{\mathcal{L}}$, p and S also include the contributions of the (mechanical and thermal) loads acting on the lateral boundaries of the three-dimensional rod, as we will see later on.

Let us consider the Helmholtz free energy function Ψ defined by

$$\Psi = \mathcal{U} - \theta\eta. \tag{21}$$

Then, the energy balance equation (18) can be written in the form

$$\rho_0\dot{\Psi} + \rho_0(\theta\dot{\eta} + \dot{\theta}\eta) - \mathcal{P} - q' - \rho_0 S = 0. \tag{22}$$

Inserting (22) into (20) we obtain the following form of the entropy inequality

$$-\rho_0\dot{\Psi} - \rho_0\dot{\theta}\eta + \mathcal{P} + \frac{\theta'}{\theta}q \geq 0. \tag{23}$$

In order to write the energy equation in a convenient form, we introduce the *vectors of deformation* defined in Zhilin (2006b). The *vector of extension-shear* $\boldsymbol{\mathcal{E}}$ and the *vector of bending-twisting* $\boldsymbol{\Phi}$ are given by

$$\boldsymbol{\mathcal{E}} = \boldsymbol{R}' - \boldsymbol{P}\cdot\boldsymbol{t}, \qquad \boldsymbol{P}' = \boldsymbol{\Phi}\times\boldsymbol{P}, \tag{24}$$

i.e. $\boldsymbol{\Phi}$ is the axial vector of the antisymmetric tensor $\boldsymbol{P}'\cdot\boldsymbol{P}^{\mathrm{T}}$ or equivalently $\boldsymbol{\Phi} = -\frac{1}{2}\left[\boldsymbol{P}'\cdot\boldsymbol{P}^{\mathrm{T}}\right]_{\times}$. The geometrical and physical interpretations of these deformation vectors have been presented in Zhilin (2007, Sect. 5). Using the relations

$$\dot{\boldsymbol{\mathcal{E}}} - \boldsymbol{\omega}\times\boldsymbol{\mathcal{E}} = \boldsymbol{V}' + \boldsymbol{R}'\times\boldsymbol{\omega}, \qquad \dot{\boldsymbol{\Phi}} - \boldsymbol{\omega}\times\boldsymbol{\Phi} = \boldsymbol{\omega}',$$

the function $\mathcal{P}$ defined by (19) can be written as

$$\mathcal{P} = \boldsymbol{N}\cdot(\dot{\boldsymbol{\mathcal{E}}} - \boldsymbol{\omega}\times\boldsymbol{\mathcal{E}}) + \boldsymbol{M}\cdot(\dot{\boldsymbol{\Phi}} - \boldsymbol{\omega}\times\boldsymbol{\Phi}) + g\dot{\nu} + h\dot{\nu}'. \tag{25}$$

We introduce the energetic vectors of deformation $\boldsymbol{\mathcal{E}}_*$ and $\boldsymbol{\Phi}_*$ given by

$$\boldsymbol{\mathcal{E}}_* = \boldsymbol{P}^{\mathrm{T}}\cdot\boldsymbol{\mathcal{E}}, \qquad \boldsymbol{\Phi}_* = \boldsymbol{P}^{\mathrm{T}}\cdot\boldsymbol{\Phi}. \tag{26}$$

Then, the equation (25) and the relations

$$\dot{\boldsymbol{\mathcal{E}}}_* = \boldsymbol{P}^{\mathrm{T}}\cdot\left(\dot{\boldsymbol{\mathcal{E}}} - \boldsymbol{\omega}\times\boldsymbol{\mathcal{E}}\right), \qquad \dot{\boldsymbol{\Phi}}_* = \boldsymbol{P}^{\mathrm{T}}\cdot\left(\dot{\boldsymbol{\Phi}} - \boldsymbol{\omega}\times\boldsymbol{\Phi}\right)$$

yield

$$\mathcal{P} = (\boldsymbol{N}\cdot\boldsymbol{P})\cdot\dot{\boldsymbol{\mathcal{E}}}_* + (\boldsymbol{M}\cdot\boldsymbol{P})\cdot\dot{\boldsymbol{\Phi}}_* + g\dot{\nu} + h\dot{\nu}'. \tag{27}$$

The reduced energy balance equation takes the form (18) with $\mathcal{P}$ given by (27).

4 Constitutive Equations for Thermoelastic Porous Rods

For thermoelastic porous rods we adopt the following constitutive assumptions: the fields Ψ, $\boldsymbol{N}$, $\boldsymbol{M}$, g, h, η and q depend only on the variables $\boldsymbol{\mathcal{E}}$, $\boldsymbol{\Phi}$, $\boldsymbol{P}$, ν, ν', θ and θ' (and on the point s). Using the requirements of invariance under superposed rigid body motions, we deduce the relations

$$\begin{aligned}&\{\Psi, g, h, \eta, q\}(\boldsymbol{\mathcal{E}}, \boldsymbol{\Phi}, \boldsymbol{P}, \nu, \nu', \theta, \theta')\\&= \{\Psi, g, h, \eta, q\}(\boldsymbol{Q}\cdot\boldsymbol{\mathcal{E}}, \boldsymbol{Q}\cdot\boldsymbol{\Phi}, \boldsymbol{Q}\cdot\boldsymbol{P}, \nu, \nu', \theta, \theta'),\\&\{\boldsymbol{Q}\cdot\boldsymbol{N}, \boldsymbol{Q}\cdot\boldsymbol{M}\}(\boldsymbol{\mathcal{E}}, \boldsymbol{\Phi}, \boldsymbol{P}, \nu, \nu', \theta, \theta')\\&= \{\boldsymbol{N}, \boldsymbol{M}\}(\boldsymbol{Q}\cdot\boldsymbol{\mathcal{E}}, \boldsymbol{Q}\cdot\boldsymbol{\Phi}, \boldsymbol{Q}\cdot\boldsymbol{P}, \nu, \nu', \theta, \theta'),\end{aligned} \tag{28}$$

which must hold true for any proper orthogonal tensor $\boldsymbol{Q}$. Taking $\boldsymbol{Q} = \boldsymbol{P}^{\mathrm{T}}$ in the relations (28) and using the definitions of $\boldsymbol{\mathcal{E}}_*$ and $\boldsymbol{\Phi}_*$ we derive that Ψ is a function of the following variables

$$\Psi = \Psi(\boldsymbol{\mathcal{E}}_*, \boldsymbol{\Phi}_*, \nu, \nu', \theta, \theta'), \tag{29}$$

and the fields $\boldsymbol{N}\cdot\boldsymbol{P}$, $\boldsymbol{M}\cdot\boldsymbol{P}$, g, h, η and q depend only on the variables $\boldsymbol{\mathcal{E}}_*$, $\boldsymbol{\Phi}_*$, ν, ν', θ and θ' (and on the point s).

By the classical procedure we insert the expression (29) into the entropy inequality (23) and using Eq. (27) we finally obtain the following restrictions on the constitutive equations

$$\begin{gathered}\Psi = \Psi(\boldsymbol{\mathcal{E}}_*, \boldsymbol{\Phi}_*, \nu, \nu', \theta), \quad \eta = -\frac{\partial \Psi}{\partial \theta},\\ \boldsymbol{N} = \frac{\partial(\rho_0\Psi)}{\partial \boldsymbol{\mathcal{E}}_*}\cdot\boldsymbol{P}^{\mathrm{T}},\ \boldsymbol{M} = \frac{\partial(\rho_0\Psi)}{\partial \boldsymbol{\Phi}_*}\cdot\boldsymbol{P}^{\mathrm{T}},\ g = \frac{\partial(\rho_0\Psi)}{\partial \nu},\ h = \frac{\partial(\rho_0\Psi)}{\partial(\nu')}\end{gathered} \tag{30}$$

and

$$q = q(\boldsymbol{\mathcal{E}}_*, \boldsymbol{\Phi}_*, \nu, \nu', \theta, \theta'). \tag{31}$$

The entropy inequality (23) reduces to

$$q\theta' \geq 0, \tag{32}$$

which must be valid for any process. In view of the constitutive equations (30) we have the relation $\rho_0\dot{\Psi} = \mathcal{P} - \rho_0\eta\dot{\theta}$, and inserting it into (22) we obtain the reduced energy balance equation in the form

$$q' + \rho_0 S = \rho_0\theta\dot{\eta}. \tag{33}$$

4.1 Free Energy Function

The presentation of the constitutive equations will be complete if we specify the expression for the free energy function $(30)_1$. In view of the developments for constitutive equations in the purely elastic case (Zhilin, 2006a, 2007), we consider the following form for Ψ

$$\begin{aligned}
\rho_0\Psi &= \Psi_0 + \boldsymbol{N}_0\cdot\boldsymbol{\mathcal{E}}_* + \boldsymbol{M}_0\cdot\boldsymbol{\Phi}_* + \frac{1}{2}\boldsymbol{\mathcal{E}}_*\cdot\boldsymbol{A}\cdot\boldsymbol{\mathcal{E}}_* + \boldsymbol{\mathcal{E}}_*\cdot\boldsymbol{B}\cdot\boldsymbol{\Phi}_* \\
&+ \frac{1}{2}\boldsymbol{\Phi}_*\cdot\boldsymbol{C}\cdot\boldsymbol{\Phi}_* + \boldsymbol{\Phi}_*\cdot(\boldsymbol{\mathcal{E}}_*\cdot\boldsymbol{D})\cdot\boldsymbol{\Phi}_* + \frac{1}{2}K_1\nu^2 + \frac{1}{2}K_2(\nu')^2 \\
&+ K_3\nu\nu' + (\boldsymbol{K}_4\cdot\boldsymbol{\mathcal{E}}_*)\nu + (\boldsymbol{K}_5\cdot\boldsymbol{\Phi}_*)\nu + (\boldsymbol{K}_6\cdot\boldsymbol{\mathcal{E}}_*)\nu' + (\boldsymbol{K}_7\cdot\boldsymbol{\Phi}_*)\nu' \\
&- (\boldsymbol{G}_1\cdot\boldsymbol{\mathcal{E}}_*)\theta - (\boldsymbol{G}_2\cdot\boldsymbol{\Phi}_*)\theta - G_3\nu\theta - G_4\nu'\theta - \frac{1}{2}G\theta^2,
\end{aligned} \tag{34}$$

where Ψ_0, K_1, K_2, K_3, G_3, G_4 and G are scalars, $\boldsymbol{N}_0$, $\boldsymbol{M}_0$, $\boldsymbol{G}_1$, $\boldsymbol{G}_2$, $\boldsymbol{K}_4$, $\ldots$, $\boldsymbol{K}_7$ are vectors, $\boldsymbol{A}$, $\boldsymbol{B}$, $\boldsymbol{C}$ are tensors of second order and $\boldsymbol{D}$ is a tensor of third order. All properties are defined in the reference configuration. The vectors $\boldsymbol{N}_0$, $\boldsymbol{M}_0$ describe the initial forces and moments acting in the reference configurations, $\boldsymbol{A}$ and $\boldsymbol{C}$ are the tensors of tangential and bending stiffness, respectively, while $\boldsymbol{B}$ is the tensor describing the coupling between stretching and bending. The presence of cubic terms involving $\boldsymbol{D}$ in (34) is useful for explaining the Poynting effect (Zhilin, 2006a, 2007). $\boldsymbol{K}_5$ and $\boldsymbol{K}_7$ describe the coupling between porosity and bending, $\boldsymbol{G}_2$ accounts for the coupling between temperature and bending deformations. On the other hand, the coupling of the extension–shear deformations with temperature is characterized by $\boldsymbol{G}_1$, while the coupling of the same type of deformations with the porosity variables is described by $\boldsymbol{K}_4$ and $\boldsymbol{K}_6$. In what follows, we present the structure of $K_1, \ldots, \boldsymbol{K}_7$ which express the poro-elastic properties of thin rods, and we analyze as well the coupling thermo-poro-elastic constitutive proprties $\boldsymbol{G}_1$, $\boldsymbol{G}_2$, and G_3, G_4 and G.

4.2 Structure of Constitutive Tensors

Let us choose the vectors $\boldsymbol{d}_1$ and $\boldsymbol{d}_2$ to be directed along the principal axes of inertia for any cross–section, i.e. we assume that

$$\int_\Sigma \rho^* x \mathrm{d}x\mathrm{d}y = 0, \quad \int_\Sigma \rho^* y \mathrm{d}x\mathrm{d}y = 0, \quad \int_\Sigma \rho^* xy \mathrm{d}x\mathrm{d}y = 0. \tag{35}$$

To determine the structure of the constitutive tensors, we employ the generalized theory of tensor symmetry (Zhilin, 2006b). We consider that the anisotropy and inhomogeneity of the material do not affect the symmetry of the cross-section with respect to $\boldsymbol{d}_1$ and $\boldsymbol{d}_2$. In other words, we assume that the orthogonal tensors

$$\boldsymbol{Q}_1 = \mathbf{1} - 2\boldsymbol{d}_1 \otimes \boldsymbol{d}_1 \qquad \text{and} \qquad \boldsymbol{Q}_2 = \mathbf{1} - 2\boldsymbol{d}_2 \otimes \boldsymbol{d}_2 \tag{36}$$

belong to the symmetry group of each constitutive tensor $\boldsymbol{f}$ and $\mathbf{1}$ is the 3D unit tensor. Since any constitutive tensor $\boldsymbol{f}$ depends on the geometry of the rod through the Darboux vector $\boldsymbol{\tau}(s)$ and the angle of natural twisting $\sigma(s)$, we decompose it in the form

$$\boldsymbol{f} = \boldsymbol{f}^0 + \boldsymbol{f}^1 \cdot \boldsymbol{\tau}, \tag{37}$$

where $\boldsymbol{f}^0$ and $\boldsymbol{f}^1$ are some tensors which depend only on σ.

Imposing that the tensors (36) belong to the symmetry groups of the tensors $\boldsymbol{A}, \boldsymbol{B}, \boldsymbol{C}$ and $\boldsymbol{D}$, the structure of the elasticity tensors $\boldsymbol{A}, \boldsymbol{B}, \boldsymbol{C}$ and $\boldsymbol{D}$ has been determined in Zhilin (2006a, 2007), where the method is presented in details.

Using the same technique, we deduce the following expressions for the poro-elastic constitutive vectors

$$\begin{aligned}
\boldsymbol{K}_4 &= K_4\boldsymbol{t} + \frac{K_4^1}{R_c}\boldsymbol{d}_1 \cos\sigma + \frac{K_4^2}{R_c}\boldsymbol{d}_2 \sin\sigma,\\
\boldsymbol{K}_5 &= \frac{K_5}{R_t}\boldsymbol{t} + \frac{K_5^1}{R_c}\boldsymbol{d}_1 \sin\sigma + \frac{K_5^2}{R_c}\boldsymbol{d}_2 \cos\sigma,\\
\boldsymbol{K}_6 &= K_6\boldsymbol{t} + \frac{K_6^1}{R_c}\boldsymbol{d}_1 \cos\sigma + \frac{K_6^2}{R_c}\boldsymbol{d}_2 \sin\sigma,\\
\boldsymbol{K}_7 &= \frac{K_7}{R_t}\boldsymbol{t} + \frac{K_7^1}{R_c}\boldsymbol{d}_1 \sin\sigma + \frac{K_7^2}{R_c}\boldsymbol{d}_2 \cos\sigma,
\end{aligned} \tag{38}$$

where K_n, K_n^1 and K_n^2 are scalars which do not depend on $\boldsymbol{\tau}$.

In the same way, since the tensors (36) belong to the symmetry groups of the vectors $\boldsymbol{G}_1$ and $\boldsymbol{G}_2$, we find that the structure of the constitutive tensors $\boldsymbol{G}_1$ and $\boldsymbol{G}_2$ is given by the expressions

$$\begin{aligned}
\boldsymbol{G}_1 &= G_1\boldsymbol{t} + \frac{1}{R_c}\big(G_1^1 \cos\sigma\, \boldsymbol{d}_1 + G_1^2 \sin\sigma\, \boldsymbol{d}_2\big),\\
\boldsymbol{G}_2 &= \frac{G_2}{R_t}\boldsymbol{t} + \frac{1}{R_c}\big(G_2^1 \sin\sigma\, \boldsymbol{d}_1 + G_2^2 \cos\sigma\, \boldsymbol{d}_2\big).
\end{aligned} \tag{39}$$

Let us turn our attention to rods without natural twisting, i.e. we have $\sigma' = 0$. In this case, we assume that the symmetry groups of all constitutive tensors contain also the orthogonal tensor $\mathbf{1} - 2\boldsymbol{t}\otimes\boldsymbol{t}$ in addition to the tensors (36), i.e. the symmetry groups contain the tensors

$$\boldsymbol{Q}_1 = \mathbf{1} - 2\boldsymbol{d}_1 \otimes \boldsymbol{d}_1, \quad \boldsymbol{Q}_2 = \mathbf{1} - 2\boldsymbol{d}_2 \otimes \boldsymbol{d}_2 \quad \text{and} \quad \boldsymbol{Q}_3 = \mathbf{1} - 2\boldsymbol{t} \otimes \boldsymbol{t}. \tag{40}$$

In view of the conditions (37) and (40), we can determine the general expressions of $K_1, ..., \boldsymbol{K}_7$. When the natural twisting of the rod is absent, the expressions (38) simplify to

$$\begin{aligned} \boldsymbol{K}_4 &= K_4\boldsymbol{t}, \qquad \boldsymbol{K}_5 = \frac{K_5}{R_t}\boldsymbol{t} + \frac{K_5^1}{R_c}\boldsymbol{d}_1 \sin\sigma + \frac{K_5^2}{R_c}\boldsymbol{d}_2 \cos\sigma, \\ \boldsymbol{K}_6 &= \frac{K_6^1}{R_c}\boldsymbol{d}_1 \cos\sigma + \frac{K_6^2}{R_c}\boldsymbol{d}_2 \sin\sigma, \qquad \boldsymbol{K}_7 = \boldsymbol{0}. \end{aligned} \tag{41}$$

In addition, $K_3 = 0$. Also, the expressions of the thermo-elastic constitutive vectors (39) simplify and we obtain

$$\boldsymbol{G}_1 = G_1\boldsymbol{t}, \quad \boldsymbol{G}_2 = \frac{G_2}{R_t}\boldsymbol{t} + \frac{1}{R_c}\big(G_2^1 \sin\sigma \boldsymbol{d}_1 + G_2^2 \cos\sigma \boldsymbol{d}_2\big) \tag{42}$$

$G_4 = 0$, G_3 and G are arbitrary, where G_n $(n = 1, ..., 4)$, G_2^α and G are scalars which do not depend on $\boldsymbol{\tau}$. The elasticity tensors for rods without natural twisting have the following structure (Zhilin, 2006a)

$$\begin{aligned} \boldsymbol{A} &= A_1\boldsymbol{d}_1\otimes\boldsymbol{d}_1 + A_2\boldsymbol{d}_2\otimes\boldsymbol{d}_2 + A_3\boldsymbol{t}\otimes\boldsymbol{t}, \\ \boldsymbol{C} &= C_1\boldsymbol{d}_1\otimes\boldsymbol{d}_1 + C_2\boldsymbol{d}_2\otimes\boldsymbol{d}_2 + C_3\boldsymbol{t}\otimes\boldsymbol{t}, \\ \boldsymbol{B} &= \frac{1}{R_t}\big(B_1\boldsymbol{d}_1 \otimes \boldsymbol{d}_1 + B_2\boldsymbol{d}_2 \otimes \boldsymbol{d}_2 + B_3\boldsymbol{t} \otimes \boldsymbol{t}\big) \\ &+ \frac{1}{R_c}\big[(B_{23}\boldsymbol{d}_2 \otimes \boldsymbol{d}_3 + B_{32}\boldsymbol{d}_3 \otimes \boldsymbol{d}_2) \cos\sigma \\ &+ (B_{13}\boldsymbol{d}_1 \otimes \boldsymbol{d}_3 + B_{31}\boldsymbol{d}_3 \otimes \boldsymbol{d}_1) \sin\sigma\big], \end{aligned} \tag{43}$$

where A_i, B_i, C_i, $B_{\alpha 3}$ and $B_{3\alpha}$ are scalars which do not depend on $\boldsymbol{\tau}$.

This completes the presentation of the governing equations in the nonlinear theory of porous thermo-elastic rods. The effective values of the constitutive coefficients can be determined by analyzing problems in the linear theory of rods. In this purpose, we focus our attention next to the linearized equations.

5 Linearized Equations of Directed Rods

Let us consider the deformation of curved rods in which the displacements, rotations and variations of temperature and volume fraction field are all infinitesimal. Then the rotation tensor can be represented as $\boldsymbol{P} = \boldsymbol{1} + \boldsymbol{\psi} \times \boldsymbol{1}$ within the approximation of the linear theory, where $\boldsymbol{\psi}(s,t)$ is the *vector of small rotations*, which also satisfies the relations $\dot{\boldsymbol{\psi}} = \boldsymbol{\omega}$ and $\boldsymbol{\psi}' = \boldsymbol{\Phi}$. In the linear theory, we assume that

$$\begin{aligned}
&\boldsymbol{u}(s,t) = \boldsymbol{R}(s,t) - \boldsymbol{r}(s) = \varepsilon\tilde{\boldsymbol{u}}(s,t),\ \boldsymbol{\psi}(s,t) = \varepsilon\tilde{\boldsymbol{\psi}}(s,t),\\
&\varphi(s,t) \equiv \nu(s,t) - \nu_0(s) = \varepsilon\tilde{\varphi}(s,t),\ T(s,t) \equiv \theta(s,t) - \theta_0 = \varepsilon\tilde{T}(s,t),
\end{aligned} \tag{44}$$

where ε is a *small* non-dimensional parameter such that the quantities of order $O(\varepsilon^2)$ are neglected. The porosity function $\varphi(s,t)$ is the variation of the volume fraction field, while $T(s,t)$ is the variation of the temperature.

5.1 Boundary-Initial-Value Problems

The geometrical equations (24) can be simplified in this case and we have

$$\boldsymbol{e} \equiv \boldsymbol{u}' + \boldsymbol{t} \times \boldsymbol{\psi} = \boldsymbol{\mathcal{E}} = \boldsymbol{\mathcal{E}}_*, \qquad \boldsymbol{\kappa} \equiv \boldsymbol{\psi}' = \boldsymbol{\Phi} = \boldsymbol{\Phi}_*, \tag{45}$$

where we denote by $\boldsymbol{e}$ the extension-shear vector and by $\boldsymbol{\kappa}$ the bending-twisting vector in the linear theory. The fields $\boldsymbol{N}$, $\boldsymbol{M}$, g, h, η and q are assumed to be quantities of order $O(\varepsilon)$, which are zero in the reference configuration. Moreover, the free energy function Ψ is a quadratic form of the arguments $\{\boldsymbol{e}, \boldsymbol{\kappa}, \varphi, \varphi', T\}$. Taking into account the entropy inequality (32), it follows that the constitutive equations for q has the form

$$q = KT' \qquad \text{with} \quad K \geq 0, \tag{46}$$

where the scalar K is the thermal conductivity of the rod. The constitutive equations (30) and (34) become

$$\begin{aligned}
&\eta = -\frac{\partial\Psi}{\partial T}, \boldsymbol{N} = \frac{\partial(\rho_0\Psi)}{\partial \boldsymbol{e}}, \boldsymbol{M} = \frac{\partial(\rho_0\Psi)}{\partial \boldsymbol{\kappa}}, g = \frac{\partial(\rho_0\Psi)}{\partial \varphi}, h = \frac{\partial(\rho_0\Psi)}{\partial(\varphi')},\\
&\quad \rho_0\Psi(\boldsymbol{e},\boldsymbol{\kappa},\varphi,\varphi',T) = \rho_0\mathcal{U}(\boldsymbol{e},\boldsymbol{\kappa},\varphi,\varphi')\\
&\qquad\qquad - \big(\boldsymbol{G}_1\cdot\boldsymbol{e} + \boldsymbol{G}_2\cdot\boldsymbol{\kappa} + G_3\varphi + G_4\varphi'\big)T - \frac{1}{2}GT^2,
\end{aligned} \tag{47}$$

where we denote by $\mathcal{U}$ the deformation energy given by

$$\begin{aligned}
\rho_0\mathcal{U}(\boldsymbol{e},\boldsymbol{\kappa},\varphi,\varphi') = &\ \frac{1}{2}\boldsymbol{e}\cdot\boldsymbol{A}\cdot\boldsymbol{e} + \boldsymbol{e}\cdot\boldsymbol{B}\cdot\boldsymbol{\kappa} + \frac{1}{2}\boldsymbol{\kappa}\cdot\boldsymbol{C}\cdot\boldsymbol{\kappa}\\
&+ \frac{1}{2}K_1\varphi^2 + \frac{1}{2}K_2(\varphi')^2 + K_3\varphi\varphi'\\
&+ (\boldsymbol{K}_4\cdot\boldsymbol{e})\varphi + (\boldsymbol{K}_5\cdot\boldsymbol{\kappa})\varphi + (\boldsymbol{K}_6\cdot\boldsymbol{e})\varphi' + (\boldsymbol{K}_7\cdot\boldsymbol{\kappa})\varphi'.
\end{aligned} \tag{48}$$

The equations of motion (16) in the linear theory are

$$\boldsymbol{N}' + \rho_0\boldsymbol{\mathcal{F}} = \rho_0(\ddot{\boldsymbol{u}} + \boldsymbol{\Theta}_1^0\cdot\ddot{\boldsymbol{\psi}}), \quad \boldsymbol{M}' + \boldsymbol{t}\times\boldsymbol{N} + \rho_0\boldsymbol{\mathcal{L}} = \rho_0\big(\ddot{\boldsymbol{u}}\cdot\boldsymbol{\Theta}_1^0 + \boldsymbol{\Theta}_2^0\cdot\ddot{\boldsymbol{\psi}}\big), \tag{49}$$

the equation of equilibrated force (17) is written as

$$h' - g + \rho_0 p = \rho_0 \varkappa \ddot{\varphi}, \tag{50}$$

while the reduced equation of energy balance (33) becomes

$$q' + \rho_0 S = \rho_0 \theta_0 \dot{\eta}. \tag{51}$$

Remark 5.1. The inertia tensors $\boldsymbol{\Theta}_\alpha^0$ which appear in the right-hand sides of the equations of motion (49) are expressed by (9). We observe that the relation $\boldsymbol{\Theta}_1^0 = \mathbf{0}$ holds true if and only if $\int_\Sigma \rho^* x \tilde{\mu} \mathrm{d}x\mathrm{d}y = \int_\Sigma \rho^* y \tilde{\mu} \mathrm{d}x\mathrm{d}y = 0$.

Let us present the boundary conditions and the initial conditions associated to the above field equations. Consider that the endpoints of the rod are characterized by the arclength coordinates $\bar{s}_1 = 0$ and $\bar{s}_2 = l$. Then we take the boundary conditions

$$\begin{array}{lcl} \boldsymbol{u}(\bar{s}_\gamma, t) = \boldsymbol{u}^{(\gamma)}(t) & \text{or} & \boldsymbol{N}(\bar{s}_\gamma, t) = \boldsymbol{N}^{(\gamma)}(t), \\ \boldsymbol{\psi}(\bar{s}_\gamma, t) = \boldsymbol{\psi}^{(\gamma)}(t) & \text{or} & \boldsymbol{M}(\bar{s}_\gamma, t) = \boldsymbol{M}^{(\gamma)}(t), \\ \varphi(\bar{s}_\gamma, t) = \varphi^{(\gamma)}(t) & \text{or} & h(\bar{s}_\gamma, t) = h^{(\gamma)}(t), \\ T(\bar{s}_\gamma, t) = T^{(\gamma)}(t) & \text{or} & q(\bar{s}_\gamma, t) = q^{(\gamma)}(t) \end{array} \tag{52}$$

for $\gamma = 1, 2$ and the initial conditions

$$\begin{array}{l} \boldsymbol{u}(s,0) = \boldsymbol{u}_0(s),\ \dot{\boldsymbol{u}}(s,0) = \boldsymbol{v}_0(s),\ \boldsymbol{\psi}(s,0) = \boldsymbol{\psi}_0(s),\ \dot{\boldsymbol{\psi}}(s,0) = \boldsymbol{\omega}_0(s), \\ \varphi(s,0) = \varphi_0(s),\ \dot{\varphi}(s,0) = \lambda_0(s),\ T(s,0) = T_0(s), \quad \text{for } s \in [0, l], \end{array} \tag{53}$$

where the functions in the right–hand sides are prescribed.

The relations (45)-(53) represent the boundary–initial–value problem for the deformation of porous thermo-elastic rods. In what follows we prove the uniqueness of solution $\{\boldsymbol{u}, \boldsymbol{\psi}, \varphi, T\}$ to the boundary-initial-value problem formulated above.

5.2 Uniqueness of Solution

Let us consider the kinetic energy $K(t)$ of the rod and the function $U(t)$ given by

$$\begin{aligned} K(t) &= \frac{1}{2}\int_{\mathcal{C}_0} \rho_0 \big(\dot{\boldsymbol{u}} \cdot \dot{\boldsymbol{u}} + 2\dot{\boldsymbol{u}} \cdot \boldsymbol{\Theta}_1^0 \cdot \dot{\boldsymbol{\psi}} + \dot{\boldsymbol{\psi}} \cdot \boldsymbol{\Theta}_2^0 \cdot \dot{\boldsymbol{\psi}} + \varkappa \dot{\varphi}^2\big) \mathrm{d}s, \\ U(t) &= \int_{\mathcal{C}_0} \rho_0 \big(\Psi + \eta T\big) \mathrm{d}s. \end{aligned} \tag{54}$$

Then we can prove the following result (Bîrsan and Altenbach, 2011b).

Theorem 5.2. *If* $\{\boldsymbol{u}, \boldsymbol{\psi}, \varphi, T\}$ *is a solution of the boundary-initial-value problem, then we have*

$$\begin{aligned}\frac{\mathrm{d}}{\mathrm{d}t}\big[K(t) + U(t)\big] \;&=\; \int_{\mathcal{C}_0} \Big[\rho_0 \Big(\boldsymbol{\mathcal{F}} \cdot \dot{\boldsymbol{u}} + \boldsymbol{\mathcal{L}} \cdot \dot{\boldsymbol{\psi}} + p\dot{\varphi} + \frac{1}{\theta_0} S T\Big) - \frac{K}{\theta_0} (T')^2\Big] \mathrm{d}s \\ &+\; \big(\boldsymbol{N} \cdot \dot{\boldsymbol{u}} + \boldsymbol{M} \cdot \dot{\boldsymbol{\psi}} + h\dot{\varphi} + \frac{1}{\theta_0} q T\big)\Big|_0^l .\end{aligned} \tag{55}$$

Proof. The proof is straightforward. Indeed, if we compute the expression $\dot{K}(t) + \dot{U}(t)$ using the definitions (54), the constitutive equations $(47)_{1-5}$, the geometrical relations (45) and the field equations (49)-(51) then we obtain them right-hand side of relation (55). □

In view of the constitutive equations (47), we deduce that

$$\rho_0 \big(\Psi + \eta T\big) = \rho_0 \mathcal{U} + \frac{1}{2} G T^2. \tag{56}$$

The last relation is useful to prove the following uniqueness theorem (Bîrsan and Altenbach, 2011a).

Theorem 5.3. *Assume that the mass density* ρ_0*, the inertia coefficient* $\varkappa$ *and the constitutive coefficient* G *are positive. If the energy of deformation* $\mathcal{U}$ *satisfies*

$$\mathcal{U} \geq 0, \tag{57}$$

then the boundary-initial-value problem for porous thermo-elastic rods has at most one solution.

The proof of this theorem is given in Bîrsan and Altenbach (2011a).

Let us prove next that the property of uniqueness of solution still holds even if we do not impose the hypothesis (57). To this aim, we present first a result which is analogue to the theorem of power and energy in the classical thermo-elasticity (see Carlson, 1972, Sect. 21). The theorem first was published in Bîrsan and Altenbach (2011b).

Theorem 5.4. *For any two moments of time* $t, z \geq 0$*, we define the function*

$$\begin{aligned}Q(t,z) \;&=\; \int_{\mathcal{C}_0} \rho_0 \Big[\boldsymbol{\mathcal{F}}(t) \cdot \dot{\boldsymbol{u}}(z) + \boldsymbol{\mathcal{L}}(t) \cdot \dot{\boldsymbol{\psi}}(z) + p(t)\dot{\varphi}(z) - \frac{1}{\theta_0} S(t) T(z)\Big] \,\mathrm{d}s \\ &+\; \Big[\boldsymbol{N}(t) \cdot \dot{\boldsymbol{u}}(z) + \boldsymbol{M}(t) \cdot \dot{\boldsymbol{\psi}}(z) + h(t)\dot{\varphi}(z) - \frac{1}{\theta_0} q(t) T(z)\Big]\Big|_0^l ,\end{aligned} \tag{58}$$

where the dependence on s for the fields in the right-hand side has been omitted for brevity. Then, the following relation holds true

$$
\begin{aligned}
U(t) - K(t) &= \frac{1}{2}\int_0^t \big[Q(t+\zeta, t-\zeta) - Q(t-\zeta, t+\zeta)\big]\mathrm{d}\zeta \\
&+ \frac{1}{2}\int_{\mathcal{C}_0} \big[\boldsymbol{N}(0)\cdot\boldsymbol{e}(2t) + \boldsymbol{M}(0)\cdot\boldsymbol{\kappa}(2t) + g(0)\varphi(2t) \\
&+ h(0)\varphi'(2t) + \rho_0\eta(2t)T(0)\big]\mathrm{d}s \\
&- \frac{1}{2}\int_{\mathcal{C}_0} \rho_0\big[\dot{\boldsymbol{u}}(2t)\cdot\big(\dot{\boldsymbol{u}}(0) + \boldsymbol{\Theta}_1^0\cdot\dot{\boldsymbol{\psi}}(0)\big) \\
&+ \dot{\boldsymbol{\psi}}(2t)\cdot\big(\dot{\boldsymbol{u}}(0)\cdot\boldsymbol{\Theta}_1^0 + \boldsymbol{\Theta}_2^0\cdot\dot{\boldsymbol{\psi}}(0)\big) + \varkappa\dot{\varphi}(2t)\dot{\varphi}(0)\big]\mathrm{d}s.
\end{aligned}
\tag{59}
$$

The proof of this theorem is given in Bîrsan and Altenbach (2011b).

Let us state the main uniqueness result Bîrsan and Altenbach (2011b).

Theorem 5.5. *Assume that the mass density ρ_0, the inertia coefficient $\varkappa$ and the constitutive coefficient G are positive. Then the boundary-initial-value problem for porous thermo-elastic rods has at most one solution.*

The proof is given in Bîrsan and Altenbach (2011a).

We mention that the results established in this section are valid for curved rods made of general anisotropic materials.

5.3 Existence Results in the Dynamical Theory

In this section we present some results concerning the existence of weak solutions to the boundary-initial value problems. In order to formulate these results precisely, we employ the components of the displacement, rotation and deformation vectors.

It is convenient to denote by $\{\boldsymbol{t}_1, \boldsymbol{t}_2, \boldsymbol{t}_3\}$ the triad $\{\boldsymbol{n}, \boldsymbol{b}, \boldsymbol{t}\}$ introduced previously, such that $\boldsymbol{t}_1 = \boldsymbol{n}, \boldsymbol{t}_2 = \boldsymbol{b}, \boldsymbol{t}_3 = \boldsymbol{t}$. Then, we shall decompose any vector $\boldsymbol{f}$ in the vector basis $\{\boldsymbol{t}_1, \boldsymbol{t}_2, \boldsymbol{t}_3\}$ and denote its components by $f_i = \boldsymbol{f}\cdot\boldsymbol{t}_i$. For the components of $\boldsymbol{u}$, $\boldsymbol{\psi}$, $\boldsymbol{e}$ and $\boldsymbol{\kappa}$ we have the relations

$$
\begin{gathered}
\boldsymbol{u} = u_i\boldsymbol{t}_i, \quad \boldsymbol{\psi} = \psi_i\boldsymbol{t}_i, \quad \boldsymbol{e} = e_i\boldsymbol{t}_i, \quad \boldsymbol{\kappa} = \kappa_i\boldsymbol{t}_i, \\
e_1 = u_1' - \frac{u_2}{R_t} + \frac{u_3}{R_c} - \psi_2, \quad e_2 = u_2' + \frac{u_1}{R_t} + \psi_1, \quad e_3 = u_3' - \frac{u_1}{R_c}, \\
\kappa_1 = \psi_1' - \frac{\psi_2}{R_t} + \frac{\psi_3}{R_c}, \qquad \kappa_2 = \psi_2' + \frac{\psi_1}{R_t}, \qquad \kappa_3 = \psi_3' - \frac{\psi_1}{R_c}.
\end{gathered}
\tag{60}
$$

We shall make distinction between the ordered set of functions $\boldsymbol{y} = (u_i, \psi_i)$ representing the set of components, and the vector fields of displacement and rotation given by $\boldsymbol{u} = \boldsymbol{u}(\boldsymbol{y}) = u_i\boldsymbol{t}_i$ and $\boldsymbol{\psi} = \boldsymbol{\psi}(\boldsymbol{y}) = \psi_i\boldsymbol{t}_i$.

Let us denote by $\|\cdot\|_1$ the usual norm on the Sobolev space $\boldsymbol{H}^1[0,l]$ and by $|\cdot|_0$ the usual norm on the space $\boldsymbol{L}^2[0,l]$. The functional spaces $\boldsymbol{H}^1$, $\boldsymbol{L}^2$ are written with bold letters whether their elements have several components. For instance, for any $\boldsymbol{y}=\big(u_i(s),\psi_i(s)\big)\in\boldsymbol{H}^1[0,l]$ we have

$$\|\boldsymbol{y}\|_1^2=|\boldsymbol{y}|_0^2+\int_0^l(u_i'u_i'+\psi_i'\psi_i')\mathrm{d}s=\int_0^l(u_iu_i+\psi_i\psi_i+u_i'u_i'+\psi_i'\psi_i')\mathrm{d}s.$$

From the constitutive equations (47) we see that Ψ can be written as

$$\rho_0\Psi\big(\boldsymbol{e},\boldsymbol{\kappa},\varphi,\varphi',T\big)=\rho_0\mathcal{U}\big(\boldsymbol{e},\boldsymbol{\kappa},\varphi,\varphi'\big)-\mathcal{H}\big(\boldsymbol{e},\boldsymbol{\kappa},\varphi,\varphi'\big)T-\frac{1}{2}GT^2, \tag{61}$$

where $\mathcal{H}$ is a linear function of its arguments given by

$$\mathcal{H}\big(\boldsymbol{e},\boldsymbol{\kappa},\varphi,\varphi'\big)=\boldsymbol{G}_1\cdot\boldsymbol{e}+\boldsymbol{G}_2\cdot\boldsymbol{\kappa}+G_3\varphi+G_4\varphi'. \tag{62}$$

All the constitutive coefficients are assumed to be bounded measurable functions of s which belong to $H^1[0,l]$. The constitutive coefficient G is assumed to satisfy the restriction

$$G>0, \tag{63}$$

while the deformation energy $\mathcal{U}$ is considered to be positive definite, i.e. there exists a constant $c_3>0$ such that

$$\mathcal{U}\big(\boldsymbol{e},\boldsymbol{\kappa},\varphi,\varphi'\big)\geq c_3\big(\boldsymbol{e}\cdot\boldsymbol{e}+\boldsymbol{\kappa}\cdot\boldsymbol{\kappa}+\varphi^2+(\varphi')^2\big). \tag{64}$$

Although more general boundary conditions could be considered, in this section we restrict our attention to the following zero boundary conditions

$$\boldsymbol{u}=\boldsymbol{0},\qquad \boldsymbol{\psi}=\boldsymbol{0},\qquad \varphi=0,\qquad T=0\qquad \text{on } \Gamma=\{0,l\}. \tag{65}$$

The initial conditions (at $t=0$) are taken in the form (53).

In what follows we show the existence of weak solutions $\{\boldsymbol{u},\boldsymbol{\psi},\varphi,T\}$ to the boundary-initial-value problem formulated above. To this aim, we first write our boundary-initial-value problem in the form of an abstract Cauchy problem in an appropriate functional framework, and then we apply the theory of semigroup of linear operators. Consider the following Banach space $\big(\mathcal{W},\|\cdot\|\big)$ endowed with the usual (product) norm

$$\begin{aligned}\mathcal{W}&=\big\{\boldsymbol{Y}=(u_i,v_i,\psi_i,\omega_i,\varphi,\lambda,T)|u_i,\psi_i,\varphi\in H_0^1[0,l],v_i,\omega_i,\lambda,T\in L^2[0,l]\big\},\\ \|Y\|^2&=\sum_{i=1}^{3}\big(\|u_i\|_1^2+\|\psi_i\|_1^2+|v_i|_0^2+|\omega_i|_0^2\big)+\|\varphi\|_1^2+|\lambda|_0^2+|T|_0^2.\end{aligned} \tag{66}$$

Let us define on the space $\mathcal{W}$ the following scalar product:
for every $\boldsymbol{Y} = (u_i, v_i, \psi_i, \omega_i, \varphi, \lambda, T)$ and $\boldsymbol{Z} = (\hat{u}_i, \hat{v}_i, \hat{\psi}_i, \hat{\omega}_i, \hat{\varphi}, \hat{\lambda}, \hat{T})$ in $\mathcal{W}$ we have

$$\langle \boldsymbol{Y}, \boldsymbol{Z}\rangle_W = \frac{1}{2}\int_C \rho_0\big(\boldsymbol{v}\cdot\hat{\boldsymbol{v}} + \boldsymbol{v}\cdot\boldsymbol{\Theta}_1\cdot\hat{\boldsymbol{\omega}} + \hat{\boldsymbol{v}}\cdot\boldsymbol{\Theta}_1\cdot\boldsymbol{\omega} + \boldsymbol{\omega}\cdot\boldsymbol{\Theta}_2\cdot\hat{\boldsymbol{\omega}} + \varkappa\lambda\hat{\lambda}\big)\mathrm{d}s$$
$$+\frac{1}{2}\int_C \big[\boldsymbol{N}(\boldsymbol{z})\cdot\boldsymbol{e}(\boldsymbol{y}) + \boldsymbol{M}(\boldsymbol{z})\cdot\boldsymbol{\kappa}(\boldsymbol{y}) + g(\boldsymbol{z})\varphi + h(\boldsymbol{z})\varphi' + \rho_0\eta(\boldsymbol{y})\hat{T}\big]\mathrm{d}s, \tag{67}$$

where for brevity we have denoted by $\boldsymbol{y} = (u_i, \psi_i, \varphi, T), \boldsymbol{z} = (\hat{u}_i, \hat{\psi}_i, \hat{\varphi}, \hat{T})$. By virtue of the inequality of Korn-type for directed curves (83), we can prove that the scalar product (67) induces a norm $\|\cdot\|_W$ on the space $\mathcal{W}$, which is equivalent to the norm $\|\cdot\|$ given by (66). Consequently, $(\mathcal{W}, \langle\cdot,\cdot\rangle_W)$ is a Hilbert space.

Let us put the balance equations (49), (50) and the energy equation (51) into an operator form. In this purpose, we introduce the second order tensor $\boldsymbol{Q}$ which is the inverse of the symmetric and positive definite tensor $\boldsymbol{\Theta}_2 - \boldsymbol{\Theta}_1^T\cdot\boldsymbol{\Theta}_1$, i.e.

$$\boldsymbol{Q} = \big(\boldsymbol{\Theta}_2 - \boldsymbol{\Theta}_1^T\cdot\boldsymbol{\Theta}_1\big)^{-1}.$$

Suggested by Eqs. (49)-(51), we define the operator $\mathcal{P} : D(\mathcal{P}) \subset \mathcal{W} \to \mathcal{W}$ as follows: for every $\boldsymbol{Y} = (u_i, v_i, \psi_i, \omega_i, \varphi, \lambda, T) \in \mathcal{W}$ we have

$$\begin{aligned}
\mathcal{P}\boldsymbol{Y} &= (v_i, B_i\boldsymbol{Y}, \omega_i, C_i\boldsymbol{Y}, \lambda, D\boldsymbol{Y}, E\boldsymbol{Y}), \forall \boldsymbol{Y} = (u_i, v_i, \psi_i, \omega_i, \varphi, \lambda, T) \in D(\mathcal{P}),\\
B_i\boldsymbol{Y} &= \frac{1}{\rho_0}\boldsymbol{t}_i\cdot\boldsymbol{Q}\cdot\big[\boldsymbol{\Theta}_2\cdot\boldsymbol{N}' - \boldsymbol{\Theta}_1\cdot(\boldsymbol{M}' + \boldsymbol{t}_3\times\boldsymbol{N})\big](u_i, \psi_i, \varphi, T),\\
C_i\boldsymbol{Y} &= \frac{1}{\rho_0}\boldsymbol{t}_i\cdot\boldsymbol{Q}\cdot\big[\boldsymbol{M}' + \boldsymbol{t}\times\boldsymbol{N} - \boldsymbol{N}'\cdot\boldsymbol{\Theta}_1\big](u_i, \psi_i, \varphi, T),\\
D\boldsymbol{Y} &= \frac{1}{\rho_0\varkappa}\big[h' - g\big](u_i, \psi_i, \varphi, T), \quad E\boldsymbol{Y} = \frac{1}{G}\left[\frac{1}{\theta_0}(KT')' - H(v_i, \omega_i, \lambda)\right].
\end{aligned} \tag{68}$$

The domain of the operator $\mathcal{P}$ is given by

$$D(\mathcal{P}) = \big\{\boldsymbol{Y}\in\mathcal{W}|\mathcal{P}\boldsymbol{Y}\in\mathcal{W}, T\in H_0^1[0,l]\big\}.$$

The function H appearing in the last equation (68) is defined by the relations (62) and

$$H(u_i, \psi_i, \varphi) = \mathcal{H}(\boldsymbol{e}, \boldsymbol{\kappa}, \varphi, \varphi'), \qquad \forall(u_i, \psi_i, \varphi) \in \boldsymbol{H}_0^1[0,l].$$

With the help of the operator $\mathcal{P}$ we write our boundary–initial–value problem as the following Cauchy problem in the Hilbert space $\mathcal{W}$:

$$\frac{\mathrm{d}\boldsymbol{Y}}{\mathrm{d}t} = \mathcal{P}\boldsymbol{Y}(t) + \boldsymbol{\Phi}(t), \qquad \boldsymbol{Y}(0) = \boldsymbol{Y}_0, \tag{69}$$

where the term $\boldsymbol{\Phi}(t)$ accounts for the external body loads and heat supply, while $\boldsymbol{Y}_0$ represents the initial data. They have the expressions

$$\boldsymbol{\Phi}(t) = \left[0_i, \boldsymbol{t}_i \cdot \boldsymbol{Q} \cdot (\boldsymbol{\Theta}_2 \cdot \boldsymbol{\mathcal{F}} + \boldsymbol{\Theta}_1 \cdot \boldsymbol{\mathcal{L}}), 0_i, \boldsymbol{t}_i \cdot \boldsymbol{Q} \cdot (\boldsymbol{\mathcal{L}} - \boldsymbol{\mathcal{F}} \cdot \boldsymbol{\Theta}_1), 0, \frac{p}{\varkappa}, \frac{\rho_0 S}{\theta_0 G}\right],$$
$$\boldsymbol{Y}_0 = (u_{0i}, v_{0i}, \psi_{0i}, \omega_{0i}, \varphi_0, \lambda_0, T_0). \tag{70}$$

Thus, the problem reduces to find $\boldsymbol{Y} = (u_i, v_i, \psi_i, \omega_i, \varphi, \lambda, T) \in D(\mathcal{P})$ which satisfies the problem (69). We establish here the existence of such weak solution $\boldsymbol{Y}$.

Theorem 5.6. *Assume that the position vector $\mathbf{r}(s)$ is of class $\boldsymbol{C}^3[0,l]$, such that the radius of curvature R_c and the radius of twisting R_t given by (2) exist at any point. Let t_0 be an arbitrary moment of time such that the external body loads and heat supply satisfy the condition $\boldsymbol{\Phi}(t) \in C^1\big([0,t_0], \boldsymbol{L}^2[0,l]\big)$. If the initial data are such that $\boldsymbol{Y}_0 \in D(\mathcal{P})$, then there exists a unique solution $\boldsymbol{Y}(t) \in C^1\big([0,t_0], \mathcal{W}\big) \cap C^0\big([0,t_0], D(\mathcal{P})\big)$ to the problem (69).*

Proof. We begin by showing that the operator $\mathcal{P}$ possesses some important properties. First, we observe that the domain $D(\mathcal{P})$ is dense in $\mathcal{W}$, in view of the relation

$$\big\{\boldsymbol{Y} = (u_i, v_i, \psi_i, \omega_i, \varphi, \lambda, T) | u_i, \psi_i, \varphi, T$$
$$\in H_0^1[0,l] \cap H^2[0,l], v_i, \omega_i, \lambda \in H_0^1[0,l]\big\} \subset D(\mathcal{P}).$$

Secondly, using the definitions (67) and (68) we can prove that the operator $\mathcal{P}$ satisfies the relation

$$\langle \mathcal{P}\boldsymbol{Y}, \boldsymbol{Y}\rangle_W = -\frac{1}{2\theta_0}\int_C K\,(T')^2 \mathrm{d}s \leq 0, \qquad \forall \boldsymbol{Y} \in D(\mathcal{P}). \tag{71}$$

The above inequality holds true since the thermal conductivity coefficient satisfies $K \geq 0$, according to the entropy principle.

Finally, let us prove that the operator $\mathcal{P}$ verifies the range condition

$$\mathrm{Range}(\mathcal{I} - \mathcal{P}) = \mathcal{W}, \tag{72}$$

where $\mathcal{I}$ designates the identity operator. For any

$$\boldsymbol{Y}^* = (u_i^*, v_i^*, \psi_i^*, \omega_i^*, \varphi^*, \lambda^*, T^*) \in \mathcal{W},$$

we have to show that the equation $\boldsymbol{Y} - \mathcal{P}\boldsymbol{Y} = \boldsymbol{Y}^*$ admits a solution $\boldsymbol{Y} \in D(\mathcal{P})$. The last equation implies that $v_i = u_i - u_i^*, \omega_i = \psi_i - \psi_i^*, \lambda = \varphi - \varphi^*$, while the functions u_i, ψ_i, φ and T satisfy the system of equations

$$\begin{aligned}
&\rho_0(\boldsymbol{u} + \boldsymbol{\Theta}_1 \cdot \boldsymbol{\psi}) - \boldsymbol{N}'(u_i, \psi_i, \varphi, T) \\
&\qquad = \rho_0\big[(\boldsymbol{u}^* + \boldsymbol{v}^*) + \boldsymbol{\Theta}_1 \cdot (\boldsymbol{\psi}^* + \boldsymbol{\omega}^*)\big], \\
&\rho_0(\boldsymbol{u}\cdot\boldsymbol{\Theta}_1 + \boldsymbol{\Theta}_2\cdot\boldsymbol{\psi}) - \big[\boldsymbol{M}' + \boldsymbol{t}\times\boldsymbol{N}\big](u_i, \psi_i, \varphi, T) \\
&\qquad = \rho_0\big[(\boldsymbol{u}^* + \boldsymbol{v}^*)\cdot\boldsymbol{\Theta}_1 + \boldsymbol{\Theta}_2\cdot(\boldsymbol{\psi}^* + \boldsymbol{\omega}^*)\big], \\
&\rho_0\varkappa\varphi - [h' - g](u_i, \psi_i, \varphi, T) = \rho_0\varkappa(\varphi^* + \lambda^*), \\
&\rho_0\eta(u_i, \psi_i, \varphi, T) - \frac{1}{\theta_0}(KT')' = \rho_0\eta(u_i^*, \psi_i^*, \varphi^*, T^*).
\end{aligned} \tag{73}$$

Using the general methods for elliptic systems and the Korn inequality for rods presented by Theorem 6.3, we can show that the equations (73) admit a solution $(u_i, \psi_i, \varphi, T) \in \boldsymbol{H}_0^1[0, l]$. Hence, the relation (72) holds true.

In view of the above properties of the operator $\mathcal{P}$, we can apply the Lumer-Phillips theorem (see Pazy, 1983, p. 14) to deduce that $\mathcal{P}$ is the infinitesimal generator of a semigroup of contractions in the Hilbert space $\mathcal{W}$. We denote by $\{R(t); t \geq 0\}$ the semigroup of contractions generated by the operator $\mathcal{P}$. From the general results of the semigroup of operators theory (see Vrabie, 2003, Sect. 8.1) it follows that there exists a unique solution $\boldsymbol{Y}(t)$ to the Cauchy problem (69), which is expressed by

$$\boldsymbol{Y}(t) = R(t)\boldsymbol{Y}_0 + \int_0^t R(t-\tau)\boldsymbol{\Phi}(\tau)\mathrm{d}\tau, \quad t \in [0, t_0]. \tag{74}$$

This completes the proof. □

Remark 5.7. The continuous dependence of the weak solution $\boldsymbol{Y}(t)$ on the initial data $\boldsymbol{Y}_0$ and the external body loads and heat supply $\boldsymbol{\Phi}(t)$ is stated by the relation

$$\|\boldsymbol{Y}(t)\|_W \leq \|\boldsymbol{Y}_0\|_W + \int_0^t \|\boldsymbol{\Phi}(\tau)\|_W \mathrm{d}\tau, \qquad t \in [0, t_0]. \tag{75}$$

The estimate is obtained from Eq. (74), by virtue of the property that $\{R(t); t \geq 0\}$ is a semigroup of contractions.

6 Statical Theory for Rods

In the case of equilibrium, the functions defined previously do not depend on time. For the static theory, we observe that the energy equation (51) decouples from the balance equations (49), (50), and it can be solved separately. Thus, the thermo–conductivity problem can be treated separately. For this reason, we consider in this section the isothermal theory for porous rods.

6.1 Inequalities of Korn-Type for Cosserat Rods

Let us present the inequalities of Korn-type corresponding to the deformation of rods. We prove first the following Korn-type inequality „without boundary conditions".

Theorem 6.1. *Assume that the position vector* $\boldsymbol{r}(s)$ *is of class* $\boldsymbol{C}^3[0,l]$, *such that the radius of curvature* R_c *and the radius of twisting* R_t *given by (2) exist at any point. For every* $\boldsymbol{y} = \big(u_i(s), \psi_i(s)\big) \in \boldsymbol{H}^1[0,l]$ *we define the components of the deformation vectors* $e_i(\boldsymbol{y})$ *and* $\kappa_i(\boldsymbol{y})$ *through the relations (60). Then, there exists a constant* $c_1 = c_1(\boldsymbol{r}) > 0$ *such that*

$$\int_C \big[u_i u_i + \psi_i \psi_i + e_i(\boldsymbol{y}) e_i(\boldsymbol{y}) + \kappa_i(\boldsymbol{y}) \kappa_i(\boldsymbol{y})\big] \mathrm{d}s \\ \geq c_1 \int_C \big(u_i u_i + \psi_i \psi_i + u_i' u_i' + \psi_i' \psi_i'\big) \mathrm{d}s, \tag{76}$$

for any $\boldsymbol{y} = \big(u_i, \psi_i\big) \in \boldsymbol{H}^1[0,l]$.

Proof. We define the following norm on the space $\boldsymbol{H}^1[0,l]$:

$$\|\boldsymbol{y}\|_H^2 = \int_0^l \big[u_i u_i + \psi_i \psi_i + e_i(\boldsymbol{y}) e_i(\boldsymbol{y}) + \kappa_i(\boldsymbol{y}) \kappa_i(\boldsymbol{y})\big] \mathrm{d}s, \forall \boldsymbol{y} = \big(u_i(s), \psi_i(s)\big). \tag{77}$$

Let us show that the space $\big(\boldsymbol{H}^1[0,l], \|\cdot\|_H\big)$ is a Banach space. Consider a Cauchy sequence $\big(\boldsymbol{y}^k\big)_{k\geq 1} \subset \boldsymbol{H}^1[0,l]$ with respect to the norm $\|\cdot\|_H$. It follows from (77) that $\big(\boldsymbol{y}^k\big)_{k\geq 1}$ is a Cauchy sequence in $\boldsymbol{L}^2[0,l]$, and also $e_i\big(\boldsymbol{y}^k\big)_{k\geq 1}$ and $\kappa_i\big(\boldsymbol{y}^k\big)_{k\geq 1}$ are Cauchy sequences in $L^2[0,l]$. Since $L^2[0,l]$ is a Banach space, we deduce that there exist the elements $\tilde{\boldsymbol{y}} \in \boldsymbol{L}^2[0,l], \tilde{e}_i, \tilde{\kappa}_i \in L^2[0,l]$ such that

$$\boldsymbol{y}^k \to \tilde{\boldsymbol{y}} \,\text{in}\, \boldsymbol{L}^2[0,l],\ e_i\big(\boldsymbol{y}^k\big) \to \tilde{e}_i, \kappa_i\big(\boldsymbol{y}^k\big) \to \tilde{\kappa}_i \,\text{in}\, L^2[0,l],\ \text{for}\, k \to \infty. \tag{78}$$

Using the definition of derivatives in the sense of distributions we can show that $\tilde{\boldsymbol{y}} \in \boldsymbol{H}^1[0,l]$ and that $\tilde{e}_i = e_i(\tilde{\boldsymbol{y}}), \tilde{\kappa}_i = \kappa_i(\tilde{\boldsymbol{y}})$. Then, from (77) and (78) it follows that $\lim_{k\to\infty} \|\boldsymbol{y}^k - \tilde{\boldsymbol{y}}\|_H = 0$, i.e. the sequence $\big(\boldsymbol{y}^k\big)_{k\geq 1}$ is convergent to $\tilde{\boldsymbol{y}}$, and hence $\big(\boldsymbol{H}^1[0,l], \|\cdot\|_H\big)$ is a Banach space.

Consider now the identity mapping $\mathcal{I} : \big(\boldsymbol{H}^1[0,l], \|\cdot\|_1\big) \to \big(\boldsymbol{H}^1[0,l], \|\cdot\|_H\big)$ which is clearly bijective and continuous. Applying a corollary of the closed graph theorem (see Brezis, 1992, Corol. II.6) we obtain that the inverse mapping $\mathcal{I}^{-1}$ is also continuous. The continuity of $\mathcal{I}^{-1}$ insures the existence of a positive constant c_1 such that the inequality (76) holds. □

The inequality of Korn-type (76) is called „without boundary conditions“ because it is valid for arbitrary functions $\boldsymbol{y} = \big(u_i(s), \psi_i(s)\big) \in \boldsymbol{H}^1[0,l]$, which are not restricted by any conditions in the end points $s = 0, l$. But as we know, in many mechanical problems the functions $u_i(s)$ and $\psi_i(s)$ have to satisfy some boundary conditions for $s = 0, l$. For this reason, we shall establish in the remaining of this section a Korn inequality "with boundary conditions".

Let us denote by $\Gamma = \{0, l\}$ the boundary of the domain $(0, l) \subset \mathbb{R}$. We denote by Γ_u and Γ_ψ the subsets of Γ where the displacements u_i and the rotations ψ_i are prescribed, respectively. Assuming zero boundary conditions for simplicity, we have the following restrictions imposed to the functions u_i and ψ_i

$$u_i(s) = 0 \quad \text{for}\, s \in \Gamma_u, \qquad \psi_i(s) = 0 \quad \text{for}\, s \in \Gamma_\psi. \tag{79}$$

We introduce the subspace $\boldsymbol{V}$ of all the displacement and rotation fields which satisfy the boundary conditions (79), i.e.

$$\boldsymbol{V} = \big\{\boldsymbol{y} = \big(u_i, \psi_i\big) \in \boldsymbol{H}^1[0,l] \,|\, u_i = 0\,\text{on}\,\Gamma_u, \quad \psi_i = 0\,\text{on}\,\Gamma_\psi\big\}. \tag{80}$$

in the sense of traces. We notice that $\boldsymbol{V}$ is a closed subspace of $\boldsymbol{H}^1[0,l]$, and hence $\big(\boldsymbol{V}, \|\cdot\|_1\big)$ is a Banach space. We present first an auxiliary result concerning infinitesimal rigid body displacements and rotations for rods.

Lemma 6.2. *Assume that the hypotheses of Theorem 6.1 are satisfied. Let* $\boldsymbol{y} = \big(u_i, \psi_i\big) \in \boldsymbol{H}^1[0,l]$ *be such that*

$$e_i(\boldsymbol{y}) = 0, \qquad \kappa_i(\boldsymbol{y}) = 0 \qquad \text{on} \quad [0,l]. \tag{81}$$

Then the displacement vector $\boldsymbol{u}(\boldsymbol{y}) = u_i \boldsymbol{t}_i$ *and the rotation vector* $\boldsymbol{\psi}(\boldsymbol{y}) = \psi_i \boldsymbol{t}_i$ *represent a rigid body displacement of the rod, i.e. there exist two constant vectors* $\boldsymbol{a}$ *and* $\boldsymbol{b}$ *such that*

$$\boldsymbol{u}(\boldsymbol{y}) = \boldsymbol{a} + \boldsymbol{b} \times \boldsymbol{r}, \qquad \boldsymbol{\psi}(\boldsymbol{y}) = \boldsymbol{b}. \tag{82}$$

Moreover, if Γ_u *and* Γ_ψ *are nonempty sets and we have* $\boldsymbol{y} = \big(u_i, \psi_i\big) \in \boldsymbol{V}$, *then the relations (81) imply that* $u_i = 0, \psi_i = 0$ *on* $[0,l]$.

Proof. In view of $(45)_2$ and $(81)_2$ it follows that $\boldsymbol{\psi}'(\boldsymbol{y}) = \boldsymbol{0}$, so that there exists a constant vector $\boldsymbol{b}$ with $\boldsymbol{\psi}(\boldsymbol{y}) = \boldsymbol{b}$. Then, from $(45)_1$ and $(81)_1$ we deduce that $\big(\boldsymbol{u}(\boldsymbol{y}) + \boldsymbol{r} \times \boldsymbol{b}\big)' = \boldsymbol{0}$, so that there exists a constant vector $\boldsymbol{a}$ which satisfies the relation $(82)_1$. If the sets Γ_u and Γ_ψ are nonempty, then from (80) and (82) we deduce $\boldsymbol{a} = \boldsymbol{0}, \boldsymbol{b} = \boldsymbol{0}$ and hence $u_i = 0, \psi_i = 0$. The proof is complete. □

On the basis of Theorem 6.1 and Lemma 6.2 we can prove now the main result of this section, i.e. the Korn inequality "with boundary conditions" (Bîrsan and Altenbach, 2012a).

Theorem 6.3. *Assume that the hypotheses of Theorem 6.1 are satisfied and that* Γ_u *and* Γ_ψ *are nonempty sets. Let* $\boldsymbol{V}$ *be the space defined by (80). Then, there exists a constant* $c_2 = c_2(\boldsymbol{r}, \Gamma_u, \Gamma_\psi) > 0$ *such that for any* $\boldsymbol{y} = \big(u_i, \psi_i\big) \in \boldsymbol{V}$ *we have*

$$\int_{\mathcal{C}} \big[e_i(\boldsymbol{y})e_i(\boldsymbol{y}) + \kappa_i(\boldsymbol{y})\kappa_i(\boldsymbol{y})\big]\mathrm{d}s \geq c_2 \int_{\mathcal{C}} \big(u_i u_i + \psi_i\psi_i + u_i' u_i' + \psi_i'\psi_i'\big)\mathrm{d}s. \quad (83)$$

Proof. Let us assume, on the contrary, that there does not exist any constant c_2 which satisfies the inequality (83). Then, we can find a sequence $\big(\boldsymbol{y}^k\big)_{k\geq 1} \subset \boldsymbol{V}$ such that

$$\|\boldsymbol{y}^k\|_1 = 1\,\forall\, k \geq 1, \text{ and } \lim_{k\to\infty} \int_0^l \big[e_i(\boldsymbol{y}^k)e_i(\boldsymbol{y}^k) + \kappa_i(\boldsymbol{y}^k)\kappa_i(\boldsymbol{y}^k)\big]\mathrm{d}s = 0. \quad (84)$$

Since the sequence $\big(\boldsymbol{y}^k\big)_{k\geq 1}$ is bounded in $\boldsymbol{H}^1[0,l]$, there exists a subsequence denoted by $\big(\boldsymbol{y}^m\big)_{m\geq 1}$ which is convergent in $\boldsymbol{L}^2[0,l]$, according to the Rellich–Kondrasov theorem (Brezis, 1992). Using $(84)_2$ we deduce that the sequences $\big(e_i(\boldsymbol{y}^m)\big)_{m\geq 1}$ and $\big(\kappa_i(\boldsymbol{y}^m)\big)_{m\geq 1}$ are convergent in $L^2[0,l]$. Consequently, the subsequence $\big(\boldsymbol{y}^m\big)_{m\geq 1} \subset \boldsymbol{H}^1[0,l]$ is a Cauchy sequence with respect to the norm $\|\cdot\|_H$ defined by (77).

By virtue of the inequality (76) stated by Theorem 6.1, we deduce that $\big(\boldsymbol{y}^m\big)_{m\geq 1}$ is a Cauchy sequence also in the space $\big(\boldsymbol{H}^1[0,l], \|\cdot\|_1\big)$. Since $\big(\boldsymbol{y}^m\big)_{m\geq 1} \subset \boldsymbol{V}$, it follows that there exists $\boldsymbol{y} \in \boldsymbol{V}$ such that

$$\boldsymbol{y}^m \to \boldsymbol{y} \quad \text{in}\big(\boldsymbol{V}, \|\cdot\|_1\big), \qquad \text{for } m \to \infty. \quad (85)$$

Next, from $(84)_2$ and (85) we obtain that

$$e_i(\boldsymbol{y}) = 0, \qquad \kappa_i(\boldsymbol{y}) = 0, \qquad i = 1, 2, 3. \quad (86)$$

Finally, we apply Lemma 6.2 and from the relations (86) we get $\boldsymbol{y} = \boldsymbol{0}$. We observe that the relation $\boldsymbol{y} = \boldsymbol{0}$ contradicts the requirement $(84)_1$. This shows that our supposition was false, i.e. there exists a positive constant c_2 which fulfils the inequality (83). □

The above results are derived using the same methods as in the case of deformable surfaces, see e.g. Ciarlet (2005); Bîrsan (2008). This Korn-type inequality is useful to prove existence results for the equations of thin thermo-elastic porous rods: it has already been employed in Sect. 5.3 in the dynamical case, and will be used also in the next section for the equilibrium equations.

6.2 Existence of Solution

In the case of isothermal equilibrium theory for porous rods, the boundary-value problem reduces to find the unknown fields $\{\boldsymbol{u}, \boldsymbol{\psi}, \varphi\}$ which satisfy the equilibrium equations

$$\boldsymbol{N}' + \rho_0 \boldsymbol{\mathcal{F}} = \boldsymbol{0}, \qquad \boldsymbol{M}' + \boldsymbol{t} \times \boldsymbol{N} + \rho_0 \boldsymbol{\mathcal{L}} = \boldsymbol{0}, \qquad h' - g + \rho_0 p = 0, \tag{87}$$

together with the constitutive equations

$$\boldsymbol{N} = \frac{\partial(\rho_0 \mathcal{U})}{\partial \boldsymbol{e}}, \qquad \boldsymbol{M} = \frac{\partial(\rho_0 \mathcal{U})}{\partial \boldsymbol{\kappa}}, \qquad g = \frac{\partial(\rho_0 \mathcal{U})}{\partial \varphi}, \qquad h = \frac{\partial(\rho_0 \mathcal{U})}{\partial(\varphi')}, \tag{88}$$

and the boundary conditions

$$\begin{array}{llll} \boldsymbol{u} = \boldsymbol{0} & \text{for } s \in \Gamma_u, & \boldsymbol{N} = \boldsymbol{0} & \text{for } s \in \Gamma \setminus \Gamma_u, \\ \boldsymbol{\psi} = \boldsymbol{0} & \text{for } s \in \Gamma_\psi, & \boldsymbol{M} = \boldsymbol{0} & \text{for } s \in \Gamma \setminus \Gamma_\psi, \\ \varphi = 0 & \text{for } s \in \Gamma_\varphi, & h = 0 & \text{for } s \in \Gamma \setminus \Gamma_\varphi. \end{array} \tag{89}$$

Let us assume that the set of functions $\boldsymbol{z} = (\tilde{u}_i, \tilde{\psi}_i, \tilde{\varphi}) \in \boldsymbol{C}^2[0, l]$ satisfies the balance equations (87), the constitutive equations (88) and the geometrical relations (45). Then, for every $\boldsymbol{y} = (u_i, \psi_i, \varphi) \in \boldsymbol{C}^1[0, l]$ the following equality holds

$$\begin{aligned} &\int_C \big[\boldsymbol{N}(\boldsymbol{z}) \cdot \boldsymbol{e}(\boldsymbol{y}) + \boldsymbol{M}(\boldsymbol{z}) \cdot \boldsymbol{\kappa}(\boldsymbol{y}) + g(\boldsymbol{z})\varphi + h(\boldsymbol{z})\varphi'\big]\mathrm{d}s \\ &\qquad = \int_C \rho_0 \big[\boldsymbol{\mathcal{F}} \cdot \boldsymbol{u}(\boldsymbol{y}) + \boldsymbol{\mathcal{L}} \cdot \boldsymbol{\psi}(\boldsymbol{y}) + p\varphi\big]\mathrm{d}s \\ &\qquad\quad + \big(\boldsymbol{N}(\boldsymbol{z}) \cdot \boldsymbol{u}(\boldsymbol{y}) + \boldsymbol{M}(\boldsymbol{z}) \cdot \boldsymbol{\psi}(\boldsymbol{y}) + h(\boldsymbol{z})\varphi\big)\Big|_0^l . \end{aligned} \tag{90}$$

To prove the relation (90), we insert the geometrical relations (45) for $\boldsymbol{e}(\boldsymbol{y})$ and $\boldsymbol{\kappa}(\boldsymbol{y})$ in the left–hand side, then we use integration by parts and finally we employ the equilibrium equations (87) written for $\boldsymbol{z}$. The identity (90) can be seen as a principle of virtual work for porous rods.

Suggested by the equality (90), we shall define next the weak solution of our boundary–value problem. We shall denote by $\boldsymbol{V}$ the subspace of $\boldsymbol{H}^1[0, l]$ given by

$$\boldsymbol{V} = \big\{\boldsymbol{y} = (u_i, \psi_i, \varphi) \in \boldsymbol{H}^1[0, l] \,|\, u_i = 0 \text{ on } \Gamma_u, \psi_i = 0 \text{ on } \Gamma_\psi, \varphi = 0 \text{ on } \Gamma_\varphi\big\}. \tag{91}$$

in the sense of traces. The body loads $\boldsymbol{\mathcal{F}}$, $\boldsymbol{\mathcal{L}}$ and p are assumed to be functions of class $L^2[0,l]$. In these conditions, we say that the set of functions $\boldsymbol{z} = (\tilde{u}_i, \tilde{\psi}_i, \tilde{\varphi})$ is a *weak solution* of the boundary-value problem for porous rods if $\boldsymbol{z} \in \boldsymbol{V}$ and the following relation holds for every $\boldsymbol{y} = (u_i, \psi_i, \varphi) \in \boldsymbol{V}$

$$\begin{aligned} &\int_C \big[N_i(\boldsymbol{z})e_i(\boldsymbol{y}) + M_i(\boldsymbol{z})\kappa_i(\boldsymbol{y}) + g(\boldsymbol{z})\varphi + h(\boldsymbol{z})\varphi'\big]\mathrm{d}s \\ &\qquad = \int_C \rho_0\big(\mathcal{F}_i u_i + \mathcal{L}_i\psi_i + p\varphi\big)\mathrm{d}s. \end{aligned} \tag{92}$$

Concerning the existence of such a solution we give the next result.

Theorem 6.4. *Assume that the hypotheses of Theorem 6.1 are satisfied, and the sets Γ_u and Γ_ψ are nonempty. Then, there exists a unique weak solution $\boldsymbol{z}$ of the boundary-value problem for porous rods. The weak solution $\boldsymbol{z}$ is characterized as the minimizer on the space $\boldsymbol{V}$ of the following functional*

$$J(\boldsymbol{y}) = \int_C \rho_0\big[\mathcal{U}\big(\boldsymbol{e}(\boldsymbol{y}), \boldsymbol{\kappa}(\boldsymbol{y}), \varphi, \varphi'\big) - \mathcal{F}_i u_i - \mathcal{L}_i\psi_i - p\varphi\big]\mathrm{d}s, \ \forall \boldsymbol{y} = (u_i, \psi_i, \varphi) \in \boldsymbol{V}. \tag{93}$$

Proof. The proof resides in the use of the Korn inequality and the application of the Lax–Milgram lemma. For the sake of brevity, we introduce the bilinear form $B(\cdot,\cdot)$ on $\boldsymbol{V} \times \boldsymbol{V}$ and the linear functional $F(\cdot)$ on $\boldsymbol{V}$ defined by

$$\begin{aligned} B(\boldsymbol{y}, \boldsymbol{z}) &= \int_C \big[N_i(\boldsymbol{z})e_i(\boldsymbol{y}) + M_i(\boldsymbol{z})\kappa_i(\boldsymbol{y}) + g(\boldsymbol{z})\varphi + h(\boldsymbol{z})\varphi'\big]\mathrm{d}s, \\ F(\boldsymbol{y}) &= \int_C \rho_0\big(\mathcal{F}_i u_i + \mathcal{L}_i\psi_i + p\varphi\big)\mathrm{d}s, \quad \forall \boldsymbol{y}, \boldsymbol{z} \in \boldsymbol{V}, \quad \boldsymbol{y} = (u_i, \psi_i, \varphi). \end{aligned} \tag{94}$$

Then, the relation (92) can be written as

$$B(\boldsymbol{y}, \boldsymbol{z}) = F(\boldsymbol{y}), \qquad \forall \boldsymbol{y} \in \boldsymbol{V}. \tag{95}$$

We remark that B and F are continuous and that B is symmetric. In order to show that B is also $\boldsymbol{V}$-elliptic, we observe that

$$B(\boldsymbol{y}, \boldsymbol{y}) = 2\int_C \rho_0 \mathcal{U}\big(\boldsymbol{e}(\boldsymbol{y}), \boldsymbol{\kappa}(\boldsymbol{y}), \varphi, \varphi'\big)\mathrm{d}s, \qquad \forall \boldsymbol{y} = (u_i, \psi_i, \varphi) \in \boldsymbol{V}. \tag{96}$$

Using the relation (64) and the Korn inequality (83), then from (96) we deduce that there exists a constant $c_4 > 0$ such that

$$B(\boldsymbol{y}, \boldsymbol{y}) \geq c_4 \|\boldsymbol{y}\|_1^2, \qquad \forall \boldsymbol{y} \in \boldsymbol{V}. \tag{97}$$

Applying the Lax-Milgram lemma (Brezis, 1992) for the problem (95) we obtain the existence and uniqueness of the solution $\boldsymbol{z}$, and its characterization as the minimizer of the functional $J(\boldsymbol{y})$ on $\boldsymbol{V}$. □

The hypothesis that Γ_u and Γ_ψ are nonempty sets excludes the possibility of a „pure traction"problem, i.e. the case when the forces $\boldsymbol{N}$ and the moments $\boldsymbol{M}$ are prescribed on both ends of the rod. Since this case is of great importance in engineering, it will be analyzed in detail in the next section.

6.3 Analysis of Pure Traction Problems

For the treatment of boundary-value problems with pure traction boundary conditions we need to establish first a Korn-type inequality „over the quotient space" $\boldsymbol{H}^1[0,l]/\boldsymbol{R}$. Here, $\boldsymbol{R}$ is the subspace of $\boldsymbol{H}^1[0,l]$ given by

$$\boldsymbol{R} = \big\{\boldsymbol{y} = (u_i, \psi_i) \in \boldsymbol{H}^1[0,l] \mid e_i(\boldsymbol{y}) = 0, \kappa_i(\boldsymbol{y}) = 0\big\}.$$

We can see that $\boldsymbol{R}$ represents the subspace of rigid–body displacements and rotations which can also be characterized by the relations $u_i\boldsymbol{t}_i = \boldsymbol{a} + \boldsymbol{b} \times \boldsymbol{r}$, $\psi_i\boldsymbol{t}_i = \boldsymbol{b}$, where $\boldsymbol{a}$ and $\boldsymbol{b}$ are two constant vectors, according to Lemma 6.2.

We introduce the quotient space $\hat{\boldsymbol{V}} = \boldsymbol{H}^1[0,l]/\boldsymbol{R}$, and we denote by $\hat{\boldsymbol{y}}$ the equivalence class of any element $\boldsymbol{y} \in \boldsymbol{H}^1[0,l]$, i.e.

$$\hat{\boldsymbol{y}} = \big\{\boldsymbol{w} \in \boldsymbol{H}^1[0,l] | (\boldsymbol{y} - \boldsymbol{w}) \in \boldsymbol{R}\big\}.$$

The space $\hat{\boldsymbol{V}}$ is a Banach space, equipped with the quotient norm $\|\cdot\|_{\hat{\boldsymbol{V}}}$ defined by

$$\|\hat{\boldsymbol{y}}\|_{\hat{\boldsymbol{V}}} = \inf_{\boldsymbol{w}\in\boldsymbol{R}} \|\boldsymbol{y} + \boldsymbol{w}\|_1, \qquad \forall \hat{\boldsymbol{y}} \in \hat{\boldsymbol{V}}. \tag{98}$$

The inequality of Korn-type „over the quotient space $\hat{\boldsymbol{V}}$" is given by the next result.

Theorem 6.5. *Assume that the hypotheses of Theorem 6.1 are satisfied. Then, there exists a constant* $\hat{c} > 0$ *such that the following inequality holds*

$$\Big\{\int_{\mathcal{C}} \big[e_i(\boldsymbol{y})e_i(\boldsymbol{y}) + \kappa_i(\boldsymbol{y})\kappa_i(\boldsymbol{y})\big]\mathrm{d}s\Big\}^{1/2} \geq \hat{c}\|\hat{\boldsymbol{y}}\|_{\hat{\boldsymbol{V}}}, \quad \forall \boldsymbol{y} \in \boldsymbol{H}^1[0,l]. \tag{99}$$

To prove the inequality (99) we can follow the same lines as in Theorem 4.3-5 of (Ciarlet, 2005), see for details (Bîrsan and Altenbach, 2012b).

The inequality of Korn-type established in Theorem 6.5 is useful to prove existence results in the case of pure traction problems. In what follows, we consider the boundary–value problem (87)-(89) for the equilibrium of porous rods in the case of pure traction boundary conditions, i.e. when $\Gamma_u = \Gamma_\psi = \emptyset$.

In this situation, the variational problem can be formulated as follows: find the weak solution $\boldsymbol{z} = (\tilde{u}_i, \tilde{\psi}_i, \tilde{\varphi}) \in \boldsymbol{H}^1[0,l]$ such that the equation (92) holds for every $\boldsymbol{y} = (u_i, \psi_i, \varphi) \in \boldsymbol{H}^1[0,l]$.

Let us consider the subspace $\tilde{\boldsymbol{R}}$ of $\boldsymbol{H}^1[0,l]$ given by

$$\tilde{\boldsymbol{R}} = \{\boldsymbol{y} = (u_i, \psi_i, \varphi) \in \boldsymbol{H}^1[0,l] | e_i(\boldsymbol{y}) = 0, \kappa_i(\boldsymbol{y}) = 0, \varphi = 0\}, \qquad (100)$$

which contains the infinitesimal rigid body displacements and rotations for porous rods. We observe that the left-hand side of the variational equation (92) vanishes for every $\boldsymbol{y} \in \tilde{\boldsymbol{R}}$. Then, from (92) and (100) we obtain the following necessary condition for the existence of weak solutions

$$\int_{\mathcal{C}} \rho_0 (\mathcal{F}_i u_i + \mathcal{L}_i \psi_i) \mathrm{d}s = 0, \qquad \forall (u_i, \psi_i) \in \boldsymbol{R}. \qquad (101)$$

Let us denote by $\tilde{\boldsymbol{V}}$ the quotient space $\tilde{\boldsymbol{V}} = \boldsymbol{H}^1[0,l]/\tilde{\boldsymbol{R}}$ and by $\hat{\boldsymbol{y}} \in \tilde{\boldsymbol{V}}$ the equivalence class of any element $\boldsymbol{y} = (u_i, \psi_i, \varphi) \in \boldsymbol{H}^1[0,l]$. The next theorem states that the condition (101) is also sufficient for the existence of solutions (Bîrsan and Altenbach, 2012b).

Theorem 6.6. *Assume that the hypotheses of Theorem 6.1 are fulfilled and that the external body loads $\mathcal{F}_i$ and $\mathcal{L}_i$ satisfy the condition (101). Then, there exists a weak solution $\boldsymbol{z} = (\tilde{u}_i, \tilde{\psi}_i, \tilde{\varphi})$ of the pure traction boundary-value problem, which is unique up to an additive (rigid–body) field $\boldsymbol{w} = (u_i, \psi_i, \varphi) \in \tilde{\boldsymbol{R}}$.*

Moreover, the equivalence class of the weak solution $\hat{\boldsymbol{z}} \in \tilde{\boldsymbol{V}}$ is also the unique solution of the minimization problem

$$J(\hat{\boldsymbol{z}}) = \inf_{\hat{\boldsymbol{y}} \in \tilde{\boldsymbol{V}}} J(\hat{\boldsymbol{y}}), \qquad (102)$$

where J is the functional defined on $\tilde{\boldsymbol{V}}$ by

$$J(\hat{\boldsymbol{y}}) = \int_{\mathcal{C}} \rho_0 \big[\mathcal{U}(\boldsymbol{e}(\hat{\boldsymbol{y}}), \boldsymbol{\kappa}(\hat{\boldsymbol{y}}), \varphi, \varphi') - \mathcal{F}_i \hat{u}_i - \mathcal{L}_i \hat{\psi}_i - p\varphi\big] \mathrm{d}s, \ \forall \hat{\boldsymbol{y}} = (\hat{u}_i, \hat{\psi}_i, \varphi) \in \tilde{\boldsymbol{V}}. \qquad (103)$$

Concerning the boundary-value problems for rods we mention that one can prove regularity results for the weak solutions, using the same mathematical methods as in the case of shells (see Ciarlet, 2005, Theorems 4.4-4 and 4.4-5).

7 Equations for Straight Rods

Let us consider the case when the middle curve $\mathcal{C}_0$ is straight, but the rod can have natural twisting and an arbitrary cross-section. In this situation our boundary-initial-value problem decouples into two problems: one for extension-torsion and the other for bending-shear.

Since the curve $\mathcal{C}_0$ is straight, the unit tangent vector $\boldsymbol{t}$ is constant (does not depend on s) and the vectors $\boldsymbol{n}, \boldsymbol{b}$ are not uniquely determined. Consider a fixed orthogonal Cartesian frame $Ox_1x_2x_3$ and denote by $\boldsymbol{e}_i$ the unit vectors along the Ox_i axes, see Figure 2. We can choose the vectors $\{\boldsymbol{e}_1, \boldsymbol{e}_2, \boldsymbol{e}_3\}$ and $\{\boldsymbol{n}, \boldsymbol{b}, \boldsymbol{t}\}$ such that

$$\boldsymbol{n} = \boldsymbol{e}_1 = \boldsymbol{d}_1(0), \qquad \boldsymbol{b} = \boldsymbol{e}_2 = \boldsymbol{d}_2(0), \qquad \boldsymbol{t} = \boldsymbol{e}_3 = \boldsymbol{d}_3. \tag{104}$$

For a straight curve $\mathcal{C}_0$ the Darboux vector is zero, but the angle of natural twisting $\sigma(s)$ is arbitrary, thus we have

$$\boldsymbol{\tau} = \boldsymbol{0}, \qquad \sigma(s) = \sphericalangle\big(\boldsymbol{e}_1, \boldsymbol{d}_1(s)\big), \qquad \boldsymbol{q}(s) = \sigma'(s)\boldsymbol{e}_1. \tag{105}$$

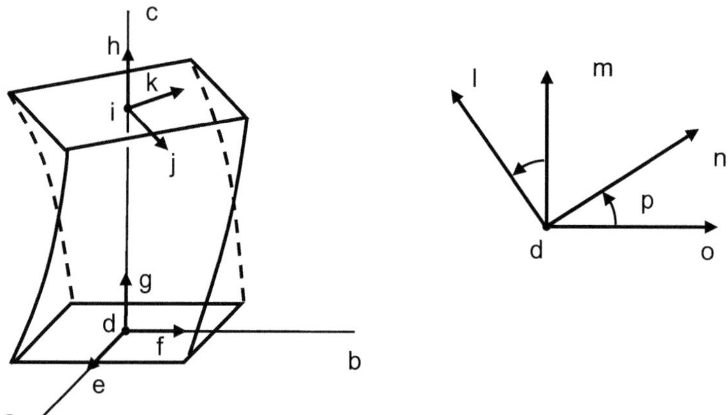

Figure 2. Reference configuration of a straight rod with natural twisting

In view of relations (9) and (35), in our case $\tilde{\mu} = 1$ and the tensors of inertia become

$$\rho_0 \boldsymbol{\Theta}_1^0 = \boldsymbol{0}, \quad \rho_0 \boldsymbol{\Theta}_2^0 = I_1 \boldsymbol{d}_1 \otimes \boldsymbol{d}_1 + I_2 \boldsymbol{d}_2 \otimes \boldsymbol{d}_2 + (I_1 + I_2)\boldsymbol{t} \otimes \boldsymbol{t}, \tag{106}$$

where we have denoted by $I_1 \equiv \int_\Sigma \rho^* y^2 \mathrm{d}x\mathrm{d}y, \quad I_2 \equiv \int_\Sigma \rho^* x^2 \mathrm{d}x\mathrm{d}y.$

The expressions of constitutive tensors simplify significantly for straight rods. Indeed, since $\boldsymbol{\tau} = \mathbf{0}$, in the decompositions of the form (37) we keep only the first term: $\boldsymbol{f} = \boldsymbol{f}^0$, for any constitutive tensor $\boldsymbol{f}$. Thus, the relations (38) reduce to

$$\boldsymbol{G}_1 = G_1\boldsymbol{t},\ \boldsymbol{G}_2 = \mathbf{0},\ \boldsymbol{K}_4 = K_4\boldsymbol{t},\ \boldsymbol{K}_5 = \mathbf{0},\ \boldsymbol{K}_6 = K_6\boldsymbol{t},\ \boldsymbol{K}_7 = \mathbf{0}, \tag{107}$$

while the elasticity tensors $\boldsymbol{A}$, $\boldsymbol{C}$ have the forms $(43)_{1,2}$ and $\boldsymbol{B} = \sigma' B_0 \boldsymbol{t} \otimes \boldsymbol{t}$ (Zhilin, 2007).

7.1 Decoupling of the Problem

In order to distinguish between the extensional, torsional, bending, and shear deformation, let us decompose the fields $\boldsymbol{u}, \boldsymbol{\psi}, \boldsymbol{N}, \boldsymbol{M}, \boldsymbol{\mathcal{F}}$ and $\boldsymbol{\mathcal{L}}$ by the tangent direction $\boldsymbol{t}$ and the normal plane $(\boldsymbol{d}_1, \boldsymbol{d}_2)$:

$$\begin{gathered} \boldsymbol{u} = u\boldsymbol{t} + \boldsymbol{w},\ \boldsymbol{\psi} = \psi\boldsymbol{t} + \boldsymbol{t} \times \boldsymbol{\vartheta},\ \ \boldsymbol{N} = F\boldsymbol{t} + \boldsymbol{Q},\ \boldsymbol{M} = H\boldsymbol{t} + \boldsymbol{t} \times \bar{\boldsymbol{L}}, \\ \boldsymbol{\mathcal{F}} = \mathcal{F}_t\boldsymbol{t} + \boldsymbol{\mathcal{F}}_n,\ \boldsymbol{\mathcal{L}} = \mathcal{L}_t\boldsymbol{t} + \boldsymbol{\mathcal{L}}_n, \end{gathered} \tag{108}$$

where the vectors $\boldsymbol{w}, \boldsymbol{\vartheta}, \boldsymbol{Q}, \bar{\boldsymbol{L}}, \boldsymbol{\mathcal{F}}_n$ and $\boldsymbol{\mathcal{L}}_n$ are orthogonal to $\boldsymbol{t}$. In relations (108) u is the longitudinal displacement, $\boldsymbol{w}$ is the vector of transversal displacement, ψ is the torsion, $\boldsymbol{\vartheta}'$ is the vector of bending deformation, F is the longitudinal force, $\boldsymbol{Q}$ is the vector of transversal force, H is the torsion moment and $\bar{\boldsymbol{L}}$ is the vector of bending moment. The geometrical equations (45) can be written as

$$\boldsymbol{e} = u'\boldsymbol{t} + \boldsymbol{\gamma}, \qquad \boldsymbol{\kappa} = \psi'\boldsymbol{t} + \boldsymbol{t} \times \boldsymbol{\vartheta}' \qquad \text{with} \quad \boldsymbol{\gamma} = \boldsymbol{w}' - \boldsymbol{\vartheta}, \tag{109}$$

where $\boldsymbol{\gamma}$ is the vector of *transverse shear* ($\boldsymbol{\gamma}$ is orthogonal to $\boldsymbol{t}$).

Considering the relations (106)-(109), the boundary-initial-value problem (45)-(53) decouples into two problems as follows: the first problem involves only the scalar unknowns u, ψ, φ, T, and represents the *extension-torsion problem*

$$\begin{aligned} &F' + \rho_0\mathcal{F}_t = \rho_0\ddot{u}, && H' + \rho_0\mathcal{L}_t = (I_1 + I_2)\ddot{\psi}, \\ &h' - g + \rho_0 p = \rho_0\varkappa\ddot{\varphi}, && q' + \rho_0 S = \rho_0\theta_0\dot{\eta}, \end{aligned} \tag{110}$$

with the constitutive equations

$$\begin{aligned} F &= A_3u' + \sigma' B_0\psi' + K_4\varphi + K_6\varphi' + G_1T, \\ H &= \sigma' B_0 u' + C_3\psi', \\ g &= K_1\varphi + K_3\varphi' + K_4u' + G_3T, \\ h &= K_2\varphi' + K_3\varphi + K_6u' + G_4T, \\ \rho_0\eta &= -GT - G_1u' - G_3\varphi - G_4\varphi', \\ q &= KT'. \end{aligned} \tag{111}$$

The second problem involves only the unknowns $\boldsymbol{w} = w_\alpha \boldsymbol{d}_\alpha, \boldsymbol{\vartheta} = \vartheta_\alpha \boldsymbol{d}_\alpha$ and represents the *bending-shear problem*

$$\boldsymbol{Q}' + \rho_0 \boldsymbol{\mathcal{F}}_n = \rho_0 \ddot{\boldsymbol{w}}, \quad \bar{\boldsymbol{L}}' + \boldsymbol{Q} - \rho_0 \boldsymbol{t} \times \boldsymbol{\mathcal{L}}_n = \big(I_2 \boldsymbol{d}_1 \otimes \boldsymbol{d}_1 + I_1 \boldsymbol{d}_2 \otimes \boldsymbol{d}_2 \big) \cdot \ddot{\boldsymbol{\vartheta}}, \quad (112)$$

with the constitutive equations

$$\boldsymbol{Q} = \big(A_1 \boldsymbol{d}_1 \otimes \boldsymbol{d}_1 + A_2 \boldsymbol{d}_2 \otimes \boldsymbol{d}_2 \big) \cdot (\boldsymbol{w}' - \boldsymbol{\vartheta}), \quad \bar{\boldsymbol{L}} = \big(C_2 \boldsymbol{d}_1 \otimes \boldsymbol{d}_1 + C_1 \boldsymbol{d}_2 \otimes \boldsymbol{d}_2 \big) \cdot \boldsymbol{\vartheta}'. \quad (113)$$

We observe that the temperature T and the porosity field φ intervene only in the extension-torsion problem (110), (111). The model of directed curves can be adapted to include the porosity and the temperature also in the bending-shear problem, but in this case we have to introduce several temperature and porosity fields, as we will show later in Sect. 10.

7.2 Solution of Simple Problems

In order to find the solution of some simple problems, let us restrict further our attention to straight rods without natural twisting. In this case we have $\sigma = 0$, $\boldsymbol{d}_i = \boldsymbol{e}_i$, and the expressions of the constitutive tensors (42), (43) become

$$\begin{gathered} \boldsymbol{B} = \boldsymbol{0}, \ \boldsymbol{G}_1 = G_1 \boldsymbol{t}, \ \boldsymbol{G}_2 = \boldsymbol{0}, \ G_4 = 0 \quad (G_3 \text{ and } G \text{ unchanged}), \\ K_3 = 0, \quad \boldsymbol{K}_4 = K_4 \boldsymbol{t}, \ \boldsymbol{K}_5 = \boldsymbol{0}, \ \boldsymbol{K}_6 = \boldsymbol{0}, \ \boldsymbol{K}_7 = \boldsymbol{0}, \end{gathered} \quad (114)$$

while $\boldsymbol{A}$ and $\boldsymbol{C}$ are given by $(43)_{1,2}$. Since $\sigma' = 0$ it follows that the torsion problem (given by $(110)_2$ and $(111)_2$) will decouple from the extensional problem. Thus, the extensional problem for straight rods without natural twisting is represented by the equations

$$F' + \rho_0 \mathcal{F}_t = \rho_0 \ddot{u}, \qquad h' - g + \rho_0 p = \rho_0 \varkappa \ddot{\varphi}, \qquad q' + \rho_0 S = \rho_0 \theta_0 \dot{\eta} \quad (115)$$

with the constitutive relations

$$\begin{gathered} F = A_3 u' + K_4 \varphi + G_1 T, \qquad g = K_1 \varphi + K_4 u' + G_3 T, \\ h = K_2 \varphi', \qquad \rho_0 \eta = -GT - G_1 u' - G_3 \varphi, \qquad q = KT'. \end{gathered} \quad (116)$$

In what follows we present 4 examples of simple problems and give their solutions, which will be used in Sect. 9 for the identification of constitutive coefficients. These solutions are exact up to an arbitrary rigid body motion $\{\bar{\boldsymbol{u}}, \bar{\boldsymbol{\psi}}, \bar{\varphi}, \bar{T}\}$. The general form for rigid body motions in the linear theory of thermo-elastic porous rods is $\bar{\boldsymbol{u}} = \bar{\boldsymbol{a}} + \bar{\boldsymbol{b}} \times \boldsymbol{r}, \bar{\boldsymbol{\psi}} = \bar{\boldsymbol{b}}, \bar{\varphi} = 0, \bar{T} = 0$, where $\bar{\boldsymbol{a}}$ and $\bar{\boldsymbol{b}}$ are constant vectors.

We consider the equilibrium deformation of straight rods made of homogeneous materials. Thus the constitutive coefficients are constant, and the requirement that the energy $\mathcal{U}$ is positive definite implies the inequalities $A_i > 0, C_i > 0, K_2 > 0$ and $A_3K_1 > (K_4)^2$. In our examples, the deformation of porous thermo-elastic rods is due to some boundary conditions on the ends, while the external body loads and heat supply are absent and the lateral surface is free, i.e. $\boldsymbol{\mathcal{F}} = \mathbf{0}$, $\boldsymbol{\mathcal{L}} = \mathbf{0}$, $p = 0$ and $S = 0$.

Problem 1: Extension Determine the equilibrium of a rod under an axial force $F^{(1)}$ applied at its ends, such that the endpoints are kept at a constant temperature $T^{(1)}$.

To solve this extensional problem, we consider the equilibrium equations given by (115) with the vanishing right–hand sides, the constitutive equations (116) and the boundary conditions

$$F(0) = F(l) = F^{(1)}, \quad T(0) = T(l) = T^{(1)}, \quad h(0) = h(l) = 0. \tag{117}$$

We observe that the equations for temperature $(115)_3$, $(116)_5$ and $(117)_3$ can be solved separately and give the solution $T(s) = T^{(1)}, \forall s \in [0, l]$. Inserting this relation into the remaining equations $(115)_{1,2}$, $(116)_{1-3}$ and $(117)_{1,2}$ we find a system of ordinary differential equations with the solution

$$\begin{aligned} u(s) &= \frac{K_1F^{(1)} + (G_1K_1 - G_3K_4)T^{(1)}}{A_3K_1 - (K_4)^2}\, s, \\ \varphi(s) &= \frac{K_4F^{(1)} + (G_1K_4 - A_3G_3)T^{(1)}}{(K_4)^2 - A_3K_1}, \end{aligned} \tag{118}$$

which gives the longitudinal displacement and the volume fraction field in the extended rod.

Problem 2: Bending–Shear Find the equilibrium of a rod subjected to transversal forces and bending moments applied to its ends.

Thus, in $s = 0$ we have the boundary conditions

$$\boldsymbol{Q}(0) = \boldsymbol{Q}^{(1)} = Q_1^{(1)}\boldsymbol{d}_1 + Q_2^{(1)}\boldsymbol{d}_2, \quad \bar{\boldsymbol{L}}(0) = \boldsymbol{L}^{(1)} = L_1^{(1)}\boldsymbol{d}_1 + L_2^{(1)}\boldsymbol{d}_2. \tag{119}$$

Using (113) and (119) we deduce the following system of equations for $w_1(s)$ and $\theta_1(s)$:

$$A_1\big(w_1''(s) - \theta_1'(s)\big) = 0, \qquad C_2\theta_1''(s) + A_1\big(w_1'(s) - \theta_1(s)\big) = 0, \quad s \in [0, l],$$

with the boundary conditions

$$A_1\big(w_1'(0) - \theta_1(0)\big) = Q_1^{(1)}, \qquad C_2\theta_1'(0) = L_1^{(1)}.$$

A similar system is obtained for the functions $w_2(s)$ and $\theta_2(s)$. Finally, we get the solution (up to a rigid–body motion)

$$w_1(s) = -\frac{Q_1^{(1)}}{6C_2}s^3 + \frac{L_1^{(1)}}{2C_2}s^2 + \frac{Q_1^{(1)}}{A_1}s, \quad w_2(s) = -\frac{Q_2^{(1)}}{6C_1}s^3 + \frac{L_2^{(1)}}{2C_1}s^2 + \frac{Q_2^{(1)}}{A_2}s,$$
$$\theta_1(s) = -\frac{Q_1^{(1)}}{2C_2}s^2 + \frac{L_1^{(1)}}{C_2}s, \qquad \theta_2(s) = -\frac{Q_2^{(1)}}{2C_1}s^2 + \frac{L_2^{(1)}}{C_1}s. \tag{120}$$

Problem 3: Torsion Determine the equilibrium of a rod under the action of a torsion moment $H^{(1)}$ applied to its ends.

The boundary conditions are $H(0) = H(l) = H^{(1)}$, and in view of the above equations for torsion we find the following solution for the function ψ (up to a rigid–body displacement)

$$\psi(s) = \frac{H^{(1)}}{C_3}s. \tag{121}$$

Problem 4: Thermal Deformation Find the equilibrium of a rod subjected to a difference of temperature $T^{(2)}$ between its two end boundaries.

This is also an extensional problem, characterized by the equations (115), (116) and we consider the end boundary conditions

$$T(l) = T^{(2)}, \; T(0) = 0, \; F(0) = F(l) = 0, \; \varphi(0) = 0, \; h(0) = h(l). \tag{122}$$

Solving first the heat equation $(115)_3$ with $(116)_5$ and $(122)_{1,2}$ we find the temperature field $T(s) = T^{(2)}s/l, \forall s \in [0,l]$. Then, we substitute this expression into the remaining equations $(115)_{1,2}$, $(116)_{1-3}$ and $(122)_{3-5}$, and by solving this boundary–value problem for ordinary differential equations we obtain the solution

$$u(s) = \frac{G_1K_1 - G_3K_4}{A_3K_1 - (K_4)^2}\frac{T^{(2)}}{2l}s^2, \qquad \varphi(s) = \frac{A_3G_3 - G_1K_4}{A_3K_1 - (K_4)^2}\frac{T^{(2)}}{l}s, \tag{123}$$

which represent the longitudinal displacement and the volume fraction field in the deformed rod.

The simple solutions obtained in this section will be used to identify the constitutive coefficients for porous thermo-elastic rods by comparison with corresponding results from the three-dimensional theory.

8 Derivation of Rods Equations from the Three-Dimensional Equations

For comparison purposes, we show in this section that one-dimensional rod equations similar to (49)-(51) can be obtained from the three-dimensional equations by integration over the cross-section.

Consider a three-dimensional rod made of a thermo-elastic material with pores which occupies the domain $\mathcal{B} = \{(x_1, x_2, x_3) | (x_1, x_2) \in \Sigma, x_3 \in [0, l]\}$. In view of $\boldsymbol{d}_i = \boldsymbol{e}_i$, we identify the coordinates $x = x_1$, $y = x_2$ and we consider that the relations (35) are satisfied. The three-dimensional governing equations of thermo-elastic material with pores are (Ieşan, 1986)

$$t^*_{ji,j} + \rho^* f^*_i = \rho^* \ddot{u}^*_i, \quad h^*_{i,i} - g^* + \rho^* p^* = \rho^* \varkappa^* \ddot{\varphi}^*, \quad q^*_{i,i} + \rho^* S^* = \rho^* \theta^*_0 \dot{\eta}^*, \tag{124}$$

where $\boldsymbol{u}^* = u^*_i \boldsymbol{e}_i$ is the displacement vector, φ^* the volume fraction field, θ^* the temperature, η^* the entropy function, $\boldsymbol{T}^* = t^*_{ij} \boldsymbol{e}_i \otimes \boldsymbol{e}_j$ is the Cauchy stress tensor, $\boldsymbol{h}^* = h^*_i \boldsymbol{e}_i$ the equilibrated stress, g^* the internal equilibrated body force, $\boldsymbol{q}^* = q^*_i \boldsymbol{e}_i$ the heat flux vector, $\boldsymbol{f}^* = f^*_i \boldsymbol{e}_i$ the body force per unit mass, p^* the assigned equilibrated body force, S^* the heat supply per unit mass and $\varkappa^*$ the equilibrated inertia. For the sake of simplicity the integration over the cross-section will be denoted by $\langle f \rangle = \int_\Sigma f \mathrm{d}x_1 \mathrm{d}x_2$ for any function f.

By integrating the three-dimensional equations (124) over the cross–section Σ of the rod $\mathcal{B}$ we obtain the following set of one-dimensional equations

$$\begin{aligned} &\hat{\boldsymbol{N}}' + \hat{\rho}_0 \hat{\boldsymbol{\mathcal{F}}} = \hat{\rho}_0 \ddot{\hat{\boldsymbol{u}}}, \qquad \hat{\boldsymbol{M}}' + \boldsymbol{t} \times \hat{\boldsymbol{N}} + \hat{\rho}_0 \hat{\boldsymbol{\mathcal{L}}} = \hat{\rho}_0 \overset{0}{\hat{\boldsymbol{\Theta}}}_2 \cdot \ddot{\hat{\boldsymbol{\psi}}}, \\ &\hat{h}' - \hat{g} + \hat{\rho}_0 \hat{p} = \hat{\rho}_0 \hat{\varkappa} \ddot{\hat{\varphi}}, \qquad \hat{q}' + \hat{\rho}_0 \hat{S} = \hat{\rho}_0 \hat{\theta}_0 \dot{\hat{\eta}}, \end{aligned} \tag{125}$$

where we have introduced the following notations

$$\hat{\boldsymbol{u}} = \frac{\langle \rho^* \boldsymbol{u}^* \rangle}{\langle \rho^* \rangle}, \quad \hat{\boldsymbol{N}} = \langle \boldsymbol{t} \cdot \boldsymbol{T}^* \rangle, \quad \hat{\boldsymbol{\mathcal{F}}} = \frac{1}{\langle \rho^* \rangle} \big(\langle \rho^* \boldsymbol{f}^* \rangle + \int_{\partial \Sigma} (\boldsymbol{n}^* \cdot \boldsymbol{T}^*) \mathrm{d}l \big),$$

$$\hat{\boldsymbol{\psi}} = \frac{\langle \rho^* x_2 u^*_3 \rangle}{I_1} \boldsymbol{e}_1 - \frac{\langle \rho^* x_1 u^*_3 \rangle}{I_2} \boldsymbol{e}_2 + \frac{\langle \rho^* (x_1 u^*_2 - x_2 u^*_1) \rangle}{I_1 + I_2} \boldsymbol{e}_3, \quad \hat{\rho}_0 = \langle \rho^* \rangle,$$

$$\hat{\boldsymbol{M}} = \langle \boldsymbol{a} \times (\boldsymbol{t} \cdot \boldsymbol{T}^*) \rangle, \quad \hat{\rho}_0 \overset{0}{\hat{\boldsymbol{\Theta}}}_2 = I_1 \boldsymbol{e}_1 \otimes \boldsymbol{e}_1 + I_2 \boldsymbol{e}_2 \otimes \boldsymbol{e}_2 + (I_1 + I_2) \boldsymbol{e}_3 \otimes \boldsymbol{e}_3,$$

$$\hat{\boldsymbol{\mathcal{L}}} = \frac{1}{\langle \rho^* \rangle} \big(\langle \rho^* \boldsymbol{a} \times \boldsymbol{f}^* \rangle + \int_{\partial \Sigma} \boldsymbol{a} \times (\boldsymbol{n}^* \cdot \boldsymbol{T}^*) \mathrm{d}l \big), \quad \hat{h} = \langle \boldsymbol{h}^* \cdot \boldsymbol{t} \rangle, \quad \hat{g} = \langle g^* \rangle,$$

$$\hat{\varphi} = \frac{\langle \rho^* \varkappa^* \varphi^* \rangle}{\langle \rho^* \varkappa^* \rangle}, \quad \hat{\varkappa} = \frac{\langle \rho^* \varkappa^* \rangle}{\langle \rho^* \rangle}, \quad \hat{p} = \frac{1}{\langle \rho^* \rangle} \big(\langle \rho^* p^* \rangle + \int_{\partial \Sigma} (\boldsymbol{h}^* \cdot \boldsymbol{n}^*) \mathrm{d}l \big),$$

$$\hat{q} = \langle \boldsymbol{q}^* \cdot \boldsymbol{t} \rangle, \quad \hat{S} = \frac{1}{\langle \rho^* \rangle}\Big(\langle \rho^* S^* \rangle + \int_{\partial\Sigma} \boldsymbol{q}^* \cdot \boldsymbol{n}^* \mathrm{d}l\Big),$$
$$\hat{T} = \frac{1}{\langle \rho^* \rangle}\langle \rho^* T^* \rangle, \quad \hat{\eta} = \frac{1}{\langle \rho^* \theta_0^* \rangle}\langle \rho^* \theta_0^* \eta^* \rangle, \quad \hat{\theta}_0 = \frac{1}{\langle \rho^* \rangle}\langle \rho^* \theta_0^* \rangle \tag{126}$$

and $\boldsymbol{n}^*$ represents the outward unit normal to the rod's lateral surface.

In view of relations (106), we remark that the one-dimensional equations (125) have exactly the same form as the governing equations in the direct approach (49)-(51). In line with the decompositions $(108)_{1,2}$, we also denote

$$\hat{w}_\alpha = \frac{\langle \rho^* u_\alpha^* \rangle}{\langle \rho^* \rangle} \quad (\alpha = 1, 2), \ \hat{u} = \frac{\langle \rho^* u_3^* \rangle}{\langle \rho^* \rangle}, \ \hat{\psi} = \frac{\langle \rho^* (x_1 u_2^* - x_2 u_1^*) \rangle}{I_1 + I_2},$$
$$\hat{\vartheta}_1 = -\frac{\langle \rho^* x_1 u_3^* \rangle}{I_2}, \ \hat{\vartheta}_2 = -\frac{\langle \rho^* x_2 u_3^* \rangle}{I_1}. \tag{127}$$

These notations are convenient to facilitate the identification of corresponding fields in the two approaches, since any field v from the direct approach will be identified with the field $\hat{v}$ from the three–dimensional approach.

Let us go further with our comparison and examine the correspondence between the constitutive equations. We will find that the two sets of constitutive equations have the same form provided we take the three–dimensional variables $\{u_i^*, \varphi^*, T^*\}$ of a certain form, appropriate for rod theory.

In what follows we turn our attention to orthotropic and homogeneous rods. The constitutive equations for orthotropic thermoelastic materials with pores are Ieşan (2009b)

$$\begin{aligned}
t_{11}^* &= c_{11}e_{11}^* + c_{12}e_{22}^* + c_{13}e_{33}^* + \beta_1\varphi^* - b_1 T^*,\\
t_{22}^* &= c_{12}e_{11}^* + c_{22}e_{22}^* + c_{23}e_{33}^* + \beta_2\varphi^* - b_2 T^*,\\
t_{33}^* &= c_{13}e_{11}^* + c_{23}e_{22}^* + c_{33}e_{33}^* + \beta_3\varphi^* - b_3 T^*,\\
t_{23}^* &= 2c_{44}e_{23}^*, \quad t_{31}^* = 2c_{55}e_{31}^*, \quad t_{12}^* = 2c_{66}e_{12}^*,\\
h_1^* &= \alpha_1\varphi_{,1}^*, \quad h_2^* = \alpha_2\varphi_{,2}^*, \quad h_3^* = \alpha_3\varphi_{,3}^*,\\
g^* &= \beta_1 e_{11}^* + \beta_2 e_{22}^* + \beta_3 e_{33}^* + \xi\varphi^* - mT^*,\\
\rho^*\eta^* &= aT^* + b_1 e_{11}^* + b_2 e_{22}^* + b_3 e_{33}^* + m\varphi^*,\\
q_i^* &= K_i^* T_{,i}^* \quad (i = 1, 2, 3 \text{ not summed}),
\end{aligned} \tag{128}$$

where $c_{rs}, \alpha_k, \beta_k, b_k, \xi, m, a$ and K_i^* are constitutive coefficients and the strain tensor is denoted by $e_{ij}^* = (u_{i,j}^* + u_{j,i}^*)/2$. The material is homogeneous and $\rho^*, \varkappa^*, \theta_0^*$ are constant and positive.

In the theory of thin rods it is customary to represent the displacement $\boldsymbol{u}^*$ as a linear function of the thickness coordinates x_1, x_2. Since in our model we have assumed that the cross-sections do not deform, then we impose the restrictions $u^*_{1,1} = u^*_{2,2} = u^*_{1,2} + u^*_{2,1} = 0$. Also, for very thin rods, it is reasonable to consider that the porosity function φ^* and the temperature field T^* do not vary across the thickness. In view of these assumptions, we represent the fields u^*_i, φ^* and T^* in the form

$$\begin{aligned} u^*_1(x_i,t) &= \hat{w}_1(x_3,t) - x_2\hat{\psi}(x_3,t), \\ u^*_2(x_i,t) &= \hat{w}_2(x_3,t) + x_1\hat{\psi}(x_3,t), \\ u^*_3(x_i,t) &= \hat{u}(x_3,t) - x_1\hat{\vartheta}_1(x_3,t) - x_2\hat{\vartheta}_2(x_3,t), \\ \varphi^*(x_i,t) &= \hat{\varphi}(x_3,t), \quad T^*(x_i,t) = \hat{T}(x_3,t), \end{aligned} \tag{129}$$

where we have used appropriate notations to conform with relations (127) and $(126)_{11,16}$.

Inserting the relations (129) into the three-dimensional equations (128) and then into the identification equalities $(126)_{2,6,9,10,14,17}$, we obtain the "resultant" constitutive equations which express $\hat{N}, \hat{M}, \hat{h}, \hat{g}, \hat{\eta}$ and $\hat{q}$ in terms of $\hat{u}, \hat{\psi}, \hat{w}_\alpha, \hat{\vartheta}_\alpha, \hat{\varphi}$ and $\hat{T}$. For instance, the "resultant" constitutive equations for the thermal and porosity fields are

$$\begin{aligned} \hat{g} &= A(\beta_3\hat{u}' + \xi\hat{\varphi} - m\hat{T}), \quad \hat{h} = A\alpha_3\hat{\varphi}', \\ \hat{\rho}_0\hat{\eta} &= A(b_3\hat{u}' + m\hat{\varphi} + a\hat{T}), \quad \hat{q} = AK^*_3\hat{T}', \end{aligned}$$

where A denotes the area of the cross–section Σ. Comparing the set of „resultant" constitutive equations with the corresponding constitutive equations (113) and (116) from the direct approach, we remark that they have the same form. This means that, from the mathematical viewpoint, we have to treat the same equations in both approaches. The comparison has been presented for straight rods, but this analysis can be extended also to curved rods.

Remark 8.1. Concerning the relations $(129)_{4,5}$ we notice that we could take into account also the variations of volume fraction φ^* and temperature T^* in the cross-section (as linear functions of x_1, x_2), but this would complicate the model by introducing three functions of porosity and three functions for temperature (see e.g. this procedure in Bîrsan (2006b), for the case of shells). In view of the identifications $(126)_{11,16}$ we deduce the interpretations of volume fraction field φ and temperature field T for directed curves, as representing the following weighted averages

$$T(s,t) = \frac{\int_\Sigma \rho^* T^*(x,y,s,t)\mathrm{d}x\mathrm{d}y}{\int_\Sigma \rho^* \mathrm{d}x\mathrm{d}y}, \quad \varphi(s,t) = \frac{\int_\Sigma \rho^* \varphi^*(x,y,s,t)\mathrm{d}x\mathrm{d}y}{\int_\Sigma \rho^* \mathrm{d}x\mathrm{d}y}. \tag{130}$$

From the relations $(126)_{3,8,13,15}$ we can observe that the external body loads and heat supply $\boldsymbol{\mathcal{F}}, \boldsymbol{\mathcal{L}}, p$ and S in the direct approach include also the contributions of loads and heat flux applied on the lateral surface of the three-dimensional rod. This comparison of the two approaches does not permit the identification of constitutive coefficients for directed curves, which is a more difficult task.

9 Identification of Constitutive Coefficients for Thermo-Elastic Porous Orthotropic Rods

In this section we determine the constitutive coefficients for straight porous rods made of an orthotropic and homogeneous material, by comparison of simple exact solutions for directed curves with the results from three-dimensional theory. To this aim, we first consider the deformation of purely elastic rods and identify the elasticity coefficients A_i and C_i in terms of the constitutive constants for orthotropic material c_{ij}. Then, we investigate the deformation of porous thermo-elastic rods, and by comparison with solutions obtained in Sect. 7.2 we determine the the thermal coefficients G_1, G_3 and the poro-elastic coefficients K_1, K_4.

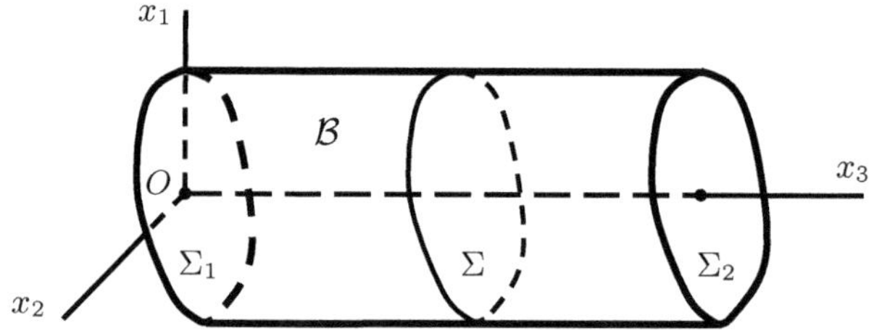

Figure 3. The three-dimensional rod in reference configuration

Let us consider a rod which occupies the domain $\mathcal{B} = \{(x_1, x_2, x_3) | (x_1, x_2) \in \Sigma, x_3 \in [0, l]\}$, made of an orthotropic and homogeneous material. We denote by Σ_1 and Σ_2 the end boundaries given by $x_3 = 0$ and $x_3 = l$, respectively (see Fig. 3). The external body loads and heat supply are zero and the lateral surface is free. We consider that the boundary conditions on the ends Σ_1 and Σ_2 are given globally (not pointwise). It is convenient to introduce the notations

$$\begin{aligned} E_0 &= \frac{\delta_2}{\delta_1}, \qquad \delta_1 = c_{11}c_{22} - c_{12}^2, \qquad \delta_2 = \det(c_{ij})_{3\times 3}, \\ \nu_1 &= \frac{c_{13}c_{22} - c_{23}c_{12}}{\delta_1}, \qquad \nu_2 = \frac{c_{23}c_{11} - c_{13}c_{12}}{\delta_1}, \end{aligned} \tag{131}$$

where we set for convenience $c_{ji} = c_{ij}$.

9.1 Bending and Extension of Orthotropic Rods

For a right cylinder $\mathcal{B}$ made of elastic material (without voids) assume that on the end boundary Σ_1 a resultant axial force of magnitude $F^{(1)}$ and a resultant bending moment $L_2^{(1)}\boldsymbol{e}_2$ are applied. Then the non-vanishing global boundary conditions on Σ_1 are

$$\int_{\Sigma_1} t_{33}^* \mathrm{d}x_1 \mathrm{d}x_2 = F^{(1)}, \qquad \int_{\Sigma_1} x_2 t_{33}^* \mathrm{d}x_1 \mathrm{d}x_2 = -L_2^{(1)},$$

and the solution for this equilibrium problem is given by Ieşan (2009a)

$$\begin{aligned}
u_1^* &= -\frac{F^{(1)}}{AE_0}\nu_1 x_1 + \frac{L_2^{(1)}}{E_0\langle x_2^2\rangle}\nu_1 x_1 x_2,\\
u_2^* &= -\frac{F^{(1)}}{AE_0}\nu_2 x_2 + \frac{L_2^{(1)}}{2E_0\langle x_2^2\rangle}(x_3^2 - \nu_1 x_1^2 + \nu_2 x_2^2),\\
u_3^* &= \frac{F^{(1)}}{AE_0}x_3 - \frac{L_2^{(1)}}{E_0\langle x_2^2\rangle}x_2 x_3.
\end{aligned} \tag{132}$$

From the three-dimensional solution (132) we obtain by integration over the cross-section

$$\begin{aligned}
&\int_\Sigma u_1^* \mathrm{d}x_1\mathrm{d}x_2 = 0, \int_\Sigma u_2^* \mathrm{d}x_1\mathrm{d}x_2 = \frac{AL_2^{(1)}}{2E_0\langle x_2^2\rangle}x_3^2 - \frac{L_2^{(1)}}{2E_0\langle x_2^2\rangle}(\nu_1\langle x_1^2\rangle - \nu_2\langle x_2^2\rangle),\\
&\int_\Sigma u_3^* \mathrm{d}x_1\mathrm{d}x_2 = \frac{F^{(1)}}{E_0}x_3, \int_\Sigma x_1 u_3^* \mathrm{d}x_1\mathrm{d}x_2 = 0, \int_\Sigma x_2 u_3^* \mathrm{d}x_1\mathrm{d}x_2 = -\frac{L_2^{(1)}}{E_0}x_3,\\
&\int_\Sigma (x_1 u_2^* - x_2 u_1^*)\mathrm{d}x_1\mathrm{d}x_2 = -\frac{L_2^{(1)}}{E_0\langle x_2^2\rangle}\big[(\nu_1 - \frac{\nu_2}{2})\langle x_1 x_2^2\rangle + \frac{\nu_1}{2}\langle x_1^3\rangle\big].
\end{aligned} \tag{133}$$

On the other hand, this problem of bending and extension can be treated in the direct approach of rods. Using the solutions of the problems solved in Sect. 7.2 we find

$$u(s) = \frac{F^{(1)}}{A_3}s, w_2(s) = \frac{L_2^{(1)}}{2C_1}s^2, \theta_2(s) = \frac{L_2^{(1)}}{C_1}s, w_1 = 0, \psi = 0, \theta_1 = 0. \tag{134}$$

Using the identifications (127) and $s = x_3$, we see that the solution in the theory of rods (134) corresponds to

$$\begin{aligned}
&\int_\Sigma u_3^* \mathrm{d}x_1\mathrm{d}x_2 = \frac{AF^{(1)}}{A_3}x_3, \int_\Sigma u_1^* \mathrm{d}x_1\mathrm{d}x_2 = 0, \int_\Sigma u_2^* \mathrm{d}x_1\mathrm{d}x_2 = \frac{AL_2^{(1)}}{2C_1}x_3^2,\\
&\int_\Sigma (x_1 u_2^* - x_2 u_1^*)\mathrm{d}x_1\mathrm{d}x_2 = \int_\Sigma x_1 u_3^* \mathrm{d}x_1\mathrm{d}x_2 = 0, \int_\Sigma x_2 u_3^* \mathrm{d}x_1\mathrm{d}x_2 = -\frac{L_2^{(1)}\langle x_2^2\rangle}{C_1}x_3.
\end{aligned} \tag{135}$$

Now we compare the solutions in the two approaches (133) and (135). First we notice that the last term in $(133)_2$ and the right-hand side of $(133)_6$ are some constants which represent rigid-body displacements and can be neglected. Then, the two solutions (133) and (135) coincide if and only if we identify the constitutive coefficients A_3 and C_1 as

$$A_3 = AE_0, \qquad C_1 = E_0\langle x_2^2\rangle = E_0 \int_\Sigma x_2^2 \mathrm{d}x_1 \mathrm{d}x_2. \tag{136}$$

Similarly, if we consider a resultant moment of the form $L_1^{(1)}\boldsymbol{e}_1$ we get

$$C_2 = E_0\langle x_1^2\rangle = E_0 \int_\Sigma x_1^2 \mathrm{d}x_1 \mathrm{d}x_2. \tag{137}$$

9.2 Torsion of Orthotropic Rods

Let us assume now that on the end boundaries of the elastic cylinder $\mathcal{B}$ we have a resultant torsion moment of magnitude $H^{(1)}$. Thus, the only non-vanishing global boundary condition on Σ_1 is

$$\int_{\Sigma_1} (x_1 t_{23}^* - x_2 t_{13}^*) \mathrm{d}x_1 \mathrm{d}x_2 = H^{(1)},$$

and the Saint-Venant solution for this equilibrium problem is (Ieşan, 2009a)

$$u_1^* = -\frac{H^{(1)}}{D_0} x_2 x_3, \qquad u_2^* = \frac{H^{(1)}}{D_0} x_1 x_3, \qquad u_3^* = \frac{H^{(1)}}{D_0} \phi(x_1, x_2), \tag{138}$$

where the torsion function $\phi(x_1, x_2)$ is the solution of the boundary–value problem

$$\begin{aligned} &c_{55}\phi_{,11} + c_{44}\phi_{,22} = 0 \qquad \text{in} \quad \Sigma, \\ &c_{55}\phi_{,1} n_1^* + c_{44}\phi_{,2} n_2^* = c_{55} x_2 n_1^* - c_{44} x_1 n_2^* \qquad \text{on} \quad \partial\Sigma, \end{aligned} \tag{139}$$

and the torsional rigidity D_0 is expressed by

$$D_0 = \int_\Sigma \big[c_{44} x_1(\phi_{,2} + x_1) - c_{55} x_2(\phi_{,1} - x_2)\big] \mathrm{d}x_1 \mathrm{d}x_2. \tag{140}$$

From the exact three-dimensional solution (138) and the conditions (35) it follows that

$$\begin{gathered} \int_\Sigma u_\alpha^* \mathrm{d}x_1 \mathrm{d}x_2 = 0 \ (\alpha = 1, 2), \quad \int_\Sigma u_3^* \mathrm{d}x_1 \mathrm{d}x_2 = \frac{H^{(1)}}{D_0} \int_\Sigma \phi(x_1, x_2) \mathrm{d}x_1 \mathrm{d}x_2, \\ \int_\Sigma (x_1 u_2^* - x_2 u_1^*) \mathrm{d}x_1 \mathrm{d}x_2 = \frac{H^{(1)}}{D_0} \langle x_1^2 + x_2^2\rangle x_3. \end{gathered} \tag{141}$$

On the other hand, the torsion of rods have been treated also in the direct approach in Sect. 7.2. In this context, we have find the solution

$$\psi(s) = \frac{H^{(1)}}{C_3} s, \qquad u(s) = w_1(s) = w_2(s) = 0.$$

In view of the identifications (127), from the last relations we find

$$\int_\Sigma (x_1 u_2^* - x_2 u_1^*) \mathrm{d}x_1 \mathrm{d}x_2 = \frac{H^{(1)}}{C_3} \langle x_1^2 + x_2^2 \rangle x_3, \; \int_\Sigma u_i^* \mathrm{d}x_1 \mathrm{d}x_2 = 0 \, (i = 1, 2, 3). \tag{142}$$

We mention that the right-hand side of $(141)_2$ is a constant which represents a rigid body displacement, so it can be neglected. By comparison of the solutions (141) and (142) obtained in the two different approaches, we conclude that C_3 must satisfy $C_3 = D_0$, which is given by (140). Using a classical procedure, see e.g. Ieşan (2009a, p. 149), we can cast the equations (139) and (140) in a more convenient form, but expressed on a modified domain Σ^*, given by

$$\Sigma^* = \{(\xi_1, \xi_2) | \xi_1 = x_1 \sqrt{\frac{c_{44} + c_{55}}{2c_{55}}}, \xi_2 = x_2 \sqrt{\frac{c_{44} + c_{55}}{2c_{44}}}, (x_1, x_2) \in \Sigma\}.$$

Thus, the constitutive coefficient C_3 can be expressed by

$$C_3 = D_0 = \frac{8(c_{44}c_{55})^{3/2}}{(c_{44} + c_{55})^2} \int_{\Sigma^*} \phi^*(\xi_1, \xi_2) \mathrm{d}\xi_1 \mathrm{d}\xi_2, \tag{143}$$

where $\phi^*(\xi_1, \xi_2)$ is the solution of the boundary-value problem on the domain Σ^*

$$\begin{aligned} &\Delta\phi^*(\xi_1, \xi_2) \equiv \frac{\partial^2 \phi^*}{\partial \xi_1^2} + \frac{\partial^2 \phi^*}{\partial \xi_2^2} = -2 \quad \text{in}\, \Sigma^*, \\ &\phi^*(\xi_1, \xi_2) = 0 \quad \text{on}\ \partial\Sigma^*. \end{aligned} \tag{144}$$

9.3 Shear Vibrations of an Orthotropic Rod

To identify the elastic coefficients A_1 and A_2, we solve here a dynamical problem for a straight rod, in the two approaches. We assume that the rod is rectangular, occupying the domain

$$\mathcal{D} = \{(x_1, x_2, x_3) | -\frac{a}{2} \leq x_1 \leq \frac{a}{2}, -\frac{b}{2} \leq x_2 \leq \frac{b}{2}, 0 \leq x_3 \leq l\},$$

and the material is orthotropic (without voids). The lateral surface is traction free, the body loads are absent and the boundary conditions on the end boundaries are given by

$$u_1^* = u_2^* = 0 \qquad \text{and} \qquad t_{33}^* = 0 \qquad \text{for} \quad x_3 = 0, l. \tag{145}$$

Under these conditions, we determine the shear vibrations of the rod. We search for the solution $\boldsymbol{u}^*$ in the form

$$\boldsymbol{u}^* = W e^{i\omega t}\sin(\lambda_k x_1)\boldsymbol{e}_3, \qquad \lambda_k = \frac{(2k+1)\pi}{a}, \quad k = 0,1,2,\dots \tag{146}$$

where W is a constant and ω is the natural frequency of the body. In view of (146), the boundary conditions (145) are satisfied and the equations of motion reduce to $t^*_{13,1} = \rho^*\ddot{u}^*_3$, which gives the relation

$$\omega^2 = \frac{c_{55}}{\rho^*}\left(\frac{(2k+1)\pi}{a}\right)^2, \qquad k = 0,1,2,\dots$$

Thus, the lowest natural frequency of shear vibrations is

$$\omega = \frac{\pi}{a}\sqrt{\frac{c_{55}}{\rho^*}}. \tag{147}$$

Let us approach the same problem from the theory of directed curves viewpoint. Consider a straight elastic rod (without natural twisting) of arclength $s \in [0, l]$ and parallel to the direction $\boldsymbol{e}_3$. Using the identifications (127) we see that the relations (145) correspond to the following boundary conditions on the ends of the rod

$$w_\alpha = 0, \quad F = 0, \quad \psi = 0, \quad L_\alpha = 0 \quad (\alpha = 1,2), \quad \text{for} s = 0, l. \tag{148}$$

We are interested in finding the shear vibrations, so we will have to solve a bending-shear problem of the form (112), (113). In view of (127), (146), we search the solution in the form

$$\theta_1 = \hat{W} e^{i\hat{\omega}t}, \qquad \theta_2 = \psi = 0, \qquad u = w_\alpha = 0, \tag{149}$$

where $\hat{W}$ and $\hat{\omega}$ are constants. By virtue of Eqs. (149) and the constitutive equations (113), (116) we see that the boundary conditions (148) are satisfied, while the equations of motion (112) reduce to $Q_1 = I_2\ddot{\theta}_1$, i.e. $-A_1\theta_1 = \rho^* A\frac{a^2}{12}\ddot{\theta}_1$, where $A = ab$ is the area of the cross–section. Inserting the expression $(149)_1$ into the last equation, we find the following value for the natural frequency

$$\hat{\omega} = \frac{1}{a}\sqrt{\frac{12A_1}{\rho^* A}}. \tag{150}$$

Finally, if we identify the natural frequencies from the two approaches ω and $\hat{\omega}$, then by relations (147) and (150) we get the expression of the elastic coefficient A_1

$$A_1 = kAc_{55}, \qquad \text{with} \quad k = \frac{\pi^2}{12}, \tag{151}$$

where k is a factor similar to the shear correction factor (Timoshenko, 1921). Analogously, we determine the expression for the elastic coefficient A_2

$$A_2 = kAc_{44} \qquad \text{with} \quad k = \frac{\pi^2}{12}. \tag{152}$$

This procedure has been used in Zhilin (2006a, 2007) to determine the coefficients A_1 and A_2 for isotropic and homogeneous elastic rods.

9.4 Problem of Thermal Deformation

Let us assume that the thermo-elastic rod deforms due to a given temperature field $T^{(2)}$ applied on the end boundary Σ_2. Then the boundary conditions on the ends are

$$\begin{gathered}\int_{\Sigma_2} T^* \mathrm{d}x_1 \mathrm{d}x_2 = AT^{(2)}, \quad \int_{\Sigma_1} T^* \mathrm{d}x_1 \mathrm{d}x_2 = 0, \quad \int_{\Sigma_\gamma} t^*_{3i} \mathrm{d}x_1 \mathrm{d}x_2 = 0,\\ \int_{\Sigma_\gamma} x_\alpha t^*_{33} \mathrm{d}x_1 \mathrm{d}x_2 = \int_{\Sigma_\gamma} (x_1 t^*_{32} - x_2 t^*_{31}) \mathrm{d}x_1 \mathrm{d}x_2 = 0, \quad (\alpha, \gamma = 1,2, i = 1,2,3)\\ \int_{\Sigma_1} \varphi^* \mathrm{d}x_1 \mathrm{d}x_2 = 0, \quad \int_{\Sigma_1} h^*_3 \mathrm{d}x_1 \mathrm{d}x_2 = \int_{\Sigma_2} h^*_3 \mathrm{d}x_1 \mathrm{d}x_2.\end{gathered} \tag{153}$$

The equilibrium equations are

$$t^*_{ji,j} = 0, \qquad h^*_{i,i} - g^* = 0, \qquad q^*_{i,i} = 0, \tag{154}$$

while the boundary conditions on the lateral surface are

$$t^*_{\alpha i} n^*_\alpha = 0, \qquad h^*_\alpha n^*_\alpha = 0, \qquad q^*_\alpha n^*_\alpha = 0. \tag{155}$$

Solving first the heat problem $(153)_{1,2}$, $(154)_3$ and $(155)_3$ we find the temperature field

$$T^*(x_1, x_2, x_3) = \frac{T^{(2)}}{l} x_3 \qquad \text{in} \quad \mathcal{B}. \tag{156}$$

If we insert (156) into the remaining Eqs. (153)-(155) we obtain an elastostatic problem for porous cylinders (Ieşan, 2009a), which admits the following solution

$$\begin{gathered}u^*_1 = \frac{T^{(2)} D_1}{lD} x_1 x_3, \qquad u^*_2 = \frac{T^{(2)} D_2}{lD} x_2 x_3,\\ u^*_3 = \frac{T^{(2)}}{2l} \frac{D_3 x_3^2 - D_1 x_1^2 - D_2 x_2^2}{D}, \qquad \varphi^* = \frac{T^{(2)} D_4}{lD} x_3,\end{gathered} \tag{157}$$

where we denote by D and D_k the determinants

$$D_1 = \begin{vmatrix} b_1 & c_{12} & c_{13} & \beta_1 \\ b_2 & c_{22} & c_{23} & \beta_2 \\ b_3 & c_{23} & c_{33} & \beta_3 \\ m & \beta_2 & \beta_3 & \xi \end{vmatrix}, \quad D_2 = \begin{vmatrix} c_{11} & b_1 & c_{13} & \beta_1 \\ c_{12} & b_2 & c_{23} & \beta_2 \\ c_{13} & b_3 & c_{33} & \beta_3 \\ \beta_1 & m & \beta_3 & \xi \end{vmatrix},$$

$$D_3 = \begin{vmatrix} c_{11} & c_{12} & b_1 & \beta_1 \\ c_{12} & c_{22} & b_2 & \beta_2 \\ c_{13} & c_{23} & b_3 & \beta_3 \\ \beta_1 & \beta_2 & m & \xi \end{vmatrix}, \quad D_4 = \begin{vmatrix} c_{ij} & b_i \\ \beta_j & m \end{vmatrix}_{4\times 4}, \quad D = \begin{vmatrix} c_{ij} & \beta_i \\ \beta_j & \xi \end{vmatrix}_{4\times 4}.$$

For comparison with results of rod theory we integrate the solution (157) over the cross–section and use the symmetry relations (35) to obtain (up to a rigid body displacement)

$$\begin{gathered} \frac{1}{A}\int_\Sigma u_3^* \mathrm{d}x_1 \mathrm{d}x_2 = \frac{T^{(2)} D_3}{2lD} x_3^2, \; \frac{1}{A}\int_\Sigma \varphi^* \mathrm{d}x_1 \mathrm{d}x_2 = \frac{T^{(2)} D_4}{lD} x_3, \; \int_\Sigma u_\alpha^* \mathrm{d}x_1 \mathrm{d}x_2 = 0, \\ \int_\Sigma (x_1 u_2^* - x_2 u_1^*) \mathrm{d}x_1 \mathrm{d}x_2 = \frac{T^{(2)}(D_2 - D_1)}{lD} x_3 \langle x_1 x_2 \rangle = 0, \\ \int_\Sigma x_\alpha u_3^* \mathrm{d}x_1 \mathrm{d}x_2 = -\frac{T^{(2)}}{2lD} \langle x_\alpha (D_1 x_1^2 + D_2 x_2^2) \rangle \simeq 0. \end{gathered} \tag{158}$$

The last integral is of order $O(d^5)$, where d is a diameter of Σ, and thus it is negligible in comparison with $\langle x_1 x_2 \rangle$ which is of order $O(d^4)$.

The same problem has been solved in Sect. 7.2 in the framework of rod theory. In view of the identifications (127) and $(130)_2$ we can compare the solutions in the two approaches (123) and (158). We see that the two solutions coincide if and only if we have the relations

$$\frac{G_1 K_1 - G_3 K_4}{A_3 K_1 - (K_4)^2} = \frac{D_3}{D}, \qquad \frac{A_3 G_3 - G_1 K_4}{A_3 K_1 - (K_4)^2} = \frac{D_4}{D}. \tag{159}$$

Thus we have obtained two equations (159) for the determination of the constants G_1, G_3, K_1 and K_4. In order to supplement the system (159) with two additional equations we turn to investigate another problem.

9.5 Extension of Porous Thermo-Elastic Rods

Let us determine the equilibrium of a three-dimensional rod under the action of an axial resultant force $F^{(1)}$ and having the temperature $T^{(1)}$ at both ends. The equilibrium equations are (154), the conditions on the lateral surfaces are (155) and the end boundary conditions are

$$\begin{gathered}\int_{\Sigma_\gamma} t^*_{33}\mathrm{d}x_1\mathrm{d}x_2 = F^{(1)},\ \int_{\Sigma_\gamma} T^*\mathrm{d}x_1\mathrm{d}x_2 = AT^{(1)},\ \int_{\Sigma_\gamma} t^*_{3\alpha}\mathrm{d}x_1\mathrm{d}x_2 = 0,\\ \int_{\Sigma_\gamma} x_\alpha t^*_{33}\mathrm{d}x_1\mathrm{d}x_2 = \int_{\Sigma_\gamma}(x_1 t^*_{32} - x_2 t^*_{31})\mathrm{d}x_1\mathrm{d}x_2 = 0,\ \int_{\Sigma_\gamma} h^*_3\mathrm{d}x_1\mathrm{d}x_2 = 0.\end{gathered} \tag{160}$$

From the equations for temperature $(154)_3$,$(155)_3$ and $(160)_2$ we obtain the constant solution $T^*(x_1,x_2,x_3) = T^{(1)}$. Using this relation into Eqs. $(154)_{1,2}$,$(155)_{1,2}$ and (160) we get an elasto-static problem for porous cylinders (Ieşan, 2009a), which has the solution

$$\begin{aligned} u^*_1 &= \frac{x_1}{D}\Big(\frac{F^{(1)}}{A}\delta_3 + T^{(1)}D_1\Big), & u^*_2 &= \frac{x_2}{D}\Big(-\frac{F^{(1)}}{A}\delta_4 + T^{(1)}D_2\Big),\\ u^*_3 &= \frac{x_3}{D}\Big(\frac{F^{(1)}}{A}\delta_5 + T^{(1)}D_3\Big), & \varphi^* &= \frac{1}{D}\Big(-\frac{F^{(1)}}{A}\delta_6 + T^{(1)}D_4\Big),\end{aligned} \tag{161}$$

where $\delta_3, ..., \delta_6$ are the determinants

$$\delta_3 = \begin{vmatrix} c_{12} & c_{13} & \beta_1\\ c_{22} & c_{23} & \beta_2\\ \beta_2 & \beta_3 & \xi\end{vmatrix}, \quad \delta_4 = \begin{vmatrix} c_{11} & c_{13} & \beta_1\\ c_{12} & c_{23} & \beta_2\\ \beta_1 & \beta_3 & \xi\end{vmatrix},$$

$$\delta_5 = \begin{vmatrix} c_{\alpha\gamma} & \beta_\alpha\\ \beta_\gamma & \xi\end{vmatrix}_{3\times 3}, \quad \delta_6 = \begin{vmatrix} c_{\alpha\gamma} & c_{\alpha 3}\\ \beta_\gamma & \beta_3\end{vmatrix}_{3\times 3}.$$

By integrating the three–dimensional solution (161) and using (35) we find

$$\begin{gathered}\frac{1}{A}\int_\Sigma u^*_3\mathrm{d}x_1\mathrm{d}x_2 = \frac{x_3}{D}\Big(\frac{F^{(1)}}{A}\delta_5 + T^{(1)}D_3\Big),\\ \frac{1}{A}\int_\Sigma \varphi^*\mathrm{d}x_1\mathrm{d}x_2 = \frac{1}{D}\Big(-\frac{F^{(1)}}{A}\delta_6 + T^{(1)}D_4\Big),\\ \int_\Sigma u^*_\alpha\mathrm{d}x_1\mathrm{d}x_2 = 0,\ \int_\Sigma(x_1u^*_2 - x_2u^*_1)\mathrm{d}x_1\mathrm{d}x_2 = 0,\ \int_\Sigma x_\alpha u^*_3\mathrm{d}x_1\mathrm{d}x_2 = 0.\end{gathered} \tag{162}$$

This extensional problem has also been solved in the direct approach in Sect. 7.2 where we obtained the solution (118). Taking into account the identifications (127) and (130), we deduce that the solutions (118) and (162) coincide if and only if the relations (159) are valid and

$$\frac{K_1}{A_3K_1 - (K_4)^2} = \frac{\delta_5}{AD}, \qquad \frac{K_4}{A_3K_1 - (K_4)^2} = \frac{\delta_6}{AD}. \tag{163}$$

From the system of four nonlinear algebraic equations (159) and (163) we can determine the constitutive coefficients G_1, G_3, K_1 and K_4. The result is

$$
\begin{aligned}
G_1 = A\frac{\delta_7}{\delta_1} = A(b_3 - b_1\nu_1 - b_2\nu_2), \quad K_4 = A\frac{\delta_6}{\delta_1} = A\big(\beta_3 - \beta_1\nu_1 - \beta_2\nu_2\big),\\
G_3 = A\frac{\delta_8}{\delta_1} = A\Big(m - \frac{c_{11}b_2\beta_2 + c_{22}b_1\beta_1 - c_{12}(b_1\beta_2 + b_2\beta_1)}{\delta_1}\Big),\\
K_1 = A\frac{\delta_5}{\delta_1} = A\Big(\xi - \frac{c_{11}\beta_2^2 + c_{22}\beta_1^2 - 2c_{12}\beta_1\beta_2}{\delta_1}\Big),
\end{aligned}
\tag{164}
$$

where we denote by $\delta_7 = \begin{vmatrix} c_{\alpha\gamma} & b_\alpha \\ c_{\gamma 3} & b_3 \end{vmatrix}_{3\times 3}, \delta_8 = \begin{vmatrix} c_{\alpha\gamma} & b_\alpha \\ \beta_\gamma & m \end{vmatrix}_{3\times 3}, \alpha, \gamma = 1, 2.$

The constitutive coefficients K_2, G and K cannot be determined from these simple solutions, but the comparison of constitutive equations suggest the values

$$
K_2 = A\alpha_3, \qquad G = Aa, \qquad K = AK_3^*. \tag{165}
$$

Remark 9.1. In the special case of isotropic and homogeneous materials, the constitutive coefficients for thermo-elastic materials with pores become (Ieşan, 2009a)

$$
\begin{aligned}
&c_{11} = c_{22} = c_{33} = \lambda + 2\mu,\ c_{12} = c_{13} = c_{23} = \lambda,\ c_{44} = c_{55} = c_{66} = \mu,\\
&\alpha_i = \alpha,\ \beta_i = \beta,\ b_i = b,\ K_i^* = K^*,\ E_0 = E,\ \nu_1 = \nu_2 = \nu,
\end{aligned}
\tag{166}
$$

where λ, μ are Lamé's constants, E is Young's modulus and ν is Poisson's ratio. Using these expressions into the relations (136), (137), (143), (151) and (152), we obtain by particularization the form of the elasticity constants A_i and C_i for isotropic and homogeneous rods

$$
\begin{aligned}
&A_1 = A_2 = k\mu A \quad (k = \frac{\pi^2}{12}), \quad A_3 = EA,\\
&C_1 = E\int_\Sigma x_2^2 \mathrm{d}x_1\mathrm{d}x_2, \quad C_2 = E\int_\Sigma x_1^2 \mathrm{d}x_1\mathrm{d}x_2,\\
&C_3 = 2\mu\int_\Sigma \phi^*(x_1, x_2)\mathrm{d}x_1\mathrm{d}x_2 \text{ with } \Delta\phi^* = -2 \text{ in } \Sigma,\ \phi^* = 0 \text{ on } \partial\Sigma.
\end{aligned}
\tag{167}
$$

The values (167) have been obtained previously by Zhilin (2006a, 2007) for the isotropic case.

Finally, if we insert (166) in the relations (164) and (165) then we obtain the expressions of the poro-thermo-elastic constitutive coefficients for isotropic homogeneous rods

$$
\begin{aligned}
G_1 = A\frac{\mu b}{\lambda + \mu}, \quad G_3 = A\Big(m - \frac{b\beta}{\lambda + \mu}\Big), \quad G = Aa, \quad K = AK^*,\\
K_1 = A\Big(\xi - \frac{\beta^2}{\lambda + \mu}\Big), \qquad K_4 = A\frac{\beta\mu}{\lambda + \mu}, \qquad K_2 = A\alpha.
\end{aligned}
\tag{168}
$$

Thus, the effective constitutive constants have been determined for rods made of isotropic or orthotropic materials.

10 Extended Thermodynamic Theory for Rods with Two Temperature Fields

As one can observe from the relation $(130)_1$, the temperature field T for rods could be identified with the averaged temperature on the cross-section of the rod. Nevertheless, in some problems it is useful to have information also about the variation of the temperature inside the cross-section. In this line of thought, we will develop in this section an extended thermodynamic theory for rods with two independent temperature fields: the absolute temperature field and the temperature deviation field.

10.1 Basic Laws

The basic laws of thermodynamics for rods with two temperature fields can deduced from Eqs. (8) assuming that there is no porosity. The balance laws of linear momentum and moment of momentum are expressed by Eqs. (11).

To describe the thermal effects in thin rods, we consider two independent temperature fields: the absolute temperature field denoted by $\theta(s,t)$ and the temperature deviation $G(s,t)$, with $\theta > 0$ and $G \geq 0$. In the reference configuration $\mathcal{C}_0$ we assume that the temperature fields θ_0, G_0 are constant. In general, the balance of energy is given by the relation

$$\frac{\mathrm{d}}{\mathrm{d}t}(\mathcal{E} + \bar{\mathcal{K}}) = \mathcal{A} + \mathcal{Q}, \tag{169}$$

where $\mathcal{E}$ and $\bar{\mathcal{K}}$ are the internal and kinetic energy functionals respectively, $\mathcal{A}$ is the mechanical power, $\mathcal{Q}$ is the heat supply. These quantities are expressed as follows:

$$\begin{aligned} &\mathcal{E} = \int_{s_1}^{s_2} U \mathrm{d}s, \quad \bar{\mathcal{K}} = \int_{s_1}^{s_2} \bar{K} \mathrm{d}s, \quad \mathcal{Q} = \int_{s_1}^{s_2} S \mathrm{d}s + q\Big|_{s_1}^{s_2}, \\ &\mathcal{A} = \int_{s_1}^{s_2} (\boldsymbol{F}\cdot\boldsymbol{V} + \boldsymbol{L}\cdot\boldsymbol{\omega})\mathrm{d}s + \big(\boldsymbol{N}\cdot\boldsymbol{V} + \boldsymbol{M}\cdot\boldsymbol{\omega}\big)\Big|_{s_1}^{s_2}, \end{aligned} \tag{170}$$

where $U(s,t)$ is the internal energy per unit length, S is the external rate of the resultant heat supply per unit length and q is the heat flux along the rod. Hence, the balance of energy is expressed by (14).

We use the second law of thermodynamics (entropy inequality) in the following form

$$\frac{\mathrm{d}}{\mathrm{d}t}\mathcal{H} \geq \mathcal{J}, \tag{171}$$

where $\mathcal{H}$ is the entropy

$$\mathcal{H} = \int_{s_1}^{s_2} \eta \mathrm{d}s,$$

$\eta(s,t)$ designates the specific entropy function, $\mathcal{J}$ is the entropy supply consisting of two parts

$$\mathcal{J} = \int_{s_1}^{s_2} j \mathrm{d}s + h\Big|_{s_1}^{s_2}, \tag{172}$$

where j is the resultant entropy supply, and h is the entropy flux along the rod. Following Simmonds (2005) we assume the following expressions for j and h

$$j = \frac{S}{\theta} - \chi G, \quad h = \frac{q}{\theta} - rG, \tag{173}$$

where χ and r are the additional heat supply and additional heat flux along the rod. Hence, the second law of thermodynamics (entropy inequality) takes the following form:

$$\int_{s_1}^{s_2} \dot{\eta} \mathrm{d}s \geq \int_{s_1}^{s_2} \Big(\frac{S}{\theta} - \chi G\Big) \mathrm{d}s + \Big(\frac{q}{\theta} - rG\Big)\Big|_{s_1}^{s_2}. \tag{174}$$

For the interpretation of various fields in terms of three–dimensional variables, we refer to the work of Simmonds (2005). For instance, the temperature field $\theta(s,t)$ and the temperature deviation $G(s,t)$ are identified in Simmonds (2005) with

$$\theta = \frac{2}{\theta_{\min}^{-1} + \theta_{\max}^{-1}}, \qquad G = \frac{1}{2}\Big(\frac{1}{\theta_{\min}} - \frac{1}{\theta_{\max}}\Big),$$

where $\theta_{\min}(s,t)$ and $\theta_{\max}(s,t)$ denote the minimum and maximum values of the absolute temperature in the cross-section of the rod characterized by the spatial coordinate s at time t. We notice that θ is the harmonic mean of $\theta_{\min}$ and $\theta_{\max}$.

Let us denote the vectors of deformation defined in (24) and (26) by: $\boldsymbol{\varepsilon} = \boldsymbol{\mathcal{E}}_*$ and $\boldsymbol{\phi} = \boldsymbol{\Phi}_*$ (for the sake of simplicity). Under suitable smoothness assumptions, we obtain in the classical manner (see Section 3) the local forms, see Eqs. (16)-(20). Let us consider the free energy function Ψ defined by

$$\Psi = U - \theta\eta - GF, \tag{175}$$

where F is a new variable called *the entropy deviation*, according to Simmonds (1984, 2005).

Then, the energy balance equation (18) can be written in the form

$$q' + S + P - \dot{\theta}\eta - \dot{G}F - \dot{\Psi} = \theta\dot{\eta} + G\dot{F}. \tag{176}$$

Inserting (176) into (20) we obtain the following form of the entropy inequality

$$\frac{\theta'}{\theta}q + P - \dot{\theta}\eta - \dot{G}F - \dot{\Psi} \geq (\dot{F} - \theta\chi - \theta r')G - \theta r G'. \tag{177}$$

10.2 Constitutive Equations

For thermo-elastic rods we adopt the following constitutive assumptions: the fields Ψ, $\boldsymbol{N}\cdot\boldsymbol{P}$, $\boldsymbol{M}\cdot\boldsymbol{P}$, η, F, r and q depend only on the variables $\boldsymbol{\varepsilon}$, $\boldsymbol{\phi}$, θ, G, θ' and G' (and on the point s), i.e.:

$$\{\Psi, \boldsymbol{N}\cdot\boldsymbol{P}, \boldsymbol{M}\cdot\boldsymbol{P}, \eta, F, r, q\} = \{\Psi, \boldsymbol{N}\cdot\boldsymbol{P}, \boldsymbol{M}\cdot\boldsymbol{P}, \eta, F, r, q\}\big(\boldsymbol{\varepsilon}, \boldsymbol{\phi}, \theta, G, \theta', G'; s\big). \tag{178}$$

In this sense we have similar assumptions as in Sect. 4 neglecting the porosity.

We use the relation (178) to express $\dot{\Psi}$ into the entropy inequality (177) and we obtain

$$\begin{aligned} &\frac{\theta'}{\theta}q + \theta G' r + \Big(\boldsymbol{N}\cdot\boldsymbol{P} - \frac{\partial\Psi}{\partial\boldsymbol{\varepsilon}}\Big)\cdot\dot{\boldsymbol{\varepsilon}} + \Big(\boldsymbol{M}\cdot\boldsymbol{P} - \frac{\partial\Psi}{\partial\boldsymbol{\phi}}\Big)\cdot\dot{\boldsymbol{\phi}} - \Big(\eta + \frac{\partial\Psi}{\partial\theta}\Big)\dot{\theta} \\ &\quad -\Big(F + \frac{\partial\Psi}{\partial G}\Big)\dot{G} - \frac{\partial\Psi}{\partial\theta'}\dot{\theta}' - \frac{\partial\Psi}{\partial G'}\dot{G}' + \Big(\theta\chi + \theta r' - \dot{F}\Big)G \geq 0. \end{aligned} \tag{179}$$

On the basis of (179) we deduce that

$$\begin{aligned} &\Big(\boldsymbol{N}\cdot\boldsymbol{P} - \frac{\partial(\Psi + FG)}{\partial\boldsymbol{\varepsilon}}\Big)\cdot\dot{\boldsymbol{\varepsilon}} + \Big(\boldsymbol{M}\cdot\boldsymbol{P} - \frac{\partial(\Psi + FG)}{\partial\boldsymbol{\phi}}\Big)\cdot\dot{\boldsymbol{\phi}} \\ &-\Big(\eta + \frac{\partial(\Psi + FG)}{\partial\theta}\Big)\dot{\theta} - \frac{\partial(\Psi + FG)}{\partial G}\dot{G} - \frac{\partial(\Psi + FG)}{\partial\theta'}\dot{\theta}' \\ &-\frac{\partial(\Psi + FG)}{\partial G'}\dot{G}' + \theta\Big(\chi + r'\Big)G + \frac{\theta'}{\theta}q + \theta G' r \geq 0. \end{aligned} \tag{180}$$

The standard Coleman-Noll procedure of analysis of the entropy inequality requires the independence of the rates of strains and temperature (see Coleman and Noll, 1963; Truesdell, 1984). In our case this means that within this procedure $\dot{\boldsymbol{\varepsilon}}$, $\dot{\boldsymbol{\phi}}$, $\dot{\theta}$, $\dot{G}$, $\dot{\theta}'$, $\dot{G}'$ have to be considered as independent quantities. Assuming for a moment this independence let us apply the

Coleman-Noll procedure to (180). Consider the following family of deformations

$$\begin{aligned} \varepsilon &= \varepsilon_0 + \varepsilon_1(t - t_0), \quad \boldsymbol{\phi} = \boldsymbol{\phi}_0 + \boldsymbol{\phi}_1(t - t_0), \\ \theta &= \theta_0 + \theta_1(t - t_0) + \theta_2(s - s_0)(t - t_0), \\ G &= G_0 + G_1(t - t_0) + G_2(s - s_0)(t - t_0), \end{aligned} \tag{181}$$

where ε_0, ε_1, $\boldsymbol{\phi}_0$, $\boldsymbol{\phi}_1$, θ_i, G_i are constants ($i = 0, 1, 2$), s_0 is any point at the rod curve and t_0 is any time instant. Substituting (181) into (180) and taking into account that at $t = t_0$, $s = s_0$ we have

$$\begin{aligned} \varepsilon = \varepsilon_0, \quad \dot{\varepsilon} = \varepsilon_1, \quad \boldsymbol{\phi} = \boldsymbol{\phi}_0, \quad \dot{\boldsymbol{\phi}} = \boldsymbol{\phi}_1, \\ \theta = \theta_0, \quad \dot{\theta} = \theta_1, \quad \dot{\theta}' = \theta_2, \\ G = G_0, \quad \dot{G} = G_1, \quad \dot{G}' = G_2, \end{aligned} \tag{182}$$

we obtain the inequality

$$\begin{aligned} &\varepsilon_1 \cdot \left(\boldsymbol{N} \cdot \boldsymbol{P} - \frac{\partial(\Psi + FG)}{\partial \varepsilon} \right) \Big|_{s=s_0, t=t_0} \\ &+ \boldsymbol{\phi}_1 \cdot \left(\boldsymbol{M} \cdot \boldsymbol{P} - \frac{\partial(\Psi + FG)}{\partial \boldsymbol{\phi}} \right) \Big|_{s=s_0, t=t_0} - \theta_1 \left(\eta + \frac{\partial(\Psi + FG)}{\partial \theta} \right) \Big|_{s=s_0, t=t_0} \\ &- G_1 \frac{\partial(\Psi + FG)}{\partial G} \Big|_{s=s_0, t=t_0} - \theta_2 \frac{\partial(\Psi + FG)}{\partial \theta'} \Big|_{s=s_0, t=t_0} \\ &- G_2 \frac{\partial(\Psi + FG)}{\partial G'} \Big|_{s=s_0, t=t_0} + \left[\left(\theta \chi + \theta r' \right) G + \frac{\theta'}{\theta} q + \theta G' r \right] \Big|_{s=s_0, t=t_0} \geq 0. \end{aligned} \tag{183}$$

The left side of the inequality (183) is a linear function of the independent variables ε_1, $\boldsymbol{\phi}_1$, θ_1, θ_2, G_1, G_2. Hence, from (183) it follows

$$\begin{aligned} \boldsymbol{N} \cdot \boldsymbol{P} = \frac{\partial(\Psi + FG)}{\partial \varepsilon}, \quad \boldsymbol{M} \cdot \boldsymbol{P} = \frac{\partial(\Psi + FG)}{\partial \boldsymbol{\phi}}, \quad \eta = -\frac{\partial(\Psi + FG)}{\partial \theta}, \\ \frac{\partial(\Psi + FG)}{\partial G} = 0, \quad \frac{\partial(\Psi + FG)}{\partial \theta'} = 0, \quad \frac{\partial(\Psi + FG)}{\partial G'} = 0, \\ \left(\theta \chi + \theta r' \right) G + \frac{\theta'}{\theta} q + \theta G' r \geq 0. \end{aligned} \tag{184}$$

From (184) it follows that the function $\widetilde{\Psi} = \Psi + FG$ does not depend on G, θ', G', i.e.

$$\widetilde{\Psi} = \widetilde{\Psi}(\varepsilon, \boldsymbol{\phi}, \theta; s),$$

and then $\boldsymbol{M}$ does not depend on G. This results seems unrealistic. Indeed, from the elementary knowledge on the strength of materials we know that non-zero through-the-thickness gradient of temperature leads to the beam bending. In other words, we have $\boldsymbol{M} \neq \mathbf{0}$ if $G \neq 0$ in general, and thus $\boldsymbol{M}$ should depend on G. In the case of non-linear thermo-elastic shells, this fact was noted in Simmonds (1984, 2011) where $\dot{F}$ was considered as an internal variable with additional constitutive equation. The similar approach was used also in Simmonds (2005) for thermo-elastic beams.

Moreover, the solution of the system of equations $(184)_{4,5,6}$ has the form

$$F = \frac{F_0(\boldsymbol{\varepsilon}, \boldsymbol{\phi}, \theta; s) - \Psi}{G},$$

where F_0 is an arbitrary function. Hence, F has a singularity when $G \to 0$.

To avoid these unsatisfactory conclusions we propose a modification of Coleman–Noll procedure assuming that not all thermodynamical processes are thermodynamically admissible. More precisely, we assume that $\boldsymbol{\varepsilon}$, $\boldsymbol{\phi}$, θ can be considered as the independent variables, while G is subjected to some constraints. Following our assumption let us consider the following family of deformations

$$\begin{gathered}\boldsymbol{\varepsilon} = \boldsymbol{\varepsilon}_0 + \boldsymbol{\varepsilon}_1(t - t_0), \qquad \boldsymbol{\phi} = \boldsymbol{\phi}_0 + \boldsymbol{\phi}_1(t - t_0), \\ \theta = \theta_0 + \theta_1(t - t_0) + \theta_2(s - s_0)(t - t_0).\end{gathered} \tag{185}$$

Substituting (185) into (179) we obtain the inequality

$$\begin{aligned}&\boldsymbol{\varepsilon}_1 \cdot \left(\boldsymbol{N}\cdot\boldsymbol{P} - \frac{\partial\Psi}{\partial\boldsymbol{\varepsilon}}\right)\Big|_{s=s_0,t=t_0} + \boldsymbol{\phi}_1 \cdot \left(\boldsymbol{M}\cdot\boldsymbol{P} - \frac{\partial\Psi}{\partial\boldsymbol{\phi}}\right)\Big|_{s=s_0,t=t_0} \\ &-\theta_1\left(\eta + \frac{\partial\Psi}{\partial\theta}\right)\Big|_{s=s_0,t=t_0} - \theta_2\frac{\partial\Psi}{\partial\theta'}\Big|_{s=s_0,t=t_0} + \left[\frac{\theta'}{\theta}q + \theta G' r\right]\Bigg|_{s=s_0,t=t_0} \\ &+\left[-\left(F + \frac{\partial\Psi}{\partial G}\right)\dot{G} - \frac{\partial\Psi}{\partial G'}\dot{G}' + \left(\theta\chi + \theta r' - \dot{F}\right)G\right]\Bigg|_{s=s_0,t=t_0} \geq 0.\end{aligned} \tag{186}$$

From Eqs. (186) it follows Eqs. $(30)_{2,3,4}$ while the free energy density Ψ is independent of θ'. The inequality (179) reduces to the relation

$$\frac{\theta'}{\theta}q + \theta G' r - \left(F + \frac{\partial\Psi}{\partial G}\right)\dot{G} - \frac{\partial\Psi}{\partial G'}\dot{G}' + \left(\theta\chi + \theta r' - \dot{F}\right)G \geq 0. \tag{187}$$

The latter inequality can be considered as the constraint for the temperature deviation function G, as well as the further constraints for admissible constitutive equations for Ψ, F, q, and r. Inequality (187) implies that

$$\frac{\partial\Psi}{\partial G'} = 0.$$

Hence, Ψ takes the form $\Psi = \Psi(\varepsilon, \phi, \theta, G)$.

In what follows we assume the constitutive relation for F in the form

$$F = -\frac{\partial \Psi}{\partial G},$$

Thus, $F = F(\varepsilon, \phi, \theta, G)$.

The entropy inequality (187) now reduces to

$$q\theta' + \theta^2 r G' + \left(\chi + r' - \theta^{-1}\dot{F}\right)\theta^2 G \geq 0. \tag{188}$$

The inequality (188) should be fulfilled for any constitutive relations and any possible values of the state variables at any point of the rod. In particular, assuming $\theta' = G' = 0$ we obtain

$$\left[\theta(\chi + r') - \dot{F}\right]\theta G \geq 0.$$

Since $\theta > 0$, $G \geq 0$, to satisfy this inequality we obtain that

$$\theta(\chi + r') - \dot{F} \geq 0.$$

Introducing a new constitutive function $c = c(\varepsilon, \phi, \theta, G, \theta', G') \geq 0$ we transform the last relation in the form

$$\theta(\chi + r') - \dot{F} = c, \tag{189}$$

where c should be determined by experiments (see Simmonds, 1984, 2005, 2011; Eremeyev and Pietraszkiewicz, 2011). The quantities χ, r, and F relate to the temperature deviation field G. It is quite natural to assume that $r = 0$, $F = 0$ in the case of homogeneous temperature and the absence of extra heat supply χ, i.e. in the case of $G = 0$ and $\chi = 0$. Hence, we assume that $c = 0$ when $G = 0$. The simplest case is $c = c_0 G$ with $c_0 = \text{const}$, which is assumed in the sequel. This case is also considered in Simmonds (2005).

Thus, the inequality (188) yields the results

$$r' + \chi - \theta^{-1}\dot{F} = c_0\theta^{-1}G, \tag{190}$$

and

$$q\theta' + \theta^2 r G' \geq 0, \tag{191}$$

which is a Fourier inequality of heat conduction augmented by a term which represents heat flux in the cross–section Σ, cf. Simmonds (2005). Equation (190) plays the role of second thermo-conductivity equation in the theory of thermo-elastic rods and extends the analogous equation obtained in Simmonds (2005) where $\dot{F}$ was considered as the internal variable.

Summarizing the constitutive equations of the thermo-elastic rods we have

$$\begin{gathered}\Psi = \Psi\big(\boldsymbol{\varepsilon}, \boldsymbol{\phi}, \theta, G; s\big),\\ \eta = -\frac{\partial \Psi}{\partial \theta}, \quad F = -\frac{\partial \Psi}{\partial G}, \quad \boldsymbol{N} = \frac{\partial \Psi}{\partial \boldsymbol{\varepsilon}} \cdot \boldsymbol{P}^T, \quad \boldsymbol{M} = \frac{\partial \Psi}{\partial \boldsymbol{\phi}} \cdot \boldsymbol{P}^T,\\ \{r, q\} = \{r, q\}\big(\boldsymbol{\varepsilon}, \boldsymbol{\phi}, \theta, G, \theta', G'; s\big).\end{gathered} \tag{192}$$

In view of the constitutive equations (192), the reduced form of the energy balance equation (176) can be written as

$$q' + S = \theta \dot{\eta} + G \dot{F}. \tag{193}$$

The similar procedure of derivation of thermo-conductivity equations for thermo-elastic and thermo-viscoelastic shells was proposed by Eremeyev and Pietraszkiewicz (2011).

To complete the presentation of the constitutive equations we have to specify the expression for the free energy function. As an example, let us consider the following quadratic form:

$$\begin{aligned}\Psi = \Psi_0 &+ \boldsymbol{N}_0 \cdot \boldsymbol{\varepsilon} + \boldsymbol{M}_0 \cdot \boldsymbol{\phi} + \frac{1}{2}\boldsymbol{\varepsilon} \cdot \boldsymbol{A} \cdot \boldsymbol{\varepsilon} + \boldsymbol{\varepsilon} \cdot \boldsymbol{B} \cdot \boldsymbol{\phi} + \frac{1}{2}\boldsymbol{\phi} \cdot \boldsymbol{C} \cdot \boldsymbol{\phi}\\ &-(\boldsymbol{k}_1 \cdot \boldsymbol{\varepsilon})(\theta - \theta_0) - (\boldsymbol{k}_2 \cdot \boldsymbol{\phi})(\theta - \theta_0) - (\boldsymbol{k}_3 \cdot \boldsymbol{\varepsilon})G - (\boldsymbol{k}_4 \cdot \boldsymbol{\phi})G\\ &-k(\theta - \theta_0)G - \frac{1}{2}a(\theta - \theta_0)^2 - \frac{1}{2}bG^2,\end{aligned} \tag{194}$$

where Ψ_0, k, a and b are scalars, θ_0 is the reference temperature, $\boldsymbol{N}_0$, $\boldsymbol{M}_0$, $\boldsymbol{k}_1$, $\boldsymbol{k}_2$, $\boldsymbol{k}_3$, $\boldsymbol{k}_4$ are vectors, and $\boldsymbol{A}$, $\boldsymbol{B}$, $\boldsymbol{C}$ are tensors of second order. All properties are defined in the reference configuration. The vectors $\boldsymbol{N}_0$, $\boldsymbol{M}_0$ describe the initial forces and moments acting in the reference configurations, $\boldsymbol{A}$ and $\boldsymbol{C}$ are the tensors of tangential and bending stiffness, respectively, $\boldsymbol{B}$ is t he tensor describing the coupling between stretching and bending, $\boldsymbol{k}_1$, $\boldsymbol{k}_2$, $\boldsymbol{k}_3$, and $\boldsymbol{k}_4$ correspond to the thermo-elastic coefficients.

To determine the structure of the constitutive tensors for thermo-elastic non-homogeneous rods, we employ the same method as in Sect. 4.2. Then, in the case of straight rods without natural twisting, we find the following structure for the constitutive tensors $\boldsymbol{k}_1$, $\boldsymbol{k}_2$, $\boldsymbol{k}_3$ and $\boldsymbol{k}_4$:

$$\boldsymbol{k}_1 = k_1 \boldsymbol{d}_3, \quad \boldsymbol{k}_2 = k_2^1 \boldsymbol{d}_1 + k_2^2 \boldsymbol{d}_2, \quad \boldsymbol{k}_3 = k_3^1 \boldsymbol{d}_1 + k_3^2 \boldsymbol{d}_2, \quad \boldsymbol{k}_4 = k_4 \boldsymbol{d}_3. \tag{195}$$

For the tensors $\boldsymbol{A}, \boldsymbol{B}$ and $\boldsymbol{C}$ we have the following expressions

$$\begin{aligned}
\boldsymbol{A} &= A_1 \boldsymbol{d}_1 \otimes \boldsymbol{d}_1 + A_2 \boldsymbol{d}_2 \otimes \boldsymbol{d}_2 + A_3 \boldsymbol{d}_3 \otimes \boldsymbol{d}_3 + A_{12}(\boldsymbol{d}_1 \otimes \boldsymbol{d}_2 + \boldsymbol{d}_2 \otimes \boldsymbol{d}_1),\\
\boldsymbol{B} &= B_{13} \boldsymbol{d}_1 \otimes \boldsymbol{d}_3 + B_{31} \boldsymbol{d}_3 \otimes \boldsymbol{d}_1 + B_{23} \boldsymbol{d}_2 \otimes \boldsymbol{d}_3 + B_{32} \boldsymbol{d}_3 \otimes \boldsymbol{d}_2,\\
\boldsymbol{C} &= C_1 \boldsymbol{d}_1 \otimes \boldsymbol{d}_1 + C_2 \boldsymbol{d}_2 \otimes \boldsymbol{d}_2 + C_3 \boldsymbol{d}_3 \otimes \boldsymbol{d}_3 + C_{12}(\boldsymbol{d}_1 \otimes \boldsymbol{d}_2 + \boldsymbol{d}_2 \otimes \boldsymbol{d}_1).
\end{aligned} \tag{196}$$

From the comparison of solutions in both the direct and the three-dimensional approach for non-homogeneous rods (Bîrsan et al., 2012) we have found that the following constitutive coefficients should be taken zero: $B_{13} = B_{23} = 0, A_{12} = 0$. Inserting (194)–(196) into (192) we find the explicit form of the constitutive equations for $\boldsymbol{N} \cdot \boldsymbol{P}$, $\boldsymbol{M} \cdot \boldsymbol{P}$, η and F. For instance, the constitutive equations for the entropy fields η and F are expressed by

$$\begin{aligned}
\eta &= k_1 \boldsymbol{\varepsilon} \cdot \boldsymbol{d}_3 + k_2^{\alpha} \boldsymbol{\phi} \cdot \boldsymbol{d}_{\alpha} + a(\theta - \theta_0) + kG,\\
F &= k_3^{\alpha} \boldsymbol{\varepsilon} \cdot \boldsymbol{d}_{\alpha} + k_4 \boldsymbol{\phi} \cdot \boldsymbol{d}_3 + k(\theta - \theta_0) + bG.
\end{aligned}$$

From these equations we can observe the coupling of the entropy fields η, F with the extension-shear and bending-twisting deformations $\boldsymbol{\varepsilon}, \boldsymbol{\phi}$.

The structure of the constitutive tensors can be derived also in the more general case of curved rods with natural twisting. In this case, the constitutive coefficients depend on the angle of natural twist (which is a function of s), and the expressions corresponding to Eqs. (195) and (196) have to be supplemented with additional terms.

In the same way as in Sect. 5, we can derive the linearized equations for thermo-elastic rods with two temperature fields. Moreover, we can prove the uniqueness of solution to the boundary-initial-value problem associated to dynamical deformations, under certain assumptions (see Altenbach et al., 2012).

11 Non-Homogeneous Rods and Composite Beams Analyzed by the Direct Approach

In this section we show that the direct approach to rods can be applied to analyze the deformation of non-homogeneous rods and composite beams. To this aim, we confine our attention to the isothermal theory and consider rods made of elastic non-homogeneous materials.

In the case of general elastic rods, the constitutive assumptions assert that the internal energy density $\mathcal{U}$ is a function of the deformation vectors

$\{\boldsymbol{\mathcal{E}}_*, \boldsymbol{\Phi}_*\}$. Thus, in this case the constitutive equations reduce to

$$\rho_0\mathcal{U}=\mathcal{U}_0+\boldsymbol{N}_0\cdot\boldsymbol{\mathcal{E}}_*+\boldsymbol{M}_0\cdot\boldsymbol{\Phi}_*+\frac{1}{2}\boldsymbol{\mathcal{E}}_*\cdot\boldsymbol{A}\cdot\boldsymbol{\mathcal{E}}_*+\boldsymbol{\mathcal{E}}_*\cdot\boldsymbol{B}\cdot\boldsymbol{\Phi}_*+\frac{1}{2}\boldsymbol{\Phi}_*\cdot\boldsymbol{C}\cdot\boldsymbol{\Phi}_*,$$
$$\boldsymbol{N}=\frac{\partial(\rho_0\mathcal{U})}{\partial\boldsymbol{\mathcal{E}}_*}\cdot\boldsymbol{P}^T,\qquad \boldsymbol{M}=\frac{\partial(\rho_0\mathcal{U})}{\partial\boldsymbol{\Phi}_*}\cdot\boldsymbol{P}^T,\tag{197}$$

where $\mathcal{U}_0$ is a scalar, $\boldsymbol{N}_0, \boldsymbol{M}_0$ are vectors, and $\boldsymbol{A}, \boldsymbol{B}, \boldsymbol{C}$ are second order tensors, defined on the reference configuration. In the same way as in Section 4.2, we deduce that the constitutive tensor $\boldsymbol{A}, \boldsymbol{B}$ and $\boldsymbol{C}$ for non-homogeneous rods have the structures as (196). We are interested in determining the expressions of the constitutive coefficients $A_i, C_i, A_{12}, C_{12}, B_{\alpha 3}$ and $B_{3\alpha}$ for functionally graded rods, in terms of the three-dimensional elastic properties. These coefficients describe the effective stiffness properties of thin rods. Since the constitutive coefficients do not depend on the deformation, their expressions can be derived by comparison of exact solutions for directed curves with the results from three-dimensional elasticity in the framework of linear theory.

Restricting ourselves to the linear theory, we consider the infinitesimal displacement vector $\boldsymbol{u}$, the vector of small rotations $\boldsymbol{\psi}$ and the deformation vectors $\boldsymbol{e}$ and $\boldsymbol{\kappa}$ defined in Section 5 by the relations (44) and (45). The equations of motion have the forms (49). The correspondence between the displacement and rotation fields $\{\boldsymbol{u}, \boldsymbol{\psi}\}$ for directed curves and the displacement vector $\boldsymbol{u}^*$ for three-dimensional rods is established by the following relations (Zhilin, 2007)

$$\rho_0\big(\boldsymbol{u}+\boldsymbol{\Theta}_1^0\cdot\boldsymbol{\psi}\big)=\langle\rho^*\boldsymbol{u}^*\tilde{\mu}\rangle,\qquad \rho_0\big(\boldsymbol{u}\cdot\boldsymbol{\Theta}_1^0+\boldsymbol{\Theta}_2^0\cdot\boldsymbol{\psi}\big)=\langle\rho^*(\boldsymbol{a}\times\boldsymbol{u}^*)\tilde{\mu}\rangle.\tag{198}$$

Also, the relations between the force and moment vector fields $\{\boldsymbol{N}, \boldsymbol{M}\}$ and the Cauchy stress tensor $\boldsymbol{T}^*$ from three-dimensional theory are given by

$$\boldsymbol{N}=\langle\boldsymbol{d}_3\cdot\boldsymbol{T}^*\rangle,\qquad \boldsymbol{M}=\langle\boldsymbol{a}\times(\boldsymbol{d}_3\cdot\boldsymbol{T}^*)\rangle.\tag{199}$$

These relations are useful when comparing the solutions of some problems in the two different approaches.

In the case of straight rods without natural twisting, we can can decompose the vector fields $\boldsymbol{u}, \boldsymbol{\psi}, \boldsymbol{e}, \boldsymbol{\kappa}, \boldsymbol{N}, \boldsymbol{M}, \boldsymbol{\mathcal{F}}$ and $\boldsymbol{\mathcal{L}}$ by the tangent direction and the normal plane in the form (108), (109). In view of (197), (196), the constitutive equations can be written in component form as

$$\begin{gathered}Q_1=A_1(w_1'-\vartheta_1)+A_{12}(w_2'-\vartheta_2)+B_{13}\psi',\\ Q_2=A_{12}(w_1'-\vartheta_1)+A_2(w_2'-\vartheta_2)+B_{23}\psi',\\ F=A_3u'-B_{31}\vartheta_2'+B_{32}\vartheta_1',\quad H=C_3\psi'+B_{13}(w_1'-\vartheta_1)+B_{23}(w_2'-\vartheta_2),\\ L_1=C_2\vartheta_1'-C_{12}\vartheta_2'+B_{32}u',\quad L_2=-C_{12}\vartheta_1'+C_1\vartheta_2'-B_{31}u'.\end{gathered}\tag{200}$$

The constitutive coefficients are constants, since we consider rods made of non-homogeneous materials which properties do not depend on the axial coordinate s.

We observe that the general boundary-initial-value problem for non-homogeneous rods does not decouple into sub-problems. This property is in contrast with the case of homogeneous rods, when the general problem decouples into the extension-torsion problem and the bending-shear problem, see Sect. 7.1.

The relations of identification (198) and (199), written for straight rods, become

$$\begin{gathered} \rho_0 w_\alpha = \langle \rho^* u_\alpha^* \rangle, \qquad \rho_0 u = \langle \rho^* u_3^* \rangle, \qquad \rho_0 = \langle \rho^* \rangle, \\ \vartheta_1 = -\frac{\langle \rho^* x_1 u_3^* \rangle}{I_2}, \quad \vartheta_2 = -\frac{\langle \rho^* x_2 u_3^* \rangle}{I_1}, \quad \psi = \frac{\langle \rho^* (x_1 u_2^* - x_2 u_1^*) \rangle}{I_1 + I_2}, \\ Q_\alpha = \langle t_{3\alpha}^* \rangle, \ F = \langle t_{33}^* \rangle, \ L_\alpha = -\langle x_\alpha t_{33}^* \rangle, \ H = \langle x_1 t_{32}^* - x_2 t_{31}^* \rangle, \end{gathered} \tag{201}$$

where u_i^* and t_{ij}^* are the components of $\boldsymbol{u}^*$ and $\boldsymbol{T}^*$, respectively.

It is not difficult to find the exact solution of the problem of extension, bending and torsion of directed curves. Consider the equilibrium of a straight rod subjected to an axial force $\bar{F}$, a torsion moment $\bar{H}$, and bending moments $\bar{L}_\alpha$ applied to both ends. The body forces and moments are absent. In our case, the equilibrium equations corresponding to (49) are

$$Q'_\alpha(s) = 0, \ F'(s) = 0, \ L'_\alpha(s) + Q_\alpha(s) = 0, \ H'(s) = 0, \ s \in (0, l), \tag{202}$$

while the boundary conditions on the ends of the rods are

$$\begin{aligned} Q_\alpha(0) = Q_\alpha(l) = 0, \quad & F(0) = F(l) = \bar{F}, \\ L_\alpha(0) = L_\alpha(l) = \bar{L}_\alpha, \quad & H(0) = H(l) = \bar{H}. \end{aligned} \tag{203}$$

Using the constitutive equations (200) we obtain a system of ordinary differential equations which yields the solution

$$w_\alpha(s) = -\frac{1}{2} a_\alpha s^2 + b_\alpha s, \ u(s) = a_3 s, \ \vartheta_\alpha(s) = -a_\alpha s, \ \psi(s) = b_3 s, \tag{204}$$

where the constants a_i and b_i are determined by the algebraic linear systems

$$\begin{bmatrix} C_2 & -C_{12} & -B_{32} \\ -C_{12} & C_1 & B_{31} \\ -B_{32} & B_{31} & A_3 \end{bmatrix} \begin{bmatrix} a_1 \\ a_2 \\ a_3 \end{bmatrix} = \begin{bmatrix} -\bar{L}_1 \\ -\bar{L}_2 \\ \bar{F} \end{bmatrix}, \quad \begin{bmatrix} A_1 & A_{12} & B_{13} \\ A_{12} & A_2 & B_{23} \\ B_{13} & B_{23} & C_3 \end{bmatrix} \begin{bmatrix} b_1 \\ b_2 \\ b_3 \end{bmatrix} = \begin{bmatrix} 0 \\ 0 \\ \bar{H} \end{bmatrix} \tag{205}$$

The force and moment vector fields corresponding to this solution are given by

$$\boldsymbol{N} = \bar{F} \boldsymbol{e}_3, \qquad \boldsymbol{M} = -\bar{L}_2 \boldsymbol{e}_1 + \bar{L}_1 \boldsymbol{e}_2 + \bar{H} \boldsymbol{e}_3. \tag{206}$$

This solution will be used later for comparison with three-dimensional solutions, in order to identify the effective stiffness coefficients for non-homogeneous thin rods.

11.1 Constitutive Coefficients for Non-Homogeneous Rods

In what follows, we determine the constitutive coefficients for non-homogeneous rods by comparison of exact solutions in the two approaches (directed curves and three-dimensional), in conjunction with the relations of identification (201).

Extension, Bending and Torsion of Non-Homogeneous Rods Let us consider a three-dimensional rod which occupies the domain

$$\mathcal{B} = \{(x_1, x_2, x_3) | (x_1, x_2) \in \Sigma, x_3 \in [0, l]\}.$$

The cross-section Σ is arbitrary and the body $\mathcal{B}$ is made of an isotropic and non-homogenous material such that the mass density ρ^* and the Lamé moduli λ, μ are independent of the axial coordinate, i.e. we have

$$\rho^* = \rho^*(x_1, x_2), \qquad \lambda = \lambda(x_1, x_2), \qquad \mu = \mu(x_1, x_2).$$

We consider the deformation of such cylinders under the action of terminal forces and moments. Assume that the body $\mathcal{B}$ is in equilibrium, in the absence of external body loads and tractions on the lateral surfaces. On the two ends of the cylinder act a resultant axial force and a resultant moment. We consider the same problem as in the previous section, but formulated in the three-dimensional setting. In view of the relations $(201)_{7-10}$ we take the boundary conditions

$$\begin{gathered} \langle t^*_{3\alpha} \rangle = 0, \qquad \langle t^*_{33} \rangle = \bar{F}, \qquad \langle x_\alpha t^*_{33} \rangle = -\bar{L}_\alpha, \\ \langle x_1 t^*_{32} - x_2 t^*_{31} \rangle = \bar{H} \text{ for } x_3 = 0, l. \end{gathered} \tag{207}$$

The solution of this three-dimensional problem for non-homogeneous rods is presented in (Ieşan, 2009a, Sects. 3.3, 3.4) where it is expressed in terms of the solutions to some auxiliary plane strain problems. We denote by $u^{(1)}_\alpha$, $u^{(2)}_\alpha$ and $u^{(3)}_\alpha$ the solutions of the 3 plane strain problems $\mathcal{D}^{(1)}$, $\mathcal{D}^{(2)}$ and $\mathcal{D}^{(3)}$ respectively, defined on the domain Σ by

$$\begin{aligned} \mathcal{D}^{(\gamma)}: &\quad t^{(\gamma)}_{\beta\alpha,\beta} + (\lambda x_\gamma)_{,\alpha} = 0 \quad \text{in } \Sigma, \quad t^{(\gamma)}_{\beta\alpha} n^*_\beta = -\lambda x_\gamma n^*_\alpha \quad \text{on } \partial\Sigma, \\ \mathcal{D}^{(3)}: &\quad t^{(3)}_{\beta\alpha,\beta} + \lambda_{,\alpha} = 0 \quad \text{in } \Sigma, \qquad t^{(3)}_{\beta\alpha} n^*_\beta = -\lambda n^*_\alpha \quad \text{on } \partial\Sigma, \end{aligned} \tag{208}$$

where obviously $t^{(k)}_{\alpha\beta} = \lambda u^{(k)}_{\rho,\rho} \delta_{\alpha\beta} + \mu(u^{(k)}_{\alpha,\beta} + u^{(k)}_{\beta,\alpha})$, $(k = 1, 2, 3$ and $\alpha, \beta = 1, 2)$ and $\boldsymbol{n}^* = n^*_\alpha \boldsymbol{e}_\alpha$ is the outward unit normal to $\partial\Sigma$. Let $\varphi(x_1, x_2)$ be the solution of the Neumann type boundary-value problem

$$\left(\mu\varphi_{,\alpha}\right)_{,\alpha} = \mu_{,1}x_2 - \mu_{,2}x_1 \quad \text{in}\Sigma, \qquad \frac{\partial\varphi}{\partial n^*} = x_2 n_1^* - x_1 n_2^* \quad \text{on}\,\partial\Sigma. \tag{209}$$

The existence of solutions to the above boundary-value problems (208) and (209) is proved in Ieşan (2009a, Sects. 3.2, 3.4). Then, the solution of our three-dimensional problem for the loads (207) is given by

$$\begin{aligned} u_1 &= -\frac{1}{2}\hat{a}_1 x_3^2 - \tau x_2 x_3 + \sum_{k=1}^{3}\hat{a}_k u_1^{(k)}(x_1,x_2), \\ u_2 &= -\frac{1}{2}\hat{a}_2 x_3^2 + \tau x_1 x_3 + \sum_{k=1}^{3}\hat{a}_k u_2^{(k)}(x_1,x_2), \\ u_3 &= (\hat{a}_1 x_1 + \hat{a}_2 x_2 + \hat{a}_3)x_3 + \tau\varphi(x_1,x_2), \end{aligned} \tag{210}$$

where the constants τ and $\hat{a}_i$ are given by the relations

$$\tau = \frac{\bar{H}}{D_*} \qquad \text{and} \qquad D_{\alpha j}\hat{a}_j = -\bar{L}_\alpha, \quad D_{3j}\hat{a}_j = \bar{F}. \tag{211}$$

Here the torsional rigidity D_* is expressed by

$$D_* = \langle \mu[x_1(x_1+\varphi_{,2}) + x_2(x_2-\varphi_{,1})]\rangle, \tag{212}$$

while the coefficients D_{ij} are given by

$$\begin{aligned} D_{\alpha\beta} &= \langle(\lambda+2\mu)x_\alpha x_\beta + \lambda x_\alpha u_{\gamma,\gamma}^{(\beta)}\rangle, \quad & D_{33} &= \langle(\lambda+2\mu) + \lambda u_{\gamma,\gamma}^{(3)}\rangle, \\ D_{\alpha 3} &= \langle(\lambda+2\mu)x_\alpha + \lambda x_\alpha u_{\gamma,\gamma}^{(3)}\rangle, \quad & D_{3\alpha} &= \langle(\lambda+2\mu)x_\alpha + \lambda u_{\gamma,\gamma}^{(\alpha)}\rangle. \end{aligned} \tag{213}$$

In Ieşan (2009a, Sect. 3.3) it is shown that $D_{ij} = D_{ji}$ and $\det(D_{ij})_{3\times 3} \neq 0$, so that we can determine the constants $\hat{a}_i$ from the system $(211)_{2,3}$.

Remark 11.1. If we introduce the stress function $\chi(x_1, x_2)$ by the relations

$$\chi_{,1} = -\mu(\varphi_{,2} + x_1), \qquad \chi_{,2} = \mu(\varphi_{,1} - x_2),$$

then the torsional rigidity is given by

$$D_* = 2\langle\chi(x_1,x_2)\rangle. \tag{214}$$

The stress function χ can be obtained from the boundary-value problem

$$\left(\frac{1}{\mu}\chi_{,\alpha}\right)_{,\alpha} = -2 \quad \text{in } \Sigma, \quad \chi = 0 \quad \text{on } \partial\Sigma,$$

provided that the domain Σ is simply-connected.

Let us compare now the three-dimensional solution (210) with the solution (204) obtained in the direct approach to rods, taking into account the relations (35) and (201). By comparison, it follows that we have to identify the constants

$$\begin{gathered} C_3 = D_*, \quad A_3 = D_{33}, \quad C_1 = D_{22}, \quad C_2 = D_{11}, \quad C_{12} = -D_{12}, \\ B_{31} = D_{23}, \quad B_{32} = -D_{13}, \quad B_{13} = B_{23} = 0. \end{gathered} \tag{215}$$

Thus, from (213)-(215) we obtain the following expressions for the constitutive coefficients

$$\begin{gathered} C_3 = 2\langle \chi(x_1,x_2)\rangle, A_3 = \langle (\lambda+2\mu) + \lambda u^{(3)}_{\gamma,\gamma}\rangle, C_1 = \langle (\lambda+2\mu)x_2^2 + \lambda x_2 u^{(2)}_{\gamma,\gamma}\rangle, \\ C_2 = \langle (\lambda+2\mu)x_1^2 + \lambda x_1 u^{(1)}_{\gamma,\gamma}\rangle, C_{12} = -\langle (\lambda+2\mu)x_1x_2 + \lambda x_1 u^{(2)}_{\gamma,\gamma}\rangle, B_{\alpha 3} = 0, \\ B_{31} = \langle (\lambda+2\mu)x_2 + \lambda x_2 u^{(3)}_{\gamma,\gamma}\rangle, B_{32} = -\langle (\lambda+2\mu)x_1 + \lambda x_1 u^{(3)}_{\gamma,\gamma}\rangle. \end{gathered} \tag{216}$$

By virtue of the identifications (201) and (216) we can verify that the fields u, w_α, ψ, $\boldsymbol{N}$ and $\boldsymbol{M}$ calculated for the solutions in the two different approaches coincide.

Remark 11.2. For the fields ϑ_α corresponding to the three-dimensional solution (210) we obtain from $(201)_{1,2}$ and (35) the expressions

$$\vartheta_\alpha = -\hat{a}_\alpha x_3 - \tau \frac{\langle \rho^* x_\alpha \varphi\rangle}{\langle \rho^* x_\alpha^2\rangle}, \qquad \alpha = 1,2, \text{not summed}.$$

Comparing this relation with the field $\vartheta_\alpha$from the solution $(204)_3$ for directed curves, we see that we need to approximate

$$\frac{\langle \rho^* x_\alpha \varphi\rangle}{\langle \rho^* x_\alpha^2\rangle} \simeq 0, \qquad \alpha = 1,2, \text{not summed},$$

where $\varphi(x_1,x_2)$ is the torsion function given by (209). For example, in the case when Σ is an elliptical domain $\Sigma = \{(x_1,x_2) | \frac{x_1^2}{a^2} + \frac{x_2^2}{b^2} < 1\}$ and μ is constant, then we have $\varphi(x_1,x_2) = \frac{b^2-a^2}{a^2+b^2} x_1 x_2$ so that the above approximation is justified.

Shear Vibrations of Rectangular Rods Due to the shear-bending coupling in the case of static problems, the effective shear stiffness coefficients A_1, A_2 and A_{12} cannot be obtained by analyzing static shear problems and using the same procedure as above. For *thin* beams, the coefficients A_1, A_2, A_{12} will not enter in the leading order terms of the solutions. For this reason, we determine next the effective shear stiffness coefficients by solving a free-vibration problem. The necessity of considering free-vibration problems for the determination of effective shear stiffness properties is also discussed in details in Zhilin (2006a, Sect. 6) and in Zhilin (2007, p. 34-38).

Consider a three-dimensional rod which occupies the domain

$$\mathcal{R} = \{(x_1, x_2, x_3) | x_1 \in (-\frac{a}{2}, \frac{a}{2}), x_2 \in (-\frac{b}{2}, \frac{b}{2}), x_3 \in (0, l)\},$$

made of a non-homogeneous isotropic material. The material parameters λ, μ and ρ^* are given functions of (x_1, x_2). Assume that the mass density ρ^* has a symmetrical distribution across the thickness: $\rho^*(x_1, x_2) = \rho^*(-x_1, x_2)$.

The body loads are zero, the lateral surfaces $x_1 = \pm\frac{a}{2}$ and $x_2 = \pm\frac{b}{2}$ are traction free, and the end boundary conditions are given by

$$u_1^* = u_2^* = 0 \qquad \text{and} \qquad t_{33}^* = 0 \qquad \text{for} \quad x_3 = 0, l. \tag{217}$$

To determine the shear vibrations of this rod, we search for solutions $\boldsymbol{u}^*$ of the form

$$\boldsymbol{u}^* = W \cos(\omega t) \sin(\frac{\pi}{a} x_1) \boldsymbol{e}_3, \tag{218}$$

where W is a constant and ω is the lowest natural frequency. We observe that all the boundary conditions are satisfied by the field (218), and the equations of motion reduce to $t_{13,1}^* = \rho^* \ddot{u}_3^*$, which by integration with respect to x_1 gives

$$t_{13}^* = -W\omega^2 \cos(\omega t) \int_{-a/2}^{x_1} \rho^*(x_1, x_2) \sin(\frac{\pi}{a} x_1) \mathrm{d}x_1.$$

Using the constitutive equation for t_{13}^* we get

$$\mu(x_1, x_2) \frac{\pi}{a} \cos(\frac{\pi}{a} x_1) = -\omega^2 \int_{-a/2}^{x_1} \rho^*(x_1, x_2) \sin(\frac{\pi}{a} x_1) \mathrm{d}x_1 \tag{219}$$

We apply the mean value theorem for the integral in (219) and we deduce that there exists a point $\alpha = \alpha(x_1, x_2) \in (-\frac{a}{2}, x_1)$ such that

$$\int_{-a/2}^{x_1} \rho^*(x_1, x_2) \sin(\frac{\pi}{a} x_1) \mathrm{d}x_1 = \rho^*(\alpha, x_2) \int_{-a/2}^{x_1} \sin(\frac{\pi}{a} x_1) \mathrm{d}x_1. \tag{220}$$

Substituting (220) into (219) and integrating over Σ we obtain

$$\omega^2 = \left(\frac{\pi}{a}\right)^2 \frac{\langle \mu(x_1, x_2) \rangle}{\langle \rho^*(\alpha, x_2) \rangle}. \tag{221}$$

Let us treat the same problem using the approach of directed curves. We consider a straight rod along the Ox_3 axis for which the arclength parameter $s \in (0, l)$. The external body loads $\boldsymbol{F}$ and $\boldsymbol{L}$ are zero. According to (201) and (217) we have the following boundary conditions on the rod ends

$$w_\alpha = 0, \quad F = 0, \quad \psi = 0, \quad L_\alpha = 0 \quad (\alpha = 1, 2), \quad \text{for } s = 0, l. \tag{222}$$

In order to study the shear vibrations, we search for solutions of the equations (49), (200) of the form

$$\vartheta_1 = \bar{W} \cos(\bar{\omega} t), \qquad \vartheta_2 = \psi = 0, \qquad u = w_\alpha = 0, \tag{223}$$

where $\bar{W}$ is a constant and $\bar{\omega}$ is the natural frequency of the rod. In view of the constitutive equations (200), we see that the boundary conditions (222) are satisfied. Imposing that the fields (223) verify the equations of motion (49) we find

$$\bar{\omega}^2 = \frac{A_1}{I_2} \qquad \text{and} \qquad A_{12} = 0. \tag{224}$$

We identify the natural frequencies ω and $\bar{\omega}$ from (221) and $(223)_1$, and we obtain the expression of the constitutive coefficient A_1 as follows

$$A_1 = k \frac{\langle \mu \rangle \langle \rho^* x_1^2 \rangle \text{Area}(\Sigma)}{\langle \rho^*(\alpha, x_2) \rangle \langle x_1^2 \rangle} \qquad \text{with} \quad k = \frac{\pi^2}{12}, \tag{225}$$

where the factor k is similar to the shear correction factor introduced first by Timoshenko (1921) in the theory of beams (note that in the original contribution of Timoshenko the value is 2/3). One can proceed analogously for the x_2 direction and find a similar expression for A_2. These relations express the transverse shear stiffness coefficients for non-homogeneous rectangular rods. The value given by (225)is verified in Bîrsan et al. (2012), where we consider the bending of cantilever functionally graded beams and make a comparison with numerical results.

Remark 11.3. In the case of *thin* rods, when ρ^* has a smooth variation across the thickness, we can employ the approximation

$$\langle \rho^*(\alpha, x_2) \rangle \simeq \langle \rho^*(x_1, x_2) \rangle. \tag{226}$$

Then, we substitute (226) into (225) and find

$$A_\gamma = k \langle \mu \rangle \frac{\langle \rho^* x_\gamma^2 \rangle \text{Area}(\Sigma)}{\langle \rho^* \rangle \langle x_\gamma^2 \rangle} \quad (\gamma = 1, 2 \text{ not summed}), \quad A_{12} = 0. \tag{227}$$

The simplified (approximate) formulas (227) can be used to estimate the transverse shear stiffness for arbitrary non-homogeneous rods (not necessarily rectangular or symmetrical) in most cases.

Remark 11.4. In the special case when the rod is made of an isotropic material with *constant Poisson ratio* ν, the above formulas can be simplified considerably. Let Young's modulus E be an arbitrary function of (x_1, x_2) and the shape of cross-section Σ is arbitrary. In this case the solutions $u_\alpha^{(k)}(x_1, x_2)$ of the problems $\mathcal{D}^{(k)}$, $k = 1, 2, 3$, defined by (208) have the simple form

$$\begin{aligned} u_1^{(1)} &= -\tfrac{1}{2}\nu(x_1^2 - x_2^2), \quad u_2^{(1)} = -\nu x_1 x_2; \\ u_1^{(2)} &= -\nu x_1 x_2, \quad u_2^{(2)} = \tfrac{1}{2}\nu(x_1^2 - x_2^2); \\ u_1^{(3)} &= -\nu x_1, \quad u_2^{(3)} = -\nu x_2. \end{aligned} \tag{228}$$

Then, from (216) we obtain the following expressions for the effective stiffness coefficients

$$\begin{gathered} A_3 = \langle E(x_1, x_2)\rangle, \quad C_{12} = -\langle x_1 x_2 E(x_1, x_2)\rangle, \quad C_1 = \langle x_2^2 E(x_1, x_2)\rangle, \\ C_2 = \langle x_1^2 E(x_1, x_2)\rangle, B_{31} = \langle x_2 E(x_1, x_2)\rangle, B_{32} = -\langle x_1 E(x_1, x_2)\rangle, B_{\alpha 3} = 0. \end{gathered} \tag{229}$$

The constitutive coefficients C_3, A_1, A_2 and A_{12} keep the same form as in the general case, given by Eqs. $(216)_1$ and (225).

We remark that in the case of a homogeneous isotropic rod, i.e. when E is also constant, from Eqs. (229) and (35) we derive the well-known formulas for the effective stiffness properties (167) obtained previously.

11.2 Effective Stiffness Properties of Composite Beams

In this section we consider rods made of two isotropic and non-homogeneous materials. The body $\mathcal{B}$ is decomposed in two regions $\mathcal{B}_1$ and $\mathcal{B}_2$ such that $\mathcal{B}_\rho = \{(x_1, x_2, x_3) | (x_1, x_2) \in S_\rho, x_3 \in (0, l)\}$. Thus, the cross-section Σ is decomposed in two domains S_1 and S_2 with $S_1 \cap S_2 = \emptyset$, see Fig. 4(a). We denote by Γ_0 the curve of separation between the domains S_1 and S_2 and by Γ_1, Γ_2 the complementary subsets of $\partial\Sigma$ such that $\partial S_\rho = \Gamma_0 \cap \Gamma_\rho$. Let $\Pi_0 = \{(x_1, x_2, x_3) | (x_1, x_2) \in \Gamma_0, x_3 \in (0, l)\}$ be the surface of separation of the two materials. We assume that the two materials are welded together along Π_0 and there is no separation of material along Π_0, so we have the conditions

$$\left[\boldsymbol{u}^*\right]_1 = \left[\boldsymbol{u}^*\right]_2, \qquad \left[\boldsymbol{T}^*\right]_1 \cdot \boldsymbol{n}^0 = \left[\boldsymbol{T}^*\right]_2 \cdot \boldsymbol{n}^0 \qquad \text{on} \quad \Pi_0, \tag{230}$$

where $\boldsymbol{n}^0 = n_\alpha^0 \boldsymbol{e}_\alpha$ is the unit normal of Π_0, outward to $\mathcal{B}_1$. The notations $\left[f\right]_1$ and $\left[f\right]_2$ represent the values of any field f on Π_0, calculated as the

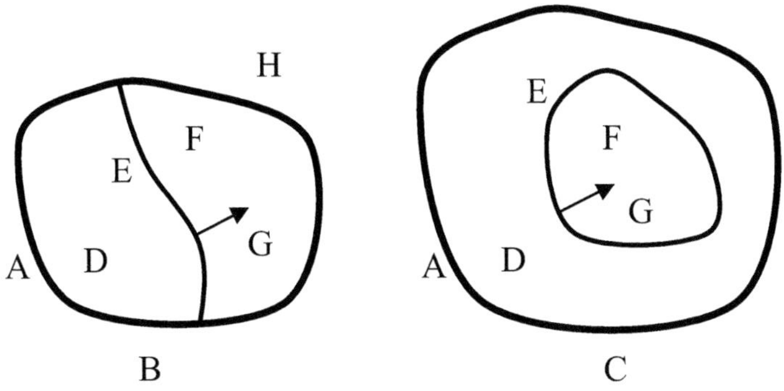

Figure 4. The cross-section of rods composed of two materials

limits of the values from the domains $\mathcal{B}_1$ and $\mathcal{B}_2$, respectively. Let us denote the Lamé moduli of the material occupying the domain $\mathcal{B}_\rho$ by $\lambda^{(\rho)}(x_1, x_2)$ and $\mu^{(\rho)}(x_1, x_2)$, with $(x_1, x_2) \in S_\rho$, $\rho = 1, 2$.

Consider the problem of extension, bending and torsion of such a compound three-dimensional rod, under the resultant forces and moments (207) acting on the ends. This problem has been treated in (Ieşan, 2009a, Sect. 3.6) and the exact solution is expressed in terms of the solutions to some auxiliary plane strain problems. Let us denote by $u_\alpha^{(1)}$, $u_\alpha^{(2)}$ and $u_\alpha^{(3)}$ the solutions of the 3 plane strain problems $\mathcal{P}^{(1)}$, $\mathcal{P}^{(2)}$ and $\mathcal{P}^{(3)}$ respectively, formulated on the domain $\Sigma = S_1 \cup S_2 \cup \Gamma_0$ by

$$\begin{aligned}
\mathcal{P}^{(\gamma)} : t_{\beta\alpha,\beta}^{(\gamma)} + (\lambda^{(\rho)} x_\gamma)_{,\alpha} = 0 \text{ in } S_\rho, \quad t_{\beta\alpha}^{(\gamma)} n_\beta^* = -\lambda^{(\rho)} x_\gamma n_\alpha^* \text{ on } \Gamma_\rho, \\
\left[u_\alpha^{(\gamma)}\right]_1 = \left[u_\alpha^{(\gamma)}\right]_2, \quad \left[t_{\beta\alpha}^{(\gamma)}\right]_1 n_\beta^0 = \left[t_{\beta\alpha}^{(\gamma)}\right]_2 n_\beta^0 + (\lambda^{(2)} - \lambda^{(1)}) x_\gamma n_\alpha^0 \text{ on } \Gamma_0, \\
\mathcal{P}^{(3)} : t_{\beta\alpha,\beta}^{(3)} + \lambda_{,\alpha}^{(\rho)} = 0 \text{ in } S_\rho, \quad t_{\beta\alpha}^{(3)} n_\beta^* = -\lambda^{(\rho)} n_\alpha^* \text{ on } \Gamma_\rho, \\
\left[u_\alpha^{(3)}\right]_1 = \left[u_\alpha^{(3)}\right]_2, \quad \left[t_{\beta\alpha}^{(3)}\right]_1 n_\beta^0 = \left[t_{\beta\alpha}^{(3)}\right]_2 n_\beta^0 + (\lambda^{(2)} - \lambda^{(1)}) n_\alpha^0 \text{ on } \Gamma_0.
\end{aligned} \tag{231}$$

We also introduce the function $\varphi(x_1, x_2)$ which is the solution of the boundary-value problem

$$\begin{aligned}
\left(\mu^{(\rho)} \varphi_{,\alpha}\right)_{,\alpha} = \mu_{,1}^{(\rho)} x_2 - \mu_{,2}^{(\rho)} x_1 \text{ in } S_\rho, \quad \frac{\partial \varphi}{\partial n^*} = x_2 n_1^* - x_1 n_2^* \text{ on } \Gamma_\rho, \\
\left[\varphi\right]_1 = \left[\varphi\right]_2, \; \mu^{(1)} \left[\frac{\partial \varphi}{\partial n^0}\right]_1 = \mu^{(2)} \left[\frac{\partial \varphi}{\partial n^0}\right]_2 + (\mu^{(1)} - \mu^{(2)})(x_2 n_1^0 - x_1 n_2^0) \text{ on } \Gamma_0.
\end{aligned} \tag{232}$$

Comparing the solution of the extension-bending-torsion problem in the direct approach given in (202)–(206) with the solution of the corresponding three-dimensional problem presented in (Ieşan, 2009a, Sect. 3.6) we deduce the following expressions for the constitutive coefficients

$$
\begin{aligned}
A_3 =&\sum_{\rho=1}^{2}\int_{S_\rho}\big(\lambda^{(\rho)}+2\mu^{(\rho)}+\lambda^{(\rho)}u^{(3)}_{\gamma,\gamma}\big)\mathrm{d}x_1\mathrm{d}x_2, \qquad B_{13}=B_{23}=0,\\
B_{31} &= \sum_{\rho=1}^{2}\int_{S_\rho} x_2\big(\lambda^{(\rho)}+2\mu^{(\rho)}+\lambda^{(\rho)}u^{(3)}_{\gamma,\gamma}\big)\mathrm{d}x_1\mathrm{d}x_2,\\
B_{32} &= -\sum_{\rho=1}^{2}\int_{S_\rho} x_1\big(\lambda^{(\rho)}+2\mu^{(\rho)}+\lambda^{(\rho)}u^{(3)}_{\gamma,\gamma}\big)\mathrm{d}x_1\mathrm{d}x_2,\\
C_1 &= \sum_{\rho=1}^{2}\int_{S_\rho} x_2\big[(\lambda^{(\rho)}+2\mu^{(\rho)})x_2+\lambda^{(\rho)}u^{(2)}_{\gamma,\gamma}\big]\mathrm{d}x_1\mathrm{d}x_2, \qquad (233)\\
C_2 &= \sum_{\rho=1}^{2}\int_{S_\rho} x_1\big[(\lambda^{(\rho)}+2\mu^{(\rho)})x_1+\lambda^{(\rho)}u^{(1)}_{\gamma,\gamma}\big]\mathrm{d}x_1\mathrm{d}x_2,\\
C_{12} &= -\sum_{\rho=1}^{2}\int_{S_\rho} x_1\big[(\lambda^{(\rho)}+2\mu^{(\rho)})x_2+\lambda^{(\rho)}u^{(2)}_{\gamma,\gamma}\big]\mathrm{d}x_1\mathrm{d}x_2,\\
C_3 &= \sum_{\rho=1}^{2}\int_{S_\rho} \mu^{(\rho)}\big[x_1(x_1+\varphi_{,2})+x_2(x_2-\varphi_{,1})\big]\mathrm{d}x_1\mathrm{d}x_2,
\end{aligned}
$$

where the functions $u^{(k)}_\alpha(x_1,x_2)$ are determined by (231) and $\varphi(x_1,x_2)$ is given by (232).

Remark 11.5. The above results (233) also hold when the distribution of the material in the rod is such that the separation curve Γ_0 is a closed curve included in Σ, see Figure 4(b). In this case we have $\Gamma_1=\partial\Sigma, \Gamma_2=\emptyset, \partial S_1 = \Gamma_1\cup\Gamma_0, \partial S_2=\Gamma_0$, and the boundary-value problems (231), (232) keep the same forms.

The results of this section can be extended to the case when the rod $\mathcal{B}$ is composed of n ($n\geq 2$) non-homogeneous and isotropic materials with different mechanical properties. Similar results can be derived also in the case of composite beams made of *orthotropic* non-homogeneous materials (see Bîrsan et al., 2012).

Example of Composite Circular Beam Let us consider a circular composite beam, such that the cross-section is decomposed as $\Sigma = S_1\cup\bar{S}_2$, where $S_1=\{(x_1,x_2)|a^2<x_1^2+x_2^2<b^2\}$ and $S_2=\{(x_1,x_2)|x_1^2+x_2^2<a^2\}$. The

first material occupies the region $S_1 \times (0,l)$ and has the Lamé moduli

$$\lambda^{(1)}(x_1,x_2) = \lambda_0 r^{-m},\ \mu^{(1)}(x_1,x_2) = \mu_0 r^{-m},\ r = \sqrt{x_1^2+x_2^2}, (x_1,x_2) \in S_1, \tag{234}$$

where $m > 0$, λ_0 and μ_0 are constants. We denote by $\nu_0 = \dfrac{\lambda_0}{2(\lambda_0+\mu_0)}$ and $E_0 = \dfrac{\mu_0(3\lambda_0+2\mu_0)}{\lambda_0+\mu_0}$. The second material occupies the region $S_2 \times (0,l)$ and its elastic properties are described by

$$E^{(2)}(x_1,x_2) = E(r),\ \nu^{(2)}(x_1,x_2) = \nu_0(\text{constant}),\ (x_1,x_2) \in S_2, \tag{235}$$

where $E(r)$ is an arbitrary given function of r.

In order to use the formulas presented above in this section, we have to solve the plane strain problems $\mathcal{P}^{(k)}$ given by (231) and the boundary-value problem (232) for the torsion function. In our case, we observe that these problems admit the following solutions

$$u_1^{(1)} = -u_2^{(2)} = \frac{1}{2}\nu_0(x_2^2 - x_1^2),\ u_2^{(1)} = u_1^{(2)} = -\nu_0 x_1 x_2,\ u_\alpha^{(3)} = -\nu_0 x_\alpha. \tag{236}$$

Inserting these functions into the general results (233) we find the effective stiffness coefficients for this compound rod

$$\begin{gathered} C_3 = \frac{\pi}{1+\nu_0}\int_0^a r^3 E(r)\mathrm{d}r + 2\pi\mu_0 d_m,\ A_3 = 2\pi\Big(\int_0^a rE(r)\mathrm{d}r + E_0 c_m\Big), \\ C_1 = C_2 = \pi\Big(\int_0^a r^3 E(r)\mathrm{d}r + E_0 d_m\Big),\ C_{12} = 0,\ B_{3\alpha} = B_{\alpha 3} = 0, \end{gathered} \tag{237}$$

where we have denoted by c_m and d_m the expressions

$$c_m = \begin{cases} \dfrac{a^{2-m} - b^{2-m}}{m-2} & \text{for } m \neq 2 \\ \log(b/a) & \text{for } m = 2 \end{cases},\ d_m = \begin{cases} \dfrac{a^{4-m} - b^{4-m}}{m-4} & \text{for } m \neq 4 \\ \log(b/a) & \text{for } m = 4 \end{cases}. \tag{238}$$

Let us find also the transverse shear stiffness coefficients A_1 and A_2. Assume that the mass density function $\rho^*(x_1,x_2)$ is given by

$$\rho^*(x_1,x_2) = \begin{cases} \rho_0^* r^{-m} & \text{for } (x_1,x_2) \in S_1 \\ \rho(r) & \text{for } (x_1,x_2) \in S_2 \end{cases}, \tag{239}$$

where $\rho_0^* > 0$ is a constant and $\rho(r)$ is an arbitrary function. Then, using the results (227) we find the expressions

$$A_1 = A_2 = \frac{\pi^3}{6b^2} \frac{\left(\int_0^a \frac{rE(r)}{1+\nu_0} \mathrm{d}r + 2\mu_0 c_m \right) \left(\int_0^a r^3 \rho(r) \mathrm{d}r + \rho_0^* d_m \right)}{\int_0^a r\rho(r) \mathrm{d}r + \rho_0^* c_m}. \tag{240}$$

The last relation expresses the transverse shear stiffness coefficients A_1 and A_2. The other effective stiffness coefficients for the composite circular beam are given by (237).

12 Conclusions

Within this chapter a theory of directed deformable curves is presented. The basic ideas coincide with the direct approach in Mechanics. The governing equations have similarities with the theory of directed surfaces (see chapter Cosserat-Type Shells in this book). In the same manner this theory combines the first theory of elastic rods elaborated by Euler using the direct approach and its generalization by the Cosserat brothers.

The theory can be applied, for example, to the case of rods

- composed of functionally graded materials and
- affected by temperature fields

For different special cases some theorems and proofs ar introduced. They are related to mathematical statements like the uniqueness of the solutions or Korn-like inequalities. These statements are important from the mathematical point of view. In addition, they are necessary for the construction of numerical algorithms.

Let us note that the theory is applied to some practical problems concerning functionally graded materials (cantilever beam with distributed load and with concentrated end force, three-point bending problem). The results are compared with finite element solutions. Details are presented in Bîrsan et al. (2012).

Acknowledgement. The authors were supported by the E.U. FP7 Programme FP7-REGPOT-2009-1 under Grant Agreement No. 245479.

Bibliography

Altenbach, H., 2000. An alternative determination of transverse shear stiffnesses for sandwich and laminated plates. Int. J. Solids Struct. 37, 3503–3520.

Altenbach J., Altenbach H., 1994. Einführung in die Kontinuumsmechanik. Teubner, Stuttgart.

Altenbach, H., Bîrsan, M., Eremeyev, V.A., 2012. On a thermodynamic theory of rods with two temperature fields, Acta Mech., in print.

Altenbach, H., Eremeyev, V.A., 2008. Direct approach-based analysis of plates composed of functionally graded materials. Arch. Appl. Mech. 78, 775–794.

Altenbach, H., Eremeyev, V.A., 2009. On the bending of viscoelastic plates made of polymer foams, Acta Mech. 204, 137–154.

Altenbach, H., Naumenko, K., Zhilin, P.A., 2006. A direct approach to the formulation of constitutive equations for rods and shells. In: Pietraszkiewicz, W., Szymczak, C. (Eds.), Shell Structures: Theory and Applications. Taylor and Francis, London, pp. 87–90.

Altenbach, H., Öchsner, A., 2010. Cellular and Porous Materials in Structures and Processes. CISM Courses and Lectures, Vol. 521, Springer Wien NewYork.

Altenbach, H., Zhilin, P.A., 1988. A general theory of elastic simple shells (in Russian). Uspekhi Mekhaniki 11, 107–148.

Antman, S.S., 1995. Nonlinear Problems of Elasticity. Series Applied Mathematical Sciences, no. 107, Springer, New York.

Berdichevsky, V.L., 2009. Variational Principles of Continuum Mechanics, Vol. II: Applications. Springer–Verlag, Berlin.

Bîrsan, M., 2006a. On the theory of elastic shells made from a material with voids. Int. J. Solids Struct. 43, 3106–3123.

Bîrsan, M., 2006b. On a thermodynamic theory of porous Cosserat elastic shells. J. Thermal Stresses 29, 879–899.

Bîrsan, M., 2008. Inequalities of Korn's type and existence results in the theory of Cosserat elastic shells. J. Elasticity 90, 227–239.

Bîrsan, M., Altenbach, H., 2011a. On the theory of porous elastic rods. Int. J. Solids Struct. 48, 910-924.

Bîrsan, M., Altenbach, H., 2011b. Theory of thin thermoelastic rods made of porous materials. Arch. Appl. Mech. 81, 1365-1391.

Bîrsan, M., Altenbach, H., 2012a. The Korn-type inequality in a Cosserat model for thin thermoelastic porous rods. Meccanica 47, 789-794.

Bîrsan, M., Altenbach, H., 2012b. On the Cosserat model for thin rods made of thermoelastic materials with voids. Discrete and Continuous Dynamical Systems - Series S, in print.

Bîrsan, M., Altenbach, H., Sadowski, T., Eremeyev, V.A., Pietras, D., 2012. Deformation analysis of functionally graded beams by the direct approach. Composites: Part B 43(3), 1315-1328.

Bîrsan, M., Bîrsan, T., 2011. An inequality of Cauchy–Schwarz type with application in the theory of elastic rods. Libertas Mathematica 31, 123–126.
Brezis, H., Analyse fonctionelle: Théorie et applications. Masson, Paris, 1992.
Capriz, G., 1989. Continua with Microstructure. Springer Tracts in Natural Philosophy, no. 35, Springer–Verlag, New York.
Capriz, G., Podio–Guidugli, P., 1981. Materials with spherical structure. Arch. Rational Mech. Anal. 75, 269–279.
Carlson, D.E., 1972. Linear Thermoelasticity. In: Flügge W (Ed.), Handbuch der Physik, vol. VI a/2, , pp. 297–346 Springer Verlag, Berlin.
Ciarlet, P.G., 2000. Mathematical Elasticity, Vol. III: Theory of Shells. North-Holland, Elsevier, Amsterdam.
Ciarlet, P.G., An Introduction to Differential Geometry with Applications to Elasticity. Springer, Dordrecht, 2005.
Ciarletta, M., Ieşan, D., 1993. Non–classical Elastic Solids. Pitman Research Notes in Mathematics, no. 293, Longman Scientific & Technical, London.
Coleman, B.D., Noll, W., 1963. The thermodynamics of elastic materials with heat conduction and viscosity. Arch. Ration. Mech. Anal. 13, 167-178.
Cosserat, E., Cosserat, F., 1909. Théorie des corps déformables. A. Herman et Fils, Paris.
Cowin, S.C., Goodman, M.A., 1976. A variational principle for granular materials. ZAMM 56, 281–286.
Cowin, S.C., Leslie, F.M., 1980. On kinetic energy and momenta in Cosserat continua. ZAMP 31, 247–260.
Cowin, S.C., Nunziato, J.W., 1983. Linear elastic materials with voids. J. Elasticity 13, 125–147.
Eremeyev, V. A., Pietraszkiewicz, W., 2011. Thermomechanics of shells undergoing phase transition. J. Mech. Phys. Solids 59, 1395-1412.
Freddi, L., Morassi, A., Paroni, R., 2007. Thin-walled beams: A derivation of Vlassov theory via Γ–convergence. J. Elasticity 86, 263–296.
Gibson, L.J., Ashby, M.F., 1997. Cellular Solids: Structure and Properties (2nd edition). Cambridge Solid State Science Series, Cambridge University Press, Cambridge.
Goodman, M.A., Cowin, S.C., 1972. A continuum theory for granular materials. Arch. Rational Mech. Anal. 44, 249–266.
Green, A.E., Naghdi, P.M., 1979. On thermal effects in the theory of rods. Int. J. Solids Struct. 15, 829–853.
Green, A.E., Naghdi, P.M., Wenner, M.L., 1974. On the theory of rods – Part II: Developments by direct approach. Proc. Royal Soc. London A337, 485–507.

Gurtin, M.E., 1972. The Linear Theory of Elasticity. In: Flügge, W. (Ed.), Handbuch der Physik, vol. VI a/2. Springer Verlag, Berlin, pp. 1–295.

Hodges, D.H., 2006. Nonlinear Composite Beam Theory. Progress in Astronautics and Aeronautics, no. 213, American Institute of Aeronautics and Astronautics Inc., Reston.

Ieşan, D., 1986. A theory of thermoelastic materials with voids. Acta Mech. 60, 67–89.

Ieşan, D., 2009a. Classical and Generalized Models of Elastic Rods. Chapman & Hall / CRC Press, Boca Raton - London - New York.

Ieşan, D., 2009b. Thermal effects in orthotropic porous elastic beams. ZAMP 60, 138–153.

Ieşan, D., Scalia, A., 2007. On the deformation of functionally graded porous elastic cylinders. J. Elasticity 87, 147-159.

Jenkins, J.T., 1975. Static equilibrium of granular materials. J. Appl. Mech. 42, 603–606.

Lebedev, L. P., Cloud, M. J., Eremeyev, V. A., 2010. Tensor Analysis with Applications in Mechanics. World Scientific, New Jersey.

Lembo, M., Podio–Guidugli, P., 2001. Internal constraints, reactive stresses, and the Timoshenko beam theory. J. Elasticity 65, 131–148.

Love, A.E.H., 1944. A Treatise in the Mathematical Theory of Elasticity. Fourth Edition, Dover Publ., New York.

Lurie, A.I., 2005. Theory of Elasticity. Springer, Berlin.

Meunier, N., 2008. Recursive derivation of one-dimensional models from three-dimensional nonlinear elasticity. Math. Mech. Solids 13, 172–194.

Mindlin, R.D., 1964. Microstructure in linear elasticity. Arch. Rational Mech. Anal. 16, 51–78.

Naumenko, K., Altenbach, H., 2007. Modeling of Creep for Structural Analysis. Springer–Verlag, Berlin.

Neff, P., 2004. A geometrically exact Cosserat shellmodel including size effects, avoiding degeneracy in the thin shell limit. Part I: Formal dimensional reduction for elastic plates and existence of minimizers for positive Cosserat couple modulus. Contin. Mech. Thermodyn. 16, 577–628.

Neff, P., 2007. A geometrically exact planar Cosserat shell-model with microstructure: Existence of minimizers for zero Cosserat couple modulus. Math. Models Meth. Appl. Sci. 17, 363–392.

Nunziato, J.W., Cowin, S.C., 1979. A nonlinear theory of elastic materials with voids. Arch. Rational Mech. Anal. 72, 175–201.

Pazy, A., 1983. Semigroups of Linear Operators and Applications to Partial Differential Equations. Springer Verlag, New York.

Podio–Guidugli, P., 2008. Validation of classical beam and plate models by variational convergence. In: Jaiani, G., Podio–Guidugli, P. (Eds.),

IUTAM Symposium on Relations of Shell, Plate, Beam, and 3D Models, IUTAM Bookseries, vol. 9. Springer Science + Business Media B.V., pp. 177–188.

Rubin, M.B., 2000. Cosserat Theories: Shells, Rods, and Points. Kluwer Academic Publishers, Dordrecht.

Simmonds, J.G., 1984. The thermodynamical theory of shells: Descent from 3-dimensions without thickness expansions. In: Axelrad, E.K., Emmerling, F.A. (eds.) Flexible Shells, Theory and Applications, pp. 1-11. Springer, Berlin.

Simmonds, J.G., 2005. A simple nonlinear thermodynamic theory of arbitrary elastic beams. J. Elasticity 81, 51-62.

Simmonds, J.G., 2011. A classical, nonlinear thermodynamic theory of elastic shells based on a single constitutive assumption. J. Elasticity 105, 305-312.

Sprekels, J., Tiba, D., 2009. The control variational approach for differential systems. SIAM J. Control Optim. 47, 3220–3226.

Svetlitsky, V.A., 2000. Statics of Rods. Springer, Berlin.

Tiba, D., Vodak, R., 2005. A general asymptotic model for Lipschitzian curved rods. Adv. Math. Sci. Appl. 15, 137–198.

Timoshenko, S.P., 1921. On the correction for shear of the differential equation for transverse vibrations of prismatic beams. Philosophical Magazine 41, 744–746.

Toupin, R.A., 1964. Theories of elasticity with couple stress. Arch. Rational Mech. Anal. 17, 85–112.

Trabucho, L., Viaño, J.M., 1996. Mathematical modelling of rods. In: Ciarlet, P.G., Lions, J.L. (Eds.), Handbook of Numerical Analysis, vol. 4. North Holland, Amsterdam, pp. 487–974.

Truesdell, C., 1984. Rational Thermodynamics (2nd edition). Springer, New York.

Vrabie, I.I., C_0–Semigroups and Applications. North–Holland, Elsevier, Amsterdam, 2003.

Zhilin, P.A., 1976. Mechanics of deformable directed surfaces. Int. J. Solids Struct. 12, 635–648.

Zhilin, P.A., 2006a. Nonlinear theory of thin rods. In: Indeitsev, D.A., Ivanova, E.A., Krivtsov, A.M. (Eds.), Advanced Problems in Mechanics, vol. 2. Instit. Problems Mech. Eng. R.A.S. Publ., St. Petersburg, pp. 227–249.

Zhilin, P.A., 2006b. Applied Mechanics – Foundations of Shell Theory (in Russian). Politekhn. Univ. Publ., St. Petersburg.

Zhilin, P.A., 2007. Applied Mechanics – Theory of Thin Elastic Rods (in Russian). Politekhn. Univ. Publ., St. Petersburg.

Micromorphic Media

Samuel Forest
MINES ParisTech, Centre des matériaux, CNRS UMR 7633
BP 87, 91003 Evry, France
samuel.forest@mines-paristech.fr

Abstract The elastoviscoplasticity theory of micromorphic media at finite deformation is presented in this chapter. Micromechanical considerations are then put forward to motivate the existence of the microdeformation degrees of freedom in the case of composite materials. Mixtures of micromorphic media are finally considered with a view to homogenising the size–dependent properties of metal polycrystals.

1 Introduction

1.1 Scope of this Chapter

A classification of generalised mechanical continuum theories is proposed in Fig. 1. The present chapter is limited to continuum media fulfilling the principle of local action, meaning that the mechanical state at a material point $\underline{X}$ depends on variables defined at this point only (Truesdell and Toupin, 1960; Truesdell and Noll, 1965). The classical Cauchy continuum is called *simple material* because its response at material point $\underline{X}$ to deformations homogeneous in a neighborhood of $\underline{X}$ determines uniquely its response to every deformation at $\underline{X}$. In *higher grade* materials, homogeneous deformations are not sufficient to characterise the material behaviour because they are sensitive to higher gradients of the displacement field. Mindlin formulated for instance the theories that include the second and third gradients of the displacement field (Mindlin, 1965). The gradient effect may be limited to the plastic part of deformation which leads to strain gradient plasticity models (Aifantis, 1984; Forest and Bertram, 2011) or, more generally, theories that include the gradient of some internal variables (Maugin, 1990). *Higher order* materials are characterised by additional degrees of freedom of the material points (Eringen, 1999). Directors can be attached to each material point that evolve in a different way from the material lines. Cosserat directors can rotate. In the micromorphic continuum designed by

Eringen and Mindlin (Eringen and Suhubi, 1964; Mindlin, 1964), the directors can also be distorted, so that a second order tensor is attributed to each material point. Tensors of higher order can even be introduced as proposed in Germain's general micromorphic theory (Germain, 1973).

Higher order media are sometimes called continua with *microstructure*. This name has now become misleading in the sense that even Cauchy material models can integrate some aspects of the underlying microstructure as illustrated by classical homogenisation methods used to derive the effective properties of composites. However generalised continua incorporate a feature of the microstructure which is not accounted for by standard homogenisation methods, namely their size–dependent material response. They involve intrinsic lengths directly stemming from the microstructure of the material. The micromorphic theory now arouses strong interest from the materials science and computational mechanics communities because of its regularisation power in the context of softening plasticity and damage and of its rather simple implementation in a finite element program. The number of degrees of freedom is not an obstacle any more with constantly increasing computer power.

The objective of this chapter is first to present the elastoviscoplasticity theory of micromorphic media at finite deformation. This presentation is based on the fundamental work of Eringen and on recent developments in the context of plasticity. The second part is dedicated to the motivation of higher order degrees of freedom by means of extended homogenisation methods. Finally the question of heterogeneous micromorphic media is addressed, with a view to applications in polycrystalline plasticity.

1.2 Notations

First, second, third, fourth and sixth order tensors are denoted by $\underline{\boldsymbol{A}}$, $\underset{\sim}{\boldsymbol{A}}$, $\underset{\simeq}{\boldsymbol{A}}$, $\underset{\approx}{\boldsymbol{A}}$ and $\underset{\underset{\sim}{\approx}}{\boldsymbol{A}}$ respectively. Their components will be considered with respect to a Cartesian basis:

$$\underline{\boldsymbol{A}} = A_i\, \underline{\boldsymbol{e}}_i, \quad \underset{\sim}{\boldsymbol{A}} = A_{ij}\, \underline{\boldsymbol{e}}_i \otimes \underline{\boldsymbol{e}}_j, \quad \underset{\simeq}{\boldsymbol{A}} = \underset{\approx}{\boldsymbol{A}} = A_{ijk}\, \underline{\boldsymbol{e}}_i \otimes \underline{\boldsymbol{e}}_j \otimes \underline{\boldsymbol{e}}_k$$

The following tensor products are defined

$$\underline{\boldsymbol{a}} \otimes \underline{\boldsymbol{b}} = a_i b_j\, \underline{\boldsymbol{e}}_i \otimes \underline{\boldsymbol{e}}_j, \quad \underset{\sim}{\boldsymbol{A}} \otimes \underset{\sim}{\boldsymbol{B}} = A_{ij} B_{kl}\, \underline{\boldsymbol{e}}_i \otimes \underline{\boldsymbol{e}}_j \otimes \underline{\boldsymbol{e}}_k \otimes \underline{\boldsymbol{e}}_l$$

$$\underset{\sim}{\boldsymbol{A}} \boxtimes \underset{\sim}{\boldsymbol{B}} = A_{ik} B_{jl}\, \underline{\boldsymbol{e}}_i \otimes \underline{\boldsymbol{e}}_j \otimes \underline{\boldsymbol{e}}_k \otimes \underline{\boldsymbol{e}}_l$$

The tensor simple and multiple contractions follow the next rules:

$$\underline{\boldsymbol{A}} \cdot \underline{\boldsymbol{B}} = A_i B_i, \quad \underset{\sim}{\boldsymbol{A}} : \underset{\sim}{\boldsymbol{B}} = A_{ij} B_{ij}, \quad \underset{\simeq}{\boldsymbol{A}} \,\vdots\, \underset{\simeq}{\boldsymbol{B}} = A_{ijk} B_{ijk}$$

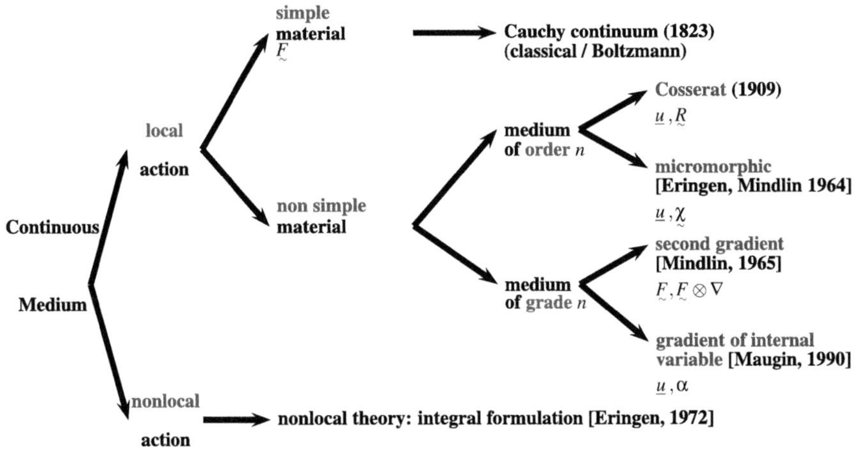

Figure 1. A classification of the mechanics of generalised continua.

For tensor analysis, nabla operators $\boldsymbol{\nabla}_X$ and $\boldsymbol{\nabla}_x$ are defined with respect to the reference and current configurations of the body, respectively.

$$\boldsymbol{\nabla}_x \;=\; _{,i}\, \underline{\boldsymbol{e}}_{\,i} = \frac{\partial}{\partial x_i}\, \underline{\boldsymbol{e}}_{\,i}, \qquad \boldsymbol{\nabla}_X \;=\; _{,J}\, \underline{\boldsymbol{E}}_{\,J} = \frac{\partial}{\partial X_J}\, \underline{\boldsymbol{E}}_{\,J}$$

where $(\underline{\boldsymbol{E}}_{\,J})_{J=1,3}$ and $(\underline{\boldsymbol{e}}_{\,i})_{i=1,3}$ denote the corresponding Cartesian bases. The comma stands for partial derivative with respect to the corresponding coordinate. The following rules are adopted

$$\underline{\boldsymbol{u}} \otimes \boldsymbol{\nabla} = u_{i,j}\, \underline{\boldsymbol{e}}_{\,i} \otimes \underline{\boldsymbol{e}}_{\,j}, \quad \underset{\sim}{\boldsymbol{\sigma}}.\boldsymbol{\nabla} = \sigma_{ij,j}\, \underline{\boldsymbol{e}}_{\,i}$$

2 Micromorphic Continua

2.1 Kinematics of Micromorphic Media

The degrees of freedom of the theory are the displacement vector $\underline{\boldsymbol{u}}$ and the microdeformation tensor $\underset{\sim}{\boldsymbol{\chi}}$:

$$DOF := \{ \;\; \underline{\boldsymbol{u}}, \quad \underset{\sim}{\boldsymbol{\chi}} \;\; \}$$

The current position of the material point is given by the transformation Φ according to $\underline{\boldsymbol{x}} \;=\; \Phi(\underline{\boldsymbol{X}}) = \underline{\boldsymbol{X}} + \underline{\boldsymbol{u}}(\underline{\boldsymbol{X}})$. The microdeformation describes

name	number of DOF	DOF	references
Cauchy	3	$\underline{\boldsymbol{u}}$	Cauchy (1822)
microdilatation	4	$\underline{\boldsymbol{u}}, \quad \chi$	Goodman and Cowin (1972) Steeb and Diebels (2003)
Cosserat	6	$\underline{\boldsymbol{u}}, \quad \underset{\sim}{\boldsymbol{R}}$	Kafadar and Eringen (1971)
microstretch	7	$\underline{\boldsymbol{u}}, \quad \chi, \quad \underset{\sim}{\boldsymbol{R}}$	Eringen (1990)
microstrain	9	$\underline{\boldsymbol{u}}, \quad \underset{\sim}{\boldsymbol{C}}^\sharp$	Forest and Sievert (2006)
micromorphic	12	$\underline{\boldsymbol{u}}, \quad \underset{\sim}{\boldsymbol{\chi}}$	Eringen and Suhubi (1964) Mindlin (1964)

Table 1. A hierarchy of higher order continua.

the deformation of a triad of directors, $\underline{\boldsymbol{\Xi}}^i$ attached to the material point

$$\underline{\boldsymbol{\xi}}^i(\underline{\boldsymbol{X}}) = \underset{\sim}{\boldsymbol{\chi}}(\underline{\boldsymbol{X}}) \cdot \underline{\boldsymbol{\Xi}}^i \tag{1}$$

The polar decomposition of the generally incompatible microdeformation field $\underline{\boldsymbol{\chi}}(\underline{\boldsymbol{X}})$ is introduced

$$\underset{\sim}{\boldsymbol{\chi}} = \underset{\sim}{\boldsymbol{R}}^\sharp \cdot \underset{\sim}{\boldsymbol{U}}^\sharp \tag{2}$$

Internal constraints can be prescribed to the microdeformation. The micromorphic medium reduces to the Cosserat medium when the microdeformation is constrained to be a pure rotation: $\underset{\sim}{\boldsymbol{\chi}} \equiv \underset{\sim}{\boldsymbol{R}}^\sharp$. The microstrain medium is obtained when $\underset{\sim}{\boldsymbol{\chi}} \equiv \underset{\sim}{\boldsymbol{U}}^\sharp$ (Forest and Sievert, 2006). Finally, the second gradient theory is retrieved when the microdeformation coincides with the deformation gradient, $\underset{\sim}{\boldsymbol{\chi}} \equiv \underset{\sim}{\boldsymbol{F}}$. A hierarchy of higher order continua can be established by specialising the micromorphic theory and depending on the targeted material class, see Table 1.

The following kinematical quantities are then introduced:

- the velocity field $\underline{\boldsymbol{v}}(\underline{\boldsymbol{x}}) := \dot{\underline{\boldsymbol{u}}}(\Phi^{-1}(\underline{\boldsymbol{x}}))$

- the deformation gradient $\underset{\sim}{F} = \underset{\sim}{1} + \underline{u} \otimes \nabla_X$
- the velocity gradient $\underline{v} \otimes \nabla_x = \dot{\underset{\sim}{F}} \cdot \underset{\sim}{F}^{-1}$
- the microdeformation rate $\dot{\underset{\sim}{\chi}} \cdot \underset{\sim}{\chi}^{-1}$
- the Lagrangean microdeformation gradient $\underset{\approx}{K} := \underset{\sim}{\chi}^{-1} \cdot \underset{\sim}{\chi} \otimes \nabla_X$
- the gradient of the microdeformation rate tensor

$$(\dot{\underset{\sim}{\chi}} \cdot \underset{\sim}{\chi}^{-1}) \otimes \nabla_x = \underset{\sim}{\chi} \cdot \dot{\underset{\approx}{K}} : (\underset{\sim}{\chi}^{-1} \boxtimes \underset{\sim}{F}^{-1}) \tag{3}$$

and the corresponding index notation:

$$(\dot{\chi}_{il}\chi_{lj}^{-1})_{,k} = \chi_{ip}\dot{K}_{pqr}\chi_{qj}^{-1}F_{rk}^{-1}$$

2.2 Principle of Virtual Power

The method of virtual power is used to introduce the generalised stress tensors and the field and boundary equations they must satisfy (Germain, 1973).
The modelling variables are introduced according to a first gradient theory:

$$MODEL = \{\ \underline{v},\quad \underline{v} \otimes \nabla_x,\quad \dot{\underset{\sim}{\chi}} \cdot \underset{\sim}{\chi}^{-1},\quad (\dot{\underset{\sim}{\chi}} \cdot \underset{\sim}{\chi}^{-1}) \otimes \nabla_x\ \}$$

The virtual power of internal forces of a subdomain $\mathcal{D} \subset \mathcal{B}$ of the body is

$$\mathcal{P}^{(i)}(\underline{v}^*, \dot{\underset{\sim}{\chi}}^* \cdot \underset{\sim}{\chi}^{*-1}) = \int_{\mathcal{D}} p^{(i)}(\underline{v}^*, \dot{\underset{\sim}{\chi}}^* \cdot \underset{\sim}{\chi}^{*-1})\, dV$$

The virtual power density of internal forces is a linear form on the fields of virtual modeling variables:

$$\begin{aligned} p^{(i)} &= \underset{\sim}{\sigma} : (\dot{\underset{\sim}{F}} \cdot \underset{\sim}{F}^{-1}) + \underset{\sim}{s} : (\dot{\underset{\sim}{F}} \cdot \underset{\sim}{F}^{-1} - \dot{\underset{\sim}{\chi}} \cdot \underset{\sim}{\chi}^{-1}) + \underset{\approx}{M} \vdots ((\dot{\underset{\sim}{\chi}} \cdot \underset{\sim}{\chi}^{-1}) \otimes \nabla_x) \\ &= \underset{\sim}{\sigma} : (\dot{\underset{\sim}{F}} \cdot \underset{\sim}{F}^{-1}) + \underset{\sim}{s} : (\underset{\sim}{\chi} \cdot (\underset{\sim}{\chi}^{-1} \cdot \underset{\sim}{F})^{\cdot} \cdot \underset{\sim}{F}^{-1}) \\ &+ \underset{\approx}{M} \vdots \left(\underset{\sim}{\chi} \cdot \dot{\underset{\approx}{K}} : (\underset{\sim}{\chi}^{-1} \boxtimes \underset{\sim}{F}^{-1})\right) \end{aligned} \tag{4}$$

where the relative deformation rate $\dot{\underset{\sim}{F}} \cdot \underset{\sim}{F}^{-1} - \dot{\underset{\sim}{\chi}} \cdot \underset{\sim}{\chi}^{-1}$ is introduced and expressed in terms of the rate of the relative deformation $\underset{\sim}{\chi}^{-1} \cdot \underset{\sim}{F}$. The virtual power density of internal forces is invariant with respect to virtual rigid body motions so that $\underset{\sim}{\sigma}$ must be symmetric. The generalised stress tensors conjugate to the velocity gradient, the relative deformation rate and the gradient of the microdeformation rate are the simple stress tensor $\underset{\sim}{\sigma}$, the relative stress tensor $\underset{\sim}{s}$ and the double stress $\underset{\approx}{M}$.

The Gauss theorem is then applied to the power of internal forces

$$\begin{aligned}\int_{\mathcal{D}} p^{(i)}\, dV &= \int_{\partial\mathcal{D}} \underline{\boldsymbol{v}}^* \cdot (\underset{\sim}{\boldsymbol{\sigma}} + \underset{\sim}{\boldsymbol{s}}) \cdot \underline{\boldsymbol{n}}\, dS + \int_{\partial\mathcal{D}} (\dot{\underset{\sim}{\boldsymbol{\chi}}}^* \cdot \underset{\sim}{\boldsymbol{\chi}}^{*-1}) : \underset{\approx}{\boldsymbol{M}} \cdot \underline{\boldsymbol{n}}\, dS \\ &- \int_{\mathcal{D}} \underline{\boldsymbol{v}}^* \cdot (\underset{\sim}{\boldsymbol{\sigma}} + \underset{\sim}{\boldsymbol{s}}) \cdot \boldsymbol{\nabla}_x\, dV - \int_{\mathcal{D}} (\dot{\underset{\sim}{\boldsymbol{\chi}}}^* \cdot \underset{\sim}{\boldsymbol{\chi}}^{*-1}) : (\underset{\approx}{\boldsymbol{M}} \cdot \boldsymbol{\nabla}_x + \underset{\sim}{\boldsymbol{s}})\, dV\end{aligned}$$

The form of the previous boundary integral dictates the form of the power of contact forces acting on the boundary $\partial\mathcal{D}$ of the subdomain $\mathcal{D} \subset \mathcal{B}$

$$\mathcal{P}^{(c)}(\underline{\boldsymbol{v}}^*, \dot{\underset{\sim}{\boldsymbol{\chi}}}^* \cdot \underset{\sim}{\boldsymbol{\chi}}^{*-1}) = \int_{\partial\mathcal{D}} p^{(c)}(\underline{\boldsymbol{v}}^*, \dot{\underset{\sim}{\boldsymbol{\chi}}}^* \cdot \underset{\sim}{\boldsymbol{\chi}}^{*-1})\, dV$$

$$p^{(c)}(\underline{\boldsymbol{v}}^*, \dot{\underset{\sim}{\boldsymbol{\chi}}}^* \cdot \underset{\sim}{\boldsymbol{\chi}}^{*-1}) = \underline{\boldsymbol{t}} \cdot \underline{\boldsymbol{v}}^* + \underset{\sim}{\boldsymbol{m}} : (\dot{\underset{\sim}{\boldsymbol{\chi}}}^* \cdot \underset{\sim}{\boldsymbol{\chi}}^{*-1})$$

where the simple traction $\underline{\boldsymbol{t}}$ and double traction $\underset{\sim}{\boldsymbol{m}}$ are introduced.

The power of forces acting at a distance is defined as

$$\mathcal{P}^{(e)}(\underline{\boldsymbol{v}}^*, \dot{\underset{\sim}{\boldsymbol{\chi}}}^* \cdot \underset{\sim}{\boldsymbol{\chi}}^{*-1}) = \int_{\mathcal{D}} p^{(e)}(\underline{\boldsymbol{v}}^*, \dot{\underset{\sim}{\boldsymbol{\chi}}}^* \cdot \underset{\sim}{\boldsymbol{\chi}}^{*-1})\, dV$$

$$p^{(e)}(\underline{\boldsymbol{v}}^*, \dot{\underset{\sim}{\boldsymbol{\chi}}}^* \cdot \underset{\sim}{\boldsymbol{\chi}}^{*-1}) = \underline{\boldsymbol{f}} \cdot \underline{\boldsymbol{v}}^* + \underset{\sim}{\boldsymbol{p}} : (\dot{\underset{\sim}{\boldsymbol{\chi}}}^* \cdot \underset{\sim}{\boldsymbol{\chi}}^{*-1})$$

including simple body forces $\underline{\boldsymbol{f}}$ and double body forces $\underset{\sim}{\boldsymbol{p}}$. More general double and triple volume forces could also be incorporated according to Germain (1973).

The principle of virtual power is now stated in the static case,

$$\begin{aligned}\forall \underline{\boldsymbol{v}}^*, \forall \underset{\sim}{\boldsymbol{\chi}}^*, \forall \mathcal{D} \subset \mathcal{B},\ \mathcal{P}^{(i)}(\underline{\boldsymbol{v}}^*, \dot{\underset{\sim}{\boldsymbol{\chi}}}^* \cdot \underset{\sim}{\boldsymbol{\chi}}^{*-1}) &= \mathcal{P}^{(c)}(\underline{\boldsymbol{v}}^*, \dot{\underset{\sim}{\boldsymbol{\chi}}}^* \cdot \underset{\sim}{\boldsymbol{\chi}}^{*-1}) \\ &+ \mathcal{P}^{(e)}(\underline{\boldsymbol{v}}^*, \dot{\underset{\sim}{\boldsymbol{\chi}}}^* \cdot \underset{\sim}{\boldsymbol{\chi}}^{*-1})\end{aligned}$$

This variational formulation leads to

$$\int_{\partial\mathcal{D}} \underline{\boldsymbol{v}}^* \cdot (\underset{\sim}{\boldsymbol{\sigma}} + \underset{\sim}{\boldsymbol{s}}) \cdot \underline{\boldsymbol{n}}\, dS + \int_{\partial\mathcal{D}} (\dot{\underset{\sim}{\boldsymbol{\chi}}}^* \cdot \underset{\sim}{\boldsymbol{\chi}}^{-1}) : \underset{\approx}{\boldsymbol{M}} \cdot \underline{\boldsymbol{n}}\, dS$$

$$- \int_{\mathcal{D}} \underline{\boldsymbol{v}}^* \cdot ((\underset{\sim}{\boldsymbol{\sigma}} + \underset{\sim}{\boldsymbol{s}}) \cdot \boldsymbol{\nabla}_x + \underline{\boldsymbol{f}})\, dV - \int_{\mathcal{D}} (\dot{\underset{\sim}{\boldsymbol{\chi}}}^* \cdot \underset{\sim}{\boldsymbol{\chi}}^{*-1}) : (\underset{\approx}{\boldsymbol{M}} \cdot \boldsymbol{\nabla}_x + \underset{\sim}{\boldsymbol{s}} + \underset{\sim}{\boldsymbol{p}})\, dV = 0$$

which delivers the field equations of the problem (Kirchner and Steinmann, 2005; Lazar and Maugin, 2007; Hirschberger et al., 2007):

- balance of momentum equation (static case)

$$(\underset{\sim}{\boldsymbol{\sigma}} + \underset{\sim}{\boldsymbol{s}}) \cdot \boldsymbol{\nabla}_x + \underline{\boldsymbol{f}} = 0, \quad \forall \underline{\boldsymbol{x}} \in \mathcal{B} \tag{5}$$

- balance of generalized moment of momentum equation (static case)

$$\underset{\approx}{\boldsymbol{M}} \cdot \boldsymbol{\nabla}_x + \underset{\sim}{\boldsymbol{s}} + \underset{\sim}{\boldsymbol{p}} = 0, \quad \forall \underline{\boldsymbol{x}} \in \mathcal{B} \tag{6}$$

- boundary conditions

$$(\underset{\sim}{\boldsymbol{\sigma}} + \underset{\sim}{\boldsymbol{s}}) \cdot \underline{\boldsymbol{n}} = \underline{\boldsymbol{t}}, \quad \forall \underline{\boldsymbol{x}} \in \partial\mathcal{B} \tag{7}$$

$$\underset{\approx}{\boldsymbol{M}} \cdot \underline{\boldsymbol{n}} = \underset{\sim}{\boldsymbol{m}}, \quad \forall \underline{\boldsymbol{x}} \in \partial\mathcal{B} \tag{8}$$

2.3 Elastoviscoplasticity of Micromorphic Media

Elastic-Plastic Decomposition of the Generalised Strain Measures

According to Eringen (1999), the following Lagrangean strain measures are adopted:

$$STRAIN = \{\underset{\sim}{\boldsymbol{C}} := \underset{\sim}{\boldsymbol{F}}^T \cdot \underset{\sim}{\boldsymbol{F}}, \quad \underset{\sim}{\boldsymbol{\Upsilon}} := \underset{\sim}{\boldsymbol{\chi}}^{-1} . \underset{\sim}{\boldsymbol{F}}, \quad \underset{\approx}{\boldsymbol{K}} := \underset{\sim}{\boldsymbol{\chi}}^{-1} . (\underset{\sim}{\boldsymbol{\chi}} \otimes \boldsymbol{\nabla}_X)\}$$

i.e. the Cauchy–Green strain tensor, the relative deformation and the microdeformation gradient.

In the presence of plastic deformation, the question arises of splitting the previous Lagrangean strain measures into elastic and plastic contributions. Following Mandel (1973), a multiplicative decomposition of the deformation gradient is postulated:

$$\underset{\sim}{\boldsymbol{F}} = \underset{\sim}{\boldsymbol{F}}^e \cdot \underset{\sim}{\boldsymbol{F}}^p = \underset{\sim}{\boldsymbol{R}}^e \cdot \underset{\sim}{\boldsymbol{U}}^e \cdot \underset{\sim}{\boldsymbol{F}}^p \tag{9}$$

which defines an intermediate local configuration at each material point, see Fig. 2. Uniqueness of the decomposition requires the suitable definition of directors. Such directors are available in any micromorphic theory.
A multiplicative decomposition of the microdeformation is also considered:

$$\underset{\sim}{\boldsymbol{\chi}} = \underset{\sim}{\boldsymbol{\chi}}^e \cdot \underset{\sim}{\boldsymbol{\chi}}^p = \underset{\sim}{\boldsymbol{R}}^{e\sharp} \cdot \underset{\sim}{\boldsymbol{U}}^{e\sharp} \cdot \underset{\sim}{\boldsymbol{\chi}}^p \tag{10}$$

according to Forest and Sievert (2003, 2006). The uniqueness of the decomposition also requires the suitable definition of directors. As an example, lattice directions in a single crystal are physically relevant directors for an elastoviscoplasticity micromorphic theory, see (Aslan et al., 2011). Finally, a partition rule must also be proposed for the third strain measure, namely the microdeformation gradient. Sansour (1998a,b) introduced an additive decomposition of curvature:

$$\underset{\approx}{\boldsymbol{K}} = \underset{\approx}{\boldsymbol{K}}^e + \underset{\approx}{\boldsymbol{K}}^p \tag{11}$$

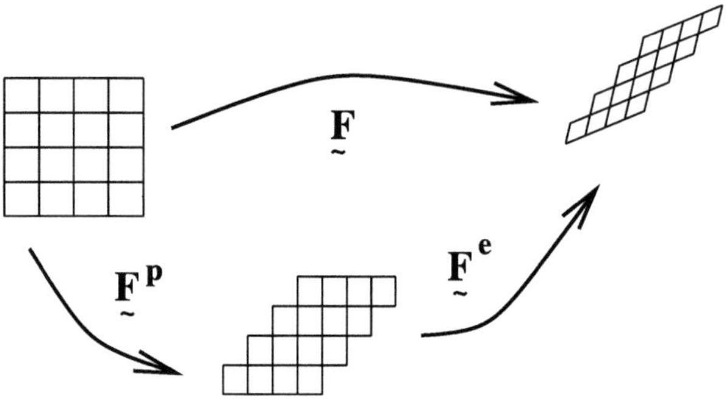

Figure 2. Multiplicative decomposition of the deformation gradient.

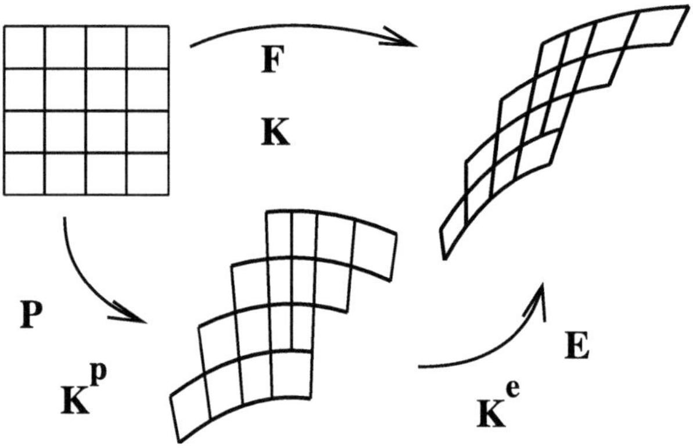

Figure 3. Definition of an intermediate local configuration for micromorphic elastoplasticity.

A quasi–additive decomposition was proposed by Forest and Sievert (2003) with the objective of defining an intermediate local configuration for which all generalised stress tensor are simultaneously released, as it will become apparent in the next section:

$$\underset{\approx}{\boldsymbol{K}} = \underset{\sim}{\boldsymbol{\chi}}^{p-1} . \underset{\simeq}{\boldsymbol{K}}^{e} : (\underset{\sim}{\boldsymbol{\chi}}^{p} \boxtimes \underset{\sim}{\boldsymbol{F}}^{p}) + \underset{\approx}{\boldsymbol{K}}^{p} \tag{12}$$

see Fig. 3.

Constitutive Equations

The continuum thermodynamic formulation is essentially unchanged in the presence of additional degrees of freedom provided that all functionals are properly extended to the new sets of variables. The local equation of energy balance is written in its usual form:

$$\rho\dot{\epsilon} = p^{(i)} - \underline{\boldsymbol{q}}.\boldsymbol{\nabla} + r$$

where ϵ is the specific internal energy density, and $p^{(i)}$ is the power density of internal forces according to Eq. (4). The heat flux vector is $\underline{\boldsymbol{q}}$ and r is a heat source term. The local form of the second principle of thermodynamics is written as

$$\rho\dot{\eta} + \left(\frac{\underline{\boldsymbol{q}}}{T}\right).\boldsymbol{\nabla} - \frac{r}{T} \geq 0$$

where η is the specific entropy density. Introducing the Helmholtz free energy function ψ, the second law becomes

$$p^{(i)} - \rho\dot{\Psi} - \eta\dot{T} - \frac{\underline{\boldsymbol{q}}}{T}.(\boldsymbol{\nabla}T) \geq 0$$

The state variables of the elastoviscoplastic micromorphic material are all the elastic strain measures and a set of internal variables q. The free energy density is a function of the state variables:

$$\Psi(\underset{\sim}{\boldsymbol{C}}^e := \underset{\sim}{\boldsymbol{F}}^{eT}.\underset{\sim}{\boldsymbol{F}}^e, \quad \underset{\sim}{\boldsymbol{\Upsilon}}^e := \underset{\sim}{\boldsymbol{\chi}}^{e-1} \cdot \underset{\sim}{\boldsymbol{F}}^e, \quad \underset{\approx}{\boldsymbol{K}}^e, \quad q)$$

The exploitation of the entropy inequality leads to the definition of the hyperelastic state laws in the form:

$$\underset{\sim}{\boldsymbol{\sigma}} = 2\underset{\sim}{\boldsymbol{F}}^e \cdot \rho\frac{\partial\Psi}{\partial\underset{\sim}{\boldsymbol{C}}^e} \cdot \underset{\sim}{\boldsymbol{F}}^{eT}, \quad \underset{\sim}{\boldsymbol{s}} = \underset{\sim}{\boldsymbol{R}}^{e\sharp} \cdot \underset{\sim}{\boldsymbol{U}}^{e\sharp-1} \cdot \rho\frac{\partial\Psi}{\partial\underset{\sim}{\boldsymbol{\Upsilon}}^e} \cdot \underset{\sim}{\boldsymbol{F}}^{eT}$$

$$\underset{\approx}{\boldsymbol{M}} = \underset{\sim}{\boldsymbol{\chi}}^{-T} \cdot \rho\frac{\partial\Psi}{\partial\underset{\approx}{\boldsymbol{K}}^e} : (\underset{\sim}{\boldsymbol{\chi}}^T \boxtimes \underset{\sim}{\boldsymbol{F}}^T) \qquad (13)$$

while the entropy density is still given by $\eta = -\frac{\partial\Psi}{\partial T}$. The thermodynamic force associated with the internal variable q is

$$R = -\rho\frac{\partial\Psi}{\partial q}$$

The hyperelasticity law (13) for the double stress tensor was derived for the additive decomposition (11). The quasi–additive decomposition (12) leads to an hyperelastic constitutive equation for the conjugate stress $\underset{\approx}{\boldsymbol{M}}$ in the current configuration, that has also the same form as for pure hyperelastic behaviour. One finds:

$$\underset{\approx}{\boldsymbol{M}} = \underset{\sim}{\boldsymbol{\chi}}^{e-T} \cdot \rho \frac{\partial \Psi}{\partial \underset{\approx}{\boldsymbol{K}}^{e}} : (\underset{\sim}{\boldsymbol{\chi}}^{eT} \boxtimes \underset{\sim}{\boldsymbol{F}}^{eT}) \tag{14}$$

The residual intrinsic dissipation is

$$D = \underset{\sim}{\boldsymbol{\Sigma}} : (\dot{\underset{\sim}{\boldsymbol{F}}}^{p}.\underset{\sim}{\boldsymbol{F}}^{p-1}) + \underset{\sim}{\boldsymbol{S}} : (\dot{\underset{\sim}{\boldsymbol{\chi}}}^{p}.\underset{\sim}{\boldsymbol{\chi}}^{p-1}) + \underset{\approx}{\boldsymbol{S}}_{0} \vdots \dot{\underset{\approx}{\boldsymbol{K}}}^{p} + R\dot{q} \geq 0$$

where generalised Mandel stress tensors have been defined

$$\underset{\sim}{\boldsymbol{\Sigma}} = \underset{\sim}{\boldsymbol{F}}^{eT}.(\underset{\sim}{\boldsymbol{\sigma}} + \underset{\sim}{\boldsymbol{s}}).\underset{\sim}{\boldsymbol{F}}^{e-T}, \quad \underset{\sim}{\boldsymbol{S}} = -\underset{\sim}{\boldsymbol{U}}^{e\sharp} \cdot \underset{\sim}{\boldsymbol{R}}^{e\sharp T} \cdot \underset{\sim}{\boldsymbol{s}} \cdot \underset{\sim}{\boldsymbol{R}}^{e\sharp} \cdot \underset{\sim}{\boldsymbol{U}}^{e\sharp-1}$$

$$\underset{\approx}{\boldsymbol{\mathcal{M}}} = \underset{\sim}{\boldsymbol{\chi}}^{T}.\underset{\simeq}{\boldsymbol{S}} : (\underset{\sim}{\boldsymbol{\chi}}^{-T} \boxtimes \underset{\sim}{\boldsymbol{F}}^{-T})$$

At this stage, one may define a dissipation potential, function of the Mandel stress tensors, from which the viscoplastic flow rule and the evolution equations for the internal variables are derived

$$\Omega(\underset{\sim}{\boldsymbol{\Sigma}}, \quad \underset{\sim}{\boldsymbol{S}}, \quad \underset{\approx}{\boldsymbol{S}}_{0})$$

$$\dot{\underset{\sim}{\boldsymbol{F}}}^{p}.\underset{\sim}{\boldsymbol{F}}^{p-1} = \frac{\partial \Omega}{\partial \underset{\sim}{\boldsymbol{\Sigma}}}, \quad \dot{\underset{\sim}{\boldsymbol{\chi}}}^{p}.\underset{\sim}{\boldsymbol{\chi}}^{p-1} = \frac{\partial \Omega}{\partial \underset{\sim}{\boldsymbol{S}}}, \quad \dot{\underset{\approx}{\boldsymbol{K}}}^{p} = \frac{\partial \Omega}{\partial \underset{\approx}{\boldsymbol{\mathcal{M}}}}, \quad \dot{q} = \frac{\partial \Omega}{\partial R}$$

The convexity of the dissipation potential with respect to its arguments ensures the positivity of the dissipation rate at each instant.

Explicit constitutive equations can be found in (Forest and Sievert, 2003; Grammenoudis and Tsakmakis, 2009; Grammenoudis et al., 2009; Regueiro, 2010; Sansour et al., 2010). Examples of application of elastoplastic micromorphic media can be found in (Dillard et al., 2006) for plasticity and failure of metallic foams.

3 From a Heterogeneous Cauchy Material to a Homogeneous Equivalent Micromorphic Medium

Two major obstacles to the use of such sophisticated continuum models are the physical interpretation of the additional degrees of freedom and the identification of the numerous additional material parameters arising in the constitutive functions of the model. Generalised continua are very often referred to as *media with microstructure* without giving precisely the link between the phenomenological constitutive equations and the detailed microstructure of the material. The mechanics of heterogeneous materials and homogenisation methods are widely used to derive the effective properties of classical Cauchy materials based on the description of a representative volume element. Extension of these methods to generalised continua would establish clear definitions of the macroscopic degrees of freedom and provide a systematic way of deriving additional macroscopic materials parameters. Homogenization techniques already exist to construct 1D Cosserat beam models and 2D Mindlin plate models (Altenbach et al., 2010). In the case of 3D generalised continua, it has been proposed in (Gologanu et al., 1997; Forest, 1998, 1999) to construct an effective generalised continuum model starting from a heterogeneous classical Cauchy material by means of extended homogenisation methods.

The present part concentrates on the construction of an overall strain gradient or micromorphic continuum from a microscopic heterogeneous Cauchy material. Such a generalised continuum approach is necessary when significantly high strain gradients develop at the macroscopic scale, more precisely, when the wave length of variation of the macroscopic fields is not sufficiently large compared to the size of the heterogeneities.

For that purpose, quadratic boundary conditions to be applied on a RVE were first proposed in (Gologanu et al., 1997; Forest and Sab, 1998) to construct an effective second gradient and Cosserat overall continuum, respectively. They represent extensions of the classical affine conditions used in classical homogenisation theory (Besson et al., 2009). They were used to identify higher order stiffness, typically bending stiffnesses, that are necessary to account for fiber size effect in composites under significant macroscopic strain gradients, in (Ostoja-Starzewski et al., 1999; Bouyge et al., 2001, 2002; Sansalone et al., 2006; Chen et al., 2009; Anthoine, 2010). Cosserat approaches are particularly well–suited to describe the effective behaviour of civil engineering and granular materials, as shown in (Trovalusci and Masiani, 2003; Goddard, 2008; Salerno and de Felice, 2009; Besdo, 2010).

Such higher order homogenisation schemes have been used in so–called FE^2 methods for which the constitutive model at each material of a computed structure is replaced by the resolution of a boundary value problem on the unit cell of the underlying heterogeneous material. The method is computationally very expensive but makes it possible to address nonlinear problems without writing explicit constitutive laws in the generalised continuum model. In (Feyel, 2003), the Cosserat model is used at the macro–level to represent a fiber matrix composite and the quadratic and cubic boundary conditions proposed by Forest and Sab (1998) are applied to each unit cell. In the references (Geers et al., 2001; Kouznetsova et al., 2002, 2004), the macroscopic medium is regarded as a strain gradient continuum so that quadratic boundary conditions are sufficient. More recently, a micromorphic overall continuum was considered in (Forest, 2002; Jänicke et al., 2009; Jänicke and Diebels, 2009) which represents currently the most general extension of classical homogenisation models.

In most cases however, the proposed extended homogenisation procedures remain heuristic and several questions are still pending: existence of a representative volume element in the presence of non–homogeneous boundary conditions, properties of the local fluctuation field in the case of a polynomial macro–field, as recently addressed by (Yuan et al., 2008; Forest and Trinh, 2011), and the contribution of this fluctuation to the extended Hill–Mandel condition.

The micromorphic theory is used in this part in the small deformation context for simplicity. Capital letters will denote variables attached to the macroscopic homogenized model, whereas small letters will characterize the microstructure level. The representative volume element of the material (RVE), a simple unit cell in the periodic case, is made of a heterogeneous Cauchy continuum characteristic of a composite material. The local coordinate in the unit cell $V(\underline{\boldsymbol{X}})$ with centre $\underline{\boldsymbol{X}}$, is denoted by $\underline{\boldsymbol{x}}$.

The degrees of freedom represented by the generally non–symmetric second order tensor field, $\underset{\sim}{\chi}(\underline{\boldsymbol{X}})$, are introduced in addition to the displacement degrees of freedom, $\underline{\boldsymbol{U}}(\underline{\boldsymbol{X}})$. It is assumed that the development of microdeformation gradient

$$\underset{\approx}{\boldsymbol{K}}(\underline{\boldsymbol{X}}) = \underset{\sim}{\chi}(\underline{\boldsymbol{X}}) \otimes \boldsymbol{\nabla}_X \tag{15}$$

is associated with internal work and energy storage. There is also an energetic price to pay for the microdeformation to depart from the macrodeformation, characterised by the relative deformation measure:

$$\underset{\sim}{e}(\underline{\boldsymbol{X}}) = \underline{\boldsymbol{U}}(\underline{\boldsymbol{X}}) \otimes \boldsymbol{\nabla}_X - \underset{\sim}{\chi}(\underline{\boldsymbol{X}}) \tag{16}$$

The micromorphic model encompasses the strain gradient theory as a limit case if the internal constraint

$$\underset{\sim}{\chi} \equiv \underline{\boldsymbol{U}} \otimes \boldsymbol{\nabla}_X \quad \Longleftrightarrow \quad \underset{\sim}{e} \equiv 0 \tag{17}$$

is enforced (Forest, 2009).

3.1 Definition of the Micromorphic Degrees of Freedom

A kinematic view of the micromorphic model has been proposed by Germain (1973), that we rephrase here. *In a theory which takes microstructure into account, from the macroscopic point of view of continuum mechanics, each particle is still represented by a material point* $\underline{\boldsymbol{X}}$*, but its kinematic* *properties are defined in a more refined way. At the microscopic level of observation, a particle appears itself as a continuum* $V(\underline{\boldsymbol{X}})$ *of small extent. Let us call* $\underline{\boldsymbol{X}}$ *its center of mass and* $\underline{\boldsymbol{x}}$ *a point of* $V(\underline{\boldsymbol{X}})$*. As* $V(\underline{\boldsymbol{X}})$ *is of small extent, it is natural to look at the Taylor expansion of the local displacement* $\underline{\boldsymbol{u}}(\underline{\boldsymbol{x}})$ *with respect to* $\underline{\boldsymbol{x}} - \underline{\boldsymbol{X}}$ *and also, as a first approximation, to stop this expansion with the terms of degree 1:*

$$\underline{\boldsymbol{u}}(\underline{\boldsymbol{x}}, \underline{\boldsymbol{X}}) = \underline{\boldsymbol{U}}(\underline{\boldsymbol{X}}) + \underset{\sim}{\chi} \cdot (\underline{\boldsymbol{x}} - \underline{\boldsymbol{X}}) \tag{18}$$

The physical significance of this assumption is clear: one postulates that one can get a sufficient description of the relative motion of the various points of the particle if one assumes that this relative motion is a homogeneous deformation. For a given local field $\underline{\boldsymbol{u}}(\underline{\boldsymbol{x}}, \underline{\boldsymbol{X}})$, in short $\underline{\boldsymbol{u}}(\underline{\boldsymbol{x}})$, instead of explicitly performing the aforementioned Taylor expansion, it has been proposed in (Forest and Sab, 1998; Forest, 2002; Jänicke et al., 2009) to determine the homogeneous deformation field (18) that is the closest to the actual displacement field, in the sense of the following minimisation problem:

$$\min_{\underline{\boldsymbol{U}}(\underline{\boldsymbol{X}}), \underset{\sim}{\chi}(\underline{\boldsymbol{X}})} \int_{V(\underline{\boldsymbol{X}})} \left|\left| \underline{\boldsymbol{u}}(\underline{\boldsymbol{x}}) - \underline{\boldsymbol{U}}(\underline{\boldsymbol{X}}) - \underset{\sim}{\chi}(\underline{\boldsymbol{X}}) \cdot (\underline{\boldsymbol{x}} - \underline{\boldsymbol{X}}) \right|\right|^2 dV \tag{19}$$

for a given material point $\underline{\boldsymbol{X}}$. The minimisation procedure is straightforward and delivers, taking $\underline{\boldsymbol{X}}$ as the centre of $V(\underline{\boldsymbol{X}})$:

$$\underline{\boldsymbol{U}}(\underline{\boldsymbol{X}}) = <\underline{\boldsymbol{u}}(\underline{\boldsymbol{x}})>_{V(\underline{\boldsymbol{X}})} \tag{20}$$

$$\begin{aligned} \underset{\sim}{\chi}(\underline{\boldsymbol{X}}) &= <(\underline{\boldsymbol{u}}(\underline{\boldsymbol{x}}) - \underline{\boldsymbol{U}}(\underline{\boldsymbol{X}})) \otimes (\underline{\boldsymbol{x}} - \underline{\boldsymbol{X}})>_{V(\underline{\boldsymbol{X}})} \cdot \underset{\sim}{\boldsymbol{A}}^{-1} \\ &= <\underline{\boldsymbol{u}}(\underline{\boldsymbol{x}}) \otimes (\underline{\boldsymbol{x}} - \underline{\boldsymbol{X}})>_{V(\underline{\boldsymbol{X}})} \cdot \underset{\sim}{\boldsymbol{A}}^{-1} \end{aligned} \tag{21}$$

with

$$\underset{\sim}{\boldsymbol{A}} = <(\underline{\boldsymbol{x}} - \underline{\boldsymbol{X}}) \otimes (\underline{\boldsymbol{x}} - \underline{\boldsymbol{X}})>_{V(\underline{\boldsymbol{X}})} \tag{22}$$

The relation (20) is known from classical homogenisation methods and defines the macroscopic displacement as the zeroth moment of the local displacement field. Formula (21) has the merit to unambiguously define the macroscopic micromorphic degrees of freedom as the first moment of the local displacement field, tensor $\underset{\sim}{\boldsymbol{A}}$ being the quadratic moment tensor of the unit cell. This represents an enhancement of the macroscopic description that incorporates additional effects of the microstructure compared to conventional schemes.

If the displacement field is a linear transformation, $\underline{u} = \underset{\sim}{E} \cdot \underline{x}$, the microdeformation is computed as

$$\chi_{ij} =< u_i x_k > A^{-1}_{kj} =< E_{il} x_l x_k > A^{-1}_{kj} = E_{il} < x_l x_k > A^{-1}_{kj} = E_{ij} \quad (23)$$

so that the microdeformation coincides with the macro–deformation $\underset{\sim}{E}$. In particular, if a rigid body motion is applied to the unit cell, the microdeformation will reduce to the applied rotation, as it should be.

3.2 Higher Order Strain Measures

The mechanical theory requires the evaluation of the macroscopic gradients of the degrees of freedom. The macroscopic gradient of the displacement field is still given by the averaging relation:

$$\underline{U} \otimes \nabla_X =< \underline{u} \otimes \nabla_x >_{V(\underline{X})} \quad (24)$$

The gradient of the microdeformation (15) is computed using the definition (21) as follows:

$$\begin{aligned} K_{ijk} &= \frac{\partial}{\partial X_k}\left(< (u_i - U_i)(x_l - X_l) > A^{-1}_{lj}\right) \\ &= < \frac{\partial}{\partial x_k}((u_i - U_i)(x_l - X_l)) > A^{-1}_{lj} \\ &+ < (u_i - U_i)(x_l - X_l) > \frac{\partial}{\partial X_k} A^{-1}_{lj} \\ &= < u_{i,k}(x_l - X_l) > A^{-1}_{lj} + < (u_i - U_i) > A^{-1}_{kj} \\ &+ < (u_i - U_i)(x_l - X_l) > A^{-1}_{lj,k} \end{aligned} \quad (25)$$

Taking (20) into account, and assuming that $\underset{\sim}{A}$ does not vary from material point to material point, the microdeformation gradient takes the simple form:

$$\underset{\approx}{K}^T(\underline{X}) =< \underline{u}(\underline{x}) \otimes \nabla_x \otimes (\underline{x} - \underline{X}) > \cdot \underset{\sim}{A}^{-1}, \quad K_{ijk} =< u_{i,k}(x_l - X_l) > A^{-1}_{lj} \quad (26)$$

where transposition of the third rank tensor is applied to the last two indices. Accordingly, the microdeformation gradient can be interpreted as the first moment of the distribution of the local displacement gradient.

The relative deformation must also be evaluated and takes the form of the difference:

$$\underset{\sim}{e}(\underline{X}) =< \underline{u}(\underline{x}) \otimes \nabla_x >_{V(\underline{X})} - < \underline{u}(\underline{x}) \otimes (\underline{x} - \underline{X}) >_{V(\underline{X})} \cdot \underset{\sim}{A}^{-1} \quad (27)$$

When the displacement field $\underline{u}$ is a linear transformation, including rigid body motions, both the relative deformation and the microdeformation gradient vanish, as it should be.

3.3 Polynomial Ansatz

Quadratic Ansätze have been initially proposed in (Gologanu et al., 1997; Forest, 1998; Kruch and Forest, 1998; Forest and Sab, 1998; Enakoutsa and Leblond, 2009) to extend the usual affine conditions of loading of the material volume element in order to incorporate strain gradient effects in the homogenisation procedure. Such polynomial developments represent an alternative to multiscale asymptotic expansions to derive effective higher order properties (Boutin, 1996), with the advantage that they can be used in a straightforward manner, irrespective of the local linear or nonlinear behaviour of the composite material.

We consider the following polynomial Ansatz of degree 4:

$$u_i^*(\underline{x}) = E_{ij}x_j + \frac{1}{2}D_{ijk}x_jx_k + \frac{1}{3}D_{ijkl}x_jx_kx_l + \frac{1}{4}D_{ijklm}x_jx_kx_lx_m, \forall \underline{x} \in V(0) \tag{28}$$

that is written in the following intrinsic form:

$$\begin{aligned} \underline{u}^*(\underline{x}) &= \underset{\sim}{E} \cdot \underline{x} + \frac{1}{2}\underset{\simeq}{D} : (\underline{x} \otimes \underline{x}) + \frac{1}{3}\underset{\approx}{D} \,\vdots\, (\underline{x} \otimes \underline{x} \otimes \underline{x}) \\ &+ \frac{1}{4}\underset{\approx}{\underline{D}} :: (\underline{x} \otimes \underline{x} \otimes \underline{x} \otimes \underline{x}), \quad \forall \underline{x} \in V(0) \end{aligned} \tag{29}$$

where the coefficients are tensors of ranks 2 to 5.

The macroscopic micromorphic strain measures are now computed successively for such a polynomial field on the reference unit cell $V(0)$:

$$< \underline{u}^* \otimes \boldsymbol{\nabla}_x >_{V(0)} = \underset{\sim}{E} + \underset{\approx}{D} : \underset{\sim}{A} \tag{30}$$

$$\underset{\sim}{\chi} = < \underline{u}^* \otimes \underline{x} >_{V(0)} \cdot \underset{\sim}{A}^{-1} = \underset{\sim}{E} + \frac{1}{3}\underset{\approx}{D} \,\vdots\, \underset{\approx}{A} \cdot \underset{\sim}{A}^{-1}, \chi_{ij} = E_{ij} + \frac{1}{3}D_{ipqr}A_{pqrk}A^{-1}_{kj} \tag{31}$$

$$\underset{\approx}{K}^T = \underset{\approx}{D} + \underset{\approx}{\underline{D}} \,\vdots\, \underset{\approx}{A} \cdot \underset{\sim}{A}^{-1}, K_{ipq} = < u_{i,q}x_r >_{V(0)} A^{-1}_{rp} = D_{iqp} + D_{iqklm}A_{klmr}A^{-1}_{rp} \tag{32}$$

These simple formula hold if the coordinate system is such that $< \underline{x} > = \underline{X} = \underline{0}$, and that the means $< x_i >$, $< x_ix_jx_k >$ and $< x_ix_jx_kx_lx_m >$ identically vanish. The fourth order geometric moment $\underset{\approx}{A}$ of the unit cell has been introduced:

$$\underset{\approx}{A} = < \underline{x} \otimes \underline{x} \otimes \underline{x} \otimes \underline{x} >_{V(0)} \tag{33}$$

It is interesting to notice that the relative deformation is related only to the third order polynomial:

$$\underset{\sim}{e} = \underset{\approx}{D} : \underset{\sim}{A} - \frac{1}{3}\underset{\approx}{D} \vdots \underset{\approx}{A} \cdot \underset{\sim}{A}^{-1} \tag{34}$$

The formula (30) to (32) set direct linear relationships between the coefficients of the polynomial and the strain measures of the effective micromorphic medium. They were used in Jänicke et al. (2009) to prescribe a given curvature $\underset{\approx}{\boldsymbol{K}}$ or relative deformation to the unit cell. However the number of coefficients in the polynomials generally differs from the number of components of the generalised strain measures. For instance, the microdeformation gradient K_{ipq} cannot be controlled solely by the coefficients D_{ipq} of the quadratic polynomial since D_{ipq} is symmetric with respect to the last two indices contrary to K_{ipq}. The selection of the relevant higher order polynomial coefficients remains to be done. In the present contribution, we will only consider the coefficients D_{ijk} and some coefficients of D_{ijkl}.

However, the polynomial (29) will usually not be applied to the whole volume but instead at the boundary ∂V of a given heterogeneous material volume element V:

$$\begin{aligned} \underline{\boldsymbol{u}}(\underline{\boldsymbol{x}}) &= \underset{\sim}{\boldsymbol{E}} \cdot \underline{\boldsymbol{x}} + \frac{1}{2}\underset{\simeq}{\boldsymbol{D}} : (\underline{\boldsymbol{x}} \otimes \underline{\boldsymbol{x}}) + \frac{1}{3}\underset{\approx}{\boldsymbol{D}} \vdots (\underline{\boldsymbol{x}} \otimes \underline{\boldsymbol{x}} \otimes \underline{\boldsymbol{x}}) \\ &+ \frac{1}{4}\underset{\approx}{\boldsymbol{D}} :: (\underline{\boldsymbol{x}} \otimes \underline{\boldsymbol{x}} \otimes \underline{\boldsymbol{x}} \otimes \underline{\boldsymbol{x}}), \quad \forall \underline{\boldsymbol{x}} \in \partial V \end{aligned} \tag{35}$$

In that case, the relation (30) is still valid but (32) must be modified. Note that the microdeformation cannot be controlled from the displacement prescribed at the boundary. The overall microdeformation gradient can be computed knowing the displacements prescribed at the boundary, using the same special coordinate system as previously, and choosing a constant translation such that $\underline{\boldsymbol{U}}(0) = 0$:

$$\begin{aligned} K_{ipq} &= \frac{1}{V}A^{-1}_{rp}\int_{V(0)} u_{i,q}x_r\, dV = \frac{1}{V}A^{-1}_{rp}\int_{V(0)} (u_i x_r)_{,q}\, dV \\ &- \frac{1}{V}A^{-1}_{qp}\int_{V(0)} u_i\, dV \\ &= \frac{1}{V}A^{-1}_{rp}\int_{V(0)} (u_i x_r)_{,q}\, dV = \frac{1}{V}A^{-1}_{rp}\int_{\partial V(0)} u_i x_r n_q\, dS \\ &= \frac{1}{V}A^{-1}_{rp}\int_{\partial V(0)} u^*_i x_r n_q\, dS \\ &= \frac{1}{V}A^{-1}_{rp}\int_{V(0)} u^*_{i,q} x_r\, dV + \frac{1}{V}A^{-1}_{qp}\int_{V(0)} u^*_i\, dV \end{aligned} \tag{36}$$

so that the expression differs from (32) by the mean value of $\underline{u}^*$. We find :

$$\underset{\simeq}{K}^T = \underset{\approx}{D} + \underset{\underset{\sim}{\approx}}{D} \vdots \underset{\approx}{A} \cdot \underset{\sim}{A}^{-1} + \underset{\simeq}{D} : \underset{\sim}{A} \otimes \underset{\sim}{A}^{-1} + \underset{\underset{\sim}{\approx}}{D} :: \underset{\approx}{A} \otimes \underset{\sim}{A}^{-1} \quad (37)$$

$$K_{ipq} = D_{iqp} + D_{iqklm} A_{klmr} A^{-1}_{rp} + D_{ijk} A_{jk} A^{-1}_{qp} + D_{ijklm} A_{jklm} A^{-1}_{qp} \quad (38)$$

Finally, the real local field will be the superposition of the polynomial and of a perturbation

$$\underline{u}(\underline{x}) = \underline{u}^*(\underline{x}) + \underline{v}(\underline{x}), \quad \forall \underline{x} \in V \quad (39)$$

The fluctuation leads to additional contributions to the micromorphic measures that are obtained by substituting $\underline{v}$ to $\underline{u}$ in the formula (21), (26) and (27). These contributions do not vanish in general.

3.4 Identification of Generalised Effective Elastic Moduli

The identification procedure of the higher order elastic moduli is now presented based on an explicit example of a peridic composite material.

Definition of the Chosen Composite Material

The chosen periodic composite material for the evaluation of the extended homogenisation methods is made of a hard isotropic linear elastic phase (h) and a soft isotropic linear elastic phase (s) :

$$E^h = 100000\,\text{MPa}, \quad \nu^h = 0.3, \quad E^s = 500\,\text{MPa}, \quad \nu^s = 0.3$$

The two phases display a contrast of 200 in their Young's modulus. The retained two–dimensional geometry of the unit cell V_0 of the periodic composite is shown in figure 4. It exhibits orthotropic symmetry. The volume fraction of the hard phase is $f^h = 0.424$. The whole microstructure is obtained by plane tessellation in the defined directions 1 and 2.

Identification of Classical Elastic Moduli

Classical periodic homogenisation is used to compute the orthotropic elastic properties of the effective Cauchy material. A constant mean deformation gradient E_{ij} is applied to the unit cell in which the displacement field is of the form :

$$\underline{u}(\underline{x}) = \underset{\sim}{E} \cdot \underline{x} + \underline{v}(\underline{x}) \quad (40)$$

where $\underline{v}$ is the periodic displacement fluctuation taking identical values at homologous points of the boundary ∂V_0 of the unit cell. The effective moduli are determined from the mean elastic energy density induced by

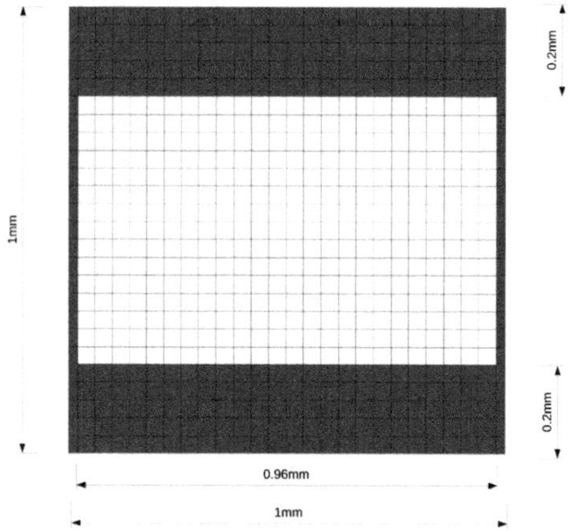

Figure 4. Unit cell V_0 of the periodic composite material. The hard phase is red and the soft phase is blue. The orthotropy axes 1 and 2 are respectively horizontal and vertical.

three successive independent loading conditions, as illustrated in figure 5. Finite element simulations are performed under plane strain conditions. The found moduli are provided in table 2. They are defined in the following matrix form:

$$\begin{bmatrix}\Sigma_{11}\\ \Sigma_{22}\\ \Sigma_{12}\end{bmatrix} = \begin{bmatrix}C_{11} & C_{12} & 0\\ C_{12} & C_{22} & 0\\ 0 & 0 & C_{44}\end{bmatrix} \begin{bmatrix}E_{11}\\ E_{22}\\ 2E_{12}\end{bmatrix} \tag{41}$$

As a comparison, we have also computed the apparent effective moduli when homogeneous deformation boundary conditions are applied to the unit cell, i.e. when the fluctuation is taken to vanish : $\underline{\boldsymbol{v}} = 0, \forall \underline{\boldsymbol{x}} \in \partial V_0$. These boundary conditions are referred to as KUBC, kinematic uniform boundary

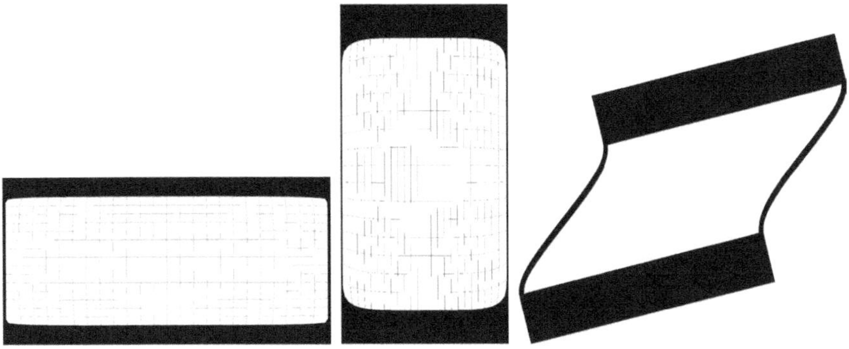

Figure 5. Loading conditions applied to the unit cell for the determination of the effective properties of the homogeneous equivalent Cauchy material. The first, second and third rows correspond to: $E_{11} = 1, E_{22} = 1, E_{12} = E_{21} = 0.25$, respectively. In each case the remaining components of E_{ij} vanish.

	C_{11} (MPa)	C_{12} (MPa)	C_{22} (MPa)	C_{44} (MPa)
periodic	44748	1579	7163	372
KUBC	45707	3181	9920	6186

Table 2. Elastic properties of the effective Cauchy material.

conditions. The corresponding apparent moduli, also given in table 2, are significantly stiffer than effective moduli from periodic homogenisation, as expected (Kanit et al., 2003).

RVE Size for Strain Gradient Overall Properties

The influence of the fluctuation type introduced in the boundary conditions in the computations of the previous section clearly shows that there is undoubtedly a boundary layer effect due to the polynomial boundary conditions, see also (Forest and Trinh, 2011). To get rid of the boundary layer effect, it is proposed to consider volume elements containing an increasing number of unit cells, typically a collection of NxN unit cells, with $N = 1, 3, 5..$ up to $N = 27$ in the following simulations. We look for the size N for which the energy distribution in the bulk of the sample, defined as a zone of fixed size MxM, does not vary any more when the polynomial

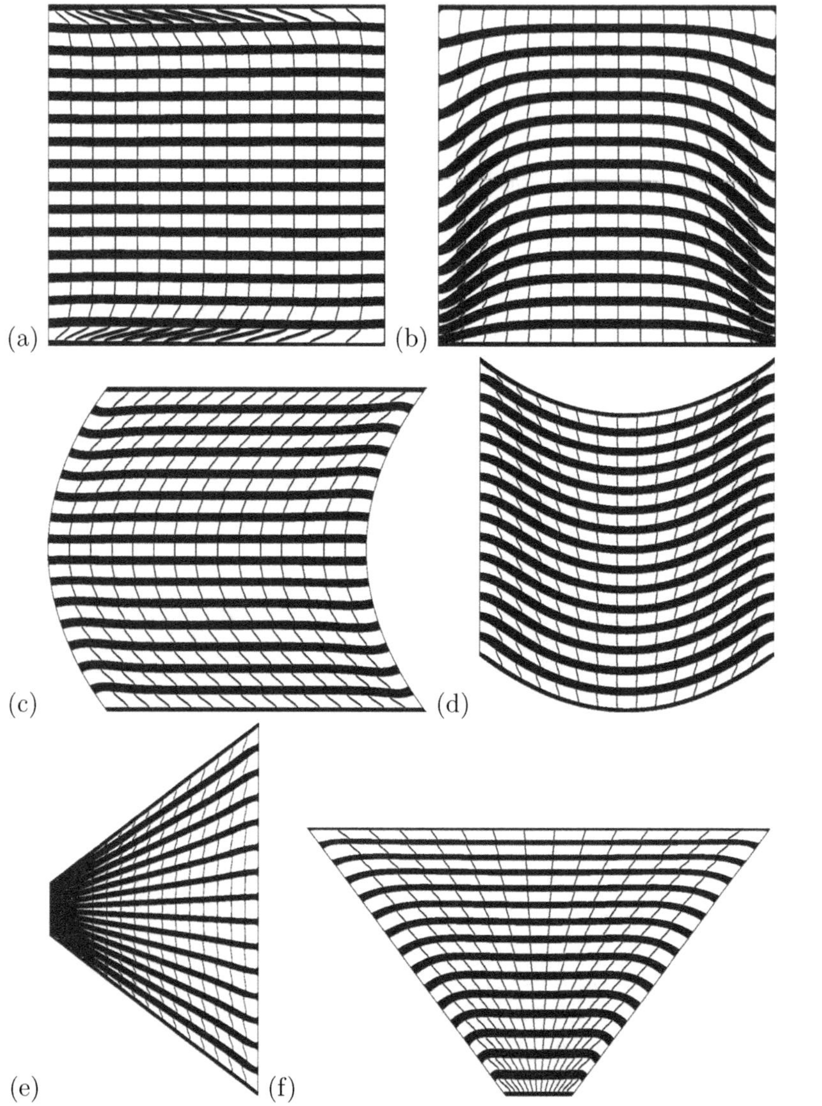

Figure 6. Deformation of a 15×15–cell volume elements corresponding to the following Dirichlet conditions at the outer boundary: (a) $D_{111} : \underline{\boldsymbol{u}} = 1/2 x_1^2 \underline{\boldsymbol{e}}_1$, (b) $D_{222} : \underline{\boldsymbol{u}} = 1/2 x_2^2 \underline{\boldsymbol{e}}_2$, (c) $D_{122} : \underline{\boldsymbol{u}} = 1/2 x_2^2 \underline{\boldsymbol{e}}_1$, (d) $D_{211} : \underline{\boldsymbol{u}} = 1/2 x_1^2 \underline{\boldsymbol{e}}_2$, (e) $D_{212} : \underline{\boldsymbol{u}} = x_1 x_2 \underline{\boldsymbol{e}}_2$, (f) $D_{112} : \underline{\boldsymbol{u}} = x_1 x_2 \underline{\boldsymbol{e}}_1$.

boundary conditions are applied at the remote boundary with the same given values of the polynomial coefficients. The obtained size will be called the RVE size for the considered polynomial conditions. In particular, the attention is focused on the energy density distribution inside the central unit cell ($M = 1$).

In the case of affine boundary conditions used for classical homogenisation, such a procedure is known to lead to a stabilized periodic stress–strain field in the bulk of the volume element. In particular, the fluctuation at the boundary of a unit cell, defined as the difference between the displacement field and the affine contribution, is then found to be periodic. This is no longer the case for quadratic conditions (Forest and Trinh, 2011).

For more general polynomial Dirichlet conditions prescribed at the outer surface, we can investigate the convergence of the mechanical fields for an increasing window size. We also define, in a similar way, the fluctuation $\underline{\boldsymbol{v}}$ and examine its properties at the boundary of the central unit cell. This program has been performed in the reference (Forest and Trinh, 2011) for a cubic grid–like composite material for quadratic polynomials. We apply it to the orthotropic microstructure of figure 4 considered in this work. We use it also to determine the corresponding overall second gradient properties and compare them with the estimations based on an *a priori* choice of the fluctuation. The analysis is limited to the quadratic polynomial term. That is why only strain gradient properties will be identified and not the full micromorphic ones.

The six 2D deformation modes corresponding to a full quadratic polynomial in equation (35): $D_{111}, D_{222}, D_{122}, D_{211}, D_{212}, D_{112}$ are considered. The associated deformed 15x15–cell volume elements are shown in figure 6. The converged shapes of the central unit cell extracted from the previous volume elements are given in figure 6, for the same magnification. The modes D_{111}, D_{222} and D_{122} induce only limited deformation in the central unit cell whereas $D_{211}, D_{212}, D_{112}$ involve significant straining. The elastic energy density levels $< \underset{\sim}{\boldsymbol{\sigma}} : \underset{\sim}{\boldsymbol{\varepsilon}} >_{V_0}$ over the central unit cell V_0 associated with these six modes are given in table 3 depending on the size N of the volume element. Convergence to finite energy values is obtained for the modes $D_{211}, D_{212}, D_{112}$ whereas the material turns out to be insensitive to the modes D_{111}, D_{222} and D_{122}.

The displayed convergence for the considered collection of cells ensures that a representative size has been reached. However, quite a large number of cells is necessary to detect the energy–free modes. Detailed analysis confirms that the fluctuation corresponding to the central unit cell response is not periodic, as pointed out by Forest and Trinh (2011).

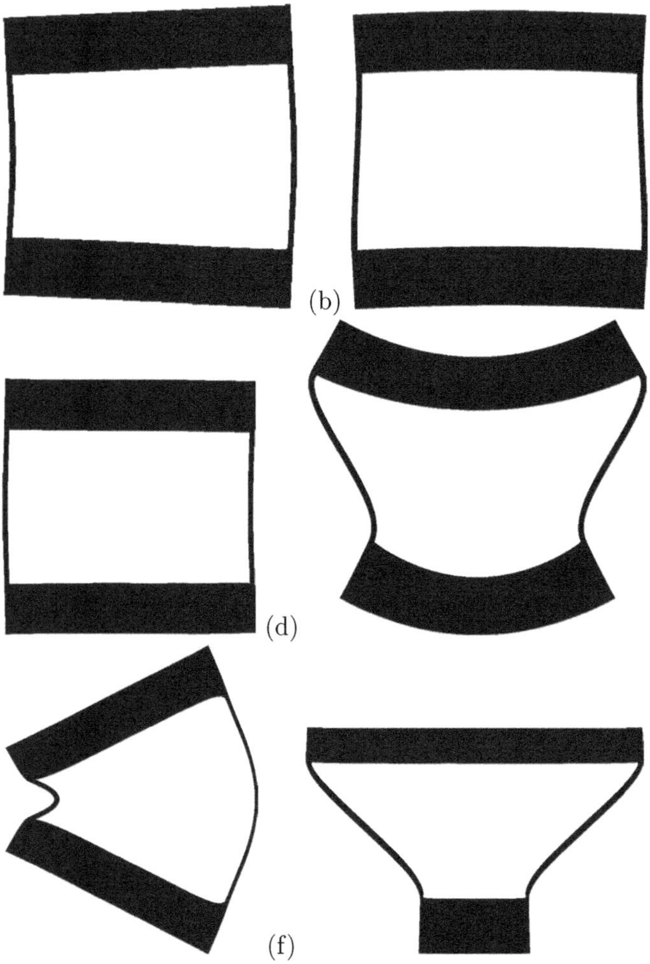

(a) (b) (c) (d) (e) (f)

Figure 7. Shape of the central cell of a 15x15 volume element subjected to the following Dirichlet boundary conditions: (a) $D_{111} : \underline{\boldsymbol{u}} = 1/2x_1^2\underline{\boldsymbol{e}}_1$, (b) $D_{222} : \underline{\boldsymbol{u}} = 1/2x_2^2\underline{\boldsymbol{e}}_2$, (c) $D_{122} : \underline{\boldsymbol{u}} = 1/2x_2^2\underline{\boldsymbol{e}}_1$, (d) $D_{211} : \underline{\boldsymbol{u}} = 1/2x_1^2\underline{\boldsymbol{e}}_2$, (e) $D_{212} : \underline{\boldsymbol{u}} = x_1x_2\underline{\boldsymbol{e}}_2$, (f) $D_{112} : \underline{\boldsymbol{u}} = x_1x_2\underline{\boldsymbol{e}}_1$.

NxN-cell	$D_{111}=1$	$D_{122}=1$	$D_{212}=1$	$D_{112}=1$	$D_{211}=1$	$D_{222}=1$
3x3	3	19	1033	527	368	324
7x7	2	0.12	789	6176	660	227
9x9	1.3	0.3	761	6079	565	89
11x11	0.9	0.4	759	5930	474	27
15x15	0.5	0.33	770	5714	371	1.8
21x21	0.4	0.32	776	5587	325	0.2
27x27	0.33	0.32	777	5548	315	0.2

Table 3. Average elastic energy density in the central unit of NxN–cell volume elements submitted to quadratic Dirichlet boundary conditions. The components D_{ijk} are given in mm^{-1} and the elastic energy values are in MPa.

Identification of Second Gradient Effective Elastic Moduli

The quadratic polynomial loading conditions D_{ijk} can be used to identify the elastic properties of an overall second gradient medium. The simple force stress tensor $\underset{\sim}{\boldsymbol{\Sigma}}$ is still related to the strain tensor $\underset{\sim}{\boldsymbol{E}}$ by the moduli (41). In a medium exhibiting point symmetry, the double stress tensor $M_{ijk}=M_{ikj}$ is linearly related to the second gradient of displacement $K_{ijk}=K_{ikj}$ by the matrix of double elasticity moduli. The structure of anisotropic six rank tensors of strain gradient elasticity was analysed by Auffray et al. (2009, 2010). In the most general situation the associated matricial representation is written:

$$\mathbf{M} = [\mathbf{A}]\mathbf{K} \tag{42}$$

with

$$\mathbf{M} = \begin{bmatrix} M_{111} \\ M_{122} \\ \sqrt{2}M_{212} \\ M_{222} \\ M_{211} \\ \sqrt{2}M_{121} \end{bmatrix}, \qquad \mathbf{K} = \begin{bmatrix} K_{111} \\ K_{122} \\ K_{212} \\ K_{222} \\ K_{211} \\ K_{121} \end{bmatrix}$$

$$[\mathbf{A}] = \begin{bmatrix} A_{111111} & A_{111122} & \sqrt{2}A_{111212} & A_{111222} & A_{111211} & \sqrt{2}A_{111121} \\ A_{122111} & A_{122122} & \sqrt{2}A_{122212} & A_{122222} & A_{122211} & \sqrt{2}A_{122121} \\ \sqrt{2}A_{212111} & \sqrt{2}A_{212122} & 2A_{212122} & \sqrt{2}A_{212222} & \sqrt{2}A_{212211} & 2A_{212121} \\ A_{222111} & A_{222122} & \sqrt{2}A_{222212} & A_{222222} & A_{222211} & \sqrt{2}A_{222121} \\ A_{211111} & A_{211122} & \sqrt{2}A_{211212} & A_{211222} & A_{211211} & \sqrt{2}A_{211121} \\ \sqrt{2}A_{121111} & \sqrt{2}A_{121122} & 2A_{121212} & \sqrt{2}A_{121222} & \sqrt{2}A_{121211} & 2A_{121121} \end{bmatrix}$$

This notation, using square root of two before K_{212} and M_{212}, defines a true second order tensorial representation of the sixth-order tensor of double elasticity. Ranking the components of the second gradient of displacement as proposed in the former matricial representation, leads, in the orthotropic case, to the uncoupled system:

$$\begin{bmatrix} M_1 \\ M_2 \\ M_3 \\ M_4 \\ M_5 \\ M_6 \end{bmatrix} = \begin{bmatrix} A_{11} & A_{12} & A_{13} & 0 & 0 & 0 \\ A_{12} & A_{22} & A_{23} & 0 & 0 & 0 \\ A_{13} & A_{23} & A_{33} & 0 & 0 & 0 \\ 0 & 0 & 0 & A_{44} & A_{45} & A_{46} \\ 0 & 0 & 0 & A_{45} & A_{55} & A_{56} \\ 0 & 0 & 0 & A_{46} & A_{56} & A_{66} \end{bmatrix} \begin{bmatrix} K_1 \\ K_2 \\ K_3 \\ K_4 \\ K_5 \\ K_6 \end{bmatrix} \tag{43}$$

with the simplified notations :

$$[K_1\ K_2\ K_3\ K_4\ K_5\ K_6] = \left[K_{111}\ K_{122}\ \sqrt{2}K_{212}\ K_{222}\ K_{211}\ \sqrt{2}K_{121}\right],$$

$$[M_1\ M_2\ M_3\ M_4\ M_5\ M_6] = \left[M_{111}\ M_{122}\ \sqrt{2}M_{212}\ M_{222}\ M_{211}\ \sqrt{2}M_{121}\right]$$

and

$$[A_{11} \quad A_{12} \quad A_{13} \quad A_{22} \quad A_{23} \quad A_{33}] =$$

$$\left[A_{111111} \quad A_{111122} \quad \sqrt{2}A_{111212} \quad A_{122122} \quad \sqrt{2}A_{122212} \quad 2A_{212212}\right],$$

$$[A_{44} \quad A_{45} \quad A_{46} \quad A_{55} \quad A_{56} \quad A_{66}] =$$

$$\left[A_{222222} \quad A_{222211} \quad \sqrt{2}A_{222121} \quad A_{211211} \quad \sqrt{2}K_{211121} \quad 2K_{121121}\right]$$

This makes 12 independent double elasticity moduli to be identified from the analysis of the response of the unit cell to non-homogeneous loading conditions. Twelve loading conditions are needed to identify them corresponding to twelve sets of the values of the coefficients D_{ijk}. The six selected loading conditions are labeled (a, b, c, d, e, f) for the identification of the first block of 6 constants in the matrix (43), taking advantage of the orthotropic symmetry of the material. Six additional ones are needed for the second block. For each loading, the post–processing procedure yields the mean energy density $2\epsilon =< \underset{\sim}{\sigma} : \underset{\sim}{\varepsilon} >_{V_0}$ in the unit cell and the overall curvature K_1, K_2 and K_3. The mean energy density is related to the overall energy density in the form:

$$2\epsilon = \begin{bmatrix} K_1 \\ K_2 \\ K_3 \\ K_4 \\ K_5 \\ K_6 \end{bmatrix}^T \begin{bmatrix} A_{11} & A_{12} & A_{13} & 0 & 0 & 0 \\ A_{12} & A_{22} & A_{23} & 0 & 0 & 0 \\ A_{13} & A_{23} & A_{33} & 0 & 0 & 0 \\ 0 & 0 & 0 & A_{44} & A_{45} & A_{46} \\ 0 & 0 & 0 & A_{45} & A_{55} & A_{56} \\ 0 & 0 & 0 & A_{46} & A_{56} & A_{66} \end{bmatrix} \begin{bmatrix} K_1 \\ K_2 \\ K_3 \\ K_4 \\ K_5 \\ K_6 \end{bmatrix} \tag{44}$$

A_{11} (MPa.mm^2)	A_{22} (MPa.mm^2)	A_{33} (MPa.mm^2)	A_{12} MPa.mm^2	A_{23} MPa.mm^2	A_{13} MPa.mm^2
134601	37436	124 548	68706	67368	127 213
A_{44}	A_{55}	A_{66}	A_{45}	A_{46}	A_{56}
69445	2801	32 175	40762	11 094	7 548

Table 4. Higher order elastic properties of the overall second-gradient material for the unit cell of figure 4(a). The fluctuation is taken to vanish at the unit cell boundary.

The found higher order moduli are listed in table 4 for a vanishing fluctuation $\underline{v}$ in (35) at the boundary of the unit cell V_0. We have not determined the effective moduli corresponding to the converged states of the unit cell embedded in a NxN–cell volume element in the sense of section 3.4, because zero–energy modes were detected as discussed above, so that the previous system of equations is undetermined. A specific procedure is necessary to determine the vanishing terms of the overall matrix.

3.5 Validation of the Extended Homogenisation Method

The performance of the generalised overall properties determined in the previous section is evaluated by considering a reference problem for a structure made of a small number of unit cells of the type of 4(a). The limitation of the Cauchy continuum is first illustrated and improvements by means of a strain gradient substitution medium are presented.

We consider the composite structure made of 10x5 cells of figure 8 (left). The following boundary value problem is considered on this structure. The left side of the structure is clamped, meaning that $U_1 = U_2 = 0$. The horizontal lower and upper sides are free of forces. The vertical displacement component $U_2 = 1$ mm is prescribed on the right side, the component U_1 being left free. The corresponding deformed shape of the structure is shown in figure 8 (right). It displays a combination of pure shear and bending modes in a boundary layer on the left side.

The same boundary value problem is considered for a homogeneous substitution Cauchy medium endowed with the elastic properties of table 2. The same clamping boundary conditions $U_1 = U_2 = 0$ are prescribed on the left side. The bottom picture of Fig. 8 shows that the Cauchy medium does not capture the bending mode of the composite structure and only provides the shearing mode. This fact had already been noticed for laminates in (Forest and Sab, 1998; Forest and Trinh, 2011).

When the structure is made of a homogeneous second gradient medium endowed with the properties of table 4, the deformed state and quantitative comparison in figure 9 show that the strain gradient effective medium fully captures the actual shear and bending modes. In figure 9, the displacement profile $U_2(x_1)$ is given along the horizontal line close to the mid–section of the structure, as drawn in figure 8.

The simulation for second gradient elasticity is made by means of a micromorphic formulation for which penalty terms ensure that the microdeformation coincides with the gradient of the displacement field. Clamping was imposed through the prescription of vanishing microdeformation on the left side. Note that the choice of the additional boundary conditions on the left side is quite heuristic, as it is the case in most beam and plate models.

Note that an effective Cosserat continuum, for which the effective moduli can be determined in the same way, performs as good as the stain gradient model as shown in Fig. 9. More elaborate examples of loading must be developed in the future to select the best–suited generalised homogeneous equivalent continuum.

4 Homogenization of Micromorphic Media

The homogenisation schemes in this section must be clearly distinguished from the one of previous section since we consider here a generalized continuum model at both the microscopic and macroscopic levels. For instance, homogenisation of Cosserat composites were considered in (Forest et al., 2001; Liu and Hu, 2003; Xun et al., 2004).

The motivation for the development of homogenisation methods for mixtures of micromorphic media is mainly related to crystal plasticity. The mechanical behaviour of metallic polycrystals is notably size dependent and the conventional crystal plasticity framework fails at convincingly predict grain and precipitate size effects (Aslan et al., 2011). Single crystals can be regarded as Cosserat, strain gradient or micromorphic continua. It follows that a polycrystal is a heterogeneous generalised continuum for which specific homogenisation methods must be designed.

4.1 Multiscale Asymptotic Expansion Method

In contrast to the previous part, the heterogeneous medium is now a mixture of micromorphic constituents, i.e. a heterogeneous micromorphic medium. One investigates the nature of the resulting homogeneous equivalent medium by means of asymptotic methods. The multiscale asymptotic method by Sanchez-Palencia (1974) is especially adequate for this purpose since, in contrast to the work done in the previous part, the nature of the

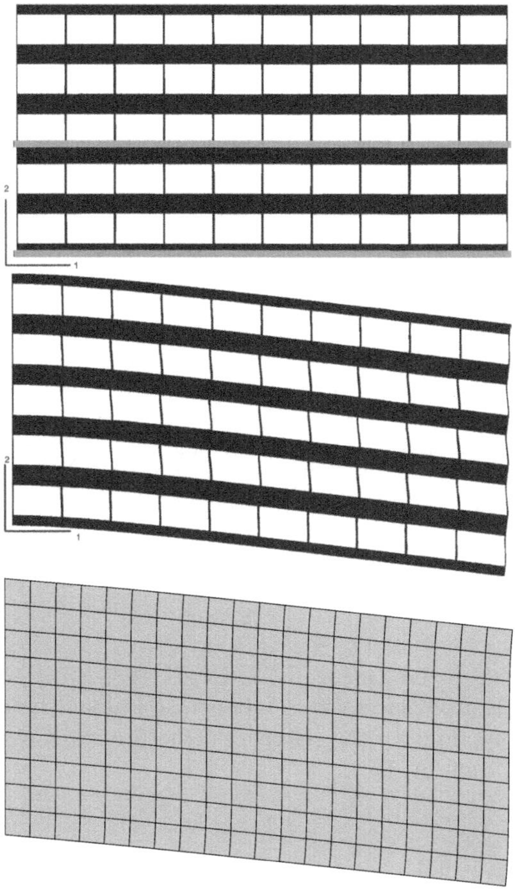

Figure 8. Reference composite structure made of 10x5 cells (left) and reference deformed shape of the structure. Two horizontal lines are shown on the structure for post–processing purposes. The bottom figure shows the deformed state predicted by the homogeneous equivalent Cauchy continuum.

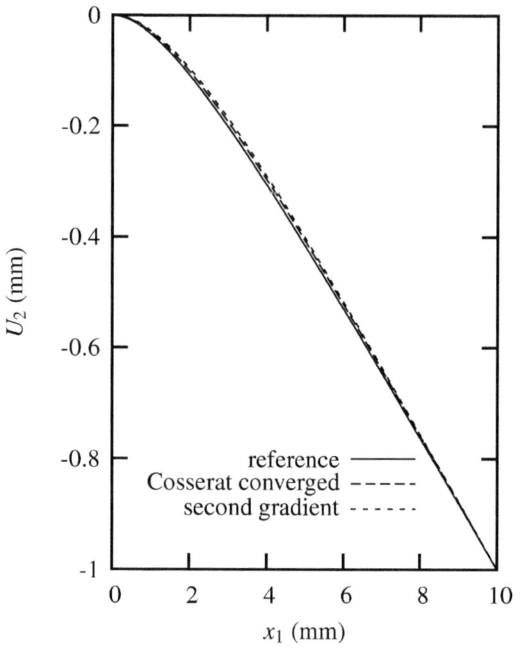

Figure 9. Vertical displacement component U_2 along the mid–line visible in figure 8 as computed for the reference structure and Cosserat and strain gradient substitution media.

effective medium is not postulated *a priori* but rather is the result of the analysis.

The balance and constitutive equations of the micromorphic continuum are recalled briefly in the linear elastic framework for which the asymptotic methods can be applied in a straightforward manner (Forest et al., 2001) The motion of a micromorphic body Ω is described by two independent sets of degrees of freedom : the displacement $\underline{u}$ and the micro–deformation $\underset{\sim}{\chi}$ attributed to each material point. The micro–deformation accounts for the rotation and distorsion of a triad associated with the underlying microstructure Eringen (1999). The micro–deformation can be split into its symmetric and skew–symmetric parts :

$$\underset{\sim}{\chi} = \underset{\sim}{\chi}^s + \underset{\sim}{\chi}^a \tag{45}$$

that are called respectively the micro–strain and the Cosserat rotation. The associated deformation fields are the classical strain tensor $\underset{\sim}{\varepsilon}$, the relative deformation $\underset{\sim}{e}$ and the micro–deformation gradient tensor $\underset{\approx}{\kappa}$ defined by :

$$\underset{\sim}{\varepsilon} = \underline{u} \overset{s}{\otimes} \nabla, \quad \underset{\sim}{e} = \underline{u} \otimes \nabla - \underset{\sim}{\chi}, \quad \underset{\approx}{\kappa} = \underset{\sim}{\chi} \otimes \nabla \tag{46}$$

The symmetric part of $\underset{\sim}{e}$ corresponds to the difference of material strain and micro–strain, whereas its skew-symmetric part accounts for the relative rotation of the material with respect to microstructure. The micro–deformation gradient can be split into two contributions :

$$\underset{\approx}{\kappa} = \underset{\approx}{\kappa}^s + \underset{\approx}{\kappa}^a, \quad \text{with} \quad \underset{\approx}{\kappa}^s = \underset{\sim}{\chi}^s \otimes \nabla, \quad \underset{\approx}{\kappa}^a = \underset{\sim}{\chi}^a \otimes \nabla \tag{47}$$

In this section, the analysis is restricted to small deformations, small micro–rotations, small micro–strains and small micro–deformation gradients. The statics of the micromorphic continuum is described by the symmetric force-stress tensor $\underset{\sim}{\sigma}$, the generally non-symmetric relative force–stress tensor $\underset{\sim}{s}$ and third–rank stress tensor $\underset{\approx}{m}$. These tensors must fulfill the local form of the balance equations in the static case, in the absence of body simple nor double forces for simplicity :

$$(\underset{\sim}{\sigma} + \underset{\sim}{s}) \cdot \nabla = 0, \quad \underset{\approx}{m} \cdot \nabla + \underset{\sim}{s} = 0 \text{ on } \Omega \tag{48}$$

The constitutive equations for linear elastic centro-symmetric micromorphic materials read :

$$\underset{\sim}{\sigma} = \underset{\approx}{a} : \underset{\sim}{\varepsilon}, \quad \underset{\sim}{s} = \underset{\approx}{b} : \underset{\sim}{e}, \quad \underset{\approx}{m} = \underset{\approx\sim}{c} \,\vdots\, \underset{\approx}{\kappa} \tag{49}$$

The elasticity tensors display the major symmetries :

$$a_{ijkl} = a_{klij}, \quad b_{ijkl} = b_{klij}, \qquad c_{ijkpqr} = c_{pqrijk} \tag{50}$$

and $\underset{\approx}{a}$ has also the usual minor symmetries. The last constitutive law can be written in the form :

$$\underset{\approx}{m} = \underset{\approx\sim}{c}^s \,\vdots\, \underset{\approx}{\kappa}^s + \underset{\approx\sim}{c}^a \,\vdots\, \underset{\approx}{\kappa}^a \tag{51}$$

For the sake of simplicity, the tensors $\underset{\approx\sim}{c}^s$ and $\underset{\approx\sim}{c}^a$ are supposed to fulfill the conditions :

$$c^s_{ijkpqr} = c^s_{jikpqr}, \quad c^a_{ijkpqr} = -c^a_{jikpqr} \tag{52}$$

thus assuming that there is no coupling between the contributions of the symmetric and skew–symmetric parts of $\underset{\sim}{\chi}$ to the third–rank stress tensor.

The setting of the boundary value problem on body Ω is then closed by the boundary conditions. In the following, Dirichlet boundary conditions are considered of the form :

$$\underline{\boldsymbol{u}}(\underline{\boldsymbol{x}}) = 0, \quad \underset{\sim}{\chi}(\underline{\boldsymbol{x}}) = 0, \quad \forall \underline{\boldsymbol{x}} \in \partial\Omega \tag{53}$$

where $\partial\Omega$ denote the boundary of Ω. The equations (46), (48), (49) and (53) define the boundary value problem $\mathcal{P}$.

The next sections of this work are restricted to micromorphic materials with periodic microstructure. The heterogeneous material is then obtained by space tesselation with cells translated from a single cell Y^l. The period of the microstructure is described by three dimensionless independent vectors $(\underline{\boldsymbol{a}}_1, \underline{\boldsymbol{a}}_2, \underline{\boldsymbol{a}}_3)$ such that :

$$Y^l = \left\{ \underline{\boldsymbol{x}} = x_i \underline{\boldsymbol{a}}_i, \ |x_i| < \frac{l}{2} \right\}$$

where l is the characteristic size of the cell. We call $\underset{\approx}{\boldsymbol{a}}^l$, $\underset{\approx}{\boldsymbol{b}}^l$ and $\underset{\approx\!\!\sim}{\boldsymbol{c}}^l$ the elasticity tensor fields of the periodic micromorphic material. They are such that :

$$\forall \underline{\boldsymbol{x}} \in \Omega, \forall (n_1, n_2, n_3) \in Z^3 / \quad \underline{\boldsymbol{x}} + l(n_1 \underline{\boldsymbol{a}}_1 + n_2 \underline{\boldsymbol{a}}_2 + n_3 \underline{\boldsymbol{a}}_3) \in \Omega,$$

$$\underset{\approx}{\boldsymbol{a}}^l(\underline{\boldsymbol{x}}) = \underset{\approx}{\boldsymbol{a}}^l(\underline{\boldsymbol{x}} + l(n_1 \underline{\boldsymbol{a}}_1 + n_2 \underline{\boldsymbol{a}}_2 + n_3 \underline{\boldsymbol{a}}_3)),$$

$$\underset{\approx}{\boldsymbol{b}}^l(\underline{\boldsymbol{x}}) = \underset{\approx}{\boldsymbol{b}}^l(\underline{\boldsymbol{x}} + l(n_1 \underline{\boldsymbol{b}}_1 + n_2 \underline{\boldsymbol{b}}_2 + n_3 \underline{\boldsymbol{b}}_3)),$$

$$\underset{\approx\!\!\sim}{\boldsymbol{c}}^l(\underline{\boldsymbol{x}}) = \underset{\approx\!\!\sim}{\boldsymbol{c}}^l(\underline{\boldsymbol{x}} + l(n_1 \underline{\boldsymbol{a}}_1 + n_2 \underline{\boldsymbol{a}}_2 + n_3 \underline{\boldsymbol{a}}_3))$$

Dimensional analysis

The size L of body Ω is defined for instance as the maximum distance between two points. Dimensionless coordinates and displacements are introduced :

$$\underline{\boldsymbol{x}}^* = \frac{\underline{\boldsymbol{x}}}{L}, \quad \underline{\boldsymbol{u}}^*(\underline{\boldsymbol{x}}^*) = \frac{\underline{\boldsymbol{u}}(\underline{\boldsymbol{x}})}{L}, \quad \underset{\sim}{\chi}^*(\underline{\boldsymbol{x}}^*) = \underset{\sim}{\chi}(\underline{\boldsymbol{x}}) \tag{54}$$

The corresponding strain measures are :

$$\underset{\sim}{\varepsilon}^*(\underline{\boldsymbol{x}}^*) = \underline{\boldsymbol{u}}^* \overset{s}{\otimes} \boldsymbol{\nabla}^* \underset{\sim}{\varepsilon}(\underline{\boldsymbol{x}}), \quad \underset{\sim}{e}^*(\underline{\boldsymbol{x}}^*) = \underline{\boldsymbol{u}}^* \otimes \boldsymbol{\nabla}^* - \underset{\sim}{\chi}^* = \underset{\sim}{e}(\underline{\boldsymbol{x}}) \tag{55}$$

$$\underset{\approx}{\boldsymbol{\kappa}}^*(\underline{\boldsymbol{x}}^*) = \underset{\sim}{\chi}^* \otimes \boldsymbol{\nabla}^* = L\, \underset{\approx}{\boldsymbol{\kappa}}(\underline{\boldsymbol{x}}) \tag{56}$$

and similarily

$$\underset{\approx}{\kappa}^{s*}(\underline{x}^*) = \underset{\sim}{\chi}^{s*} \otimes \nabla^* = L\,\underset{\approx}{\kappa}^{s}(\underline{x}), \quad \underset{\approx}{\kappa}^{a*}(\underline{x}^*) = \underset{\sim}{\chi}^{a*} \otimes \nabla^* = L\,\underset{\approx}{\kappa}^{a}(\underline{x}) \quad (57)$$

with $\underline{\nabla}^* = \left(\dfrac{\partial\,\cdot}{\partial x_i^*}\right)\underline{e}_i = L\nabla$. It is necessary to introduce next a norm of the elasticity tensors :

$$A = \underset{\underline{x}\in Y^l}{\mathrm{Max}} \left(\left|a^l_{ijkl}(\underline{x})\right|, \left|b^l_{ijkl}(\underline{x})\right|\right)$$

$$C^s = \underset{\underline{x}\in Y^l}{\mathrm{Max}} \left|c^{sl}_{ijkpqr}(\underline{x})\right|, \quad C^a = \underset{\underline{x}\in Y^l}{\mathrm{Max}} \left|c^{al}_{ijkpqr}(\underline{x})\right| \quad (58)$$

whereby characteristic lengths l_s and l_a can be defined as:

$$C^s = A l_s^2, \quad C^a = A l_a^2 \quad (59)$$

The definition of dimensionless stress and elasticity tensors follows :

$$\underset{\sim}{\sigma}^*(\underline{x}^*) = A^{-1}\underset{\sim}{\sigma}(\underline{x}), \quad \underset{\sim}{s}^*(\underline{x}^*) = A^{-1}\underset{\sim}{s}(\underline{x}), \quad \underset{\approx}{m}^*(\underline{x}^*) = (AL)^{-1}\underset{\approx}{m}(\underline{x}) \quad (60)$$

$$\underset{\approx}{a}^*(\underline{x}^*) = A^{-1}\underset{\approx}{a}^l(\underline{x}), \quad \underset{\approx}{b}^*(\underline{x}^*) = A^{-1}\underset{\approx}{b}^l(\underline{x}), \quad (61)$$

$$\underset{\approx\approx}{c}^{s*}(\underline{x}^*) = (A l_c^2)^{-1}\underset{\approx\approx}{c}^{sl}(\underline{x}), \quad \underset{\approx\approx}{c}^{a*}(\underline{x}^*) = (A l_c^2)^{-1}\underset{\approx\approx}{c}^{al}(\underline{x}) \quad (62)$$

Since the initial tensors $\underset{\approx}{a}^l, \underset{\approx}{b}^l$ and $\underset{\approx\approx}{c}^l$ are Y^l-periodic, the dimensionless counterparts are Y^*-periodic :

$$Y^* = \frac{l}{L}Y, \quad Y = \left\{\underline{y} = y_i\underline{a}_i,\ |y_i| < \frac{1}{2}\right\} \quad (63)$$

Y is the (dimensionless) unit cell used in the present asymptotic analyses. As a result, the dimensionless stress and strain tensors are related by the following constitutive equations :

$$\underset{\sim}{\sigma}^* = \underset{\approx}{a}^* : \underset{\sim}{\varepsilon}^*, \quad \underset{\sim}{s}^* = \underset{\approx}{b}^* : \underset{\sim}{e}^*, \quad \underset{\approx}{m}^* = \left(\frac{l_s}{L}\right)^2 \underset{\approx\approx}{c}^{s*} \,\vdots\, \underset{\approx}{\kappa}^{s*} + \left(\frac{l_a}{L}\right)^2 \underset{\approx\approx}{c}^{a*} \,\vdots\, \underset{\approx}{\kappa}^{a*} \quad (64)$$

The dimensionless balance equations read :

$$\forall \underline{x}^* \in \Omega^*, \quad (\underset{\sim}{\sigma}^* + \underset{\sim}{s}^*)\cdot\nabla^* = 0, \quad \underset{\approx}{m}^*\cdot\nabla^* + \underset{\sim}{s}^* = 0 \quad (65)$$

A boundary value problem $\mathcal{P}^*$ can be defined using equations (56), (64) and (65), complemented by the boundary conditions :

$$\forall \underline{x}^* \in \partial\Omega^*, \quad \underline{u}^*(\underline{x}^*) = 0, \quad \underset{\sim}{\chi}^*(\underline{x}^*) = 0 \quad (66)$$

The homogenisation problem

The boundary value problem $\mathcal{P}^*$ is treated here as an element of a series of problems $(\mathcal{P}_\epsilon)_{\epsilon>0}$ on Ω^*. The homogenisation problem consists in the determination of the limit of this series when the dimensionless parameter ϵ, regarded as small, tends towards 0. The series is chosen such that

$$\mathcal{P}_{\epsilon=\frac{l}{L}} = \mathcal{P}^*$$

The unknowns of boundary value problem $\mathcal{P}_\epsilon$ are the displacement and micro–deformation fields $\underline{\boldsymbol{u}}^{\,\epsilon}$ and $\underset{\sim}{\boldsymbol{\chi}}^\epsilon$ satisfying the following field equations on Ω^* :

$$\underset{\sim}{\boldsymbol{\sigma}}^\epsilon = \underset{\approx}{\boldsymbol{a}}^\epsilon : (\underline{\boldsymbol{u}}^{\,\epsilon} \overset{s}{\otimes} \boldsymbol{\nabla}^*), \quad \underset{\sim}{\boldsymbol{s}}^\epsilon = \underset{\approx}{\boldsymbol{b}}^\epsilon : (\underline{\boldsymbol{u}}^{\,\epsilon} \otimes \boldsymbol{\nabla}^* - \underset{\sim}{\boldsymbol{\chi}}^\epsilon), \quad \underset{\sim}{\boldsymbol{m}}^\epsilon = \underset{\approx}{\boldsymbol{c}}^\epsilon \vdots (\underset{\sim}{\boldsymbol{\chi}}^\epsilon \otimes \boldsymbol{\nabla}^*) \tag{67}$$

$$(\underset{\sim}{\boldsymbol{\sigma}}^\epsilon + \underset{\sim}{\boldsymbol{s}}^\epsilon) \cdot \boldsymbol{\nabla}^* = 0, \quad \underset{\sim}{\boldsymbol{m}}^\epsilon \cdot \boldsymbol{\nabla}^* + \underset{\sim}{\boldsymbol{s}}^\epsilon = 0 \tag{68}$$

Different cases must now be distinguished depending on the relative position of the constitutive lengths l_s and l_a with respect to the characteristic lengths l and L of the problem. Four special cases are relevant for the present asymptotic analysis. The first case corresponds to a limiting process for which l_s/l and l_a/l remain constant when l/L goes to zero. The second case corresponds to the situation for which l_s/L and l_a/L remain constant when l/L goes to zero. The third (resp. fourth) situation assumes that l_s/l and l_a/L (resp. l_s/L and l_a/l) remain constant when l/L goes to zero. These assumptions lead to four different homogenisation schemes labelled $HS1$ to $HS4$ in the sequel. The homogenisation scheme 1 (resp. 2) will be relevant when the ratio l/L is small enough and when l_s, l_a and l (resp. L) have the same order of magnitude.
Accordingly, the following tensors of elastic moduli can be defined :

$$\underset{\approx}{\boldsymbol{a}}^{(0)}(\underline{\boldsymbol{y}}) = \underset{\approx}{\boldsymbol{a}}^*(\frac{l}{L}\underline{\boldsymbol{y}}), \quad \underset{\approx}{\boldsymbol{b}}^{(0)}(\underline{\boldsymbol{y}}) = \underset{\approx}{\boldsymbol{b}}^*(\frac{l}{L}\underline{\boldsymbol{y}}), \tag{69}$$

$$\underset{\approx}{\boldsymbol{c}}^{(1)}(\underline{\boldsymbol{y}}) = \left(\frac{l_s}{l}\right)^2 \underset{\approx}{\boldsymbol{c}}^*(\frac{l}{L}\underline{\boldsymbol{y}}), \quad \underset{\approx}{\boldsymbol{c}}^{(2)}(\underline{\boldsymbol{y}}) = \left(\frac{l_s}{L}\right)^2 \underset{\approx}{\boldsymbol{c}}^*(\frac{l}{L}\underline{\boldsymbol{y}}), \tag{70}$$

$$\underset{\approx}{\boldsymbol{c}}^{s(1)}(\underline{\boldsymbol{y}}) = \left(\frac{l_s}{l}\right)^2 \underset{\approx}{\boldsymbol{c}}^{s*}(\frac{l}{L}\underline{\boldsymbol{y}}), \quad \underset{\approx}{\boldsymbol{c}}^{a(1)}(\underline{\boldsymbol{y}}) = \left(\frac{l_a}{l}\right)^2 \underset{\approx}{\boldsymbol{c}}^{a*}(\frac{l}{L}\underline{\boldsymbol{y}}), \tag{71}$$

$$\underset{\approx}{\boldsymbol{c}}^{s(2)}(\underline{\boldsymbol{y}}) = \left(\frac{l_s}{L}\right)^2 \underset{\approx}{\boldsymbol{c}}^{s*}(\frac{l}{L}\underline{\boldsymbol{y}}), \quad \underset{\approx}{\boldsymbol{c}}^{a(2)}(\underline{\boldsymbol{y}}) = \left(\frac{l_a}{L}\right)^2 \underset{\approx}{\boldsymbol{c}}^{a*}(\frac{l}{L}\underline{\boldsymbol{y}}) \tag{72}$$

They are Y-periodic since $\underset{\approx}{\boldsymbol{a}}^*, \underset{\approx}{\boldsymbol{b}}^*$ and $\underset{\approx}{\boldsymbol{c}}^*$ are Y^*-periodic. Four different hypotheses will be made concerning the constitutive tensors of problem $\mathcal{P}_\epsilon$:

Assumption 1: $\underset{\approx}{\boldsymbol{a}}^\epsilon(\underline{\boldsymbol{x}}^*) = \underset{\approx}{\boldsymbol{a}}^{(0)}(\epsilon^{-1}\underline{\boldsymbol{x}}^*)$, $\underset{\approx}{\boldsymbol{b}}^\epsilon(\underline{\boldsymbol{x}}^*) = \underset{\approx}{\boldsymbol{b}}^{(0)}(\epsilon^{-1}\underline{\boldsymbol{x}}^*)$ and
$\underset{\approx}{\boldsymbol{c}}^\epsilon(\underline{\boldsymbol{x}}^*) = \epsilon^2\underset{\approx}{\boldsymbol{c}}^{(1)}(\epsilon^{-1}\underline{\boldsymbol{x}}^*)$;

Assumption 2: $\underset{\approx}{\boldsymbol{a}}^\epsilon(\underline{\boldsymbol{x}}^*) = \underset{\approx}{\boldsymbol{a}}^{(0)}(\epsilon^{-1}\underline{\boldsymbol{x}}^*)$, $\underset{\approx}{\boldsymbol{b}}^\epsilon(\underline{\boldsymbol{x}}^*) = \underset{\approx}{\boldsymbol{b}}^{(0)}(\epsilon^{-1}\underline{\boldsymbol{x}}^*)$ and
$\underset{\approx}{\boldsymbol{c}}^\epsilon(\underline{\boldsymbol{x}}^*) = \underset{\approx}{\boldsymbol{c}}^{(2)}(\epsilon^{-1}\underline{\boldsymbol{x}}^*)$;

Assumption 3: $\underset{\approx}{\boldsymbol{a}}^\epsilon(\underline{\boldsymbol{x}}^*) = \underset{\approx}{\boldsymbol{a}}^{(0)}(\epsilon^{-1}\underline{\boldsymbol{x}}^*)$, $\underset{\approx}{\boldsymbol{b}}^\epsilon(\underline{\boldsymbol{x}}^*) = \underset{\approx}{\boldsymbol{b}}^{(0)}(\epsilon^{-1}\underline{\boldsymbol{x}}^*)$ and
$\underset{\approx}{\boldsymbol{c}}^{s\epsilon}(\underline{\boldsymbol{x}}^*) = \epsilon^2\underset{\approx}{\boldsymbol{c}}^{s(1)}(\epsilon^{-1}\underline{\boldsymbol{x}}^*)$, $\underset{\approx}{\boldsymbol{c}}^{a\epsilon}(\underline{\boldsymbol{x}}^*) = \underset{\approx}{\boldsymbol{c}}^{a(2)}(\epsilon^{-1}\underline{\boldsymbol{x}}^*)$;

Assumption 4: $\underset{\approx}{\boldsymbol{a}}^\epsilon(\underline{\boldsymbol{x}}^*) = \underset{\approx}{\boldsymbol{a}}^{(0)}(\epsilon^{-1}\underline{\boldsymbol{x}}^*)$, $\underset{\approx}{\boldsymbol{b}}^\epsilon(\underline{\boldsymbol{x}}^*) = \underset{\approx}{\boldsymbol{b}}^{(0)}(\epsilon^{-1}\underline{\boldsymbol{x}}^*)$ and
$\underset{\approx}{\boldsymbol{c}}^{s\epsilon}(\underline{\boldsymbol{x}}^*) = \underset{\approx}{\boldsymbol{c}}^{s(2)}(\epsilon^{-1}\underline{\boldsymbol{x}}^*)$, $\underset{\approx}{\boldsymbol{c}}^{a\epsilon}(\underline{\boldsymbol{x}}^*) = \epsilon^2\underset{\approx}{\boldsymbol{c}}^{a(1)}(\epsilon^{-1}\underline{\boldsymbol{x}}^*)$.

Assumptions 1 and 2 respectively correspond to the homogenisation schemes HS1 and HS2. Both choices meet the requirement that

$$(\epsilon = \frac{l}{L}) \Rightarrow (\underset{\approx}{\boldsymbol{a}}^\epsilon = \underset{\approx}{\boldsymbol{a}}^* \quad \text{and} \quad \underset{\approx}{\boldsymbol{c}}^\epsilon = (\frac{l_s}{L})^2 \underset{\approx}{\boldsymbol{c}}^*)$$

Assumptions 3 and 4 respectively correspond to the homogenisation schemes HS3 and HS4. Both choices meet the requirement that

$$(\epsilon = \frac{l}{L}) \Rightarrow (\underset{\approx}{\boldsymbol{a}}^\epsilon = \underset{\approx}{\boldsymbol{a}}^*, \quad \underset{\approx}{\boldsymbol{c}}^{s\epsilon} = (\frac{l_s}{L})^2 \underset{\approx}{\boldsymbol{c}}^{s*} \quad \text{and} \quad \underset{\approx}{\boldsymbol{c}}^{a\epsilon} = (\frac{l_a}{L})^2 \underset{\approx}{\boldsymbol{c}}^{a*})$$

It must be noted that, in our presentation of the asymptotic analysis, the lengths l, l_s, l_a and L are given and fixed, whereas parameter ϵ is allowed to tend to zero in the limiting process.

In the sequel, the stars $*$ are dropped for conciseness.

Multiscale asymptotic method

In the setting of the homogenisation problems two space variables have been distinguished : $\underline{\boldsymbol{x}}$ describes the macroscopic scale and $\underline{\boldsymbol{y}}$ is the local variable in the unit cell Y. To solve the homogenisation problem, it is resorted to the method of multiscale asymptotic developments initially introduced in Sanchez-Palencia (1974). According to this method, all fields are regarded as functions of both variables $\underline{\boldsymbol{x}}$ and $\underline{\boldsymbol{y}}$. It is assumed that they can be expanded in a series of powers of small parameter ϵ. In particular, the displacement, micro-deformation, force and double stress fields are supposed to take the form :

$$
\begin{aligned}
\underline{u}^{\epsilon}(\underline{x}) &= \underline{u}_0(\underline{x},\underline{y}) + \epsilon \underline{u}_1(\underline{x},\underline{y}) + \epsilon^2 \underline{u}_2(\underline{x},\underline{y}) + \ldots \\
\underset{\sim}{\chi}^{\epsilon}(\underline{x}) &= \underset{\sim}{\chi}_1(\underline{x},\underline{y}) + \epsilon \underset{\sim}{\chi}_2(\underline{x},\underline{y}) + \epsilon^2 \underset{\sim}{\chi}_3(\underline{x},\underline{y}) + \ldots \\
\underset{\sim}{\sigma}^{\epsilon}(\underline{x}) &= \underset{\sim}{\sigma}_0(\underline{x},\underline{y}) + \epsilon \underset{\sim}{\sigma}_1(\underline{x},\underline{y}) + \epsilon^2 \underset{\sim}{\sigma}_2(\underline{x},\underline{y}) + \ldots \\
\underset{\sim}{s}^{\epsilon}(\underline{x}) &= \underset{\sim}{s}_0(\underline{x},\underline{y}) + \epsilon \underset{\sim}{s}_1(\underline{x},\underline{y}) + \epsilon^2 \underset{\sim}{s}_2(\underline{x},\underline{y}) + \ldots \\
\underset{\approx}{m}^{\epsilon}(\underline{x}) &= \underset{\approx}{m}_0(\underline{x},\underline{y}) + \epsilon \underset{\approx}{m}_1(\underline{x},\underline{y}) + \epsilon^2 \underset{\approx}{m}_2(\underline{x},\underline{y}) + \ldots
\end{aligned}
\qquad (73)
$$

where the coefficients $\underline{u}_i(\underline{x},\underline{y})$, $\underset{\sim}{\chi}_i(\underline{x},\underline{y})$, $\underset{\sim}{\sigma}_i(\underline{x},\underline{y})$, $\underset{\sim}{s}_i(\underline{x},\underline{y})$ and $\underset{\approx}{m}_i(\underline{x},\underline{y})$ are assumed to have the same order of magnitude and to be Y-periodic with respect to variable $\underline{y}$ ($\underline{y} = \underline{x}/\epsilon$). The average operator over the unit cell Y is denoted by

$$
\langle \cdot \rangle = \frac{1}{|Y|} \int_Y \cdot \, dV
$$

As a result,

$$
< \underline{u}^{\epsilon} >= \underline{U}_0 + \epsilon \underline{U}_1 + \ldots \quad \text{and} \quad < \underset{\sim}{\chi}^{\epsilon} >= \underset{\sim}{\Xi}_1 + \epsilon \underset{\sim}{\Xi}_2 + \ldots \qquad (74)
$$

where $\underline{U}_i =< \underline{u}_i >$ and $\underset{\sim}{\Xi}_i =< \underset{\sim}{\chi}_i >$. The gradient operator can be split into partial derivatives with respect to $\underline{x}$ and $\underline{y}$:

$$
\boldsymbol{\nabla} = \boldsymbol{\nabla}_x + \frac{1}{\epsilon} \boldsymbol{\nabla}_y \qquad (75)
$$

This operator is used to compute the strain measures and balance equations :

$$
\begin{aligned}
\underset{\sim}{\varepsilon}^{\epsilon} &= \epsilon^{-1} \underset{\sim}{\varepsilon}_{-1} + \underset{\sim}{\varepsilon}_0 + \epsilon^1 \underset{\sim}{\varepsilon}_1 + \ldots \\
&= \epsilon^{-1} \underline{u}_0 \overset{s}{\otimes} \boldsymbol{\nabla}_y + (\underline{u}_0 \overset{s}{\otimes} \boldsymbol{\nabla}_x + \underline{u}_1 \overset{s}{\otimes} \boldsymbol{\nabla}_y) \\
&+ \epsilon (\underline{u}_1 \overset{s}{\otimes} \boldsymbol{\nabla}_x + \underline{u}_2 \overset{s}{\otimes} \boldsymbol{\nabla}_y) + \ldots \\
\underset{\sim}{e}^{\epsilon} &= \epsilon^{-1} \underset{\sim}{e}_{-1} + \underset{\sim}{e}_0 + \epsilon^1 \underset{\sim}{e}_1 + \ldots \\
&= \epsilon^{-1} \underline{u}_0 \otimes \boldsymbol{\nabla}_y + (\underline{u}_0 \otimes \boldsymbol{\nabla}_x + \underline{u}_1 \otimes \boldsymbol{\nabla}_y - \underset{\sim}{\chi}_1) \\
&+ \epsilon (\underline{u}_1 \otimes \boldsymbol{\nabla}_x + \underline{u}_2 \otimes \boldsymbol{\nabla}_y - \underset{\sim}{\chi}_2) + \ldots \\
\underset{\approx}{\kappa}^{\epsilon} &= \epsilon^{-1} \underset{\approx}{\kappa}_{-1} + \underset{\approx}{\kappa}_0 + \epsilon^1 \underset{\approx}{\kappa}_1 + \ldots \\
&= \epsilon^{-1} \underset{\sim}{\chi}_1 \otimes \boldsymbol{\nabla}_y + (\underset{\sim}{\chi}_1 \otimes \boldsymbol{\nabla}_x + \underset{\sim}{\chi}_2 \otimes \boldsymbol{\nabla}_y) \\
&+ \epsilon (\underset{\sim}{\chi}_2 \otimes \boldsymbol{\nabla}_x + \underset{\sim}{\chi}_3 \otimes \boldsymbol{\nabla}_y) + \ldots
\end{aligned}
$$

$$
(\underset{\sim}{\sigma}^{\epsilon} + \underset{\sim}{s}^{\epsilon}) \cdot \boldsymbol{\nabla}_x + \epsilon^{-1} (\underset{\sim}{\sigma}^{\epsilon} + \underset{\sim}{s}^{\epsilon}) \cdot \boldsymbol{\nabla}_y = 0, \quad \underset{\approx}{m}^{\epsilon} \cdot \boldsymbol{\nabla}_x + \epsilon^{-1} \underset{\approx}{m}^{\epsilon} \cdot \boldsymbol{\nabla}_y + \underset{\sim}{s}^{\epsilon} = 0 \quad (76)
$$

Similar expansions are valid for the tensors $\underset{\approx}{\kappa}^s, \underset{\approx}{\kappa}^a$. The expansions of the stress tensors are then introduced in the balance equations (76) and the terms can be ordered with respect to the powers of ϵ. Identifying the terms of same order, we are lead to the following set of equations :

- *order* ϵ^{-1},

$$(\underset{\sim}{\boldsymbol{\sigma}}_0 + \underset{\sim}{\boldsymbol{s}}_0) \cdot \boldsymbol{\nabla}_y = 0 \quad \text{and} \quad \underset{\approx}{\boldsymbol{m}}_0 \cdot \boldsymbol{\nabla}_y = 0 \tag{77}$$

- *order* ϵ^{0},

$$(\underset{\sim}{\boldsymbol{\sigma}}_0 + \underset{\sim}{\boldsymbol{s}}_0) \cdot \boldsymbol{\nabla}_x + (\underset{\sim}{\boldsymbol{\sigma}}_1 + \underset{\sim}{\boldsymbol{s}}_1) \cdot \boldsymbol{\nabla}_y = 0 \quad \text{and} \quad \underset{\approx}{\boldsymbol{S}}_0 \cdot \boldsymbol{\nabla}_x + \underset{\approx}{\boldsymbol{S}}_1 \cdot \boldsymbol{\nabla}_y + \underset{\sim}{\boldsymbol{s}}_1 = 0 \tag{78}$$

The effective balance equations follow (78) by averaging over the unit cell Y and, at the order ϵ^0 one gets :

$$(\underset{\sim}{\boldsymbol{\Sigma}}_0 + \underset{\sim}{\boldsymbol{S}}_0) \cdot \boldsymbol{\nabla} = 0 \quad \text{and} \quad \underset{\approx}{\boldsymbol{M}}_0 \cdot \boldsymbol{\nabla} + \underset{\sim}{\boldsymbol{S}}_0 = 0 \tag{79}$$

where $\underset{\sim}{\boldsymbol{\Sigma}}_0 =< \underset{\sim}{\boldsymbol{\sigma}}_0 >$, $\underset{\sim}{\boldsymbol{S}}_0 =< \underset{\sim}{\boldsymbol{s}}_0 >$ and $\underset{\approx}{\boldsymbol{M}}_0 =< \underset{\approx}{\boldsymbol{m}}_0 >$.

Homogenization scheme HS1

For the first homogenisation scheme defined in section 4.1, the equations describing the local behaviour are :

$$\underset{\sim}{\boldsymbol{\sigma}}^\epsilon = \underset{\approx}{\boldsymbol{a}}^{(0)}(\underline{\boldsymbol{y}}) : \underset{\sim}{\boldsymbol{\varepsilon}}^\epsilon, \quad \underset{\sim}{\boldsymbol{s}}^\epsilon = \underset{\approx}{\boldsymbol{b}}^{(0)}(\underline{\boldsymbol{y}}) : \underset{\sim}{\boldsymbol{e}}^\epsilon \quad \text{and} \quad \underset{\approx}{\boldsymbol{m}}^\epsilon = \epsilon^2 \underset{\approx\approx}{\boldsymbol{c}}^{(1)}(\underline{\boldsymbol{y}}) \vdots \underset{\approx}{\boldsymbol{\kappa}}^\epsilon \tag{80}$$

At this stage, the expansion (76) can be substituted into the constitutive equations (80). Identifying the terms of same order, we get :

- *order* ϵ^{-1},

$$\underset{\approx}{\boldsymbol{a}}^{(0)} : \underset{\sim}{\boldsymbol{\varepsilon}}_{-1} = \underset{\approx}{\boldsymbol{a}}^{(0)} : (\underline{\boldsymbol{u}}_0 \overset{s}{\otimes} \boldsymbol{\nabla}_y) = 0, \quad \underset{\approx}{\boldsymbol{b}}^{(0)} : \underset{\sim}{\boldsymbol{e}}_0 = \underset{\approx}{\boldsymbol{b}}^{(0)} : (\underline{\boldsymbol{u}}_0 \otimes \boldsymbol{\nabla}_y) = 0 \tag{81}$$

- *order* ϵ^{0},

$$\underset{\sim}{\boldsymbol{\sigma}}_0 = \underset{\approx}{\boldsymbol{a}}^{(0)} : \underset{\sim}{\boldsymbol{\varepsilon}}_0, \quad \underset{\sim}{\boldsymbol{s}}_0 = \underset{\approx}{\boldsymbol{b}}^{(0)} : \underset{\sim}{\boldsymbol{e}}_0, \quad \underset{\approx}{\boldsymbol{m}}_0 = 0 \tag{82}$$

- *order* ϵ^{1},

$$\underset{\sim}{\boldsymbol{\sigma}}_1 = \underset{\approx}{\boldsymbol{a}}^{(0)} : \underset{\sim}{\boldsymbol{\varepsilon}}_1, \quad \underset{\sim}{\boldsymbol{s}}_1 = \underset{\approx}{\boldsymbol{b}}^{(0)} : \underset{\sim}{\boldsymbol{e}}_1, \quad \underset{\approx}{\boldsymbol{m}}_1 = \underset{\approx\approx}{\boldsymbol{c}}^{(1)} \vdots \underset{\approx}{\boldsymbol{\kappa}}_{-1} \tag{83}$$

The equation (81) implies that $\underline{\boldsymbol{u}}_0$ does not depend on the local variable $\underline{\boldsymbol{y}}$:

$$\underline{\boldsymbol{u}}_0(\underline{\boldsymbol{x}}, \underline{\boldsymbol{y}}) = \underline{\boldsymbol{U}}_0(\underline{\boldsymbol{x}})$$

At the order ϵ^0, the higher order stress tensor vanishes,

$$\underset{\approx}{\boldsymbol{M}}_0 =< \underset{\approx}{\boldsymbol{m}}_0 >= 0$$

Finally, the fields $(\underline{u}_1, \underset{\sim}{\chi}_1, \underset{\sim}{\sigma}_0, \underset{\sim}{s}_0, \underset{\approx}{m}_1)$ are solutions of the following auxiliary boundary value problem defined on the unit cell :

$$\begin{cases} \underset{\sim}{\varepsilon}_0 = \underline{U}_0 \overset{s}{\otimes} \nabla_x + \underline{u}_1 \overset{s}{\otimes} \nabla_y, & \underset{\sim}{e}_0 = \underline{U}_0 \otimes \nabla_x + \underline{u}_1 \otimes \nabla_y - \underset{\sim}{\chi}_1 \\ \underset{\approx}{\kappa}_{-1} = \underset{\sim}{\chi}_1 \otimes \nabla_y & \\ \underset{\sim}{\sigma}_0 = \underset{\approx}{a}^{(0)} : \underset{\sim}{\varepsilon}_0, \quad \underset{\sim}{s}_0 = \underset{\approx}{b}^{(0)} : \underset{\sim}{e}_0, & \underset{\approx}{m}_1 = \underset{\approx}{\underset{\sim}{c}}^{(1)} : \underset{\sim}{\kappa}_{-1} \\ (\underset{\sim}{\sigma}_0 + \underset{\sim}{s}_0) \cdot \nabla_y = 0, & \underset{\approx}{m}_1 \cdot \nabla_y + \underset{\sim}{s}_0 = 0 \end{cases} \tag{84}$$

The boundary conditions of this problem are given by the periodicity requirements for the unknown fields. A series of auxiliary problems similar to (84) can be defined to obtain the solutions at higher orders. It must be noted that these problems must be solved in cascade since, for instance, the solution of (84) requires the knowledge of $\underline{U}_0$. A particular solution $\underset{\sim}{\chi}$ for a vanishing prescribed $\underline{U}_0 \overset{s}{\otimes} \nabla_x$ is $\underset{\sim}{\chi} = \underline{U}_0 \overset{a}{\otimes} \nabla_x$. It follows that the solution $(\underline{u}_1, \underline{U}_0 \overset{a}{\otimes} \nabla_x - \underset{\sim}{\chi}_1)$ to problem (84) depends linearly on $\underline{U}_0 \overset{s}{\otimes} \nabla_x$, up to a translation term, so that :

$$\underline{u}^{\epsilon} = \underline{U}_0(\underline{x}) + \epsilon(\underline{U}_1(\underline{x}) + \underset{\simeq}{X}^{(1)}_u(\underline{y}) : (\underline{U}_0 \overset{s}{\otimes} \nabla)) + \dots \tag{85}$$

$$\underset{\sim}{\chi}^{\epsilon} = \underline{U}_0 \overset{a}{\otimes} \nabla_x + \underset{\approx}{X}^{(1)}_{\chi}(\underline{y}) : \underline{U}_0 \overset{s}{\otimes} \nabla + \dots \tag{86}$$

where concentration tensors $\underset{\simeq}{X}^{(1)}_u$ and $\underset{\approx}{X}^{(1)}_{\chi}$ have been introduced, the components of which are determined by the successive solutions of the auxiliary problem for unit values of the components of $\underline{U}_0 \overset{s}{\otimes} \nabla$. Concentration tensor $\underset{\simeq}{X}^{(1)}_u$ is such that its mean value over the unit cell vanishes.

The macroscopic stress tensor is given by :

$$\underset{\sim}{\Sigma}_0 = < \underset{\sim}{\sigma}_0 > = < \underset{\approx}{a}^{(0)} : (\underset{\approx}{1} + \nabla_x \overset{s}{\otimes} \underset{\simeq}{X}^{(1)}_u) > : (\underline{U}_0 \overset{s}{\otimes} \nabla) = \underset{\approx}{A}^{(1)}_0 : (\underline{U}_0 \overset{s}{\otimes} \nabla) \tag{87}$$

Accordingly, the tensor of effective moduli possesses all symmetries of classical elastic moduli for a Cauchy medium :

$$A^{(1)}_{0\,ijkl} = A^{(1)}_{0\,klij} = A^{(1)}_{0\,jikl} = A^{(1)}_{0\,ijlk}$$

The additional second rank stress tensor can be shown to vanish :

$$\underset{\sim}{S}_0 = < \underset{\sim}{s}_0 > = < -\underset{\approx}{m}_1 \cdot \nabla_y > = 0 \tag{88}$$

The effective medium is therefore governed by the single equation :

$$\underset{\sim}{\Sigma}_0 \cdot \nabla = 0 \tag{89}$$

The effective medium turns out to be a Cauchy continuum with symmetric stress tensor.

Homogenization scheme HS2

For the second homogenisation scheme defined in section 4.1, the equations describing the local behaviour are :

$$\underset{\sim}{\sigma}^\epsilon = \underset{\approx}{a}^{(0)}(\underline{y}) : \underset{\sim}{\varepsilon}^\epsilon, \quad \underset{\sim}{s}^\epsilon = \underset{\approx}{b}^{(0)}(\underline{y}) : \underset{\sim}{e}^\epsilon, \quad \text{and} \quad \underset{\approx}{m}^\epsilon = \underset{\approx\approx}{c}^{(2)}(\underline{y}) \,\vdots\, \underset{\approx}{\kappa}^\epsilon \tag{90}$$

The different steps of the asymptotic analysis are the same as in the previous section for HS1. We will only focus here on the main results. At the order ϵ^{-1}, one gets

$$\underset{\approx}{a}^{(0)} : \underset{\sim}{\varepsilon}_{-1} = 0, \quad \underset{\approx}{b}^{(0)} : \underset{\sim}{e}_{-1} = 0, \quad \underset{\approx\approx}{c}^{(2)} \,\vdots\, \underset{\approx}{\kappa}_{-1} = 0 \tag{91}$$

This implies that the gradients of $\underline{u}_0$ and $\underset{\sim}{\chi}_1$ with respect to $\underline{y}$ vanish, so that :

$$\underline{u}_0(\underline{x}, \underline{y}) = \underline{U}_0(\underline{x}), \quad \underset{\sim}{\chi}_1(\underline{x}, \underline{y}) = \underset{\sim}{\Xi}_1(\underline{x}) \tag{92}$$

The fields $(\underline{u}_1, \underset{\sim}{\chi}_1, \underset{\sim}{\sigma}_0, \underset{\sim}{s}_0, \underset{\approx}{m}_0)$ are solutions of the two following auxiliary boundary value problems defined on the unit cell :

$$\begin{cases} \underset{\sim}{\varepsilon}_0 = \underline{U}_0 \overset{s}{\otimes} \nabla_x + \underline{u}_1 \overset{s}{\otimes} \nabla_y, \quad \underset{\sim}{e}_0 = \underline{U}_0 \otimes \nabla_x + \underline{u}_1 \otimes \nabla_y - \underset{\sim}{\Xi}_1 \\ \underset{\sim}{\sigma}_0 = \underset{\approx}{a}^{(0)} : \underset{\sim}{\varepsilon}_0, \quad \underset{\sim}{s}_0 = \underset{\approx}{b}^{(0)} : \underset{\sim}{e}_0 \\ (\underset{\sim}{\sigma}_0 + \underset{\sim}{s}_0) \cdot \nabla_y = 0 \end{cases}$$

$$\begin{cases} \underset{\approx}{\kappa}_0 = \underset{\sim}{\Xi}_1 \otimes \nabla_x + \underset{\sim}{\chi}_2 \otimes \nabla_y \\ \underset{\approx}{m}_0 = \underset{\approx\approx}{c}^{(2)} \vdots \underset{\approx}{\kappa}_0, \quad \underset{\approx}{m}_0 \cdot \nabla_y = 0 \end{cases}$$

We are therefore left with two decoupled boundary value problems : the first one with main unknown $\underline{u}_1$ depends linearly on $\underline{U}_0 \overset{s}{\otimes} \nabla_x$ and $\underline{U}_0 \otimes \nabla_x - \underset{\sim}{\Xi}_1$, whereas the second one with unknown $\underset{\sim}{\chi}_2$ is linear in $\underset{\sim}{\Xi}_1 \otimes \nabla_x$. The solutions take the form :

$$\underline{u}^\epsilon = \underline{U}_0(\underline{x}) + \epsilon(\underline{U}_1(\underline{x}) + \underset{\approx}{X}_u^{(2)}(\underline{y}) : (\underline{U}_0 \overset{s}{\otimes} \nabla) + \underset{\approx}{X}_e^{(2)}(\underline{y}) : (\underline{U}_0 \otimes \nabla - \underset{\sim}{\Xi}_1))$$

$$+ \ldots, \quad \underset{\sim}{\chi}^\epsilon = \underset{\sim}{\Xi}_1(\underline{x}) + \epsilon(\underset{\sim}{\Xi}_2(\underline{x}) + \underset{\approx\approx}{X}_\kappa^{(2)}(\underline{y}) \vdots (\underset{\sim}{\Xi}_1 \otimes \nabla)) + \ldots \tag{93}$$

where concentration tensors $\underset{\simeq}{\boldsymbol{X}}^{(2)}_u, \underset{\simeq}{\boldsymbol{X}}^{(2)}_e$ and $\underset{\approx}{\boldsymbol{X}}^{(1)}_\kappa$ have been introduced. Their components are determined by the successive solutions of the auxiliary problem for unit values of the components of $\underline{\boldsymbol{U}}_0 \overset{s}{\otimes} \boldsymbol{\nabla}, \underline{\boldsymbol{U}}_0 \otimes \boldsymbol{\nabla} - \underset{\sim}{\Xi}_1$ and $\underset{\sim}{\Xi}_1 \otimes \boldsymbol{\nabla}_y$. They are such that their mean value over the unit cell vanishes.

The macroscopic stress tensors and effective elastic properties are given by :

$$
\begin{aligned}
\underset{\sim}{\boldsymbol{\Sigma}}_0 &= \; < \underset{\approx}{\boldsymbol{a}}^{(0)} : (\underset{\approx}{\boldsymbol{1}} + \boldsymbol{\nabla}_y \overset{s}{\otimes} \underset{\simeq}{\boldsymbol{X}}^{(2)}_u) >: (\underline{\boldsymbol{U}}_0 \overset{s}{\otimes} \boldsymbol{\nabla}) \\
&+ \; < \underset{\approx}{\boldsymbol{a}}^{(0)} : (\boldsymbol{\nabla}_y \overset{s}{\otimes} \underset{\simeq}{\boldsymbol{X}}^{(2)}_e) >: (\underline{\boldsymbol{U}}_0 \otimes \boldsymbol{\nabla} - \underset{\sim}{\Xi}_1)
\end{aligned} \tag{94}
$$

$$
\begin{aligned}
\underset{\sim}{\boldsymbol{S}}_0 &= \; < \underset{\sim}{\boldsymbol{s}}_0 > = < \underset{\approx}{\boldsymbol{b}}^{(0)} : (\boldsymbol{\nabla}_y \otimes \underset{\simeq}{\boldsymbol{X}}^{(2)}_u) >: (\underline{\boldsymbol{U}}_0 \overset{s}{\otimes} \boldsymbol{\nabla}) \\
&+ \; < \underset{\approx}{\boldsymbol{b}}^{(0)} : (\boldsymbol{\nabla}_y \otimes \underset{\simeq}{\boldsymbol{X}}^{(2)}_e) >: (\underline{\boldsymbol{U}}_0 \otimes \boldsymbol{\nabla} - \underset{\sim}{\Xi}_1)
\end{aligned} \tag{95}
$$

$$
\underset{\approx}{\boldsymbol{M}}_0 = < \underset{\approx}{\boldsymbol{m}}_0 > = < \underset{\approx}{\boldsymbol{c}}^{(2)} \vdots (\underset{\approx}{\boldsymbol{1}} + \boldsymbol{\nabla}_y \otimes \underset{\approx}{\boldsymbol{X}}^{(2)}_\kappa) > \vdots \, \underset{\sim}{\Xi}_1 \otimes \boldsymbol{\nabla} \tag{96}
$$

None of these tensors vanishes in general, which means that the effective medium is a full micromorphic continuum governed by the balance equations (79).

Homogenization scheme HS3

For the third homogenisation scheme defined in section 4.1, the equations describing the local behaviour are :

$$
\underset{\sim}{\boldsymbol{\sigma}}^\epsilon = \underset{\approx}{\boldsymbol{a}}^{(0)}(\underline{\boldsymbol{y}}) : \underset{\sim}{\boldsymbol{\varepsilon}}^\epsilon, \quad \underset{\sim}{\boldsymbol{s}}^\epsilon = \underset{\approx}{\boldsymbol{b}}^{(0)}(\underline{\boldsymbol{y}}) : \underset{\sim}{\boldsymbol{e}}^\epsilon, \tag{97}
$$

$$
\underset{\approx}{\boldsymbol{m}}^\epsilon = \epsilon^2 \underset{\approx}{\boldsymbol{c}}^{s(1)}(\underline{\boldsymbol{y}}) \vdots \underset{\approx}{\boldsymbol{\kappa}}^{s\epsilon} + \underset{\approx}{\boldsymbol{c}}^{a(2)}(\underline{\boldsymbol{y}}) \vdots \underset{\approx}{\boldsymbol{\kappa}}^{a\epsilon} \tag{98}
$$

At the order ϵ^{-1}, one gets

$$
\underset{\approx}{\boldsymbol{a}}^{(0)} : \underset{\sim}{\boldsymbol{\varepsilon}}_{-1} = 0, \quad \underset{\approx}{\boldsymbol{b}}^{(0)} : \underset{\sim}{\boldsymbol{e}}_{-1} = 0, \quad \underset{\approx}{\boldsymbol{c}}^{a(2)} \vdots \underset{\approx}{\boldsymbol{\kappa}}^a_{-1} = 0 \tag{99}
$$

This implies that the gradients of $\underline{\boldsymbol{u}}_0$ and $\underset{\sim}{\chi}^a_1$ with respect to $\underline{\boldsymbol{y}}$ vanish, so that :

$$
\underline{\boldsymbol{u}}_0(\underline{\boldsymbol{x}}, \underline{\boldsymbol{y}}) = \underline{\boldsymbol{U}}_0(\underline{\boldsymbol{x}}), \quad \underset{\sim}{\chi}^a_1(\underline{\boldsymbol{x}}, \underline{\boldsymbol{y}}) = \underset{\sim}{\Xi}^a_1(\underline{\boldsymbol{x}}) \tag{100}
$$

The fields $(\underline{u}_1, \underset{\sim}{\chi}_1^s, \underset{\sim}{\chi}_2^a, \underset{\sim}{\chi}_3^a, \underset{\sim}{\sigma}_0, \underset{\sim}{s}_0, \underset{\approx}{m}_0, \underset{\approx}{m}_1)$ are solutions of the following auxiliary boundary value problem defined on the unit cell :

$$\begin{cases} \underset{\sim}{\varepsilon}_0 = \underline{U}_0 \overset{s}{\otimes} \nabla_x + \underline{u}_1 \overset{s}{\otimes} \nabla_y, \quad \underset{\sim}{e}_0 = \underline{U}_0 \otimes \nabla_x + \underline{u}_1 \otimes \nabla_y - \underset{\sim}{\Xi}_1^a - \underset{\sim}{\chi}_1^s \\ \underset{\approx}{\kappa}_{-1}^s = \underset{\sim}{\chi}_1^s \otimes \nabla_y, \ \underset{\approx}{\kappa}_0^a = \underset{\sim}{\Xi}_1^a \otimes \nabla_x + \underset{\sim}{\chi}_2^a \otimes \nabla_y, \ \underset{\approx}{\kappa}_1^a = \underset{\sim}{\chi}_2^a \otimes \nabla_x + \underset{\sim}{\chi}_3^a \otimes \nabla_y \\ \underset{\sim}{\sigma}_0 = \underset{\approx}{a}^{(0)} : \underset{\sim}{\varepsilon}_0, \quad \underset{\sim}{s}_0 = \underset{\approx}{b}^{(0)} : \underset{\sim}{e}_0 \\ \underset{\approx}{m}_0 = \underset{\approx\approx}{c}^{a(2)} \vdots \underset{\approx}{\kappa}_0^a, \quad \underset{\approx}{m}_1 = \underset{\approx\approx}{c}^{s(1)} \vdots \underset{\approx}{\kappa}_{-1}^s + \underset{\approx\approx}{c}^{a(2)} \vdots \underset{\approx}{\kappa}_1^a \\ (\underset{\sim}{\sigma}_0 + \underset{\sim}{s}_0) \cdot \nabla_y = 0, \quad \underset{\approx}{m}_0 . \nabla_y = 0, \quad \underset{\approx}{m}_0 \cdot \nabla_x + \underset{\approx}{m}_1 \cdot \nabla_y + \underset{\sim}{s}_0 = 0 \end{cases}$$

This complex problem can be seen to depend linearly on $\underline{U}_0 \overset{s}{\otimes} \nabla, \underline{U}_0 \overset{a}{\otimes} \nabla - \underset{\sim}{\Xi}_1^a$ and $\underset{\sim}{\Xi}_1^a \otimes \nabla$. The solutions take the form :

$$\begin{aligned} \underline{u}^\epsilon &= \underline{U}_0(\underline{x}) + \epsilon(\underline{U}_1(\underline{x}) + \underset{\simeq}{X}_u^{(3)}(\underline{y}) : (\underline{U}_0 \overset{s}{\otimes} \nabla) \\ &+ \underset{\simeq}{X}_e^{(3)}(\underline{y}) : (\underline{U}_0 \overset{a}{\otimes} \nabla - \underset{\sim}{\Xi}_1^a)) + \dots \end{aligned} \tag{101}$$

$$\underset{\sim}{\chi}^\epsilon = \underset{\sim}{\Xi}_1(\underline{x}) + \epsilon(\underset{\sim}{\Xi}_2(\underline{x}) + \underset{\approx\approx}{X}_\kappa^{(3)}(\underline{y}) \vdots (\underset{\sim}{\Xi}_1^a \otimes \nabla)) + \dots \tag{102}$$

where concentration tensors $\underset{\simeq}{X}_u^{(3)}, \underset{\simeq}{X}_e^{(3)}$ and $\underset{\approx\approx}{X}_\kappa^{(3)}$ have been introduced. Their components are determined by the successive solutions of the auxiliary problem for unit values of the components of $\underline{U}_0 \overset{s}{\otimes} \nabla, \underline{U}_0 \overset{a}{\otimes} \nabla - \underset{\sim}{\Xi}_1^a$ and $\underset{\sim}{\Xi}_1^a \otimes \nabla_y$. They are such that their mean value over the unit cell vanishes.

The macroscopic stress tensors and effective elastic properties are given by :

$$\begin{aligned} \underset{\sim}{\Sigma}_0 &= < \underset{\approx}{a}^{(0)} : (\underset{\approx}{1} + \nabla_x \overset{s}{\otimes} \underset{\simeq}{X}_u^{(3)}) >: (\underline{U}_0 \overset{s}{\otimes} \nabla) \\ &+ < \underset{\approx}{a}^{(0)} : (\nabla_y \overset{s}{\otimes} \underset{\simeq}{X}_e^{(3)}) >: (\underline{U}_0 \overset{a}{\otimes} \nabla - \underset{\sim}{\Xi}_1^a) \end{aligned} \tag{103}$$

$$\begin{aligned} \underset{\sim}{S}_0 &= < \underset{\sim}{s}_0 >=< \underset{\approx}{b}^{(0)} : (\nabla_y \otimes \underset{\simeq}{X}_u^{(3)}) >: (\underline{U}_0 \overset{s}{\otimes} \nabla) \\ &+ < \underset{\approx}{b}^{(0)} : (\nabla_y \otimes \underset{\simeq}{X}_e^{(3)}) >: (\underline{U}_0 \overset{a}{\otimes} \nabla - \underset{\sim}{\Xi}_1^a) \end{aligned} \tag{104}$$

$$\underset{\approx}{M}_0 =< \underset{\approx}{m}_0 >=< \underset{\approx\approx}{c}^{a(2)} \vdots (\underset{\approx\approx}{1} + \nabla_y \otimes \underset{\approx\approx}{X}_\kappa^{(3)}) > \vdots \underset{\sim}{\Xi}_1^a \otimes \nabla \tag{105}$$

They must fulfill the balance equations (79). Note that $\underset{\approx}{m}_0$ and therefore $\underset{\approx}{M}_0$ are skew–symmetric with respect to their first two indices. The averaged equation of balance of moment of momentum implies then that $\underset{\sim}{S}_0$ is skew–symmetric. The macroscopic degrees of freedom are the displacement field $\underline{U}_0$ and the rotation associated to $\underset{\sim}{\Xi}^a_1$. The found balance and constitutive equations are therefore that of a Cosserat effective medium. The more classical form of the Cosserat theory is retrieved once one rewrites the previous equations using the axial vector associated to $\underset{\sim}{\Xi}^a$ (Forest, 2001).

Homogenization scheme HS4

For the last homogenisation scheme defined in section 4.1, the equations describing the local behaviour are :

$$\underset{\sim}{\sigma}^\epsilon = \underset{\approx}{a}^{(0)}(\underline{y}) : \underset{\sim}{\varepsilon}^\epsilon, \quad \underset{\sim}{s}^\epsilon = \underset{\approx}{b}^{(0)}(\underline{y}) : \underset{\sim}{e}^\epsilon \tag{106}$$

$$\underset{\sim}{m}^\epsilon = \underset{\approx}{c}^{s(2)}(\underline{y}) \vdots \underset{\sim}{\kappa}^{s\epsilon} + \epsilon^2 \underset{\approx}{c}^{a(1)}(\underline{y}) \vdots \underset{\sim}{\kappa}^{a\epsilon} \tag{107}$$

At the order ϵ^{-1}, one gets

$$\underset{\approx}{a}^{(0)} : \underset{\sim}{\varepsilon}_{-1} = 0, \quad \underset{\approx}{b}^{(0)} : \underset{\sim}{e}_{-1} = 0, \quad \underset{\approx}{c}^{s(2)} \vdots \underset{\sim}{\kappa}^s_{-1} = 0 \tag{108}$$

This implies that the gradients of $\underline{u}_0$ and $\underset{\sim}{\chi}^s_1$ with respect to $\underline{y}$ vanish, so that :

$$\underline{u}_0(\underline{x}, \underline{y}) = \underline{U}_0(\underline{x}), \quad \underset{\sim}{\chi}^s_1(\underline{x}, \underline{y}) = \underset{\sim}{\Xi}^s_1(\underline{x}) \tag{109}$$

The fields $(\underline{u}_1, \underset{\sim}{\chi}^a_1, \underset{\sim}{\chi}^s_2, \underset{\sim}{\chi}^s_3, \underset{\sim}{\sigma}_0, \underset{\sim}{s}_0, \underset{\sim}{m}_0, \underset{\sim}{m}_1)$ are solutions of the following auxiliary boundary value problem defined on the unit cell :

$$\begin{cases} \underset{\sim}{\varepsilon}_0 = \underline{U}_0 \overset{s}{\otimes} \nabla_x + \underline{u}_1 \overset{s}{\otimes} \nabla_y, \quad \underset{\sim}{e}_0 = \underline{U}_0 \otimes \nabla_x + \underline{u}_1 \otimes \nabla_y - \underset{\sim}{\Xi}^s_1 - \underset{\sim}{\chi}^a_1 \\ \underset{\sim}{\kappa}^a_{-1} = \underset{\sim}{\chi}^a_1 \otimes \nabla_y, \ \underset{\sim}{\kappa}^s_0 = \underset{\sim}{\Xi}^s_1 \otimes \nabla_x + \underset{\sim}{\chi}^s_2 \otimes \nabla_y, \ \underset{\sim}{\kappa}^a_1 = \underset{\sim}{\chi}^a_2 \otimes \nabla_x + \underset{\sim}{\chi}^a_3 \otimes \nabla_y \\ \underset{\sim}{\sigma}_0 = \underset{\approx}{a}^{(0)} : \underset{\sim}{\varepsilon}_0, \quad \underset{\sim}{s}_0 = \underset{\approx}{b}^{(0)} : \underset{\sim}{e}_0 \\ \underset{\sim}{m}_0 = \underset{\approx}{c}^{s(2)} \vdots \underset{\sim}{\kappa}^s_0, \quad \underset{\sim}{m}_1 = \underset{\approx}{c}^{a(1)} \vdots \underset{\sim}{\kappa}^a_{-1} + \underset{\approx}{c}^{s(2)} \vdots \underset{\sim}{\kappa}^s_1 \\ (\underset{\sim}{\sigma}_0 + \underset{\sim}{s}_0) \cdot \nabla_y = 0, \quad \underset{\sim}{m}_0 . \nabla_y = 0, \quad \underset{\sim}{m}_0 \cdot \nabla_x + \underset{\sim}{m}_1 \cdot \nabla_y + \underset{\sim}{s}_0 = 0 \end{cases}$$

This complex problem can be seen to depend linearly on $\underline{U}_0 \overset{s}{\otimes} \nabla, \underline{U}_0 \overset{s}{\otimes} \nabla - \underset{\sim}{\Xi}^s_1$ and $\underset{\sim}{\Xi}^s_1 \otimes \nabla$. The solutions take the form :

$$\begin{aligned} \underline{u}^\epsilon &= \underline{U}_0(\underline{x}) + \epsilon(\underline{U}_1(\underline{x}) + \underset{\approx}{X}^{(4)}_u(\underline{y}) : (\underline{U}_0 \overset{s}{\otimes} \nabla) \\ &+ \underset{\approx}{X}^{(4)}_e(\underline{y}) : (\underline{U}_0 \overset{s}{\otimes} \nabla - \underset{\sim}{\Xi}^s_1)) + \ldots \end{aligned} \tag{110}$$

$$\underset{\sim}{\chi}^\epsilon = \underset{\sim}{\Xi}_1(\underline{x}) + \epsilon(\underset{\sim}{\Xi}_2(\underline{x}) + \underset{\approx}{X}^{(4)}_\kappa(\underline{y}) \vdots (\underset{\sim}{\Xi}^s_1 \otimes \nabla)) + \ldots \tag{111}$$

where concentration tensors $\underset{\approx}{X}_u^{(4)}, \underset{\approx}{X}_e^{(4)}$ and $\underset{\approx}{X}_\kappa^{(4)}$ have been introduced. Their components are determined by the successive solutions of the auxiliary problem for unit values of the components of $\underline{U}_0 \overset{s}{\otimes} \nabla, \underline{U}_0 \overset{s}{\otimes} \nabla - \underset{\sim}{\Xi}_1^s$ and $\underset{\sim}{\Xi}_1^s \otimes \nabla_y$. They are such that their mean value over the unit cell vanishes.

The macroscopic stress tensors and effective elastic properties are given by :

$$\begin{aligned}\underset{\sim}{\Sigma}_0 &= < \underset{\approx}{a}^{(0)} : (\underset{\approx}{1} + \nabla_x \overset{s}{\otimes} \underset{\approx}{X}_u^{(4)}) >: (\underline{U}_0 \overset{s}{\otimes} \nabla) \\ &+ < \underset{\approx}{a}^{(0)} : (\nabla_y \overset{s}{\otimes} \underset{\approx}{X}_e^{(4)}) >: (\underline{U}_0 \overset{s}{\otimes} \nabla - \underset{\sim}{\Xi}_1^s)\end{aligned} \tag{112}$$

$$\begin{aligned}\underset{\sim}{S}_0 =< \underset{\sim}{s}_0 > &= < \underset{\approx}{b}^{(0)} : (\nabla_y \otimes \underset{\approx}{X}_u^{(4)}) >: (\underline{U}_0 \overset{s}{\otimes} \nabla) \\ &+ < \underset{\approx}{b}^{(0)} >: (\underline{U}_0 \overset{s}{\otimes} \nabla - \underset{\sim}{\Xi}_1^s)\end{aligned} \tag{113}$$

$$\underset{\approx}{M}_0 =< \underset{\approx}{m}_0 >=< \underset{\approx}{c}^{s(2)} \vdots (\underset{\approx}{1} + \nabla_y \otimes \underset{\approx}{X}_\kappa^{(4)}) > \vdots (\underset{\sim}{\Xi}_1^s \otimes \nabla) \tag{114}$$

They must fulfill the balance equations (79). Note that $\underset{\approx}{m}_0$ and therefore $\underset{\approx}{M}_0$ are symmetric with respect to their first two indices. The averaged equation of balance of moment of momentum implies then that $\underset{\sim}{S}_0 = - < \underset{\approx}{m}_0 > \cdot \nabla$ is symmetric. The macroscopic degrees of freedom are the displacement field $\underline{U}_0$ and the symmetric strain tensor $\underset{\sim}{\Xi}_1^s$. Such a continuum is called a microstrain medium (Forest and Sievert, 2006).

As a conclusion, depending on the relative contributions of the various intrinsic length scales of the micromorphic continuum, different effective media are obtained, as summarised in table 5. The effective medium can be of micromorphic, microstrain, Cosserat or Cauchy type.

homogenisation scheme	characteristic lengths	effective medium
HS1	$l_s \sim l, l_a \sim l$	Cauchy
HS2	$l_s \sim L, l_a \sim L$	micromorphic
HS3	$l_s \sim l, l_a \sim L$	Cosserat
HS4	$l_s \sim L, l_a \sim l$	microstrain

Table 5. Homogenization of heterogenous micromorphic media : Nature of the homogeneous equivalent medium depending on the values of the intrinsic lengths of the constituents.

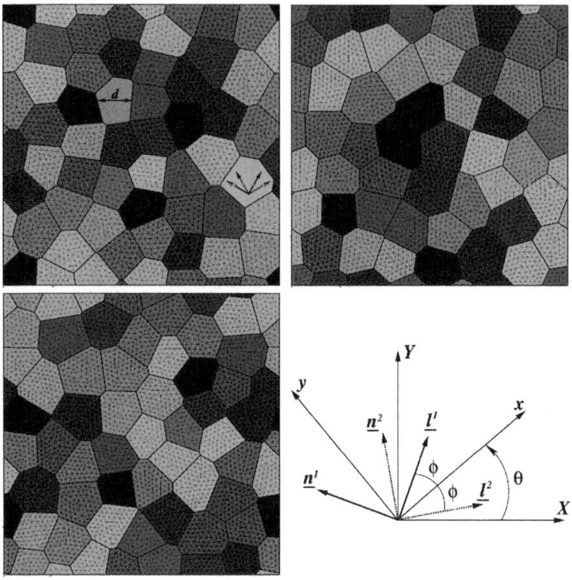

Figure 10. Periodic meshes of the 2D periodic aggregates used in the finite element simulations: (a) 52 grains, (b) 47 grains and (c) 55 grains. Two slip systems are taken into account in each randomly oriented grain. Various mean grain sizes, d, ranging from tens of nanometers to hundreds of microns, are investigated. (d) Description of the two effective slip systems for 2D planar double slip.

4.2 Application to Polycrystalline Plasticity

The previous homogenisation method is extended to non linear micromorphic constitutive equations in order to predict size effects in the plasticity of polycrystals. The micromorphic single crystal model is not presented here and the reader is referred to (Cordero et al., 2010, 2012) for a detailed presentation of the model and a more complete description of polycrystal homogenisation.

Periodic Homogenisation of Micromorphic Polycrystals

The computation of polycrystalline aggregates based on standard crystal plasticity models follows the rule of classical homogenisation theory in the sense that a mean strain is prescribed to a volume element of poly-

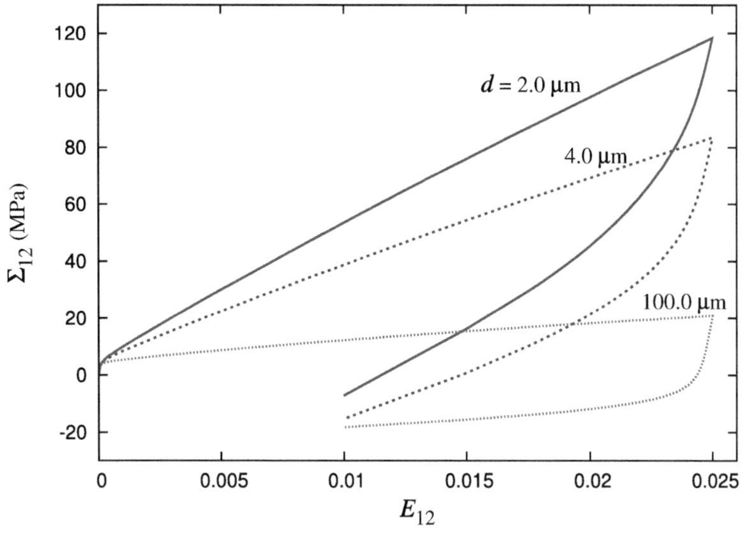

Figure 11. Macroscopic stress–strain response of the 52–grain aggregate of Fig. 10(a) under simple shear loading conditions including unloading for three different grain sizes.

crystalline materials using suitable boundary conditions like strain–based, stress–based or periodic ones. The structure of the boundary value problem is modified if a generalized continuum approach is used inside the considered volume element. In the present work, we are considering the computational homogenisation of a heterogeneous micromorphic medium and suitable boundary conditions for displacement and plastic microdeformation must be defined. In the case of linear material behavior, the structure of the unit cell problem to be solved can be derived from multiscale asymptotic expansion analysis, as shown in section 4.1. The obtained boundary conditions are then assumed to hold also for non–linear material responses. We look for the displacement field $\underline{\boldsymbol{u}}$ and the plastic microdeformation field $\underset{\sim}{\boldsymbol{\chi}}^p$ in the polycrystal volume element such that

$$\underline{\boldsymbol{u}}(\underline{\boldsymbol{x}}) = \underset{\sim}{\boldsymbol{E}} \cdot \underline{\boldsymbol{x}} + \underline{\boldsymbol{v}}(\underline{\boldsymbol{x}}), \quad \forall \underline{\boldsymbol{x}} \in V \tag{115}$$

the fluctuation $\underline{\boldsymbol{v}}$ being periodic at homologous points of the boundary ∂V. Under these conditions the prescribed average strain is the symmetric second order tensor $\underset{\sim}{\boldsymbol{E}}$. The plastic microdeformation $\underset{\sim}{\boldsymbol{\chi}}^p$ is periodic at ho-

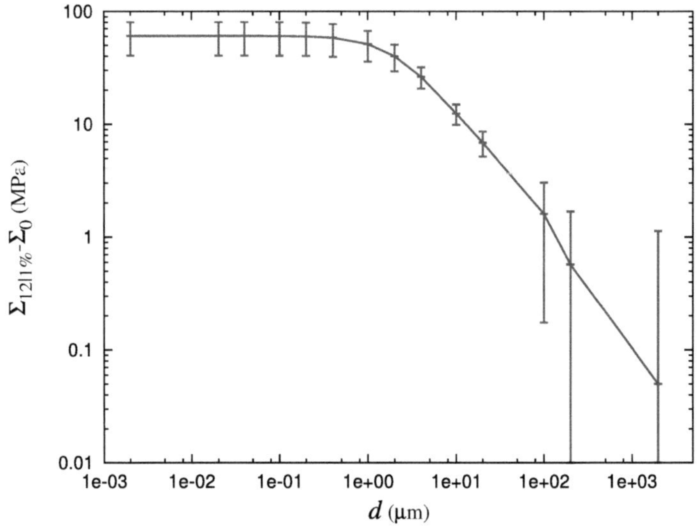

Figure 12. Effect of the mean grain size, d, on the macroscopic flow stress $\Sigma_{12|1\%}$ at 1% mean plastic strain. The results are obtained with the three aggregates of Fig. 10 under simple shear. The error bars give the standard deviation.

mologous points of ∂V. As a result, the mean value of the microdeformation gradient $\underset{\sim}{\mathbf{\Gamma}}$ vanishes.

The grain boundary conditions must now be discussed. At any interface of a micromorphic continuum, there may exist some jump conditions for the degrees of freedom of the theory and the associated reactions, namely the simple and double traction vectors. We consider in this work that such jumps do not exist. Instead, the displacement vector and the plastic microdeformation tensor are assumed to be continuous at grain boundaries. As a result, the simple and double tractions also are continuous. The continuity of plastic microdeformation is a new grain boundary condition that does not exist in classical crystal plasticity. It will generate boundary layers at grain boundaries which are essential for the observed size effects (Cordero et al., 2010). Let us imagine a grain boundary between a plastically deforming grain and an elastic grain where plasticity is not triggered. The condition of continuity of plastic microdeformation implies that the plastic microdeformation should vanish at this grain boundary, thus leading to a decrease of plastic slip close to the grain boundary, associated with pile–up formation. More generally continuity of plastic microdeformation is enforced at grain boundaries. Also grain boundaries are assumed to transmit simple and double tractions. Jump conditions or more specific interface laws may well be more realistic or physically motivated but they would require additional considerable computational effort. We think that the continuity conditions carry the main physical ingredient to capture the targeted size effects in plasticity.

The simulations are limited to two–dimensional crystals under plane strain conditions and a mean shear strain E_{12} is prescribed to the volume elements. Three volume elements are considered, made of 52, 47 and 55 grains, respectively, according to a 2D Voronoi tessellation with periodic constraints. They are shown in Fig. 10. The three realisations of the material have different grain shapes and different orientations chosen randomly. Homothetic volumes constructed from the three previous volume elements will be considered, thus having different mean grain sizes but the same grain morphology and crystallographic texture.

Only two planar slip systems are considered in most simulations of this work. The slip directions and normal to the slip planes are contained in the considered plane. They are separated by the angle 2ϕ with $\phi = 35.1°$ following Bennett and McDowell (2003). The sets of material parameters used for the simulations are given in (Cordero et al., 2012).

The micromorphic constitutive model contains a characteristic length equal here to $l_\omega = 450$ nm. Note that this intrinsic length is defined from the ratio of two constitutive moduli for reason of dimensionality. The resulting characteristic thickness of boundary layers affected by the strain gradient effects, especially close to grain boundary, will generally be proportional to l_ω with a factor depending on other constitutive parameters and on the type of boundary value problem. Such characteristic lengths have been derived from analytical solutions in some simple boundary value problems in (Cordero et al., 2010). For polycrystals, they will emerge from the computational analysis.

Evidence of Size-Dependent Kinematic Hardening

Fig. 11 gives the mean shear stress as a function of mean shear strain as a result of a finite element simulation of the 52–grain aggregate of Fig. 10(a) for three different grain sizes. One shear loading branch up to 0.025 mean shear strain followed by the unloading branch are presented. The stress–strain curves clearly exhibit an overall kinematic hardening effect induced by the local contributions of the double stress tensor, as proved in (Cordero et al., 2010). The kinematic hardening vanishes for large grains and is all the stronger as the grain size is smaller.

From the overall shear curves, the shear stress value $\Sigma_{12|p_0}$ was recorded at a given level of mean plastic microstrain $\chi_{12}^{ps} = p_0$ where χ_{12}^{ps} is defined as:

$$\chi_{12}^{ps} = (\chi_{12}^p + \chi_{21}^p)/2 \tag{116}$$

The mean shear stress $\Sigma_{12|p_0}$ was plotted as a function of grain size. It turns out that the shear stress value converges toward a fixed value Σ_0 for large grain sizes. This limit depends only on the value of the critical resolved shear stress entering the Schmid law and on the specific geometry and orientations of the considered polycrystalline aggregates. It is therefore possible to draw a Hall–Petch diagram which is a log–log plot of $\Sigma_{12|p_0} - \Sigma_0$ vs. the grain size d. Such a plot is given in Fig. 12 for $p_0 = 0.01$. The continuous line gives the mean value of the shear stress level for the three realizations of the microstructure considered in Fig. 10. Error bars are also provided showing the scatter of the results which is rather strong due to the small number of grains in each microstructure and the small number of considered aggregates. The diagram of Fig. 12 clearly shows two regimes in the relation between stress level and grain size. For grain sizes smaller than 1 μm, no dependence of the overall stress on grain size is observed. For grain sizes larger than 1 μm, a power law is found in the form

$$\Sigma_{12|p_0} - \Sigma_0 \propto d^m \tag{117}$$

with an exponent m of the order of -0.9 for the mean curve in Fig. 12. The micromorphic model therefore can account for grain size effects with a saturation for too small grain sizes.

The dislocation density tensor $\underset{\sim}{\mathbf{\Gamma}} = \text{curl}\, \underset{\sim}{\boldsymbol{\chi}}^p$ does not only impact the overall polycrystal behavior but also the way plastic deformation develops inside the grains. An example of the spreading of plastic deformation in a polycrystal depending on the grain size is shown in Fig. 13 for the 52–grain aggregate of Fig. 10(a). The shown maps are the contour plots of the field of equivalent plastic deformation p. At the onset of plastic deformation, plasticity starts in the same grains and at the same locations in 100μm–grains as in 1μm–grains, as shown by the pictures of Fig. 13(a) and (b). This is due to the fact that the same critical resolved shear stress is adopted for both grain sizes, corresponding to the same initial dislocation density. In contrast, at higher mean plastic strain levels, the strongly different values of the plastic microdeformation gradients lead to significantly different plastic strain fields. Two main features are evidenced by the Fig. 13(c) to (f). First, a tendency to strain localization in bands is observed for small grain sizes. The observed bands cross several grains whereas plastic strain is more diffuse at larger grain sizes. This fact was already observed in the

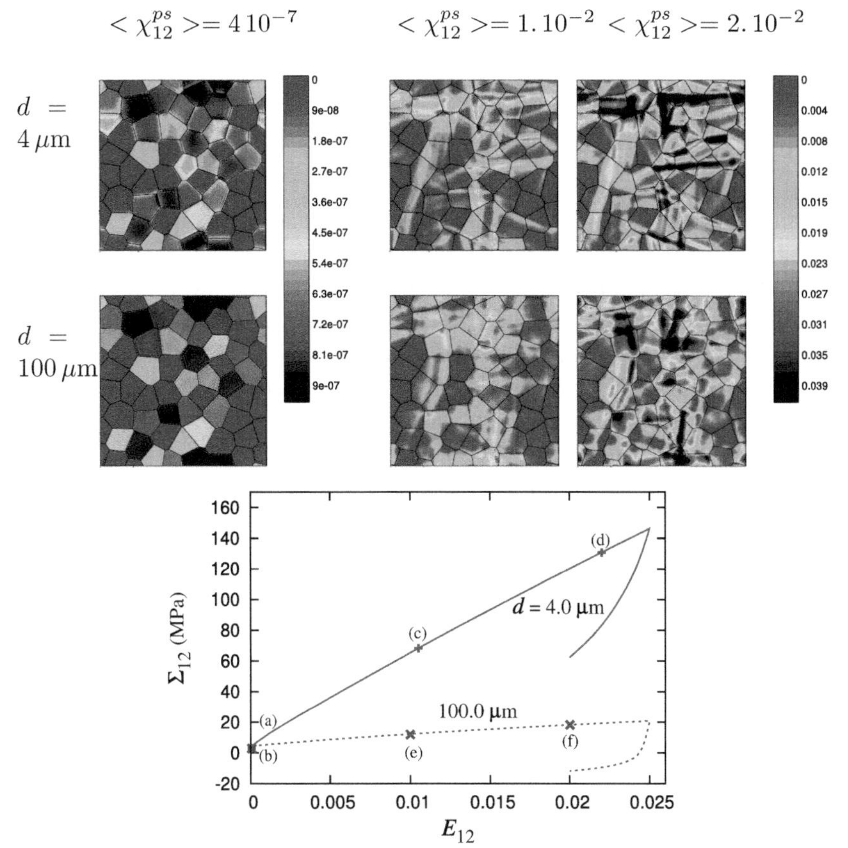

Figure 13. Contour plots of the accumulative plastic strain p for two grain sizes, $d = 100$ and $4\,\mu$m, and for three different mean values of the plastic strain : $< \chi_{12}^{ps} >\approx 0.$, $< \chi_{12}^{ps} >= 0.01$ and $< \chi_{12}^{ps} >= 0.02$, obtained with the 55–grain aggregate of Fig. 10(c) under simple shear. (g) Macroscopic stress–strain response of the corresponding aggregate, letters indicating the loading steps for which the maps are shown.

simulations presented in Cordero et al. (2011, 2012). Second, a consequence of this localization is that some small grains are significantly less deformed that the larger ones.

Bibliography

E.C. Aifantis. On the microstructural origin of certain inelastic models. *Journal of Engineering Materials and Technology*, 106:326–330, 1984.

J. Altenbach, H. Altenbach, and V.A. Eremeyev. On generalized cosserat-type theories of plates and shells: a short review and bibliography. *Arch. Appl. Mech.*, 80:73–92, 2010.

A. Anthoine. Second-order homogenisation of functionally graded materials. *Int. J. Solids Structures*, 47:1477–1489, 2010.

O. Aslan, N. M. Cordero, A. Gaubert, and S. Forest. Micromorphic approach to single crystal plasticity and damage. *International Journal of Engineering Science*, 49:1311–1325, 2011.

N. Auffray, R. Bouchet, and Y. Bréchet. Derivation of anisotropic matrix for bi–dimensional strain–gradient elasticity behavior. *International Journal of Solids and Structures*, 46:440–454, 2009.

N. Auffray, R. Bouchet, and Y. Bréchet. Strain gradient elastic homogenization of bidimensional cellular media. *International Journal of Solids and Structures*, 47:1698–1710, 2010.

V.P. Bennett and D.L. McDowell. Crack tip displacements of microstructurally small surface cracks in single phase ductile polycrystals. *Engineering Fracture Mechanics*, 70(2):185–207, 2003.

D. Besdo. Towards a Cosserat-theory describing motion of an originally rectangular structure of blocks. *Arch. Appl. Mech.*, 80:25–45, 2010.

J. Besson, G. Cailletaud, J.-L. Chaboche, S. Forest, and M. Blétry. *Non–Linear Mechanics of Materials.* Series: Solid Mechanics and Its Applications, Vol. 167, Springer, ISBN: 978-90-481-3355-0, 433 p., 2009.

C. Boutin. Microstructural effects in elastic composites. *Int. J. Solids Structures*, 33:1023–1051, 1996.

F. Bouyge, I. Jasiuk, and M. Ostoja-Starzewski. A micromechanically based couple-stress model of an elastic two-phase composite. *Int. J. Solids Structures*, 38:1721–1735, 2001.

F. Bouyge, I. Jasiuk, S. Boccara, and M. Ostoja-Starzewski. A micromechanically based couple-stress model of an elastic orthotropic two-phase composite. *European Journal of Mechanics A/solids*, 21:465–481, 2002.

H. Chen, X. Liu, G. Hu, and H. Yuan. Identification of material parameters of micropolar theory for composites by homogenization method. *Computational Materials Science*, 46:733–737, 2009.

N. M. Cordero, S. Forest, E. P. Busso, S. Berbenni, and M. Cherkaoui. Grain size effect in generalized continuum crystal plasticity. In I. R. Ionescu, S. Bouvier, O. Cazacu, and P. Franciosi, editors, *Plasticity of crystalline materials*, pages 101–122. ISTE-Wiley, 2011.

N. M. Cordero, S. Forest, E. P. Busso, S. Berbenni, and M. Cherkaoui. Grain size effects on plastic strain and dislocation density tensor fields in metal polycrystals. *Computational Materials Science*, 52:7–13, 2012.

N.M. Cordero, A. Gaubert, S. Forest, E. Busso, F. Gallerneau, and S. Kruch. Size effects in generalised continuum crystal plasticity for two–phase laminates. *Journal of the Mechanics and Physics of Solids*, 58:1963–1994, 2010.

T. Dillard, S. Forest, and P. Ienny. Micromorphic continuum modelling of the deformation and fracture behaviour of nickel foams. *European Journal of Mechanics A/Solids*, 25:526–549, 2006.

K. Enakoutsa and J.B. Leblond. Numerical implementation and assessment of the glpd micromorphic model of ductile rupture. *European Journal of Mechanics A/solids*, 28:445–460, 2009.

A.C. Eringen. Theory of thermo-microstretch elastic solids. *Int. J. Engng Sci.*, 28:1291–1301, 1990.

A.C. Eringen. *Microcontinuum field theories.* Springer, New York, 1999.

A.C. Eringen and E.S. Suhubi. Nonlinear theory of simple microelastic solids. *Int. J. Engng Sci.*, 2:189–203, 389–404, 1964.

F. Feyel. A multilevel finite element method (FE2) to describe the response of highly non-linear structures using generalized continua. *Comp. Meth. Appl. Mech. Engng*, 192:3233–3244, 2003.

S. Forest. The micromorphic approach for gradient elasticity, viscoplasticity and damage. *ASCE Journal of Engineering Mechanics*, 135:117–131, 2009.

S. Forest. Mechanics of generalized continua : Construction by homogenization. *Journal de Physique IV*, 8:Pr4–39–48, 1998.

S. Forest. Homogenization methods and the mechanics of generalized continua–Part 2. *Theoretical and Applied Mechanics*, 28–29:113–143, 2002.

S. Forest. Aufbau und Identifikation von Stoffgleichungen für höhere Kontinua mittels Homogenisierungsmethoden. *Technische Mechanik*, Band 19, Heft 4:297–306, 1999.

S. Forest. Cosserat media. In K.H.J. Buschow, R.W. Cahn, M.C. Flemings, B. Ilschner, E.J. Kramer, and S. Mahajan, editors, *Encyclopedia of Materials : Science and Technology*, pages 1715–1718. Elsevier, 2001.

S. Forest and A. Bertram. Formulations of strain gradient plasticity. In H. Altenbach, G. A. Maugin, and V. Erofeev, editors, *Mechanics of Generalized Continua*, pages 137–150. Advanced Structured Materials vol. 7, Springer, 2011.

S. Forest and K. Sab. Cosserat overall modeling of heterogeneous materials. *Mechanics Research Communications*, 25(4):449–454, 1998.

S. Forest and R. Sievert. Elastoviscoplastic constitutive frameworks for generalized continua. *Acta Mechanica*, 160:71–111, 2003.

S. Forest and R. Sievert. Nonlinear microstrain theories. *International Journal of Solids and Structures*, 43:7224–7245, 2006.

S. Forest and D. K. Trinh. Generalized continua and non–homogeneous boundary conditions in homogenization methods. *ZAMM*, 91:90 109, 2011.

S. Forest, F. Pradel, and K. Sab. Asymptotic analysis of heterogeneous Cosserat media. *International Journal of Solids and Structures*, 38:4585–4608, 2001.

M.G.D. Geers, V.G. Kouznetsova, and W.A.M Brekelmans. Gradient–enhanced computational homogenization for the micro–macro scale transition. *Journal de Physique IV*, 11:Pr5–145–152, 2001.

P. Germain. The method of virtual power in continuum mechanics. part 2 : Microstructure. *SIAM J. Appl. Math.*, 25:556–575, 1973.

J.D. Goddard. *Mathematical models of granular matter, ed. by P. Mariano, G. Capriz, and P. Giovine*, chapter From Granular Matter to Generalized Continuum, pages 1–20. vol. 1937 of Lecture Notes in Mathematics, Springer, Berlin, 2008.

M. Gologanu, J. B. Leblond, and J. Devaux. *Continuum micromechanics*, volume 377, chapter Recent extensions of Gurson's model for porous ductile metals, pages 61–130. Springer Verlag, CISM Courses and Lectures No. 377, 1997.

M.A. Goodman and S.C. Cowin. A continuum theory for granular materials. *Arch. Rational Mech. and Anal.*, 44:249–266, 1972.

P. Grammenoudis and Ch. Tsakmakis. Micromorphic continuum Part I: Strain and stress tensors and their associated rates. *International Journal of Non–Linear Mechanics*, 44:943–956, 2009.

P. Grammenoudis, Ch. Tsakmakis, and D. Hofer. Micromorphic continuum Part II: Finite deformation plasticity coupled with damage. *International Journal of Non–Linear Mechanics*, 44:957–974, 2009.

C.B. Hirschberger, E. Kuhl, and P. Steinmann. On deformational and configurational mechanics of micromorphic hyperelasticity - theory and computation. *Computer Methods in Applied Mechanics and Engineering*, 196:4027–4044, 2007.

R. Jänicke and Diebels. A numerical homogenisation strategy for micromorphic continua. *Nuovo Cimento della Societa Italiana di Fisica C-Geophysics and Space Physics*, 32:121–132, 2009.

R. Jänicke, S. Diebels, H.-G. Sehlhorst, and A. Düster. Two–scale modelling of micromorphic continua. *Continuum Mechanics and Thermodynamics*, 21:297–315, 2009.

C.B. Kafadar and A.C. Eringen. Micropolar media: I the classical theory. *Int. J. Engng Sci.*, 9:271–305, 1971.

T. Kanit, S. Forest, I. Galliet, V. Mounoury, and D. Jeulin. Determination of the size of the representative volume element for random composites : statistical and numerical approach. *International Journal of Solids and Structures*, 40:3647–3679, 2003.

N. Kirchner and P. Steinmann. A unifying treatise on variational principles for gradient and micromorphic continua. *Philosophical Magazine*, 85: 3875–3895, 2005.

V. G. Kouznetsova, M. G. D. Geers, and W. A. M. Brekelmans. Multi-scale constitutive modelling of heterogeneous materials with a gradient-enhanced computational homogenization scheme. *Int. J. Numer. Meth. Engng*, 54:1235–1260, 2002.

V. G. Kouznetsova, M. G. D. Geers, and W. A. M. Brekelmans. Multi-scale second-order computational homogenization of multi-phase materials : A nested finite element solution strategy. *Computer Methods in Applied Mechanics and Engineering*, 193:5525–5550, 2004.

S. Kruch and S. Forest. Computation of coarse grain structures using a homogeneous equivalent medium. *Journal de Physique IV*, 8:Pr8–197–205, 1998.

M. Lazar and G.A. Maugin. On microcontinuum field theories: the eshelby stress tensor and incompatibility conditions. *Philosophical Magazine*, 87: 3853–3870, 2007.

X. Liu and G. Hu. Inclusion problem of microstretch continuum. *International Journal of Engineering Science*, 42:849–860, 2003.

J. Mandel. Equations constitutives et directeurs dans les milieux plastiques et viscoplastiques. *Int. J. Solids Structures*, 9:725–740, 1973.

G.A. Maugin. Internal variables and dissipative structures. *J. Non–Equilib. Thermodyn.*, 15:173–192, 1990.

R.D. Mindlin. Micro–structure in linear elasticity. *Arch. Rat. Mech. Anal.*, 16:51–78, 1964.

R.D. Mindlin. Second gradient of strain and surface–tension in linear elasticity. *Int. J. Solids Structures*, 1:417–438, 1965.

M. Ostoja-Starzewski, S. D. Boccara, and I. Jasiuk. Couple-stress moduli and characteristic length of two-phase composite. *Mechanics Research Communication*, 26:387–396, 1999.

R.A. Regueiro. On finite strain micromorphic elastoplasticity. *International Journal of Solids and Structures*, 47:786–800, 2010.

G. Salerno and F. de Felice. Continuum modeling of periodic brickwork. *International Journal of Solids and Structures*, 46:1251–1267, 2009.

E. Sanchez-Palencia. Comportement local et macroscopique d'un type de milieux physiques hétérogènes. *International Journal of Engineering Science*, 12:331–351, 1974.

V. Sansalone, P. Trovalusci, and F. Cleri. Multiscale modeling of materials by a multifield approach : microscopic stress and strain distribution in fiber-matrix composites. *Acta Materialia*, 54:3485–3492, 2006.

C. Sansour. A unified concept of elastic–viscoplastic Cosserat and micromorphic continua. *Journal de Physique IV*, 8:Pr8–341–348, 1998a.

C. Sansour. A theory of the elastic–viscoplastic cosserat continuum. *Archives of Mechanics*, 50:577–597, 1998b.

C. Sansour, S. Skatulla, and H. Zbib. A formulation for the micromorphic continuum at finite inelastic strains. *Int. J. Solids Structures*, 47:1546–1554, 2010.

H. Steeb and S. Diebels. A thermodynamic–consistent model describing growth and remodeling phenomena. *Computational Materials Science*, 28:597–607, 2003.

P. Trovalusci and R. Masiani. Non-linear micropolar and classical continua for anisotropic discontinuous materials. *International Journal of Solids and Structures*, 40:1281–1297, 2003.

C. Truesdell and W. Noll. *The non-linear field theories of mechanics*. Handbuch der Physik, edited by S. Flügge, reedition Springer Verlag 2004, 1965.

C.A. Truesdell and R.A. Toupin. *Handbuch der Physik, S. Flügge, editor, vol. 3*, chapter The classical field theories, pages 226–793. Springer verlag, Berlin, 1960.

F. Xun, G. Hu, and Z. Huang. Size–dependence of overall in–plane plasticity for fiber composites. *International Journal of Solids and Structures*, 41: 4713–4730, 2004.

X. Yuan, Y. Tomita, and T. Andou. A micromechanical approach of nonlocal modeling for media with periodic microstructures. *Mechanics Research Communications*, 35:126–133, 2008.

Electromagnetism and Generalized Continua

Gérard A. Maugin
Institut Jean Le Rond d'Alembert
Université Pierre et Marie Curie, Paris, France
E-mail: gerard.maugin@upmc.fr

Abstract In this series of lectures, after an introduction of the most basic elements and some historical perspective on the matter, an exposition of electromagnetic source terms to be taken into account in the Galilean invariant continuum thermodynamics of deformable continua is first given. The emphasis here is placed on the notions of ponderomotive force and couple, of which the latter already hints at some generalization of usual continua by the necessity to envisage nonsymmetric Cauchy stress tensors. Then the notion of magnetic continua endowed with a dynamic magnetic structure, such as in ferromagnets and antiferromagnets, is envisioned. The corresponding modelling can be made by exploiting different methods, among these a direct model of magneto-mechanical interactions in the manner of H. F. Tiersten but also an application of a generalized version of the principle of virtual power, or that of a Hamiltonian variational principle in the absence of dissipation. Such approaches allow one to exhibit a strict analogy with the equations that govern generalized, purely mechanical, continua such as micropolar or oriented media in the manner of the Cosserat brothers or Eringen, by introducing the notions of spin and couple stress. A parallel approach is given for electro-deformable media endowed with permanent electric polarization, e.g., ferroelectrics. Then analogies are established between the resonance couplings arising in certain structures (plates, shells) as shown by Mindlin in classical studies, and those existing for coupled magnetoelastic and electroelastic waves of different types. Finally, the contribution of these electromagnetic microstructures in the computation of configurational forces (e.g., driving forces acting on cracks) is shown to be quite similar to the terms due to a mechanical microstructure.

1 Introduction and Historical Perspective

An introduction to basic properties of electromagnetic properties is to be found in Chapter 1, pp. 1–61, of Maugin (1988).

Placed in an electric field $\mathbf{E}$, a point-like electric charge (monopole, a scalar) q is subjected to a force

$$\mathbf{F}^e = q\,\mathbf{E}. \tag{1}$$

In the same condition an electric dipole $\mathbf{p}$ (a polar vector) is acted upon by a couple (axial vector)

$$\mathbf{C}^e = \mathbf{p} \times \mathbf{E}. \tag{2}$$

There exists, so far, no evidence for the existence of magnetic monopoles (magnetic charges). But a magnetic dipole $\mathbf{m}$ (an axial vector) placed in a magnetic field $\mathbf{H}$ is the object of a mechanical couple

$$\mathbf{C}^m = \mathbf{m} \times \mathbf{H}. \tag{3}$$

It is this concept that explains the alignment of the needle of a compass with the local earth magnetic field. Full alignment after a transient period nullifies the couple (3).

In a continuous body where one assumes a continuous distribution of electric charges, electric dipoles, magnetic dipoles, etc., the force expression per unit volume generalizing (1) is much more complicated and requires a specific derivation (see Section 2). But the generalizations of (2) and (3) are relatively simple. One defines the electric polarization $\mathbf{P}$ and magnetization $\mathbf{M}$ per unit volume in such a way that (2) and (3) are replaced by

$$\mathbf{C}^m = \mathbf{P} \times \mathbf{E} + \mathbf{M} \times \mathbf{H}. \tag{4}$$

Here $\mathbf{P}$ and $\mathbf{M}$ are primarily determined by $\mathbf{E}$ and $\mathbf{H}$, i.e., we can write symbolically

$$\mathbf{P} = \mathbf{P}(\mathbf{E};.), \qquad \mathbf{M} = \mathbf{M}(\mathbf{H};.) \tag{5}$$

the missing arguments being temperature, strain, etc. Equations (5) are *constitutive equations* that characterize a specific material. The fields $\mathbf{P}$ and $\mathbf{M}$ are "material" fields (they vanish in a vacuum) and are classically defined per unit of matter. They introduce the difference between $\mathbf{E}$ and the electric displacement vector $\mathbf{D}$ on the one hand, and between $\mathbf{H}$ and the magnetic induction $\mathbf{B}$, on the other hand, so that — in so-called Lorentz–Heaviside electromagnetic units —

$$\mathbf{H} = \mathbf{B} - \mathbf{M}, \qquad \mathbf{D} = \mathbf{E} + \mathbf{P}. \tag{6}$$

This, in turn, means that $\mathbf{D}$ can be used instead of $\mathbf{E}$ in equation (4). Similarly, $\mathbf{B}$ can be used in place of $\mathbf{H}$ in that equation.

Equation (4) is important in the present context for the following reason. It has long been difficult to conceive of a purely mechanical means to produce a couple per unit volume. Accordingly, the possible existence of couples such as (4) has regularly been advanced as a justification to consider **nonsymmetric** stress tensors (that are classically shown to be symmetric in the absence of body couple) and this, as we know (cf. Maugin, 2010, 2011a), was the primary reason to introduce the first and simplest generalization of continuum mechanics.

In truth the interaction between electromagnetic behavior and microstructure and the mechanical, statical or dynamical, response of a material may be much more farfetched than simply through the couple (4).

First of all, while (4) may be referred to as the *ponderomotive couple* (the naming is misleading as it still smells of its "point-like" origin), the corresponding *ponderomotive force* (a force per unit volume of the material) may be much more involved than that given by (1). A justification for a rather reasonable form of this force will be given in Section 2.

Second, and that is most important as quite often these are the only remaining effects accounted for by engineers, there may exist couplings of *energetic origin* such as of piezoelectric, electrostrictive, electroelastic, piezomagnetic and magnetostrictive types. It might be a surprise to most readers to learn that *magnetostriction* (a longitudinal strain produced in a mechanically free body by a magnetic field in the length direction of the specimen, and that goes like the *square* of this field) was discovered by James Joule (of electric-conduction and thermodynamics fame) in 1842 in Nickel. This magneto-mechanical coupling exists in all ferromagnetic bodies to a greater or lesser extent. There is no symmetry restriction. The analogous electro-mechanical coupling is called *electrostriction* and also exists in all electro-deformable bodies to a greater or lesser extent for the same reason. Like magnetostriction it is an effect of the *second order* in the field. Because of its smallness the effect was in fact observed and modelled only in the 1920s. The energy electro-magneto-mechanical effects of the *first order* are of different nature and complexity for they necessarily involve appropriate symmetry properties of the considered material. Thus, (inverse) *piezoelectricity*, discovered in 1881 by the Curie brothers after their discovery of the "direct" piezoelectric effect (1880; appearance of electric charges at the boundary of a deformed body), provides a strain that is linear in the applied electric field but only for certain allowed material symmetries of crystals (see Katzir, 2003). It requires the absence of a center of symmetry to allow for a direct coupling between a stress (essentially a second-order tensor variable) and an electric field (essentially a polar vector). *Piezomagnetism*, the somewhat equivalent magneto-mechanical coupling first observed in Russia in 1960 (Borovik-Romanov), is even rarer in that the axial nature of a magnetic field requires a specific magnetic symmetry of the material (e.g., some antiferromagnetic fluorides). Higher order electro- and magneto-mechanical couplings of energy origin are simply referred to as *electroelastic couplings* and *magnetoelastic couplings* of higher order (see, e.g., Maugin et al., 1992). It must be noted that Pierre Curie paid special attention to the differences between polar vectors (e.g., electric polarization) and axial vectors (e.g., magnetization) as this plays a fundamental role in his well known statement of his principle of symmetry.

The just mentioned energetic couplings do not modify before hand the standard balance equations of mechanics (e.g., the balance of moment of momentum: *the stress remains symmetric* as it still is the derivative of the internal or free energy with respect to a symmetric strain). With much more drastic consequences is the fact that some electromagnetic materials are endowed with an electric or magnetic microstructure which, in spite of being of microscopic origin, does influence the mechanical behavior and, in particular, yields a non symmetric stress tensor in an equation of moment of momentum that acquires additional contributions. This is the case in materials exhibiting so-called ferroic states (for a classification of these, see Aizu, 1970). Here, because of the different vectorial natures of electric polarization and magnetization we must distinguish between the magnetic case and the electric case.

Examining first the electric case, we note that electric polarization is akin to a *mechanical displacement* up to an electric charge (cf. Maugin, 1988, chap. 1). That is why in some electric materials we can associate a kind of classical inertia with the *electric microstructure* provided by a network of permanent electric dipoles. That is, $\dot{\mathbf{p}}$ denoting the time derivative of an electric dipole density, we may have to consider an associated "kinetic energy" of the type

$$K(\dot{\mathbf{p}}) = \frac{1}{2} d_E \, \dot{\mathbf{p}} \cdot \dot{\mathbf{p}}, \tag{7}$$

where d_E is a kind of inertia to be evaluated from a microscopic model. Such an evaluation was given in a work by Pouget et al. (1986a,b) for ferroelectric materials of the molecular-group type (e.g., $NaNO_2$). An expression such as (7) is reminiscent of the kinetic energy formally associated with so-called directors in Ericksen's (1960) theory of "anisotropic" (microstructured, liquid-crystal) fluids. In addition, because of a prevalent electric ordering in the network of electric dipoles in ferroelectrics, the interaction between neighbouring electric dipoles leads to considering the presence of the gradient of electric polarization in the internal or free density energy of the material considered as a continuum, a kind of *weak nonlocality*.

In the magnetic case of ferroic states — in ferromagnetism, antiferromagnetism and ferrimagnetism — the magnetic dipole density is akin to an angular momentum — a spin, as of a particle in rotation about an axis; see the microscopic definition of magnetization in Maugin (1988). Accordigly, there exists the celebrated gyromagnetic relation of quantum-mechanical origin between a spin density and a magnetization density, e.g., per unit of mass of the material,

$$\mathbf{s} = \gamma^{-1}\,\mu, \tag{8}$$

where γ is the so-called gyromagnetic ratio and μ denotes the magnetization per unit mass (cf. Van Vleck, 1932).

The existence of the relation (8) makes that inertia in magnetic ferroic states is completely different in form from that in ferroelectrics — equation (7). As a matter of fact, since (8) obviously relates to rotational-precessional dynamical properties, there does not exist a closed form for the continuum magnetic kinetic energy in such bodies. The reason is that, magnetization having reached saturation in small domains of the material, we have the important identity

$$\dot{\mathbf{s}} \cdot \omega \equiv 0, \tag{9}$$

where ω is the precessional velocity of the magnetic spin. We say that $\dot{\mathbf{s}}$ is a *d'Alembertian inertia couple* (i.e., like a minute gyroscope, it does not produce any power in a real precessional velocity; see Tiersten, 1964). Only a quantum-mechanical formulation using the appropriate formalism (spinors, Pauli matrices) allows one to introduce an integrated form of the kinetic energy (cf. Nelson and Chen, 1994). Authors not aware of these facts were tempted to introduce a magnetization kinetic energy by analogy with that of electric dipoles (7) and Ericksen's (1960) director theory. This is the case of Lenz (1972) and more recent works that ignore the physical bases (e.g., De Simone and Podio-Guidugli, 1996). This is pitiful because it yields wrong results concerning wave propagation phenomena. But similarly to the ferroelectric case, the interaction between neighboring magnetic spins induces a dependency of the internal or free energy on the gradient of magnetization.

On account of the above-made remarks, we can cite a selection of those papers that have contributed much to the development of the equations, and their dynamical exploitation thereof, of deformable ferroelectrics and ferromagnets in a continuum landscape:

- For ferroelectrics and ionic crystals: Voigt (1928), Tiersten (1971), Mindlin (1972), Maugin (1976c), Maugin and Pouget (1980);
- For ferromagnets and their generalizations: Kittel (1958), Tiersten (1964, 1965), Brown (1966), Akhiezer et al. (1968), Maugin and Eringen (1972a), Maugin (1971, 1976a,b, 1988), Soumahoro and Pouget (1994), Maugin et al. (1992).

From the above exposed arguments, we deduce that *a microstructure related to electric polarization properties* will be primarily described dynamically by an equation that governs a "displacement", i.e., an equation that will resemble a standard equation of motion governing a *linear momentum* in continuum physics. In contrast, *a microstructure related to magnetic properties* will be described dynamically by an equation that will essentially be an equation governing a *moment of momentum.* As to the means of establishing such equations, we identify three basic ones:

(*i*) In the absence of dissipation but knowing the functional dependence of the energy density, a variational principle may be appropriate, but caution should be taken in the magnetic case because of the property (9) — see Tiersten (1965); Maugin and Eringen (1972a). For the electric case see Suhubi (1969).

(*ii*) For a general thermodynamical behavior and basing on an a priori statement of balance laws, one needs a model for introducing the balance equation related to the relevant electromagnetic microstructure. For this approach see Askar et al. (1970), Tiersten (1971), Mindlin (1972), Maugin (1976c) for electric properties; Tiersten (1964), Maugin and Eringen (1972a), Maugin (1976a, p. 1737), Maugin (1988, Figure 6.2.1), Eringen and Maugin (1990, chap. 9) for magnetic properties.

(*iii*) For an equivalent state of generality but a more formal structural algebraic approach, one may exploit a modern formulation of the principle of virtual power in which new electromagnetic degrees of freedom are granted a status equivalent to that of standard deformation processes; for this see Collet and Maugin (1974), Maugin and Pouget (1980), Soumahoro and Pouget (1994) for electric properties, and Maugin (1974, 1976a,b) for magnetic ones; Maugin (1980a) for a general comprehensive presentation.

Before proceeding to the construction of these equations, we need recall the expression of source terms due to electromagnetic fields that appear in standard balance laws of continuum mechanics. This itself is based on the consideration of microscopic fields and a meaningful modelling.

2 Electromagnetic Sources in Galilean Invariant Continuum Physics

2.1 Maxwell's Equations

In magnetized, electrically polarized, and electrically conducting matter, the celebrated set of Maxwell's equations *in a fixed laboratory frame* reads in full generality — and according to Heaviside's synthetic formulation — as

$$\nabla \times \mathbf{E} + \frac{1}{c}\frac{\partial \mathbf{B}}{\partial t} = \mathbf{0}, \qquad \nabla \cdot \mathbf{B} = 0, \tag{10}$$

and

$$\nabla \times \mathbf{H} - \frac{1}{c}\frac{\partial \mathbf{D}}{\partial t} = \frac{1}{c}\mathbf{J}, \qquad \nabla \cdot \mathbf{D} = q_f, \tag{11}$$

where c is the velocity of light in vacuum, $\mathbf{J}$ is the electric current vector, and q_f is the density of free electric charges. The first set (10) is valid everywhere and yields the notion of electromagnetic potentials. In order to close the system of field equations (10)–(11), we must be given electromagnetic constitutive equations (5) together with an equation for electric current, e.g., $\mathbf{J} = \mathbf{J}(\mathbf{E}, ..)$ and the relations (6).

Taking the divergence of $(11)_1$ and accounting for $(11)_2$, we obtain the law of *conservation of electric charge*:

$$\frac{\partial q_f}{\partial t} + \nabla \cdot \mathbf{J} = 0, \tag{12}$$

a *strict* conservation law. Second, by a usual manipulation, one also deduces from (10)–(11) an *energy identity* called the "Poynting–Umov theorem", such that

$$\mathbf{H} \cdot \frac{\partial B}{\partial t} + \mathbf{E} \cdot \frac{\partial D}{\partial t} = -\mathbf{J} \cdot \mathbf{E} - \nabla \cdot \mathbf{S}, \qquad \mathbf{S} \equiv c\,\mathbf{E} \times \mathbf{H}, \tag{13}$$

without any hypothesis concerning the electromagnetic constitutive equations. Note that the Joule term $\mathbf{J}\cdot\mathbf{E}$ can be interpreted as a power expended by an electric force. Indeed, we can write as an example

$$\mathbf{J} \cdot \mathbf{E} = (q\,\mathbf{v}) \cdot \mathbf{E} = (q\,\mathbf{E}) \cdot \mathbf{v} = \mathbf{f} \cdot \mathbf{v}, \tag{14}$$

where $\mathbf{f} = q\,\mathbf{E}$ is seen in statics, according to Lorentz, as the elementary mechanical force acting on a point particle of electric charge q in an electric field $\mathbf{E}$. For a particle moving at velocity $\dot{\mathbf{x}} = \mathbf{v}$, we have the *Lorentz force*

$$\mathbf{f} = q\,\mathbf{E} + \frac{q}{c}\mathbf{v} \times \mathbf{B} = q\,\widetilde{\mathbf{E}}, \qquad \widetilde{\mathbf{E}} = \mathbf{E} + \frac{1}{c}\mathbf{v} \times \mathbf{B}, \tag{15}$$

where the electric field $\widetilde{\mathbf{E}}$ is called the *electromotive intensity.* In addition to this field, we will be led to introducing other fields in a so-called co-moving frame:

$$\widetilde{\mathbf{J}} = \mathbf{J} - q_f\,\mathbf{v}, \qquad \widetilde{\mathbf{M}} = \mathbf{M} + \frac{1}{c}\mathbf{v} \times \mathbf{P} \tag{16}$$

where the first is the conduction current per se, and the last is the magnetization per unit volume in this frame. The large classes of electromagnetic materials are defined thus:

- *Insulators*:

$$q_f = 0, \qquad \widetilde{\mathbf{J}} = \mathbf{0}; \tag{17}$$

- *Non-polarized materials* [Galilean approximation; compare to $(16)_2$, see Maugin (1988)]:

$$\widetilde{\mathbf{P}} \equiv \mathbf{P} = \mathbf{0}. \tag{18}$$

- *Non-magnetized materials*:

$$\widetilde{\mathbf{M}} = \mathbf{0}. \tag{19}$$

  This reduces to $\mathbf{M} = \mathbf{0}$ in statics or if the material simultaneously is non polarized electrically.
- *Dielectric materials*: These are insulators in which, generally,

$$\mathbf{P} \neq \mathbf{0}. \tag{20}$$

- In *quasi-electrostatics* (true electro-magnetic effects discarded but the remaining fields are still time-dependent), (10) and (11) reduce to

$$\nabla \times \mathbf{E} = \mathbf{0}, \qquad \nabla \cdot \mathbf{D} = q_f, \qquad \mathbf{D} = \mathbf{E} + \mathbf{P}; \tag{21}$$

  The first of these tells us that we can introduce a scalar electric potential φ such that $\mathbf{E} = -\nabla\varphi$.
- In *quasi-magnetostatics* (true electro-magnetic effects discarded but the remaining fields are still time-dependent), (10) and (11) reduce to

$$\nabla \times \mathbf{H} = \frac{1}{c}\mathbf{J}, \qquad \nabla \cdot \mathbf{B} = 0, \qquad \mathbf{H} = \mathbf{B} - \mathbf{M}. \tag{22}$$

- For *dielectrics* one has to take $q_f = 0$ in the second of (21), while for *insulators* one has to take $\mathbf{J} = \mathbf{0}$ in the first of (22). In this case $\nabla \times \mathbf{H} = \mathbf{0}$ and one can introduce a scalar magneto-static potential ϕ such that $\mathbf{H} = -\nabla\phi$.

Equation (12) is fully discarded in the case of dielectrics. The Poynting–Umov theorem (13) takes the form

$$\mathbf{E} \cdot \frac{\partial \mathbf{D}}{\partial t} = -\nabla \cdot \mathbf{S} \tag{23}$$

in non-magnetized dielectrics in quasi-electrostatics, while it reduces to

$$\mathbf{H} \cdot \frac{\partial \mathbf{B}}{\partial t} = -\nabla \cdot \mathbf{S}, \tag{24}$$

in insulating non-polarized magnetic materials in quasi-magnetostatics.

These two equations show that without loss in generality we can define the Poynting–Umov energy flux as

$$\mathbf{S} = \varphi \frac{\partial \mathbf{D}}{\partial t} \quad \text{and} \quad \mathbf{S} = \phi \frac{\partial \mathbf{B}}{\partial t} \tag{25}$$

in the corresponding approximations because $\nabla \cdot \mathbf{D} = 0$ in dielectrics and $\nabla \cdot \mathbf{B} = 0$ always, respectively.

2.2 Ponderomotive Force and Couple in a Continuum

The generalization of the force expression (15) and the corresponding couple in a general electromagnetic continuum is a difficult matter that was pondered for a long time. Rather than postulating a form (on what bases?) we prefer to follow the line of H. A. Lorentz (1909, 1952) already followed by Dixon and Eringen (1965), Nelson (1979) and Maugin and Eringen (1977). This involves a type of homogenization (passing from the discrete to the continuum) introducing the approximations of multipoles, a truncation of these at a certain order, and effecting a volume or phase-space average. Lorentz's vision is essentially that of a free space containing charged point particles (Figure 1). We report here only the general traits of this derivation, and give the resulting expressions in the comprehensive form given by Maugin

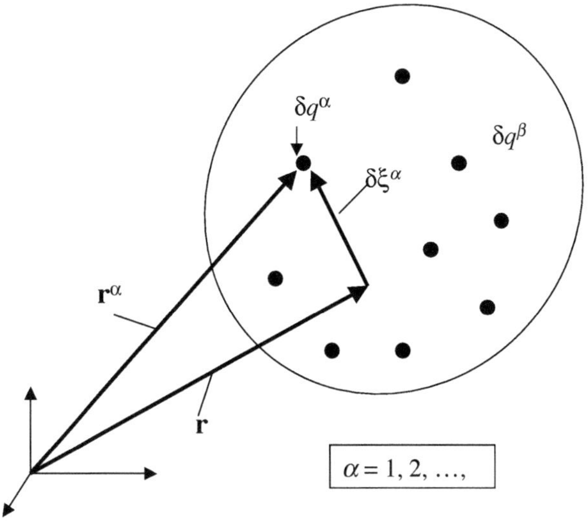

Figure 1. Stable group of elementary electric charges in the Lorentz's averaging approach.

and Eringen (1977). Each elementary electric charge δq_α, $\alpha = 1, 2, \ldots$ contained in a representative volume element ΔV is acted upon by a *Lorentz force* (compare (15))

$$\delta \mathbf{f}_\alpha = \delta q_\alpha \left(\mathbf{e}(\mathbf{r}_\alpha) + \frac{1}{c}\, \dot{\mathbf{x}}_\alpha \times \mathbf{b}(\mathbf{r}_\alpha) \right), \tag{26}$$

where $\mathbf{e}$ and $\mathbf{b}$ are the electric field and magnetic induction at the actual placement $\mathbf{r}_\alpha$ of the charge δq_α. The computation consists then in evaluating the quantities (here, for the sake of simplicity, we adopt a simple volume average, while De Groot and Suttorp (1972) use a relativistically invariant phase average):

$$\sum_{\alpha \in \Delta V} \delta \mathbf{f}_\alpha, \qquad \sum_{\alpha \in \Delta V} (\mathbf{r}_\alpha \times \delta \mathbf{f}_\alpha), \qquad \sum_{\alpha \in \Delta V} \delta \mathbf{f}_\alpha \cdot \dot{\mathbf{x}}_\alpha, \tag{27}$$

and then dividing by ΔV. On neglecting quadrupole contributions and higher order multipoles, lengthy calculations (cf. Maugin and Eringen, 1977) lead to electromagnetic source terms of force, couple **and** energy per unit continuum volume:

$$\mathbf{f}^{em} = q_f\, \widetilde{\mathbf{E}} + \frac{1}{c}\, (\widetilde{\mathbf{J}} + \mathbf{P}^*) \times \mathbf{B} + (\mathbf{P} \cdot \nabla)\, \mathbf{E} + (\nabla \mathbf{B}) \cdot \widetilde{\mathbf{M}}, \tag{28}$$

$$\mathbf{c}^{em} = \mathbf{r} \times \mathbf{f}^{em} + \tilde{\mathbf{c}}^{em}, \tag{29}$$

$$w^{em} = \mathbf{f}^{em} \cdot \mathbf{v} + \tilde{\mathbf{c}}^{em} \cdot \Omega + \rho\, h^{em}, \tag{30}$$

where $\mathbf{r}$ refers to the center of charges of the volume element, ρ is the matter density, and $\mathbf{v}$ is the physical velocity of the continuum, Ω is the vorticity $\Omega = (\nabla \times \mathbf{v})/2$, and we have set

$$q_f(\mathbf{x}, t) = (\Delta V)^{-1} \sum_{\alpha \in \Delta V} \delta q_\alpha, \tag{31}$$

$$\mathbf{P}(\mathbf{x}, t) = (\Delta V)^{-1} \sum_{\alpha \in \Delta V} \delta q_\alpha\, \xi_\alpha(\mathbf{x}, t), \tag{32}$$

$$\mathbf{M}(\mathbf{x}, t) = (\Delta V)^{-1} \sum_{\alpha \in \Delta V} \frac{1}{2\,c}\, \delta q_\alpha\, \xi_\alpha \times \dot{\xi}_\alpha, \tag{33}$$

where $\xi_\alpha = \mathbf{x}_\alpha(t) - \mathbf{x}$ are internal coordinates vectors in ΔV. Note the lack of symmetry between polarization and magnetization effects which clearly demonstrates the "displacement" nature of the polarization and the axial nature (involving a rotation) of the magnetization — cf. section 1 above.

Various contributions in (28) are easily identified. The first term is a continuum generalization of (1). The third term means that $\mathbf{P}$, essentially a continuum density of electric dipoles, is the object of a force when placed in a spatially nonuniform electric field. The last term is quite similar but it relates to a continuum density of magnetic dipoles that feels a nonuniform magnetic induction. As to the second term, it can be understood since a time varying electric polarization creates a displacement current according to Maxwell's equations. We have also defined the intrinsic electromagnetic sources of couple, energy and stress by (here tr = trace; subscript s stands for symmetrization)

$$\tilde{\mathbf{c}}^{em} = \mathbf{P} \times \widetilde{\mathbf{E}} + \widetilde{\mathbf{M}} \times \mathbf{B}, \tag{34}$$

$$\rho\, h^{em} = \widetilde{\mathbf{J}} \cdot \widetilde{\mathbf{E}} + \widetilde{\mathbf{E}} \cdot \mathbf{P}^{*} - \widetilde{\mathbf{M}} \cdot \mathbf{B}^{*} + \operatorname{tr}\left(\tilde{\mathbf{t}}^{em}\,(\nabla \mathbf{v})_S\right), \tag{35}$$

and

$$\tilde{\mathbf{t}}^{em} = \mathbf{P} \otimes \widetilde{\mathbf{E}} - \mathbf{B} \otimes \widetilde{\mathbf{M}} + \left(\widetilde{\mathbf{M}} \cdot \mathbf{B}\right) \mathbf{1}. \tag{36}$$

We note that $\tilde{\mathbf{c}}^{em}$ is the axial vector dual to the skewsymmetric part of this last tensor. Electromagnetic fields in a co-moving frame have already been defined while $\mathbf{E}$ and $\mathbf{B}$ are simple volume averages of $\mathbf{e}$ and $\mathbf{b}$. The first contribution in the *r-h-s* of (28) is none other than a "Lorentz force" per unit volume (compare (15)).

Finally, a right asterisk denotes a so-called convected time derivative such that

$$\mathbf{P}^{*} = \frac{\partial \mathbf{P}}{\partial t} + \nabla \times (\mathbf{P} \times \mathbf{v}) + \mathbf{v}\,(\nabla \cdot \mathbf{P}) = \frac{\mathrm{d}\mathbf{P}}{dt} - (\mathbf{P} \cdot \nabla)\mathbf{v} + \mathbf{P}\,(\nabla \cdot \mathbf{v}). \tag{37}$$

In principle, the above obtained *source terms*, once their origin forgotten, have to be carried into the classical balance laws of a continuum (allowing for a possibly *non symmetric* Cauchy stress), leaving however the internal/free energy of the medium to depend on the electromagnetic fields. A remarkable fact is that in spite of their farfetched outlook, some may be given a form that reminds us of some standard expression (such as in (5)). For instance, Maugin and Eringen (1977) have shown that (30) can also be written as

$$\begin{aligned} w^{em} &= \mathbf{J} \cdot \mathbf{E} + \mathbf{E} \cdot \frac{\partial \mathbf{P}}{\partial t} - \mathbf{M} \cdot \frac{\partial \mathbf{B}}{\partial t} + \nabla\big(\mathbf{v}\,(\mathbf{E} \cdot \mathbf{P})\big) \\ &= -\frac{\partial u^{em.f}}{\partial t} - \nabla \cdot \big(\mathbf{S} - \mathbf{v}\,(\mathbf{E} \cdot \mathbf{P})\big), \end{aligned} \tag{38}$$

in which we identify some of the terms in (13) or a possible direct combination with some of them. The volume energy density of *free* electromagnetic fields $u^{em.f}$ is given by

$$u^{em.f} = \frac{1}{2}\left(\mathbf{E}^2 + \mathbf{B}^2\right). \tag{39}$$

Another interesting equivalent form of (38) is given by (cf. Maugin and Eringen, 1977, a superimposed dot denotes the material time derivative):

$$w^{em} = \mathbf{f}^{em} \cdot \mathbf{v} + \rho\, \dot{e}^{mag} + \rho\, \widetilde{\mathbf{E}} \cdot \dot{\pi} + \rho\, \mathbf{B} \cdot \dot{\mu} + \widetilde{\mathbf{J}} \cdot \widetilde{\mathbf{E}}, \tag{40}$$

where π and μ are the electric polarization and magnetization per unit mass

$$\pi = \mathbf{P}/\rho, \qquad \mu = \widetilde{\mathbf{M}}/\rho, \tag{41}$$

and

$$e^{mag} = -\mu \cdot \mathbf{B} \tag{42}$$

is the energy of magnetic doublets per unit mass.

Particular Cases

We have the following obvious reductions for the ponderomotive force and couple and the accompagning energy source:

In the quasi-electrostatics of dielectrics:

$$\begin{aligned} \mathbf{f}^{em} &= (\mathbf{P} \cdot \nabla)\,\mathbf{E} \equiv (\nabla \mathbf{E}) \cdot \mathbf{P}, \\ \tilde{\mathbf{c}}^{em} &= \mathbf{P} \times \mathbf{E} = \mathbf{P} \times \mathbf{D}, \end{aligned} \tag{43}$$

$$w^{e} = \mathbf{f}^{em} \cdot \mathbf{v} + \rho\, \mathbf{E} \cdot \dot{\pi}, \tag{44}$$

In the quasi-magnetostatics of insulators:

$$\begin{aligned} \mathbf{f}^{em} &= (\nabla \mathbf{B}) \cdot \mathbf{M} \equiv (\mathbf{M} \cdot \nabla)\,\mathbf{H} + \tfrac{1}{2}\,\nabla \mathbf{M}^{2}, \\ \tilde{\mathbf{c}}^{em} &= \mathbf{M} \times \mathbf{B} \equiv \mathbf{M} \times \mathbf{H}, \end{aligned} \tag{45}$$

$$w^{em} = \mathbf{f}^{em} \cdot \mathbf{v} + \rho\, \dot{e}^{mag} + \rho\, \mathbf{B} \cdot \dot{\mu}. \tag{46}$$

3 Deformable Magnetized Bodies with Magnetic Microstructure

3.1 Model of Interactions

For the sake of simplicity we consider the *quasi-magnetostatics* of non-electrically polarized *insulators*.

Using the standard notation of nonlinear continuum mechanics we may consider to start with the following *generalized motion* for deformable magnetized bodies of the ferroïc type:

$$\mathbf{x} = \bar{\mathbf{x}}(\mathbf{X}, t); \qquad \mu = \bar{\mu}(\mathbf{X}, t), \tag{47}$$

where the first of these denotes the classical finite deformation at Newtonian time t between the reference configuration K_R and the actual configuration K_t. Here $\mathbf{x}$ is the placement of Euclidean coordinates x_i, $i = 1, 2, 3$, and $\mathbf{X}$ denotes the material point of coordinates X_K, $K = 1, 2, 3$ in material space. The second of (47) means that the magnetization per unit mass here is considered as a primary quantity. The reason for this is that in ferroïc states in small regions of the bodies (so-called domains), one may have a nonvanishing magnetization in the absence of applied magnetic field. This in fact is the very definition of such a state. Borrowing the denomination introduced by Tiersten (1964), we may say that the first of (47) describes the time evolution of the **lattice continuum** or LC (standard matter in the macroscopic description), while — because of the relation (8) — the second of (47) provides the time evolution of the (magnetic or electronic) **spin continuum** or SC. These two continua should be treated on an equal footing in the vision of **generalized continua**. But they do not respond exactly to the same kind of loads while we must also envisage interactions between these two "continua".

In particular, the spin continuum cannot translate with respect to the lattice continuum. It, therefore, "expands" and "contracts" with the lattice continuum and, accordingly, its volumetric behavior is governed by the usual continuity equation. As usual, the lattice continuum is assumed to be able to respond to volume and surface forces, hence exhibits stresses, and to volume couples, so that stress is not expected to be symmetric. We assume that it is not equipped with any mechanism to respond to surface couples, so that it does not exhibit couple stresses of mechanical origin. The balance of linear (physical) momentum simply says that whatever force of magnetic origin — e.g., the reduced ponderomotive force in (45) — is applied to a point in the spin continuum, it is directly transferred to the lattice continuum at the same point. The spin continuum, by its very nature, can respond only to **couples**, which may be either of the volume or of the surface type. Accordingly, we consider that the ponderomotive couple (cf. Equations (45))

$$\mathbf{c}^{em} = \widetilde{\mathbf{M}} \times \mathbf{B} = \rho\,\mu \times \mathbf{B} \tag{48}$$

is directly applied to the spin continuum.

In so far as the interactions between lattice and spin continua are concerned, they must necessarily be of the couple type since the spin continuum is sensitive only to that type of interaction. Following Tiersten (1964), we naturally assume that this couple is due to a *local magnetic induction* $\mathbf{B}^L$ — to be given a constitutive equation —, so that we can apply the "recipe" (compare (48))

$$\mathbf{c}_{(LC/SC)} = \mathbf{M} \times \mathbf{B}^L = \rho\,\mu \times \mathbf{B}^L. \tag{49}$$

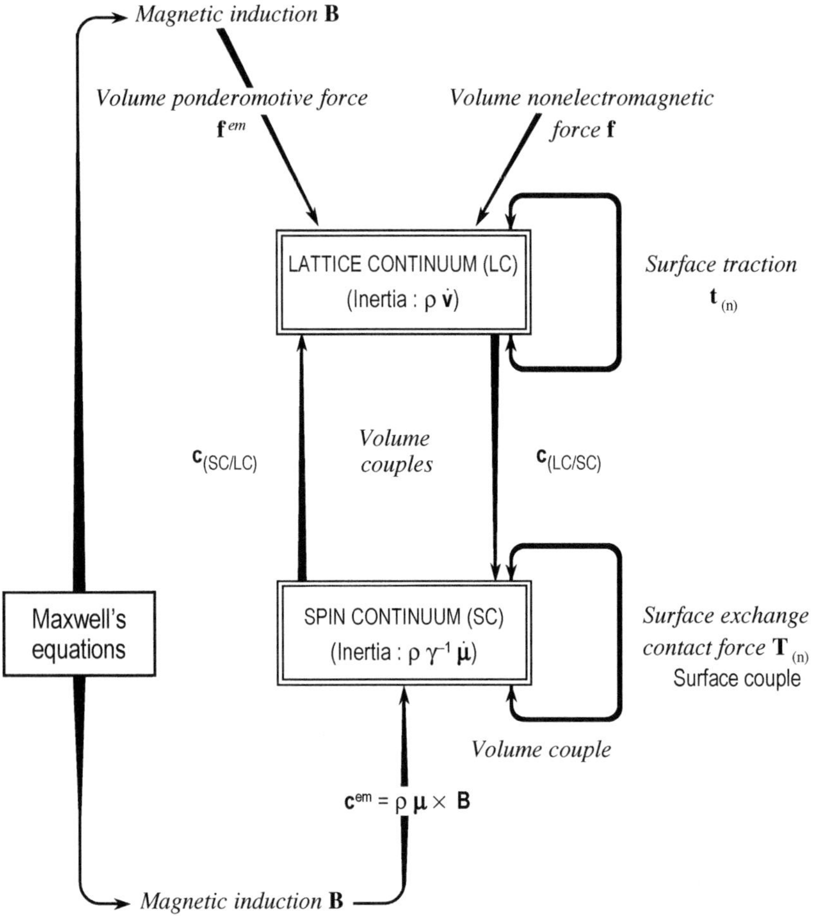

Figure 2. Interactions in deformable ferromagnets (after Maugin, 1979).

Angular momentum being conserved (exchanged) between the two continua, an equal and opposite couple (see Figure 2)

$$\mathbf{c}_{(SC/LC)} = -\mathbf{c}_{(LC/SC)} = \mathbf{B}^L \times \mathbf{M} = \rho \mathbf{B}^L \times \mu \tag{50}$$

is exerted on the unit volume of the lattice continuum.

Finally, in order to account for ferromagnetic (Heisenberg) exchange forces of quantum origin (interactions between neighbouring spins) that fall off rapidly with distance, we can represent these "forces" in a continuum description by a *contact action* in much the same manner as the stress vector for a Cauchy deformable continuum, except that this "surface exchange contact force" must also obey the "recipe" of a magnetic couple (the density ρ is included in $\mathbf{A}$):

$$\mathbf{c}_{(LC-surface)} = \mu \times \mathbf{A}, \tag{51}$$

where $\mathbf{A}$ is an axial vector that depends on the local unit normal $\mathbf{n}$ to the surface and can be written in the same way as the classical stress vector (Cauchy principle), i.e.,

$$\mathbf{A} = \mathbf{A}_{(n)} = \mathbf{A}(\mathbf{x}, t; \mathbf{n}) = \mathbf{n} \cdot \widehat{\mathbf{B}}(\mathbf{x}, t), \tag{52}$$

so that (51) yields the following surface couple density acting on the spin continuum:

$$\mathbf{c}_{(SC-surface)} = \mu \times \mathbf{A} = \mu \times (\mathbf{n} \cdot \widehat{\mathbf{B}}). \tag{53}$$

Because of this very expression we can surmise that only the portion of $\mathbf{A}$ orthogonal to $\mathbf{M}$ is effectively defined. Thus, without loss in generality we can set forth the following condition:

$$\mathbf{A} \cdot \mathbf{M} = \mathbf{n} \cdot \widehat{\mathbf{B}} \cdot \mathbf{M} = 0 \tag{54}$$

at any point at the surface of the body. A similar orthogonality condition can be imposed on $\mathbf{B}^L$ but at any point inside the body.

3.2 Statement of Global Balance Laws

Collecting now the various proposed expressions according to the scheme shown in Figure 2, we can write down the global balance laws at time t in K_t for a magnetic body of volume B_t and regular bounding surface ∂B_t (for the sake of simplicity we ignore any discontinuity surface within the body; for the equations at discontinuity surfaces, see Maugin (1988)):

Balance of mass for the combined continuum:

$$\frac{\mathrm{d}}{\mathrm{d}t} \int_B \rho \, \mathrm{d}v = 0; \tag{55}$$

Balance of linear momentum for the LC:

$$\frac{\mathrm{d}}{\mathrm{d}t} \int_B \rho \mathbf{v} \, \mathrm{d}v = \int_B (\mathbf{f} + \mathbf{f}^{em}) \, \mathrm{d}v + \int_{\partial B} \mathbf{t}_{(n)} \, \mathrm{d}a; \tag{56}$$

Balance of angular momentum for the LC:

$$\begin{aligned} \frac{\mathrm{d}}{\mathrm{d}t} \int_B (\mathbf{r} \times \rho \mathbf{v}) \, \mathrm{d}v = & \int_B (\mathbf{r} \times (\mathbf{f} + \mathbf{f}^{em}) + \mathbf{c}_{(SC/LC)}) \, \mathrm{d}v \\ & + \int_{\partial B} (\mathbf{r} \times \mathbf{t}_{(n)}) \, \mathrm{d}a; \end{aligned} \tag{57}$$

Balance of angular momentum for the SC:

$$\frac{\mathrm{d}}{\mathrm{d}t}\int_B \rho\gamma^{-1}\mu\,\mathrm{d}v = \int_B \left(\mathbf{c}^{em} + \mathbf{c}_{(LC/SC)}\right)\mathrm{d}v + \int_{\partial B} \mu \times \mathbf{A}_{(n)}\,\mathrm{d}a; \tag{58}$$

First law of thermodynamics for the combined continuum:

$$\begin{aligned}\frac{\mathrm{d}}{\mathrm{d}t}\int_B \rho\left(\frac{1}{2}\mathbf{v}^2 + e\right)\mathrm{d}v &= \int_B (\mathbf{f}\cdot\mathbf{v} + w^{em} + \rho\,h)\,\mathrm{d}v \\ &\quad + \int_{\partial B}\left(\mathbf{t}_{(n)}\cdot\mathbf{v} + \mathbf{A}_{(n)}\cdot\dot{\mu} + q_{(n)}\right)\mathrm{d}a;\end{aligned} \tag{59}$$

Second law of thermodynamics for the combined continuum:

$$\frac{\mathrm{d}}{\mathrm{d}t}\int_B \rho\,\eta\,\mathrm{d}v \geq \int_B \rho\,\theta^{-1}\,h\,\mathrm{d}v + \int_{\partial B}\theta^{-1}\,q_{(n)}\,\mathrm{d}a. \tag{60}$$

In these equations, $\mathbf{t}_{(n)}$ is the surface traction, $\mathbf{f}$ is a body mechanical force (e.g., gravity), e is the internal energy per unit mass, η is the entropy per unit mass, h is the body heat source, q is the heat influx. Classically (compare (52)),

$$\mathbf{t}_{(n)} = \mathbf{n}\cdot\mathbf{t}, \qquad q_{(n)} = -\mathbf{n}\cdot\mathbf{q}, \tag{61}$$

where $\mathbf{t}$ is the Cauchy stress and $\mathbf{q}$ is the heat-flux vector.

Standard localization of these global equations on account of the assumed continuity of all fields yields the following *local equations* at any point in B:

$$\dot{\rho} + \rho\,\nabla\cdot\mathbf{v} = 0, \tag{62}$$

$$\rho\,\dot{\mathbf{v}} = \mathrm{div}\;\mathbf{t} + \mathbf{f} + \mathbf{f}^{em}, \tag{63}$$

$$\varepsilon_{ijk}\left(t_{jk} + \rho\,B_j^L\,\mu_k\right) = 0, \tag{64}$$

$$\gamma^{-1}\,\dot{\mu}_i = \left[\mu\times\left(\mathbf{B} + \mathbf{B}^L + \rho^{-1}\,\mathrm{div}\;\widehat{\mathbf{B}}\right)\right]_i + \rho^{-1}\,\varepsilon_{ijk}\,\widehat{B}_{pk}\,\mu_{j,p}, \tag{65}$$

$$\begin{aligned}\rho\left(\dot{e} + \frac{\mathrm{d}}{\mathrm{d}t}\left(\frac{1}{2}\mathbf{v}^2\right)\right) &= t_{ji}\,v_{i,j} + (t_{kj,k}\,v_j + \mathbf{f}\cdot\mathbf{v}) + \widehat{B}_{kj}\,\dot{\mu}_{j,k} \\ &\quad + \widehat{B}_{kj,k}\,\dot{\mu}_j + w^{em} + \rho\,h - \nabla\cdot\mathbf{q},\end{aligned} \tag{66}$$

and

$$\rho\,\dot{\eta} \geq \theta^{-1}\,\rho\,h - \theta^{-1}\,\nabla\cdot\mathbf{q} - \mathbf{q}\cdot\nabla(\theta^{-1}), \tag{67}$$

where the divergence of nonsymmetric tensors is to be taken on the first index, and a superimposed dot denotes the classical material time derivative.

Equations (66), (64), (65) and (67) are transformed thus. In (66), we must account for the kinetic energy theorem obtained by taking the inner product of the motion equation (63) by $\mathbf{v}$:

$$\rho \frac{\mathrm{d}}{\mathrm{d}t}\left(\frac{1}{2}\mathbf{v}^2\right) = (t_{kj,k}\, v_j + \mathbf{f}\cdot\mathbf{v}) + \mathbf{f}^{em}\cdot\mathbf{v}, \tag{68}$$

while w^{em} is given by the reduced form (46). Therefore, (66) reads

$$\begin{aligned} \rho\,\dot{\hat{e}} &= t_{ji}\, v_{i,j} + \rho\,\mathbf{B}\cdot\dot{\mu} + \widehat{B}_{kj}\,\dot{\mu}_{jk} + \widehat{B}_{kj,k}\,\dot{\mu}_j + \rho\, h - \nabla\cdot\mathbf{q}, \\ \hat{e} &= e - e^{mag} \equiv e + \mu\cdot\mathbf{B}. \end{aligned} \tag{69}$$

But if (8) and (9) hold good, $\dot{\mu}$ must be of the purely *precessional* form

$$\dot{\mu} = \omega\times\mu. \tag{70}$$

This corresponds to saturation of the magnetization in each magnetic domain. As a consequence the last contribution in (65) must vanish:

$$\widehat{B}_{k[j}\,\mu_{i],k} = 0. \tag{71}$$

On account of this we check that

$$(\mathrm{div}\,\widehat{\mathbf{B}})\cdot\dot{\mu} = -\rho\,(\mathbf{B}+\mathbf{B}^L)\cdot\dot{\mu}, \tag{72}$$

because

$$\omega = -\gamma\,\mathbf{B}^{\mathit{eff}}, \qquad \mathbf{B}^{\mathit{eff}} = \mathbf{B} + \mathbf{B}^L + \rho^{-1}\,\mathrm{div}\,\widehat{\mathbf{B}}. \tag{73}$$

This may be viewed as a continuum generalization of the celebrated Larmor precession equation $\omega_{\mathrm{Larmor}} = -\gamma\,\mathbf{B}$ for an isolated electron in a magnetic induction $\mathbf{B}$.

Finally, (69) transforms to the following form using an intrinsic notation (T = transpose):

$$\rho\,\dot{\hat{e}} = \mathrm{tr}\big[\mathbf{t}\,(\nabla\mathbf{v})^T\big] - \rho\,\mathbf{B}^L\cdot\dot{\mu} + \mathrm{tr}\big[\widehat{\mathbf{B}}\,(\nabla\dot{\mu})^T\big] - \nabla\cdot\mathbf{q} + \rho\, h. \tag{74}$$

We let the reader show that in the same conditions (67) provides the following *Clausius–Duhem inequality*:

$$-\rho\,(\dot{\widehat{\psi}} + \eta\,\dot{\theta}) + \mathrm{tr}\big[\mathbf{t}\,(\nabla\mathbf{v})^T\big] - \rho\,\mathbf{B}^L\cdot\dot{\mu} + \mathrm{tr}\big[\widehat{\mathbf{B}}\,(\nabla\dot{\mu})^T\big] - \theta^{-1}\,\mathbf{q}\cdot\nabla\theta \geq 0, \tag{75}$$

wherein the free energy density has been defined by

$$\widehat{\psi} = \hat{e} - \eta\,\theta. \tag{76}$$

Equation (75) nowadays plays an essential role in the construction of constitutive equations that we need for the set of quantities

$$\left\{\widehat{\psi}, \eta, \mathbf{t}, \mathbf{B}^L, \widehat{\mathbf{B}}, \mathbf{q}\right\}. \tag{77}$$

Equation (75) represents a constraint imposed by thermodynamic irreversibility (in particular in the so-called Coleman–Noll exploitation that we shall adopt here). In this formulation we usually look for the expression of so-called objective (or materially indifferent) entities. To that purpose we should rewrite (75) in terms of such quantities. This is achieved as follows. On the one hand we note from (64) that the skewsymmetric part of $\mathbf{t}$ is given by

$$t_{[ji]} = \rho\, \mu_{[j}\, B^L_{i]} \tag{78}$$

and we can write

$$\mathbf{t} = \mathbf{t}^S + \mathbf{t}^A, \quad \text{i.e.,} \quad t_{ji} = t_{(ji)} + t_{[ji]}. \tag{79}$$

We introduce the following objective time rates (Maugin, 1974):

$$D_{ij} = \frac{1}{2}\,(v_{i,j} + v_{j,i}) \tag{80}$$

and

$$\widehat{m}_i = (D_J\, \mu)_i \equiv \dot{\mu}_i - \Omega_{ij}\, \mu_j, \qquad \widehat{M}_{ij} = (\dot{\mu}_i)_{,j} - \Omega_{ik}\, \mu_{k,j}, \tag{81}$$

with

$$\Omega_{ij} = \frac{1}{2}\,(v_{i,j} - v_{j,i}). \tag{82}$$

We let the reader prove that the quantities defined in (80) and (81) are indeed *objective.*

The first of (81) is none other than a so-called *Jaumann* derivative. The second of (81) is not exactly the Jaumann derivative of the gradient of μ, but it is closely related to it modulo a term involving the rate of strain (81). On account of these we show that

$$\begin{aligned} &\mathrm{tr}\left[\mathbf{t}\,(\nabla\mathbf{v})^T\right] - \rho\,\mathbf{B}^L \cdot \dot{\mu} + \mathrm{tr}\left[\widehat{\mathbf{B}}\,(\nabla\dot{\mu})^T\right] \\ &\equiv \mathrm{tr}\left(\mathbf{t}^S\,\mathbf{D}\right) - \rho\,\mathbf{B}^L \cdot \widehat{\mathbf{m}} + \mathrm{tr}\left(\widehat{\mathbf{B}}\,\widehat{\mathbf{M}}^T\right), \end{aligned} \tag{83}$$

whence the looked for reduced useful expression for (74) and (75).

In summary, for the present modeling the local field equations at any regular material point in the body B are provided by equations (62), (63), (70), (74) and the reduced form of Maxwell's equations

$$\nabla \times \mathbf{H} = \mathbf{0}, \quad \nabla \cdot \mathbf{B} = 0, \quad \mathbf{H} = \mathbf{B} - \rho\,\mu, \tag{84}$$

in which the Cauchy stress $\mathbf{t}$ is given by equations (78) and (79), and the effective magnetic induction $\mathbf{B}^{eff}$ is given by (73). Equation (74) will eventually provide the heat-propagation equation while the inequality (75) constrains the constitutive behavior. Interestingly enough, we note that the energy equation (59) does not contain any contribution due to the spin lattice by virtue of the d'Alembertian nature of this quantity (cf. (9)) — more on this point in the following two paragraphs.

3.3 Approach via the Principle of Virtual Power

In modern continuum mechanics, an elegant and powerful means of constructing field equations and associated natural boundary conditions is provided by an algebraically structured formulation of the (d'Alembert) principle of virtual power as exposed at length in Maugin (1980a). In this somewhat abstract formulation this principle is enunciated in the following form for global powers over the body B and its boundary ∂B: *The virtual power of inertial forces is, at each instant of time, balanced by the total virtual power of "internal forces" and that of externally applied forces both in the bulk and at the surface*, the word "force" being understood in a generalized manner. Inertial forces have an expression provided by physics, internal forces need to be given a constitutive equation, and external forces are prescribed in form and perhaps in value. In mathematical terms:

$$P^*_{inert}(B) = P^*_{int}(B) + P^*_{extern}(B, \partial B), \tag{85}$$

where an asterisk will denote the value of an expression is a so-called virtual velocity field (itself noted with an asterisk). In the present case, the generalized kinematical description of the model (47) from which the basic virtual velocity field is given by

$$v^* = \{v_i^*, (\dot{\mu}_i)^* = (\omega^* \times \mu)_i\}, \tag{86}$$

where ω^* is a virtual precessional velocity of the SC. Thus

$$P^*_{inert}(B) = \int_B (\rho\,\dot{\mathbf{v}} \cdot \mathbf{v}^* + \gamma^{-1}\,\rho\,\dot{\mu} \cdot \omega^*)\,\mathrm{d}v, \tag{87}$$

where we clearly distinguish between real fields (no asterisks; actual solutions of a problem) and virtual ones (noted with an asterisk; at our disposal in this type of variational formulation). In particular, for real fields, because of (8) and (9), (86) yields

$$P_{inert}(B) = \frac{\mathrm{d}}{\mathrm{d}t}\int_B \left(\frac{1}{2}\,\rho\,\mathbf{v}^2\right)\,\mathrm{d}v = \frac{\mathrm{d}}{\mathrm{d}t}K(B), \tag{88}$$

where $K(B)$ is the total kinetic energy of the traditional motion.

The total power of external forces is obviously given by the following expression:

$$P^*_{extern}(B, \partial B) = P^*(B) + P^*(\partial B) \tag{89}$$

wherein

$$P^*(B) = \int_B \left((\mathbf{f} + \mathbf{f}^{em}) \cdot \mathbf{v}^* + \rho\, \mathbf{B} \cdot (\dot{\mu})^* \right) \mathrm{d}v, \tag{90}$$

and

$$P^*(\partial B) = \int_{\partial B} \left((\mathbf{t}_{(n)} + \mathbf{t}^{em}_{(n)}) \cdot \mathbf{v}^* + \mathbf{A} \cdot (\dot{\mu})^* \right) \mathrm{d}a, \tag{91}$$

where $\mathbf{t}^{em}_{(n)}$ is an eventual magnetic surface traction related to the possible existence of a magnetic field outside B (see Maugin, 1988, chap. 6).

Finally, the global virtual power of internal forces is constructed as follows. First a "gradient order" is selected for the kinematics associated with internal forces. Generalizing classical continuum mechanics (which is a first order gradient theory of displacement) we consider a first-order gradient theory based on (86). That is,

$$V = \{v_i,\ v_{i,j},\ \dot{\mu}_i,\ \dot{\mu}_{i,j}\}. \tag{92}$$

But internal forces must be *objective*, i.e., frame indifferent, or invariant under changes of observer in the actual configuration (superimposition of a rigid body motion of dimension 6). Accordingly, one must extract from the 24-dimensional space spanned by (92) a set of objective quantities, this set V_{obj}, a quotient space, being necessarily of dimension $24 - 6 = 18$. We have shown elsewhere (Maugin, 1980a) how to systematically construct such quotient spaces. In the present case a good set is given by

$$V_{obj} = \{D_{ij},\ \widehat{m}_i,\ \widehat{M}_{ij}\}, \tag{93}$$

where it happens that the quantities thus formally introduced have already been defined in (80)–(82). Then the power P^*_{inter} is written as a continuous linear form on the set V^*_{obj}, introducing thus formally internal forces $\{\mathbf{t}^S, -\rho\,\mathbf{B}^L, \widehat{\mathbf{B}}\}$ as co-factors of the elements of V^*_{obj}. That is (signs are chosen for convenience),

$$P^*_{int}(B) = -\int_B \left(t^S_{ji}\, v^*_{i,j} - \rho\, B^L_i\, \widehat{m}^*_i + \widehat{B}_{ji}\, \widehat{M}^*_{ij} \right) \mathrm{d}v. \tag{94}$$

Collecting the various contributions and assuming that the obtained global expression is valid for any element of volume and surface and any virtual velocity field (86) we obtain the local equations

$$\rho\, \dot{\mathbf{v}} = \mathrm{div}\ \mathbf{t} + \mathbf{f} + \mathbf{f}^{em} \quad \text{in } B, \tag{95}$$

and

$$\gamma^{-1}\,\dot{\mu} = -(\mathbf{B}^{\mathit{eff}} \times \mu) \quad \text{in } B, \tag{96}$$

with the nonsymmetric stress $\mathbf{t}$ given by

$$t_{ji} = t^{S}_{ji} - \rho\, B^{L}_{[j}\,\mu_{i]} \tag{97}$$

on account of the constraint (71), and $\mathbf{B}^{\mathit{eff}}$ given by the second of (73). Simultaneously, we obtain the natural boundary conditions (not given here — see Maugin, 1988, chap. 6 —) for $\mathbf{t}$ and $\mathbf{A}$.

It is readily checked that equations (95) and (96), together with (97) and (73) are identical to the equations deduced in the foregoing paragraph. Pursuing along the same line, and considering the principle (85) for real velocity fields, on account of (88) we obtain the *global equation of kinetic energy* in the form:

$$\frac{\mathrm{d}}{\mathrm{d}t}K(B) = P_{int}(B) + P_{extern}(B, \partial B). \tag{98}$$

This is to be combined with the global statement of the first law of thermodynamics (59) to deduce the global form of the *internal-energy theorem*. By localization this will yield (74) with the already transformed expression involving the objective internal forces. The exploitation of the inequality (75) is unchanged.

The present formulation (introduced in Maugin, 1974) — formal as it is — has certain advantages, one of which being the account of the d'Alembert-inertia couple in the expression (86). But more interestingly, it provides a direct modelling of more general ferroïc cases such as in ferrimagnets and antiferromagnets (see Paragraph 3.5).

3.4 Hamiltonian Variational Formulation

The above-given formulation is valid for both deformable solid and fluid behaviors and also in the presence of dissipative processes such as *viscosity* (via $\mathbf{D}$) and *spin-lattice relaxation* (via $\widehat{\mathbf{m}}$). In the absence of dissipative processes and for an a priori known behavior — e.g., elasticity — it is possible to approach the present theory via a Hamiltonian variational principle. Such an approach to *elastic ferromagnets* is to be found in Tiersten (1965), Brown (1966), and Maugin and Eringen (1972a). We base the present exposition on the latter. Again, we must account for the d'alembertian nature of the magnetic spin inertia and, therefore, introduce an *already varied term* for this effect. That is, we shall write down the *variational formulation* as follows:

$$\delta W_{spin} + \delta A + \delta W_{data} + \delta W_{constr} = 0. \tag{99}$$

Here δW_{spin} is the mentioned already varied term

$$\delta W_{spin} = \int_t \mathrm{d}t \int_{B_R} \rho_R \gamma^{-1} \dot{\mu} \cdot \delta\Theta \, \mathrm{d}v_R, \tag{100}$$

the scalar A is the *action* such that

$$A = \int_t \mathrm{d}t \int_{B_R} L \, \mathrm{d}v_R, \tag{101}$$

$$L = \frac{1}{2} \rho_R \mathbf{v}^2 - \Psi(\mathbf{F} = \nabla_R \overline{\mathbf{x}}, \, \mu, \, \nabla_R \bar{\mu}), \tag{102}$$

δW_{data} accounts for the external loads in such a way that

$$\begin{aligned} \delta W_{data} = & \int_t \mathrm{d}t \int_{B_R} \left((\mathbf{f} + \mathbf{f}^{em}) \cdot \delta\mathbf{x} + \rho_R \, \mathbf{B} \cdot \delta\mu \right) \mathrm{d}v_R \\ & + \int_t \mathrm{d}t \int_{\partial B_R} \left((\mathbf{t}_{(n)} + \mathbf{t}^{em}_{(n)}) \cdot \delta\mathbf{x} + \mathbf{A} \cdot \delta\mu \right) \mathrm{d}a_R, \end{aligned} \tag{103}$$

and δW_{constr} is possibly introduced to account, via the introduction of appropriate Lagrange multipliers, for the constraints provided by the constancy of the modulus of μ i.e., $\mu_i \, \mu_i = \mu_S^2 = const.$, and the derived relation $(\nabla_R \mu) \cdot \mu = 0$ for its spatial uniformity within a domain, where ∇_R denotes the material gradient that commutes with the partial time derivative. But these are accounted for systematically in the variation of the other terms. Here the variations are Lagrangian (taken at fixed material coordinates so that they commute with the material gradient), ρ_R is the matter density in the reference configuration K_R of the body of volume B_R and regular boundary ∂B_R. The Lagrangian variation $\delta\mu$ respects the constraint $\mu \cdot \delta\mu = 0$ and therefore is such that it involves the infinitesimal (vectorial) angular variation $\delta\Theta$ through the relation $\delta\mu = \delta\Theta \times \mu$. The Lagrangian density L involves the standard kinetic energy per unit material volume and a magneto-elastic energy Ψ that accounts for a first-order gradient theory with respect to the two basic elements of the generalized motion. We let the reader exploit the variational formulation (99) for arbitrary variations $(\delta\mathbf{x}, \delta\Theta)$ as the above expressions are given only for the sake of comparison with the previous formulation exploiting the principle of virtual power. In particular, we note the introduction of the term (100) that compares to the spin contribution in $P^*_{inertia}$. Just the same, we emphasize the similarity between the expression of δW_{data} and that of P^*_{extern}. We also note the following that is of interest compared to the Cosserats'work of 1909 on "Cosserat" continua. This is the possible exploitation of the so-called

Euclidean invariance — a first attempt to use group theory in continuum mechanics taken over by Toupin (1964) and Maugin (1970). This consists in applying (99) with special variations corresponding to *pure spatial translations*:

$$\delta \mathbf{x} = \varepsilon\, \mathbf{d}, \qquad \delta \mu = 0 \tag{104}$$

and then simultaneous *infinitesimal rotation* of both *LC* and *SC* such as

$$\delta x_i = \varepsilon\, \varpi_{ij}\, x_j, \qquad \delta \mu_i = \varepsilon\, \varpi_{ij}\, \mu_j, \tag{105}$$

where ε is an infinitesimally small parameter, $\mathbf{d}$ is a fixed finite vector, and $\varpi_{ij} = -\varpi_{ji}$ is a fixed skewsymmetric tensor. Application of (104) and (105) directly yields the local balance of linear and angular momentum in the form of the theory of Cosserat continua (Maugin, 1971; Maugin and Eringen, 1972a). We shall return to this point in Paragraph 3.6.

3.5 Ferrimagnetic and Antiferromagnetic Materials

The magnetic description considered in the foregoing paragraphs often is insufficient and not realistic enough for many magnetic materials such as ferrites. Louis Néel (Nobel prize in physics 1970 for this matter) introduced in the early 1940s a model in which the most general description of the magnetization field in a magnetically ordered crystal below its magnetic-phase-transition temperature consists in the vectorial resultant of the sum of n magnetization fields μ_α, $\alpha = 1, 2, \ldots, n$ per unit mass — referred to as *magnetic sub-lattices* — arising at each point from n different ionic species having different spectroscopic splitting factors, thus various gyromagnetic ratios γ_α, so that the total magnetic spin per unit mass is not necessarily aligned with the total magnetization. This model proved to be efficient in accounting for the unusual magnetic properties (e.g., susceptibility) of ferrites — iron oxides — for which Néel coined the behavior name *ferrimagnetism.* Simple *antiferromagnetism* is the special case for which only two magnetic sub-lattices subsist, of equal magnitude and opposite direction, allowing for the absence of global magnetization in the absence of applied magnetic field. But the magnetic response is quite different from that of classical ferromagnetism when a magnetic field is applied (see Eringen and Maugin, 1990, Vol. I, pp. 110–111). The resulting dynamics is also much more involved yielding a multiplicity of magnon branches in the case of ferrimagnetism.

A rational modelling of deformable ferrimagnetic bodies in the spirit of the model of Paragraph 3.1 would be somewhat messy, although a scheme generalizing that of Figure 2 can easily be drawn for antiferromagnetic deformable bodies equipped with two co-existing interacting magnetic sub-

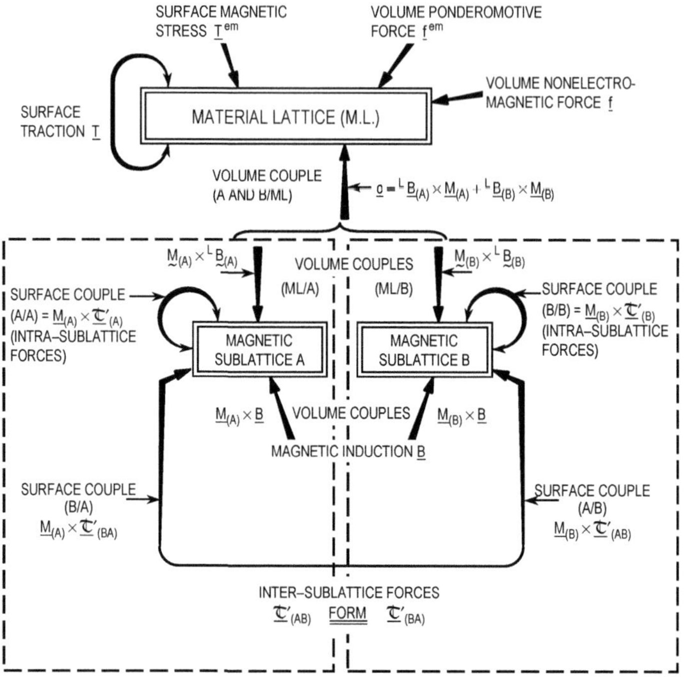

Figure 3. Interactions in a deformable antiferromagnet (after Maugin, 1976a,b, JMP).

lattices (see Figure 3 — from Maugin, 1976a). It is much more convenient and safe to exploit a modelling generalizing that of the principle of virtual power in the manner of Paragraph 3.3 in which it is "sufficient" to enlarge the initial set by replacing μ by the series of $\mu_\alpha's$ and constructing the various sets of velocities for the corresponding power of internal forces once a gradient order has been selected (first-order is sufficient). A new type of interactions will appear, that between different magnetic sub-lattices, having in turn strong consequences for the coupling between elastic waves (phonons) and magnetic oscillations (magnons) — see Eringen and Maugin, 1990, Vol. II, pp. 492–493. Such a modelling was initially proposed by Maugin (1976a,b) and its dynamical consequences examined in detail by Sioké-Rainaldy and Maugin (1983); Maugin and Sioké-Rainaldy (1983).

3.6 Analogy with Cosserat Continua

Returning of the ferromagnetic case, we note that the spin-precession equation (96) deals with axial vectors. Accordingly, we can introduce dual skewsymmetric tensors by applying the alternation symbol ε_{kli} to its i-component. On account of the well known formula

$$\varepsilon_{kli}\,\varepsilon_{ipq} = \delta_{kp}\,\delta_{lq} - \delta_{kq}\,\delta_{lp}, \tag{106}$$

this operation results in the equation

$$\frac{1}{2}\,\rho\gamma^{-1}\,\varepsilon_{kli}\,\dot{\mu}_i = \rho\,\mu_{[k}\,B_{l]} + \rho\,\mu_{[k}\,B^L_{l]} + \left(\mu_{[k}\,\widehat{B}_{ml]}\right)_{,m} - \mu_{[k,m}\,\widehat{B}_{ml]}. \tag{107}$$

But we note that the last term in the right-hand of this equation vanishes identically because of the constraint (71), while the resulting penultimate term is none other that the skew part of the Cauchy stress according to (97), and the first term is none other than the ponderomotive couple written as a skew tensor (dual of the axial vector $\mathbf{c}^{em}$). Thus equation (107) reads

$$\rho\,\dot{S}_{kl} = M_{pkl,p} + t_{[kl]} + C_{kl}, \tag{108}$$

wherein

$$S_{kl} = \frac{1}{2}\,\gamma^{-1}\,\varepsilon_{kli}\,\mu_i, \quad M_{pkl} = \mu_{[k}\,\widehat{B}_{pl]}, \quad C_{kl} = C^{em}_{kl} = \varepsilon_{kli}\,\tilde{c}^{em}_i. \tag{109}$$

Equation (108) is in the canonical form of the balance equation of angular momentum in *Cosserat or micropolar continua* — compare Eringen (1999) and the present Appendix A — except that all contributions have a magnetic origin, the gyromagnetic relation for the inertial term S_{kl}, Heisenberg exchange forces for the couple stress tensor M_{pkl}, the applied coupled C_{ij}, and the skew part of the nonsymmetric Cauchy stress. Equation (108) and the accompanying boundary condition were deduced by the author in his PhD thesis (Maugin, 1971). A formally similar result can easily be obtained for the model of ferrimagnetic deformable bodies mentioned in the foregoing paragraph with the appropriate summation over the various magnetic sub-lattices. In the case of an exploitation of a Hamiltonian variational formulation, application of the rotational part (105) of the Euclidean invariance directly yields (108) (Maugin, 1971).

3.7 Reduction to a Model without Microstructure (Paramagnetic and Soft-Ferromagnetic Bodies)

When true ferromagnetic effects (gyromagnetic effect, Heisenberg exchange forces) are discarded, equation (108), reduces to

$$t_{[kl]} = C^{em}_{[kl]} = -M_{[k}\,B_{l]} = B_{[k}\,M_{l]}. \tag{110}$$

This applies to the simpler cases of nonlinear paramagnetic and soft-ferromagnetic bodies. The resulting theory applies, in particular, to magnetoelastic polymers as recently developed (on this subject, one can refer to the Udine course of 2009; Ogden and Steigmann, 2010, Editors; and also, Eringen and Maugin, 1990, chap. 8, Vol. I). Whenever field and magnetization are aligned (case of magnetically isotropic bodies) or in a purely linear theory in which one discards the right-hand side of (110) as being second order in the fields, the skew part of the stress is zero. The only remaining magneto-mechanical coupling in the first case remains magnetostriction for any symmetry, while in the second case, only piezomagnetism may exist, under severe symmetry conditions however.

4 Deformable Dielectrics with Electric-Polarization Microstructure

4.1 Model of Interactions

We consider the case of the quasi-electrostatics of deformable dielectrics for the sake of simplicity. We can envisage a generalized motion described by the functions

$$\mathbf{x} = \overline{\mathbf{x}}(\mathbf{X}, t), \qquad \boldsymbol{\pi} = \overline{\boldsymbol{\pi}}(\mathbf{X}, t), \tag{111}$$

where $\boldsymbol{\pi}$ is an electric polarization (polar vector) per unit mass in the deformed configuration. The second function defines a **polarization continuum**, PC. According to the contents of Section 1, PC responds to electric fields only. Accounting for the inertia introduced in Section 1, we are tempted to write down a balance equation for PC for the whole body B in a more or less standard form:

$$\frac{\mathrm{d}}{\mathrm{d}t} \int_B \rho\, d_E\, \dot{\pi}\, \mathrm{d}v = \int_B \rho\, (\mathbf{E} + \mathbf{E}^L)\, \mathrm{d}v + \int_{\partial B} \mathbf{A}\, \mathrm{d}a, \tag{112}$$

where $\mathbf{E}$ is the Maxwellian electrostatic field, $\mathbf{E}^L$ is a quantity akin to an electric field and due to the possible interaction with the lattice continuum LC, whose deformation is described by the first of (111). Finally, $\mathbf{A}$, also akin to an electric field or a surface electric polarization, accounts in the form of a contact action for interactions between neighbouring electric dipoles. Applying to this the Cauchy principle, we can introduce a second order — nonsymmetric — tensor $\widehat{\mathbf{E}}$ such that

$$\mathbf{A} = \mathbf{n} \cdot \widehat{\mathbf{E}} \quad \text{at } \partial B. \tag{113}$$

Localization of (112) therefore yields the balance equation

$$d_E\, \ddot{\pi} = \mathbf{E} + \mathbf{E}^L + \rho^{-1} \operatorname{div} \widehat{\mathbf{E}} \quad \text{in } B. \tag{114}$$

Although considered by some authors (e.g., Maugin, 1976c), it is difficult to grant a true physical meaning to the balance (112) which strongly resembles the balance law postulated, with the same degree of arbitrariness, in anisotropic fluids (nematic liquid crystals) by Ericksen (1960). One possibility of interpretation is that (114) is a standard equation of motion for a unit (hypothetical) electric charge (but the medium considered is a dielectric free of charges). Indeed, as exemplified by equation (1) the product of a charge and an electric field is a classical force. In this interpretation *LC* and *PC* may be viewed as two interpenetrating continua. Such an a priori interpretation was advanced by Tiersten (1971). As to the surface condition (113) we can write it more explicitly as

$$\rho^{-1}\,\mathbf{n}\cdot\widehat{\mathbf{E}} = \pi_S, \tag{115}$$

where π_S is a density of surface electric polarization (a polar vector).

The global balances of linear and angular momenta for the lattice continuum naturally read as

$$\frac{\mathrm{d}}{\mathrm{d}t}\int_B \rho\,\mathbf{v}\,\mathrm{d}v = \int_B (\mathbf{f}+\mathbf{f}^{em})\,\mathrm{d}v + \int_{\partial B}\mathbf{t}_{(n)}\,\mathrm{d}a; \tag{116}$$

and

$$\begin{aligned}\frac{\mathrm{d}}{\mathrm{d}t}\int_B (\mathbf{r}\times\rho\,\mathbf{v})\,\mathrm{d}v = &\int_B \big(\mathbf{r}\times(\mathbf{f}+\mathbf{f}^{em}) + \mathbf{c}_{(PC/LC)}\big)\,\mathrm{d}v\\ &+\int_{\partial B}(\mathbf{r}\times\mathbf{t}_{(n)})\,\mathrm{d}a.\end{aligned} \tag{117}$$

Then, artificial as this may look, the balance of angular momentum for the *PC* is given by:

$$\frac{\mathrm{d}}{\mathrm{d}t}\int_B \rho\,\pi\times\dot{\pi}\,\mathrm{d}v = \int_B \big(\mathbf{c}^{em} + \mathbf{c}_{(LC/PC)}\big)\,\mathrm{d}v + \int_{\partial B}\pi\times\mathbf{A}_{(n)}\,\mathrm{d}a. \tag{118}$$

This is complemented by the first law of thermodynamics for the combined continuum:

$$\begin{aligned}\frac{\mathrm{d}}{\mathrm{d}t}\int_B \rho\left(\frac{1}{2}\mathbf{v}^2 + e\right)\,\mathrm{d}v = &\int_B \big(\mathbf{f}\cdot\mathbf{v} + w^{em} + \rho\,h\big)\,\mathrm{d}v\\ &+\int_{\partial B}\big(\mathbf{t}_{(n)}\cdot\mathbf{v} + \mathbf{A}_{(n)}\cdot\dot{\pi} + q_{(n)}\big)\,\mathrm{d}a,\end{aligned} \tag{119}$$

and the second law of thermodynamics for the combined continuum:

$$\frac{\mathrm{d}}{\mathrm{d}t}\int_B \rho\,\eta\,\mathrm{d}v \geq \int_B \rho\,\theta^{-1}\,h\,\mathrm{d}v + \int_{\partial B}\theta^{-1}\,q_{(n)}\,\mathrm{d}a. \tag{120}$$

In these equations,

$$\mathbf{c}^{em} = \rho\pi \times \mathbf{E}, \quad \mathbf{c}_{LC/PC} = \rho\pi \times \mathbf{E}^L, \quad \mathbf{A}_{(n)} = \mathbf{n} \cdot \widehat{\mathbf{E}}. \tag{121}$$

On account of (114) and the local form of (116) the local forms of (117) and (118) are easily established as

$$t_{[ji]} = \rho\, E^L_{[j}\, \pi_{i]} - \widehat{E}_{p[j}\, \pi_{i],p} \tag{122}$$

and

$$\rho\, \frac{\mathrm{d}}{\mathrm{d}t}(\pi \times \dot{\pi})_i = c_i^{em} + c_i^L + \varepsilon_{ijk}\left(\pi_j\, \widehat{E}_{pk}\right)_{,p}. \tag{123}$$

The rest of this approach consists in expressing (119) and (120). In introducing objective time rates such as (80) and (compare (81))

$$\begin{aligned} \hat{p}_i &= (D_J\, \pi)_i \equiv \dot{\pi}_i - \Omega_{ij}\, \pi_j, \\ \widehat{\Pi}_{ij} &= (\dot{\pi}_i)_{,j} - \Omega_{ik}\, \pi_{k,j} \end{aligned} \tag{124}$$

we can show that (119) and (120) lead to the following local forms of the energy equation and of the Clausius–Duhem inequality:

$$\rho\, \dot{e} = \mathrm{tr}(\mathbf{t}^S\, \mathbf{D}) - \rho\, \mathbf{E}^L \cdot \hat{\mathbf{p}} + \mathrm{tr}\big(\widehat{\mathbf{E}}^L\, \widehat{\Pi}^T\big) - \nabla \cdot \mathbf{q} + \rho\, h \tag{125}$$

and

$$-\rho\left(\dot{\psi} + \eta\, \dot{\theta}\right) + \mathrm{tr}(\mathbf{t}^S\, \mathbf{D}) - \rho\, \mathbf{E}^L \cdot \hat{\mathbf{p}} + \mathrm{tr}\big(\widehat{\mathbf{E}}\, \widehat{\Pi}^T\big) - \theta^{-1}\, \mathbf{q} \cdot \nabla\theta \geq 0, \tag{126}$$

with

$$t_{ji} = t^S_{ji} + t_{[ji]}, \qquad t_{[ji]} = \rho\, E^L_{[j}\, \pi_{i]} - \widehat{E}_{p[j}\, \pi_{i],p}. \tag{127}$$

Together with the Maxwell's electrostatic equations for dielectrics,

$$\nabla \times \mathbf{E} = \mathbf{0}, \qquad \nabla \cdot \mathbf{D} = 0, \qquad \mathbf{D} = \mathbf{E} + \rho\,\pi, \tag{128}$$

this concludes the formal construction of the theory before establishing constitutive equations constrained by the inequality (126).

4.2 Approach via the Principle of Virtual Power

It is now clear that an approach exploiting directly the principle of virtual power for the present theory will be very much like what was achieved in Paragraph 3.4 for ferromagnets except for the essential difference regarding the inertial force of the polarization lattice PC. That is, we shall a priori write

$$P^*_{inert}(B) = \int_B \left(\rho\, \dot{\mathbf{v}} \cdot \mathbf{v}^* + \rho\, d_E \ddot{\pi}_i\, \dot{\pi}^*_i\right)\, \mathrm{d}v, \tag{129}$$

where we clearly distinguish between real fields (actual solutions of a problem) and virtual ones (at our disposal in this type of variational formulation). In particular, for real fields, this yields

$$P_{inert}(B) = \frac{\mathrm{d}}{\mathrm{d}t}\int_B \left(\frac{1}{2}\,\rho\,\mathbf{v}^2 + \frac{1}{2}\,\rho\, d_E\,\dot{\pi}^2\right)\,\mathrm{d}v = \frac{\mathrm{d}}{\mathrm{d}t}K(B). \tag{130}$$

The other global virtual powers are directly written down as

$$P^*(B) = \int_B \left((\mathbf{f} + \mathbf{f}^{em})\cdot\mathbf{v}^* + \rho\,\mathbf{E}\cdot(\dot{\pi})^*\right)\,\mathrm{d}v, \tag{131}$$

$$P^*(\partial B) = \int_{\partial B} \left((\mathbf{t}_{(n)} + \mathbf{t}_{(n)}^{em})\cdot\mathbf{v}^* + \rho\,\pi_S\cdot(\dot{\pi})^*\right)\,\mathrm{d}a, \tag{132}$$

and

$$P^*_{int}(B) = -\int_B \left(t^S_{ji}\,v^*_{i,j} - \rho\,E^L_i\,\hat{p}^*_i + \widehat{E}_{ji}\,\widehat{\Pi}^*_{ij}\right)\,\mathrm{d}v. \tag{133}$$

From the standard application of the principle of virtual power for any volume and surface elements and for arbitrary members of the set $\{\mathbf{v}^*, (\dot{\pi})^*\}$, one deduces the local equations of linear momentum of the LC and the governing equation (114) of the PC, together with the accompanying natural boundary conditions. Then equations (115) and (126) follow in the usual way, using the result (130).

4.3 Hamiltonian Variational Principle

Again this strategy applies when one knows a priori the functional dependence of the internal energy. There is no special problem with the kinetic energy that is given by the expression appearing in (130). For instance, for a first-order gradient theory of electroelasticity one would consider a Lagrangian density per unit volume of the reference configuration K_R

$$L = \frac{1}{2}\,\rho_R\,\mathbf{v}^2 + \frac{1}{2}\,\rho_R\,\dot{\pi}^2 - \Psi(\mathbf{F} = \nabla_R\,\overline{\mathbf{x}},\ \pi,\ \nabla_R\,\overline{\pi}), \tag{134}$$

but one must add to this the electrostatic energy including both free-field and electric-dipole energies:

$$e^{elec} = \frac{1}{2}\,\mathbf{E}^2 + \rho_R\,\pi\cdot\mathbf{E}. \tag{135}$$

This formulation applies to both elastic ferroelectrics (theory of Pouget and Maugin) and elastic ionic crystals (theory of Mindlin). Suhubi (1969) gave such a formulation for Mindlin's theory of elastic dielectrics with polarization gradients. The true aficionados will find in Maugin and Eringen (1972b) a relativistically invariant variational formulation containing simultaneously both ferromagnetic and ferroelectric descriptions (see also Maugin, 1978). Similarly, he will find in Collet and Maugin (1974) and Maugin (1980a), a formulation using the principle of virtual power for these two descriptions simultaneously.

4.4 Antiferroelectric Materials

It is easily imagined that a theory of deformable antiferroelectrics (e.g., lead zirconate or sodium niobate) in which an antiparallel arrangement of permanent electric dipoles can be devised by analogy with the theory of antiferromagnetics, i.e., by considering the polarization density π as arising from the vectorial sum of two opposite polarization sub-lattices of equal magnitude. Such a model was constructed by Soumahoro and Pouget (1994) who also studied in detail its dynamical consequences.

4.5 Analogy with Cosserat Continua

Applying the alternation symbol to equation (123) and using the identity (106) or, equivalently, taking the tensor product of (114) and then the skew part of the result we obtain

$$\rho \frac{\mathrm{d}}{\mathrm{d}t}(d_E \, \dot{\pi}_{[i} \, \pi_{j]}) = E_{[i} \, P_{j]} + \left(\rho \, E^L_{[i} \, \pi_{j]} - \widehat{E}_{k[i} \, \pi_{j],k}\right) + \left(\widehat{E}_{k[i} \, \pi_{j]}\right)_{,k}, \tag{136}$$

or

$$\rho \, \dot{S}_{ij} = C^{em}_{ij} + t_{[ji]} + M_{kij,k}, \tag{137}$$

where we accounted for (127) and we set

$$S_{ij} = d_E \, \dot{\pi}_{[i} \, \pi_{j]}, \qquad C^{em}_{ij} = E_{[i} \, P_{j]}, \qquad M_{kij} = \widehat{E}_{k[i} \, \pi_{j]}. \tag{138}$$

Simultaneously, (115) yields the associated natural boundary condition at ∂B:

$$n_k \, M_{kij} = M_{(n)ij} \equiv \pi_{S[i} \, P_{j]}. \tag{139}$$

Equations (137) and (139) are in the canonical form of the local balance of angular momentum for a Cosserat or micropolar continuum in Eringen's classification, but all terms have an electric origin. These equations were obtained by the author (Maugin, 1971, 1980a).

4.6 Reduction to a Model without Microstructure

When pure ferroelectric features are ignored or neglecting polarization inertia and polarization-gradient effects in Mindlin's theory, equation (137) reduces to

$$t_{[ij]} = E_{[i}\, P_{j]}. \tag{140}$$

This corresponds to the classical theory of nonlinear dielectrics as originally built by Toupin (1956, 1963) and Eringen (1963) — see Eringen and Maugin, 1990, Vol. 1, chap. 7). This nonlinear theory in finite strains applies in particular to electroelastic polymers. The skewsymmetric part of the stress vanishes when polarization and electric fields are aligned. This occurs in isotropic bodies. Still the ponderomotive force is present. However, if quadratic effects in the electric field are discarded altogether, corresponding to a fully linear theory, then both ponderomotive force and couple disappear leaving for only possible electromechanical couplings piezoelectricity, material symmetry permitting (no center of symmetry).

4.7 Remark on Electric Quadrupoles

The microscopic electric description considered in Section 2 and at the beginning of this section views electric macroscopoic polarization as a polar vector. Its thermodynamical dual is akin to an electric field ($\mathbf{E}^L$). Its gradient has for thermodynamical dual $\widehat{\mathbf{E}}$ (dimensionally, an elecric field multiplied by a length). However, another view consists, while making the construct recalled in Section 2, to consider macroscopic electric polarization as made of an electric dipole density $\overline{\mathbf{P}}$, per se, and an electric quadrupole density, $\overline{\mathbf{Q}}$, so that $\mathbf{P} = \overline{\mathbf{P}} - \operatorname{div} \overline{\mathbf{Q}}$ (a natural outcome of the Lorentz modelling), and then considering $\overline{\mathbf{P}}$ and $\overline{\mathbf{Q}}$ as independent electric independent variables. The thermodynamical dual of $\overline{\mathbf{Q}}$ will then be a gradient of electric field. Such a description, envisaged by the author in the early 1970s, also yields an electric continuum endowed with a microstructure involving couple stresses. For instance, in quasi-electrostatics, we would have instead of (28), (34), (36) and (44),

$$\begin{aligned} f_i^{em} &= \overline{P}_j\, E_{i,j} + \overline{Q}_{jp}\, E_{i,jp}, \\ c_i^{em} &= \varepsilon_{ijk}\left(\overline{P}_j\, E_k + \overline{Q}_{mj}\, E_{k,m}\right) \end{aligned} \tag{141}$$

and

$$\begin{aligned} t_{ji}^{em} &= D_j\, E_i + \overline{Q}_{jk}\, E_{i,k} - \frac{1}{2}\,\mathbf{E}^2\,\delta_{ji}, \\ w^{em} &= f_j^{em}\, v_j + \rho\, E_i\, \dot{\overline{\pi}}_i + \rho\, E_{i,j}\, \mathrm{d}\big(\overline{Q}_{ji}/\rho\big)/\mathrm{d}t. \end{aligned} \tag{142}$$

More on these in the form of problems in pp. 87–89 in Eringen and Maugin, 1990, Vol. 1.

5 Dynamical Couplings between Deformation and Electromagnetic Microstructure

5.1 Introductory Note: Resonance Coupling between Wave Modes

A. General Features

In linear or linearized dynamical theories of continua expressed by a system of partial differential equations with derivatives in terms of space and time, one is often interested in knowing the possible *travelling wave modes*, functions of a space-time phase variable $\varphi = \mathbf{k} \cdot \mathbf{x} - \omega t$, where $\mathbf{k} = k\,\mathbf{m}$ is the wave vector, k is the wave number, $\mathbf{m}$ denotes the director cosines, and ω is the circular frequency. In substituting for trigonometric functions of φ or exponential functions of $i\,\varphi$ in the system of field equations one is led for nonzero amplitudes to a relation between the components of $\mathbf{k}$ and ω known as the *dispersion relation* written as

$$D(k, \omega) = 0. \tag{143}$$

The quantity $v_\varphi = \omega/k$ is called the phase velocity while the quantity $\mathbf{v}_g = \partial\omega/\partial\mathbf{k}$ is the group velocity. The wavelength is defined as $\lambda = 2\,\pi/k$. When v_φ does not depend on λ or k, the studied system is said to be *nondispersive*. Then the corresponding group velocity equals the phase velocity for any ω and k (or λ). If this is not the case, then the system is said to be *dispersive*. In the latter case the Fourier components of a non-monochromatic signal travel at different velocities – causing the *dispersion* of the signal –, all this independently of the amplitude (that does not depend on these properties for a linear system). Dispersive systems are characterized by systems of partial differential equations that do not admit a polynomial of differentiation that is homogeneous (a homogeneous polynomial of differentiation has all terms with space-time partial derivatives of the same order). Thus the classical *d'Alembert equation*

$$\frac{\partial^2 u}{\partial t^2} - c^2 \frac{\partial^2 u}{\partial x^2} = 0, \tag{144}$$

where c is a constant provides a *nondispersive* system, while the *Klein–Gordon equation*

$$\frac{\partial^2 u}{\partial t^2} - c^2 \frac{\partial^2 u}{\partial x^2} + m\,u = 0 \tag{145}$$

where m is also a constant, yields a *dispersive* system. The so-called *sine-Gordon equation*

$$\frac{\partial^2 u}{\partial t^2} - c^2 \frac{\partial^2 u}{\partial x^2} + m \sin u = 0, \tag{146}$$

is both *dispersive* **and** *nonlinear* (since obviously trigonometric functions are not linear functions of their argument). An equation such as the *Boussinesq equation* of crystal physics,

$$\frac{\partial^2 u}{\partial t^2} - c^2 \frac{\partial^2 u}{\partial x^2} - c^2 \delta^2 \frac{\partial^4 u}{\partial x^4} - c^2 \beta \frac{\partial u}{\partial x} \frac{\partial^2 u}{\partial x^2} = 0 \tag{147}$$

where δ is a characteristic length and β a nondimensional nonlinearity parameter, also is both *dispersive* **and** *nonlinear*. We shall have the opportunity later on to deal with equations such as (146) or (147). For the time being we are concerned with equations of the simpler types (144) and (145).

B. Resonance Coupling between Modes

We are interested in the following exemplary situation. Consider a linear physical system in which three bulk modes, A_α, $\alpha = 1, 2, 3$, of the plane time-harmonic type may propagate:

$$A_\alpha = \widehat{A}_\alpha \exp[i\,(\mathbf{k} \cdot \mathbf{x} - \omega\, t)]. \tag{148}$$

Assume that the system of considered field equations is such that with trial solutions (148) it yields a dispersion relation of the type (cf. Maugin, 1980b)

$$D(\omega, k) = [\omega - \omega_3(k)] \left([\omega - \omega_1(k)]\, [\omega - \omega_2(k)] - \varepsilon\, \varpi^2 \right) = 0 \tag{149}$$

where the $\omega_\alpha{}'s$ are known functions of k and of material parameters, ϖ is a characteristic (eventually wave number dependent) angular frequency, and ε is an infinitesimally small parameter. The relations

$$D_\alpha(\omega, k) = \omega - \omega_\alpha(k) = 0, \qquad \alpha = 1, 2, 3, \tag{150}$$

are the dispersion relations for *uncoupled* modes.

Of course, equation (149) tells us that the component A_3 is not coupled with the other two, and its associated dispersion relation is influenced only by material parameters that may appear in $\omega_3(k)$. The remaining two solutions of (149) are coupled via ε. Let (ω_0, k_0) denote the intersection point of the two curves $D_\alpha(\omega, k) = 0$, $\alpha = 1, 2$, in the positive quarter of the (ω, k) plane. In the neighbourhood of this critical point C, which is called a *crossover region* for the coupled modes, we have

$$\omega_\alpha(k) \cong \omega_0 + v_\alpha(k - k_0), \qquad \omega_0 = \omega_1(k_0) = \omega_2(k_0), \tag{151}$$

and the $v_\alpha{}'s$ are the group velocities of the uncoupled modes of C.

In the neighbourhood of C the remaining factor in (149) yields two approximate coupled solutions

$$\omega_{\pm}(k)-\omega_0=\frac{1}{2}\Big((v_1+v_2)\,(k-k_0)\pm\big((k-k_0)^2\,(v_1-v_2)^2+4\,\varepsilon\,\varpi^2\big)^{1/2}\Big) \quad (152)$$

or

$$k_{\pm}(\omega)-k_0=\frac{1}{2\,v_1\,v_2}\Big((v_1+v_2)\,(\omega-\omega_0) \pm\big((\omega-\omega_0)^2\,(v_1-v_2)^2+4\,\varepsilon\,\varpi^2\,v_1\,v_2\big)^{1/2}\Big). \quad (153)$$

Depending on the sign of ε and of the product $v_1\,v_2$, four different pictures of the crossover region can be sketched out. But we single out the case $\varepsilon>0$, $v_1\,v_2>0$, both $v_\alpha>0$, that is typical of what happens in problems of mechanics. The overall behavior of the remaining two coupled solutions therefore is as follows. For $k\in[0,+\infty)=\mathbb{R}^+$, we have

$$\begin{aligned}\omega_I(k)&=f_I\big(\omega_1(k),\omega_2(k);\varepsilon\big),\\ \omega_{II}(k)&=f_{II}\big(\omega_1(k),\omega_2(k);\varepsilon\big),\end{aligned} \quad (154)$$

with

$$\omega_I\cong\omega_1,\quad \omega_{II}\cong\omega_2,\quad \text{for } k\ll k_0, \quad (155)$$

and the reverse situation for $k\gg k_0$, while in the neighbourhood of point C,

$$\omega_I\cong\omega_0+\omega_+,\qquad \omega_{II}\cong\omega_0+\omega_-.$$

We see that the critical point C, in fact, no longer belongs to the coupled dispersion diagram. It is said that a *repulsion* of the dispersion curves has occurred at point C. This repulsion has the essential property to be such that

$$\frac{\Delta\omega}{\varpi}=\left|\frac{\omega_I^2-\omega_{II}^2}{\varpi\,(\omega_I+\omega_{II})}\right|\Bigg|_{k=k_0}\cong\left|\frac{\omega_I^2-\omega_{II}^2}{2\,\omega_0\,\varpi}\right|(k_0)=\mathcal{O}(\sqrt{\varepsilon}). \quad (156)$$

Simultaneously, a *resonance effect* takes place in the crossover region since it can be shown that the amplitudes satisfy a relation of the type

$$\left|\frac{\widehat{A}_1}{\widehat{A}_2}\right|\propto\sqrt{\varepsilon}\,\varpi\,\big|\omega(k)-\omega_2(k)\big|^{-1}, \quad (157)$$

which blows up for $\omega(k)$ approaching $\omega_2(k)$ at C in the absence of damping. Furthermore, in following continuously one of the coupled dispersion curves with increasing k we observe an *energy conversion* from one type of oscillation to the other type. This is typical of some wave systems in the mechanics of deformable solids as studied in depth by R. D. Mindlin (e.g., in a three-term model of rectangular rod or in the dynamics of plates; cf. Mindlin, 1955, 1960; Graff, 1975).

Two essential remarks are in order concerning this resonance phenomenon:

- First, in the case where a dissipative process is associated with each of the $A_\alpha{}'s$, the resonance effect is smoothed out (it presents a maximum instead of a divergence), and the $\omega's$ becoming complex, a relaxation time accompanies each real-frequency solution and an interchange of relaxation is observed in the cross over region.
- Second, in the case of surface waves the requirement that the amplitudes decrease with depth in the substrate implies that the allowed domain of dispersion may be reduced and that some of the coupled branches (154) in fact are not attainable.

The subsequent sections illustrate this phenomenon in the models of coupled fields sketched out in previous sections.

5.2 The Case of Magnetoelasticity in Ferromagnets

A. Magnon

Magnons are the quasi-particles quantum mechanically associated with spin waves that are oscillations in the ordered array of magnetic spins such as described by equations (70) in the absence of couplings with elasticity. The corresponding frequency mode is shown to be parabolic ($\omega \approx k^2$) with a cut-off defined by the initial non-zero static magnetization M_0, i.e., typically for such a mode

$$\omega_S(k) = \omega_M\,(\alpha\,k^2 + \beta), \qquad \omega_M = \gamma\,M_0, \tag{158}$$

where α and β are reduced exchange and magnetic anisotropy constants.

In the magnetoelastic case the mode (158) will couple with elastic modes that typically have a linear dispersion relation (subscript P for "phonon" = elastic vibrations)

$$\omega_P(k) = c_T\,k. \tag{159}$$

The coupling between these two via *magnetostriction* (or piezomagnetism induced by magnetostriction in the presence of a bias magnetization) is of the resonance type discussed in the preceding paragraph.

B. Bulk Magnetoelastic Modes

It is not the purpose here to establish in detail the coupling between "magnons" and "phonons" on the basis of the coupled continuum equations recalled in Paragraph 3.2. The reader will find this dealt with at length in Maugin (1988, chap. 6), and Eringen and Maugin (1990, chap. 9). We shall rather exhibit exemplary pictures of this dynamical magnetoelastic (or "magnetoacoustic") coupling. What in fact happens is the resonance coupling between one elastic transverse mode and the spin mode while the other elastic transverse mode is practically uncoupled, and we are in the dispersion situation sketched out in equation (149), so that the comments in Paragraph 3.1 apply. This is true for a propagation at zero angle with respect to the direction of the initial magnetization (see Figure 4). For a nonzero angle an additional coupling with the elastic longitudinal mode occurs (see Figure 5). Figures 6 and 7 show numerically computed coupled dispersion relations for two good deformable ferromagnets, Cobalt and Yttrium-Iron-Garnet (YIG). These figures also exhibit a so-called magnetoelastic **Faraday effect** (the fact that the polarization plane of magnetoelastic waves rotates as right-polarized and left-polarized waves propagate at different speeds; Part (c) in these figures). Furthermore, if viscosity and spin-relaxation are taken into account, according to the modelling developed by the author, an exchange of relaxation between modes take place in the cross-over region (Part (d) in

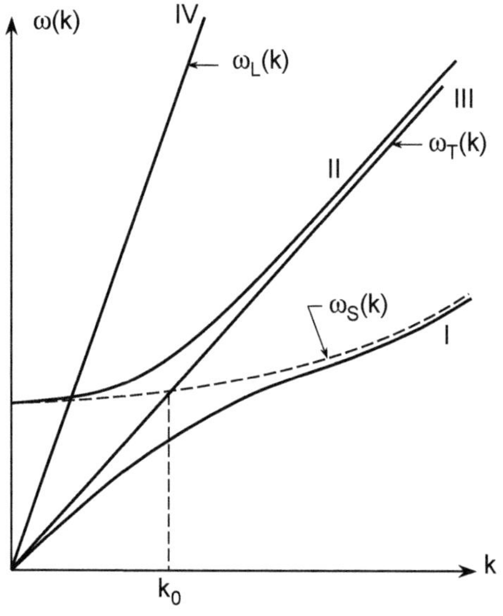

Figure 4. Dispersion diagram for coupled magnetoelastic waves in ferromagnets (after Maugin, 1981, IJES).

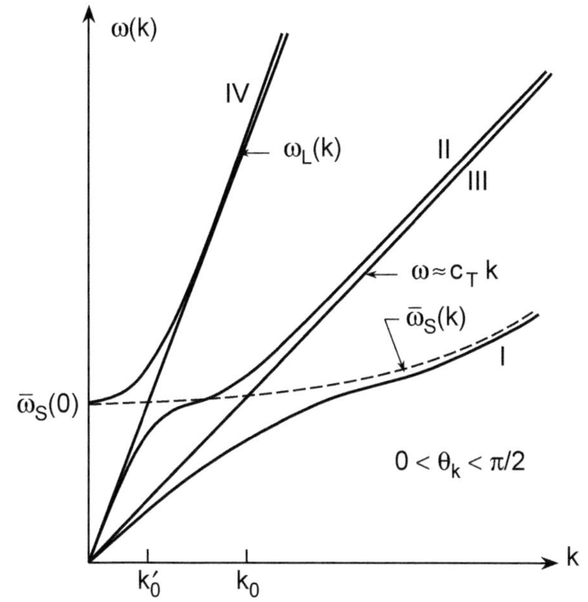

Figure 5. Dispersion diagram for coupled magnetoelastic waves in ferromagnets (after Maugin, 1981, IJES).

these figures).

C. Surface Magnetoelastic Modes

This case is much more subtle because we must account for the condition of existence of surface waves while the symmetry between right and left propagation is broken (so called "non-reciprocity" of propagation to the right and the left). With an initial setting of the static magnetization orthogonal the sagittal plane, and neglect of the curvature (magnetization gradients) of the spin-wave mode, one obtains dispersion curves such as sketched out in Figure 8. Here hatched regions are forbidden (surface modes with amplitude attenuation with depth do not exist), the low branches are limited to small wave numbers, while for forward travelling waves the upper branch tends to a shear-horizontal elastic mode, which is not the case for the backward travelling mode where the dynamic upper branch tends towards a so-called "magnetostatic" mode (as obtained in a nondeformable half space).

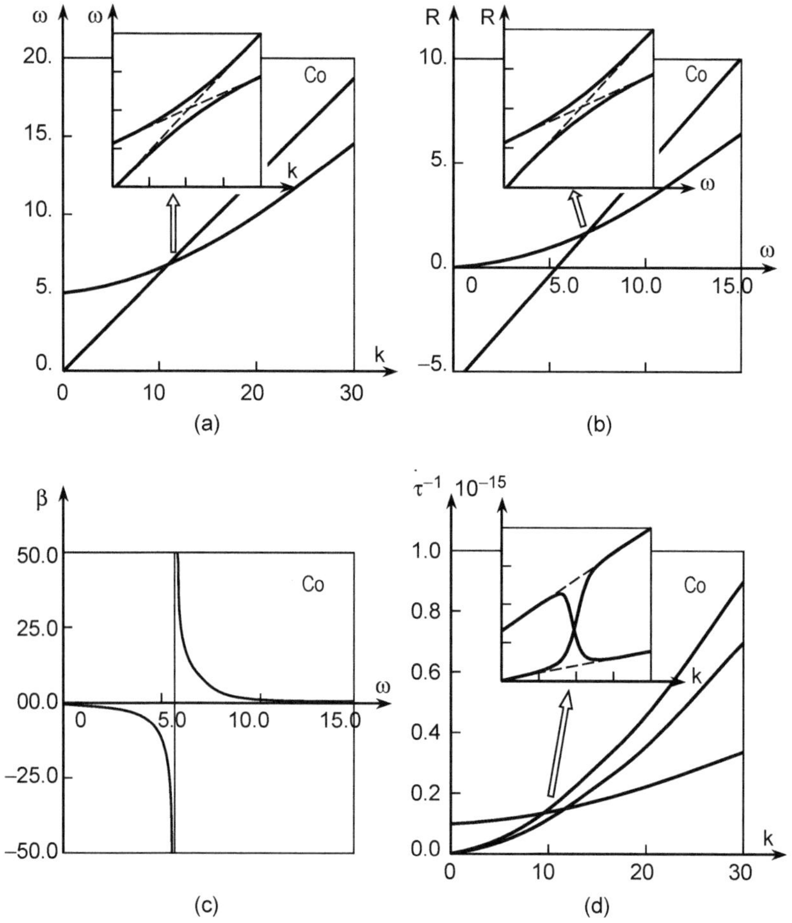

Figure 6. Magnetoaoustic resonance in Cobalt: (a) dimensionless real dispersion relation for coupled magnons and transverse phonons (ω versus k); (b) dimensionless real dispersion relation ($R = k^2$ versus ω); (c) magnetoacoustic Faraday effect; (d) exchange of relaxation between modes. (after Maugin and Pouget, 1981, IJES).

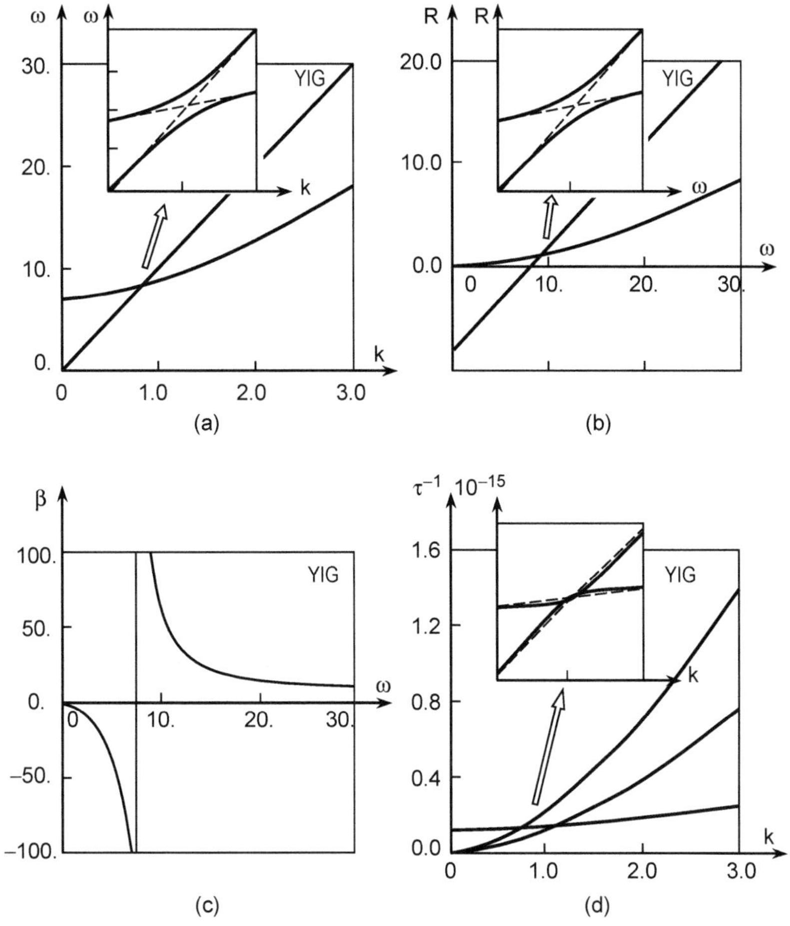

Figure 7. Magnetacoustic resonance in Yttrium-Iron-Garnet (YIG): (a) dimensionless real dispersion relation for coupled magnons and transverse phonons (ω versus k); (b) dimensionless real dispersion relation ($R = k^2$ versus ω); (c) magneoacoustic Faraday effect; (d) exchange of relaxation between modes. (after Maugin and Pouget, 1981, IJES).

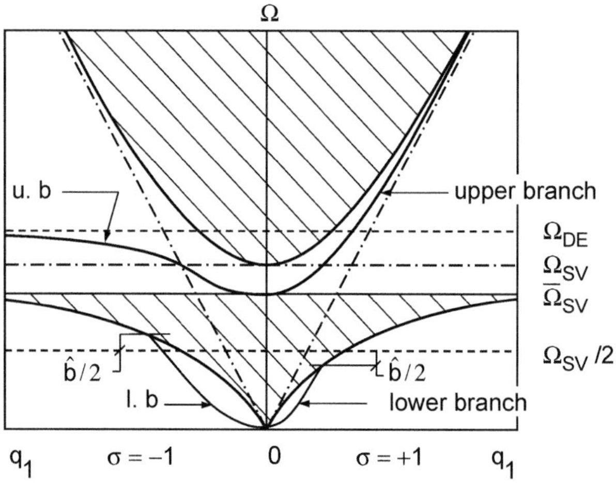

Figure 8. Dispersion curves for coupled surface magnetoelastic waves for an orthogonal setting of the bias magnetic field (exchange forces neglected): Difference between right and left propagation (after Maugin and Hakmi, 1985, JASA).

D. Case of Elastic Antiferromagnets

In the case of ferrimagnets, the multiplicity of magnetic lattices has for consequence the existence of several magnon branches, two in the case of antiferromagnets equipped with two magnetic sublattices. The magnetoelastic couplings are exhibited in the model sketched out in Figure 3. Examples of possible couplings between transverse elastic modes and two (lower and upper) spin-wave branches are shown in Figure 9 after the author and co-workers. The coupling scheme becomes complicated but reminds us exactly of what happens in the pure mechanical wave modes in some structures according to Mindlin. Figure 10 reproduces experimental results of such couplings in antiferromagnetic $FeCl_2$ (energy versus reduced wave number).

E. Magnetoacoustic Solitons

Solitons is the name given to strongly localized dynamical solutions that propagate undeformed and interact just like elastic particles during collisions. They exist because of a strict compensation between nonlinearity (that tends to make a signal getting more and more steep like in the formation of a shock wave) and dispersion (that tends to spread out a sig-

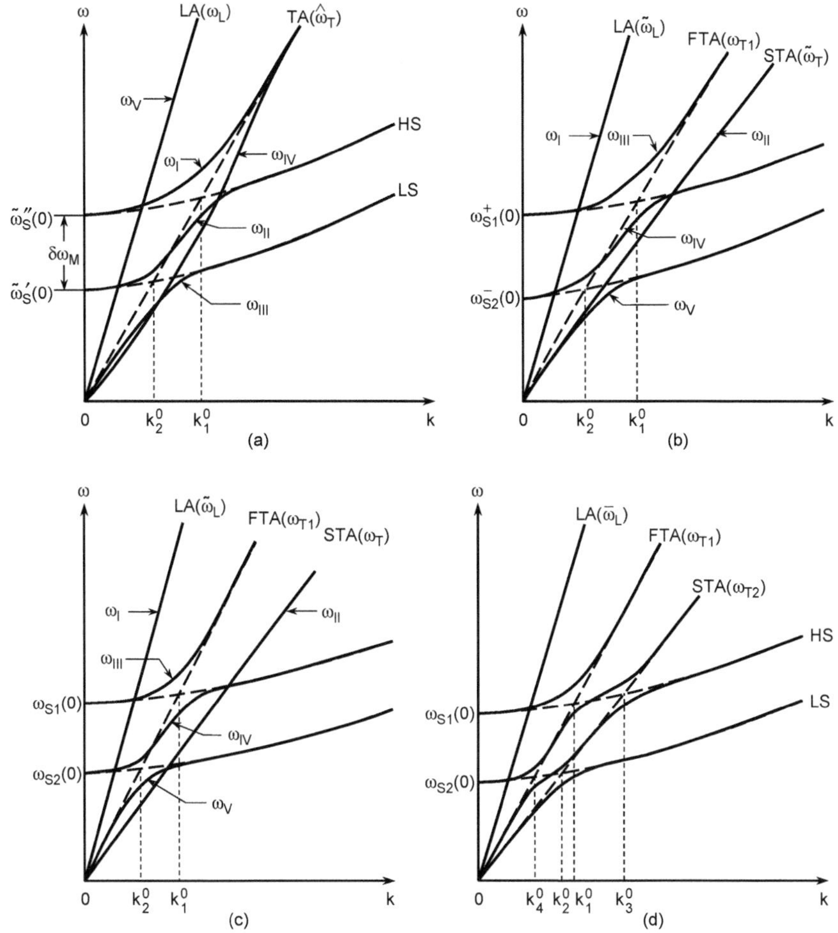

Figure 9. Qualitative dispersion relation for coupled magnetoelastic waves in simple antiferromagnets (after Maugin and Sioké-Rainaldy, 1983, 1985, JAP): (a) Longitudinal setting for a moderate bias field for nonzero global magnetization initially; (b) orthogonal setting for a moderate bias field, same initial configuration; (c) orthogonal setting for a strong bias field with nonzero global magnetization initially; (d) longitudinal setting for a strong bias field.

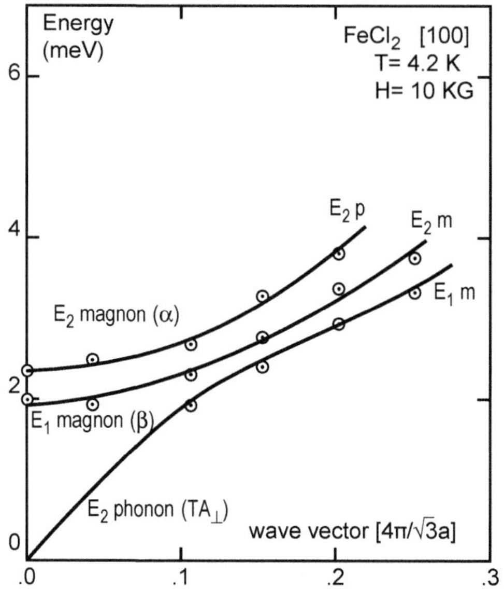

Figure 10. Real dispersion relation in antiferromagnetic $FeCl_2$ (experimental results of Ziebeck and Houmann; after Maugin, 1980b).

nal by making the various Fourier components of the signal to travel at different speeds). Equations (146) and (147) exhibit such solutions.

In the foregoing paragraphs only small amplitude oscillations of both the mechanical displacement, and the magnetic spin orientation were considered. But we may also contemplate large angular deviations of the magnetic spin allowed by the essentially non-linear (gyroscopic-like) spin equation, still coupled via magnetostriction with small elastic strains. This situation in which the simultaneous presence of nonlinearity and dispersion favours the existence of **solitons** was first studied by the author and Miled (1986a). They have shown that the coupled system of Section 3.2 can yield a system of partial differential equations now called the *sine-Gordon-D'Alembert* system that coupled via magnetostriction an equation such as (146) with a linear elastic mode. That is, in appropriate units,

$$\frac{\partial^2 \phi}{\partial t^2} - \frac{\partial^2 \phi}{\partial x^2} + \sin\phi = -\beta \frac{\partial u}{\partial x} (\cos\phi), \tag{160}$$

$$\frac{\partial^2 u}{\partial t^2} - c_T^2 \frac{\partial^2 u}{\partial x^2} = \beta \frac{\partial}{\partial x} (\sin\phi), \tag{161}$$

where u is a transverse elastic displacement, ϕ is the only remaining angle describing the spin precession, and β stands for the magnetostriction coupling. System (160)–(161) exhibits soliton-like solutions which are not exactly soliton solutions in the mathematical sense because it exhibits some radiations due to the essentially linear equation (161). Such solutions represent the dynamics of magnetoelastic domain walls of the Néel type (in-plane 180 degrees rotation of the local magnetization through the wall) in thin magnetic films. Remarkably enough, a similar nonlinear dispersive wave system was exhibited in the purely mechanical cases of micropolar elasticity — in the sense of Eringen — (Maugin and Miled, 1986b; also Eringen, 1999) and elastic media endowed with a microstructure described by a set of rigid directors – i.e., oriented media in the sense of Duhem & Ericksen – (Pouget and Maugin, 1989).

5.3 The Case of Electroelasticity in Ferroelectrics

A. Polaritons

Polaritons refer to the waves that result from a coupling between oscillations in the system of ordered electric dipoles — governed by equation of (114) — and the full set of Maxwell electromagnetic equations. The polarization mode is slightly dispersive and presents a cut-off, while the electromagnetic one is characterized by its high (light) velocity. This results in Figure 11a in a large split between the two modes 2 and 3.

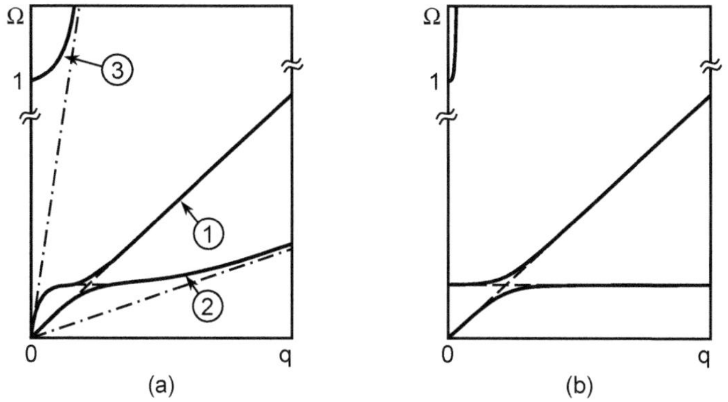

Figure 11. Dispersion relation curves for coupled transverse modes in elastic ferroelectrics: (a) mixed acoustic-polariton branches; (b) schematic view with Maxwell electrostatic equations and neglect of polarization gradients (after Maugin, 1988, p. 531).

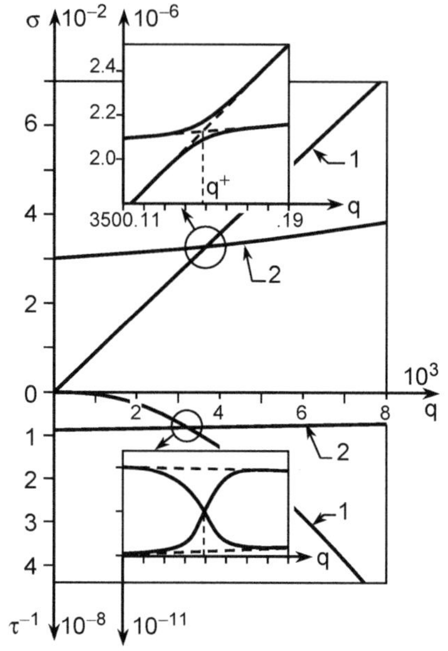

Figure 12. Numerical dispersion curves for $BaTiO_3$ (mixed transverse acoustic and polariton modes in presence of dissipation (after Pouget and Maugin, 1980, JASA).

B. Bulk and Surface Coupled Mode

For our purpose, however, the coupling with the electromagnetic mode can be discarded (quasi-electrostatic hypothesis) and we retain only the possible resonance coupling between an elastic mode and a practically flat polarization mode such as shown in Figure 11b. Such coupled modes calculated on the basis of the model developed by Maugin and Pouget (1980) are sketched in Figure 12 for a good ferroelectric material such as $BaTiO_3$ (after Pouget and Maugin, 1980). The eventual exchange of relaxation accompanying this coupling is also shown in the bottom part of the figure. Coupled surface wave modes localized in the vicinity of a limiting plane surface and characterized by a decrease of amplitude with depth can also be placed in evidence. Figure 13 shows the corresponding coupling between the polarization mode (c) and a transverse elastic mode (a) for a wave of the Rayleigh surface type (elastic displacement polarized in the sagittal plane) with a hatched forbidden dispersion zone (after Pouget and Maugin, 1981).

Finally, quite similar to the magnetoelastic solitons of the previous para-

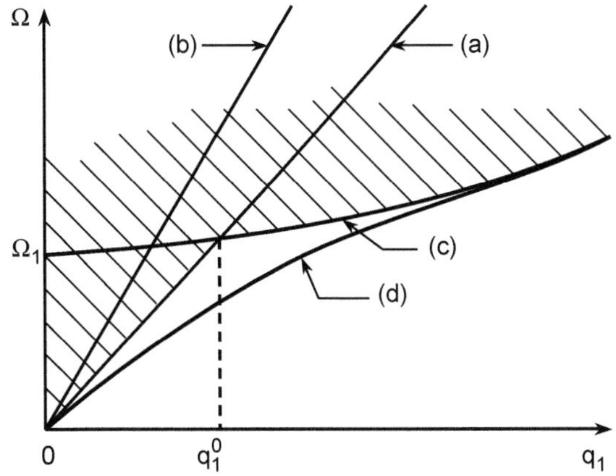

Figure 13. Piezoelectric Rayleigh waves in elastic ferroelectrics: qualitative sketch of the dispersion relation (hatched region is forbidden) (after Maugin, 1988, p. 552).

graph, **electroelastic solitons** can be shown to exist for ferroelectrics of the molecular group type (e.g., $NaCl_2$). They are governed by a nonlinear dispersive system of equations of the type of (160)–(161); cf. Pouget and Maugin, 1984.

6 Configurational Forces in Presence of an Electromagnetic Microstructure

6.1 Definition

The theory of *configurational forces* has recently become a rapidly developing active chapter of continuum thermomechanics (see the book Maugin, 2011b for an overview for professionals). We remind the reader that configurational forces are those forces of thermodynamical nature that are associated with changes of the reference configuration while traditional Newtonian-Eulerian forces are those that appear in the actual configuration of the body (standard applied forces or couples, Cauchy stress). Accordingly, configurational forces are the driving forces behind the evolution of structural defects on, and topological changes of, the material manifold. The theories of fracture, dislocation and disclination motions, structural changes such as in plasticity, damage and material growth, and the progress of phase-transition fronts belong in the general theory of configurational forces (Maugin, 2011b). In the absence of dissipation — a case which is sufficient in the present context — the relevant thermomechanical equations of the theory of configurational forces are those equations of conservation that follow from the application of Noether's theorem in a Hamiltonian-Lagrangian variational formulation. They are the local conservation equations of energy and material momentum that account for the invariance of the considered physical system under changes of time and space parametrization (the system is then said to be homogeneous in time and material space). These two equations can also be obtained by direct manipulation of the field equations by appropriate multiplication by time derivative and spatial gradient of the fields and rearrangement of the results on account of the known constitutive equations. We shall only give a flavour of the matter by recalling first what happens for a pure mechanical system, e.g., the case of micropolar solids in small deformation and small micro-rotations.

6.2 Reminder of a Purely Mechanical Case

In the dynamical case of small deformation and small micro-rotation the standard balance laws of linear and angular momenta are given in Cartesian tensorial components by (cf. present Appendix and Eringen, 1968; no applied force and couple for the sake of simplicity):

$$\rho_0 \frac{\partial^2 u_i}{\partial t^2} - \frac{\partial}{\partial x_j} \sigma_{ji} = 0, \tag{162}$$

and

$$\rho_0 \, j_{ij} \frac{\partial^2 \phi_j}{\partial t^2} - \frac{\partial m_{ji}}{\partial x_j} - \varepsilon_{ipq} \, \sigma_{pq} = 0. \tag{163}$$

Isotropic microinertia, $\mathbf{j} = I \, \mathbf{1}$, i.e., $j_{ij} = I \, \delta_{ij}$ is often assumed for the sake of simplicity or as an evident conclusion from a true micro-analysis. In terms of the displacement of components u_i and the microrotation of components ϕ_i, the relevant measures of generalized deformations for elastic solids are defined by

$$\begin{aligned} \mathbf{e} &:= (\nabla \mathbf{u})^T + \text{dual } \phi = \{ e_{ji} = u_{i,j} - \varepsilon_{jik} \, \phi_k \}, \\ \gamma &:= \nabla \phi = \{ \gamma_{ji} = \phi_{i,j} \}, \end{aligned} \tag{164}$$

and the constitutive equations of interest are given by

$$\sigma = \frac{\partial \widehat{W}}{\partial \mathbf{e}}, \quad \mathbf{m} = \frac{\partial \widehat{W}}{\partial \gamma}, \quad S = -\frac{\partial \widehat{W}}{\partial \theta}; \quad W = \widehat{W}(\mathbf{e}, \gamma, \theta; \mathbf{x}) \tag{165}$$

for the usual (but here nonsymmetric) *Cauchy stress* tensor σ and the *couple-stress* tensor $\mathbf{m}$. Here θ is the thermodynamical temperature and

S is the entropy per unit volume. The presence of $\mathbf{x}$ among the dependence of the free energy W indicates a possible material inhomogeneity via the constitutive equations. Such a dependence is also a priori considered for the density ρ_0 in the reference configuration. Let us define a "Lagrangian" density per unit volume by

$$\widehat{L} = K - W, \tag{166}$$

with

$$K = \frac{1}{2}\rho_0\left(\dot{u}_i\,\dot{u}_i + \dot{\phi}_i\,j_{ij}\,\dot{\phi}_j\right), \qquad W = \widehat{W}. \tag{167}$$

We encourage the reader to prove the following two equations by multiplying equations (162) and (163), respectively by $\dot{u}_i$ and $\dot{\phi}_i$, adding the two results, and rearranging terms on account of (165), and performing similar operations but by applying $u_{i,k}$ and $\phi_{i,k}$ to (162) and (163), adding the two results, and accounting for (165). This results in obtaining the local canonical balance equations of *energy* and *momentum* in the following form (e = internal energy per unit mass, $\mathbf{q}$ is a possible heat flux)

$$\frac{\partial}{\partial t}\left(\rho_0\left(e + \frac{1}{2}\,\dot{\mathbf{u}}^2 + \frac{1}{2}\,\dot{\phi}_i\,j_{ij}\,\dot{\phi}_j\right)\right) - \frac{\partial}{\partial x_j}\left(\sigma_{ji}\,\dot{u}_i + m_{ji}\,\dot{\phi}_i - q_j\right) = 0, \tag{168}$$

and

$$\frac{\partial}{\partial t}P_i^{tot.f} - \frac{\partial}{\partial x_j}b_{ji} = f_i^{inh} + f_i^{th}, \tag{169}$$

wherein

$$P_i^{tot.f} = -\rho_0\left(\dot{u}_j\,u_{j,i} + \dot{\phi}_k\,j_{kj}\,\phi_{j,i}\right),\ f_i^{inh} = \left(\frac{\partial\widehat{L}}{\partial x_i}\right)_{expl},\ f_i^{th} = S\,\frac{\partial\theta}{\partial x_i}, \tag{170}$$

and

$$\begin{aligned} &\mathbf{b} = -L\,\mathbf{1} - \sigma\cdot(\nabla\mathbf{u})^T - m\cdot(\nabla\phi)^T \\ \text{or}\ \ &b_{ji} = -(L\,\delta_{ji} + \sigma_{jk}\,u_{k,i} + m_{jk}\,\phi_{k,i}). \end{aligned} \tag{171}$$

The nonsymmetric stress tensor $\mathbf{b}$ is referred to as the Eshelby stress tensor; it is a true fully material stress tensor on the material manifold in the original finite-strain formulation.

Here the notation used in $(170)_2$ means (assuming j_{ij} is the same at all material points)

$$\left(\frac{\partial\widehat{L}}{\partial x_k}\right)_{expl} = \frac{1}{2}\,\frac{\partial\rho_0}{\partial x_k}\left(\dot{u}_i\,\dot{u}_i + \dot{\phi}_i\,j_{ij}\,\dot{\phi}_j\right) - \frac{\partial}{\partial x_k}\widehat{W}\bigg|_{fixed\ fields}. \tag{172}$$

In the absence of thermal effects and for materially homogeneous bodies, equations (168) and (169) reduce to the following strict conservation laws:

$$\frac{\partial}{\partial t}H - \frac{\partial}{\partial x_j}\left(\sigma_{ji}\,\dot{u}_i + m_{ji}\,\dot{\phi}_i\right) = 0, \tag{173}$$

and

$$\frac{\partial}{\partial t}P_i^{tot.f} - \frac{\partial}{\partial x_j}b_{ji} = 0, \tag{174}$$

wherein

$$H = K + W, \qquad W = \widehat{W}(e_{ij}, \gamma_{ij}). \tag{175}$$

Equations (173) and (174) and the associated jump relations across a discontinuity surface provide the basis for most calculations of driving forces on defects. In particular, for a straight through crack in the direction x_1 (see Figure 14) the celebrated J-integral of fracture theory for a quasi-static progress is given by

$$J = \oint_\Gamma \left(W\,n_1 - n_j \left(\sigma_{ji}\,\frac{\partial u_i}{\partial x_1} + m_{ji}\,\frac{\partial \phi_i}{\partial x_1} \right) \right) \mathrm{d}\Gamma, \tag{176}$$

where Γ is a circuit in the (x_1, x_2)-plane starting form the bottom stress-free face of the crack and ending on its tip stress-free face (hence in a counterclockwise circuit with unit outward normal of components n_i; component n_1 along the direction x_1). Formula (176), generalizing the standard (no microstructure) elastic case of Rice (1968), was first given by Atkinson and Leppington (1974), and reformulated since then by various authors (see

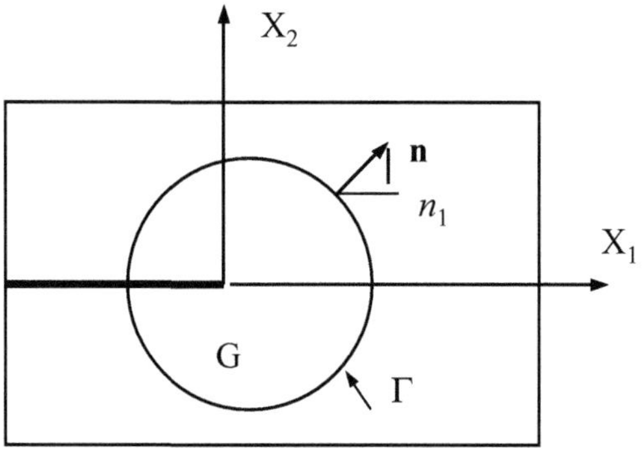

Figure 14. Straight through crack with integration contour for the J-integral.

Maugin, 2011b, p. 268). Equation (176) is obtained by integrating (174) in a domain of the (x_1, x_2)-plane encircling the crack tip, ignoring the inertia terms, invoking the divergence theorem, and projecting the result onto the x_1 direction. In the same approximation, the expression of the accompanying energy-release rate is obtained by performing similar manipulations on equation (173).

In a similar manner, the configurational force driving a phase-transition front Σ of unit oriented normal $\mathbf{n}$ is obtained by the formula

$$f_\Sigma = \mathbf{n} \cdot [\mathbf{b}] \cdot \mathbf{n}, \tag{177}$$

where $\mathbf{b}$ is reduced to its quasi-static part and the brackets denote the jump of the enclosed quantity. The details of the proof of (177) are given in Maugin (2011b, pp. 264–267) in the case of finite fields.

6.3 The Ferroelectric Case

It is clear that the ferroelectric elastic case sketched out in Section 4 is very similar to the micropolar case recalled in Paragraph 6.2. In particular, in parallel with (166) and (167) we have the following expressions:

$$\widehat{L} = K - W, \tag{178}$$

with

$$K = \frac{1}{2}\rho_0 \, (\dot{u}_i \, \dot{u}_i + \dot{\pi}_i \, d_E \, \dot{\pi}_i), \qquad W = \widehat{W}, \tag{179}$$

but here

$$\widehat{W} = \overline{W}(e_{ij} = u_{(i,j)}, \pi_i, \pi_{i,j}) - \left(\frac{1}{2}\,\mathbf{E}^2 + \rho_0 \, \pi_i \, E_i\right), \tag{180}$$

where E_i are the components of the Maxwellian quasi-static electric field $\mathbf{E}$, and we consider only small strains and electric fields in the absence of thermal and other dissipative effects, and of material inhomogeneities, the whole in the quasi-electrostatics of dielectrics. The relevant constitutive equations are given by (see, e.g., Maugin, 1988, chap. 7)

$$\begin{aligned} \sigma_{ji} &= \frac{\partial \overline{W}}{\partial e_{ij}}, \\ E_i^L &= -\rho_0^{-1} \frac{\partial \overline{W}}{\partial \pi_i}, \\ \widehat{E}_{ji} &= \frac{\partial \overline{W}}{\partial \pi_{i,j}}, \end{aligned} \tag{181}$$

while the field equations are the equation of motion, the equation governing the electric polarization, and the remaining Maxwell's equations for dielectrics. That is,

$$\rho_0 \frac{\partial^2 u_i}{\partial t^2} = \frac{\partial \sigma_{ji}}{\partial x_j} + f_i^{em}, \qquad f_i^{em} = \rho_0 \, \pi_j \frac{\partial E_i}{\partial x_j}, \tag{182}$$

$$d_E \frac{\partial^2 \pi_i}{\partial t^2} = E_i + E_i^L + \rho_0^{-1} \frac{\partial \widehat{E}_{ji}}{\partial x_j}, \tag{183}$$

and

$$\begin{aligned} &\nabla \cdot \mathbf{D} = 0, \\ &\nabla \times \mathbf{E} = \mathbf{0} \Rightarrow \mathbf{E} = -\nabla \phi, \\ &D_i = E_i + \rho_0 \, \pi_i. \end{aligned} \tag{184}$$

Here ϕ is the electrostatic potential not to be mistaken for the microrotation of the previous paragraph.

Because the conservation laws of energy and material momentum are canonical (i.e., their formal expression is independent of the true physical interpretation of the variables; see Maugin, 2011b) we could write them down at once by analogy with equations (168) through (171). The thermomechanics of the corresponding configurational forces was given by Restuccia and Maugin (2008) in the finite-strain framework. We could exploit Noether's theorem — as there is no dissipation — since all equations are derivable from a unique Hamiltonian-Lagrangian variational principle. More naively, for example, in order to obtain the conservation equation of canonical (material) momentum, we can simply combine the three co-vectorial equations obtained by applying the field $u_{i,k}$, $\pi_{i,k}$ and $\nabla \phi$, respectively to (182), (183) and (184) and accounting for (179) through (181), to arrive at the equation

$$\frac{\partial}{\partial t} P_i^{tot.f} - \frac{\partial}{\partial x_j} b_{ji} = 0 \tag{185}$$

wherein

$$P_i^{tot.f} = -\rho_0 \left(\dot{u}_j \, u_{j,i} + d_E \, \dot{\pi}_j \, \pi_{j,i} \right), \tag{186}$$

and

$$b_{ji} = -\big(L \, \delta_{ji} + \sigma_{jk} \, u_{k,i} + D_j \, \phi_{,i} + \widehat{E}_{jk} \, \pi_{k,i} \big). \tag{187}$$

Just the same as (174), equation (185) provides the basis for the construction of configurational forces acting on defects in elastic ferroelectrics. In particular, in quasi-statics, the integral of (185) in the proper plane (see Figure 14) yields the generalized J-integral useful in fracture studies:

$$J = \oint_\Gamma \left(W \, n_1 - n_j \left(\sigma_{ji} \frac{\partial u_i}{\partial x_1} + D_j \frac{\partial \phi}{\partial x_1} + \widehat{E}_{ji} \frac{\partial \pi_i}{\partial x_1} \right) \right) \mathrm{d}\Gamma. \tag{188}$$

If the body is piezoelectric but not ferroelectric, then $d_E = 0$, $\widehat{E}_{ij} = 0$, and equation (183) reduces to

$$\mathbf{E} + \mathbf{E}^L = 0, \tag{189}$$

while

$$\overline{W} = \overline{W}(e_{ij}, \pi_i), \qquad E_i^L = -\rho_0^{-1} \frac{\partial \overline{W}}{\partial \pi_i}. \tag{190}$$

We can perform a partial Legendre transformation on W such as, on account of (189):

$$\begin{aligned} E_i &= \frac{\partial \overline{W}}{\partial P_i}, \\ P_i &= \rho_0 \, \pi_i = -\frac{\partial \widetilde{W}}{\partial E_i}, \\ \overline{W}(e_{ij}, P_i) &- \widetilde{W}(e_{ij}, E_j) = E_k \, P_k = \rho_0 \, \pi_k \, E_k. \end{aligned} \tag{191}$$

Thus, in this approximation,

$$D_i = E_i + P_i = -\frac{\partial \breve{W}}{\partial E_i}, \qquad \breve{W}(e_{ji}, E_i) = -\frac{1}{2} \mathbf{E}^2 + \widetilde{W}(e_{ij}, E_j). \tag{192}$$

This is the standard theory of electroelasticity of piezoelectrics for which (188) reduces to the known formula (given by several authors, see Maugin, 2011b, sec. 11.9)

$$J = \oint_\Gamma \left(W \, n_1 - n_j \left(\sigma_{ji} \, \frac{\partial u_i}{\partial x_1} + D_j \, \frac{\partial \phi}{\partial x_1} \right) \right) \, \mathrm{d}\Gamma. \tag{193}$$

For the formulation of the driving force acting on phase-transition fronts, we refer the reader to our book (Maugin, 2011b, sec. 11.9).

6.4 The Ferromagnetic Case

This case is peculiar in the dynamic framework because of the special nature of the magnetic spin for which there is no kinetic energy expressed in the traditional form. First of all it can be shown (Maugin, 2011b, p. 370) that the local energy equation (66) for small strains can be rewritten in a more traditional (canonical) form as

$$\frac{\partial}{\partial t} \left(\rho_0 \left(\frac{1}{2} \, \dot{\mathbf{u}}^2 + e - \mathbf{B} \cdot \mu \right) \right) - \frac{\partial}{\partial x_j} (\sigma_{ji} \, \dot{u}_i + \widehat{B}_{ji} \, \dot{\mu}_i - q_j) = 0, \tag{194}$$

where e is the internal energy per unit mass and q_j stands for the heat flux. In the absence of thermal and dissipative processes this is rewritten as

$$\frac{\partial}{\partial t} \left(\frac{1}{2} \, \rho_0 \, \dot{\mathbf{u}}^2 + W - \rho_0 \, B_j \, \mu_j \right) - \frac{\partial}{\partial x_j} (\sigma_{ji} \, \dot{u}_i + \widehat{B}_{ji} \, \dot{\mu}_i) = 0, \tag{195}$$

where there is no apparent kinetic energy for the magnetic spin, while the volume energy W remains a function of the set of variables $(e_{ji}, \mu_i, \mu_{i,j})$. The associated canonical conservation law of (material) momentum was formulated by Fomethe and Maugin (1996). In small strains and materially homogeneous materials it reads

$$\frac{\partial}{\partial t} P_i^{mech} - \frac{\partial}{\partial x_j} b_{ji} = f_i^{ferro}, \tag{196}$$

wherein

$$P_i^{mech} = -\rho_0 \, \dot{u}_k \, u_{k,i}, \tag{197}$$

$$b_{ji} = -\big(L \, \delta_{ji} + \sigma_{jk} \, u_{k,i} + \widehat{B}_{jk} \, \mu_{k,i}\big), \tag{198}$$

$$L = \frac{1}{2} \rho_0 \, \dot{\mathbf{u}}^2 + \rho_0 \, B_j \, \mu_j - W(e_{ji}, \mu_i, \mu_{i,j}), \tag{199}$$

and

$$f_i^{ferro} = -\gamma^{-1} \, \omega_k \, \mu_{k,i}, \tag{200}$$

where ω_i are the components of the precessional velocity of the magnetic spin, i.e.,

$$\omega_i = -\gamma \left(B_i + B_i^L + \rho_0^{-1} \, \widehat{B}_{ji,j}\right). \tag{201}$$

The generally nonvanishing right-hand side in (196) is the print left by the peculiar gyroscopic nature of the magnetic spin. Whatever we do, we cannot incorporate it in any of the two terms in the left-hand side. This "material" force (200) can also be written in the self-speaking form

$$f_i^{ferro} = \rho_0 \, \frac{\delta L}{\delta \mu_k} \, \mu_{k,i}, \tag{202}$$

where the Euler–Lagrange variational derivative is given by

$$\frac{\delta L}{\delta \mu_k} = \frac{\partial L}{\partial \mu_k} - \frac{\partial}{\partial x_p} \frac{\partial L}{\partial \mu_{k,p}} = B_k - \frac{\delta W}{\delta \mu_k}. \tag{203}$$

It is verified that the "ferromagnetic" material force (200) or (202) has no dissipative contents by computing the (identically nil) power that it expends in a material velocity field on the material manifold by virtue of equation (9), in full agreement with (195). Thus it will not contribute any term in the energy-release rate that we could deduce from (196) in a study of fracture. We refer the reader to Chapter 11 of Maugin (2011b) for further applications to fracture and the progress of phase-transition fronts and magnetic domain walls.

7 Conclusive Remark

In the early developments of three-dimensional generalized continuum mechanics the difficulty of physically realizing a volume density of distributed mechanical couples was noticed. An early justification for considering magnetized and electrically polarized materials was the possibility to induce such couples by electromagnetic means: the general non-alignment of magnetic field and magnetization in the magnetic case, that of electric field and electric dipoles in the electric case. What we have shown in the foregoing sections is that a further electromagnetic microstructure, whether magnetic or electric in nature, induces the presence of other fields that theoretically exist in purely mechanical theories, those of hyperstress and intrinsic spin. As shown in this set of lectures, a consequence of the presence of such fields makes that some of the results and expressions of pure continuum mechanics are translated into new electromagnetically-based quantities. This is true in most of the applications such as coupled-wave propagation, driving forces on cracks and phase-transformation. This does not come as a surprise in the field-theoretical approach presented in these lectures where many of the expressions are indeed canonical, and thus formally independent of the precise physical meaning of the involved fields.

A Reminder of Basic Equations of Generalized Mechanical Continua

Here we remind the reader of the basic local equations of balance of now standard generalized continua (see, e.g., Eringen, 1999; Maugin, 2011a).

Let σ the Cauchy stress of continuum mechanics, i.e., the stress tensor of Cartesian tensor components σ_{ji} in the actual configuration of a body B at Newtonian time t. In classical continuum mechanics (no applied couple, no internal structure) this is symmetric satisfying the two local balance equations of linear momentum and moment of momentum:

$$\frac{\partial}{\partial t}(\rho_0\,\mathbf{v}) - \mathrm{div}\,\sigma = 0 \qquad \text{or} \qquad \frac{\partial}{\partial t}(\rho_0\,\dot{u}_i) - \sigma_{ji,j} = 0, \tag{204}$$

and

$$\sigma = \sigma^T \quad \text{or} \quad \sigma_{ji} = \sigma_{ij} \iff \sigma_{[ji]} = 0, \tag{205}$$

in the absence of body force. Here ρ_0 is the constant matter density (for a homogeneous body) and $v_i := \dot{u}_i$ denotes the velocity.

In the most popular (purely mechanical) generalized continuum mechanics, Equation (205) generalizes to the following ones (written in quasi-statics for the sake of simplicity):

Micromorphic Bodies (Eringen, Mindlin; 1964):

$$\begin{aligned} &\mu_{kij,k} + \sigma_{ji} - s_{ji} + l_{ij} = 0, \\ &\sigma_{ji} = \sigma_{(ji)} + \sigma_{[ji]}, \; s_{[ji]} = 0, \\ &l_{ji} = C_{ji} + l_{(ji)}. \end{aligned} \tag{206}$$

Micropolar Bodies (Cosserat brothers, etc.):

$$\mu_{k[ji],k} + \sigma_{[ji]} + C_{ij} = 0 \tag{207}$$

or

$$m_{ji,j} + \varepsilon_{ikj}\,\sigma_{kj} + C_i = 0. \tag{208}$$

Bodies with Microstretch (Eringen, 1969):

$$\mu_{klm} = \frac{1}{3}\,m_k\,\delta_{lm} - \frac{1}{2}\,\varepsilon_{lmr}\,m_{kr} \tag{209}$$

so that

$$\begin{aligned} m_{kl,k} + \varepsilon_{lmn}\,\sigma_{mn} + C_l &= 0, \\ m_{k,k} + \sigma - s + l &= 0. \end{aligned} \tag{210}$$

Dilatational Elasticity (Cowin and Nunziato, 1983):

$$m_{k,k} + \sigma - s + l = 0. \tag{211}$$

In these equations given in Cartesian components in order to avoid any misunderstanding (note that the divergence is always taken on the first index of the tensorial object to which it applies), μ_{kij} is a new internal force having the nature of a third-order tensor. It has no specific symmetry in Equation (206) and it may be referred to as a *hyperstress*. In the case of Equations (207) this quantity μ_{kij} is skewsymmetric in its last two indices and a dual second order tensor — called a *couple stress* — of components m_{ji} can be introduced having *axial* nature with respect to its second index. The fields s_{ji} and l_{ij} are, respectively, a symmetric second order tensor and a general second order tensor. The former is an *intrinsic interaction stress*, while the latter refers to an external source of *both* stress and couple according to the last of Equations (206). Only the skew part of the later remains in the special case of micropolar materials (Equations (207) in which C_i represents the components of an *applied couple*, an axial vector associated with the skewsymmetric tensor C_{ji}). The latter can be of electromagnetic origin, and more rarely of pure mechanical origin. Equations (209) and (210) represent a kind of intermediate case between micromorphic and micropolar materials. The case of dilatational elasticity in Equation (211) appears as a further reduction of that in Equation (210). This can be useful in describing the mechanical behavior of media exhibiting a distribution of holes or cavities in evolution.

In a fully dynamical case, a dynamic (inertia) term is present in the right-hand side of equations $(206)_1$, $(207)_1$, $(210)_1$ and (211). For instance, in the case of $(207)_1$, its dynamical generalization reads

$$\mu_{k[ji],k} + \sigma_{[ji]} + C_{ij} = \rho_0 \, S_{ji} \tag{212}$$

or

$$m_{ji,j} + \varepsilon_{ikj} \, \sigma_{kj} + C_i = \rho_0 \, S_i, \tag{213}$$

where the intrinsic spin tensor of components S_{ji} (and its dual axial vector S_k) are given by

$$S_{ji} = -\varepsilon_{jik} \, S_k, \qquad S_k = \frac{\mathrm{d}}{\mathrm{d}t}(j_{kp} \, \nu_p), \tag{214}$$

where j_{kp} stands for a symmetric tensor of rotational (or micro) "inertia" and ν_p denotes the vector components of an intrinsic rotational velocity (generally different from standard vorticity $\omega_i = \frac{1}{2}(\nabla \times \mathbf{v})_i$). Eringen (1966) has shown that j_{kp} satisfies a "balance law of micro-inertia" in perfect analogy with the standard conservation of mass for the macroscopic motion. For the purpose of analogy with electromagnetically micro-structured media, we note that the spin tensor that will appear in the right-hand side of equation $(206)_1$ would be defined microscopically by an average of the type

$$S_{ji} \equiv \langle \ddot{\xi}_j \, \xi_i \rangle, \tag{215}$$

where ξ_j refers to internal coordinates in a micro-element defined at point $\mathbf{X}$ in the body. Only the skewsymmetric (i.e., antisymmetric) part of this tensor is involved in a *micropolar* body so that we have the reduction

$$S_{ji} = \langle \ddot{\xi}_{[j} \, \xi_{i]} \rangle = \left\langle \frac{\mathrm{d}}{\mathrm{d}t} (\dot{\xi}_{[j} \, \xi_{i]}) \right\rangle = \frac{\mathrm{d}}{\mathrm{d}t} \langle \dot{\xi}_{[j} \, \xi_{i]} \rangle. \tag{216}$$

This is to be compared to expressions such as that in the left-hand side of (123).

Bibliography

K. Aizu. Possible species of ferromagnetic, ferroelectric and ferroelastic crystals. *Phys. Rev. B*, 2:754–772, 1970.

A. I. Akhiezer, V. G. Bar'yakhtar, and S. V. Peletmiskii. *Spin Waves*. North-Holland, Amsterdam, 1968. (translation from the Russian).

A. Askar, P. C. Y. Lee, and E. S. Cakmak. Lattice-dynamics approach to the theory of elastic dielectrics with polarization gradients. *Phys. Rev. B*, 1:3525–3537, 1970.

C. Atkinson and F. G. Leppington. Some calculations of the energy-release rate G for cracks in micropolar and couple-stress elastic media. *Int. J. Fract.*, 10:599–602, 1974.

A. S. Borovik-Romanov. Piezomagnetism in the antiferromagnetic fluorides of cobalt and manganese. *Soviet Phys. JETP*, 11:786, 1960.

W. F. Jr Brown. *Magnetoelastic Interactions*. Springer, New York, 1966.

B. Collet and G. A. Maugin. Sur l'électrodynamique des milieux continus avec interactions. *C. R. Acad. Sci. Paris B*, 279:379–382, 1974.

E. Cosserat and F. Cosserat. *Théorie des corps déformables*. Hermann, Paris, 1909. (Reprint, Gabay, Paris, 2008).

S. C. Cowin and J. W. Nunziato. Linear elastic materials with voids. *J. Elasticity*, 13:125–147, 1983.

P. Curie and J. Curie. Développement par pression de l'électricité polaire dans les cristaux hémièdres à faces inclinées (This concerns the direct effect.). *C. R. Acad. Sci. Paris*, 91:294–295, 1880. (also, Bulletin de Minéralogie de France, 3:90–93, 1880; also, *Œuvres de Pierre Curie* (pp. 6–9). Société Française de Physique, editor. Gauthier-Villars, Paris, 1908).

P. Curie and J. Curie. Contractions et dilatations produites par des tensions électriques dans les cristaux hémièdres à faces inclinées (This concerns the inverse effect.). *C. R. Acad. Sci. Paris*, 93:1137–1139, 1881.

R. De Groot and L. G. Suttorp. *Foundations of electrodynamics*. North-Holland, Amsterdam, 1972.

A. De Simone and P. Podio-Guidugli. On the continuum theory of deformable ferromagnetic solids. *Arch. Rat. Mech. Anal.*, 136(3):201–233, 1996.

R. C. Dixon and A. C. Eringen. A dynamical theory of polar elastic materials, I, II. *Int. J. Engng. Sci.*, 3:359–377, 379–398, 1965.

J. L. Ericksen. Anisotropic fluids. *Arch. Rat. Mech. Anal.*, 4:231–237, 1960.

A. C. Eringen. On the foundations of electrostatics. *Int. J. Engng. Sci.*, 1: 127–153, 1963.

A. C. Eringen. Theory of micropolar fluids. *J. Math. Mech.*, 16:1–18, 1966.

A. C. Eringen. Theory of micropolar elasticity. In H. Liebowitz, editor, *Fracture: A Treatise, Vol. 2*, pages 621–729. Academic Press, New York, 1968.

A. C. Eringen. Micropolar fluids with stretch. *Int. J. Engng. Sci.*, 7:115–127, 1969.

A. C. Eringen. *Microcontinuum field theories, I - Foundations and solids*. Springer, New York, 1999.

A. C. Eringen and G. A. Maugin. *Electrodynamics of continua*, volume I and II. Springer, New York, 1990.

A. Fomethe and G. A. Maugin. Material forces in thermoelastic ferromagnets. *Continum Mech. Thermodyn.*, 8:275–292, 1996.

K. F. Graff. *Wave motion in elastic solids*. Ohio State University Press, Colombus, 1975.

J. P. Joule. On a new class of magnetic forces. *Sturgeon's Annals of Electricity*, 8:219, 1842. [also, On the effects of magnetism upon the dimensions of iron and steel bars. *Phil. Mag.*, 30(3):76–87, 225–241, 1847.].

S. Katzir. The discovery of the piezoelectric effect. *Arch. Hist. Exact Sciences*, 57:61–91, 2003.

C. Kittel. Interactions of spin waves and ultrasonic waves in ferromagnetic crystals. *Phys. Rev.*, 110:836–841, 1958.

J. Lenz. Magnetoelastic interactions in solids. *Zeit. Angew. Math. Mechanik*, 52(4):T139–T141, 1972. [also, Spin precession in deformable ferromagnetic crystals produced entirely by mechanical induction. *Zeit. Angew. Math. Mechanik*, 68(4):T208–T210, 1988.].

H. A. Lorentz. *The theory of electrons and its applications to the phenomena of lights and radiant heat*. Teubner, Leipzig, 1909. (reprinted by Dover, New York, 1952).

G. A. Maugin. Un principe variationnel pour des milieux micromorphiques non dissipatifs. *C. R. Acad. Sci. Paris A*, 271:807–810, 1970.

G. A. Maugin. *Micromagnetism and polar media*. PhD thesis, AMS Dept, Princeton University, Princeton, N.J., USA, 1971.

G. A. Maugin. Sur la dynamique des milieux déformables magnétisés avec spin – Théorie classique. *J. Mécanique (Paris)*, 13:75–96, 1974.

G. A. Maugin. A continuum theory of deformable ferrimagnetic bodies: Field equations. *J. Math. Phys.*, 17:1727–1738, 1976a.

G. A. Maugin. A continuum theory of deformable ferrimagnetic bodies: Thermodynamics, constitutive theory. *J. Math. Phys.*, 17:1739–1751, 1976b.

G. A. Maugin. Deformable dielectrics - I - Equations for a dielectric made of several molecular species. *Ach. Mechanics (PL)*, 28:679–692, 1976c.

G. A. Maugin. On the covariant equations of the relativistic electrodynamics of continua. *J. Math. Phys.*, 19:1212–1219, 1978.

G. A. Maugin. Classical magnetoelasticity in ferromagnets with defects. In H. Parkus, editor, *Electromagnetic interaction in elastic solids*, CISM Courses and Lectures no. 257, Udine 1977, pages 243–324. Springer, Wien, 1979.

G. A. Maugin. The method of virtual power in continuum mechanics: Application to coupled fields. *Acta Mechanica*, 35:1–70, 1980a.

G. A. Maugin. Elastic-electromagnetic resonance couplings in electromagnetically ordered media. In F. P. J. Rimrott and B. Tabarrok, editors, *Theoretical and Applied Mechanics (Proc. 15th ICTAM, Toronto, 1980)*, pages 345–355. North-Holland, Amsterdam, 1980b.

G. A. Maugin. Wave motion in magnetizable deformable solids. *Int. J. Engng. Sci.*, 19:321–388, 1981.

G. A. Maugin. *Continuum Mechanics of Electromagnetic Solids*. North-Holland, Amsterdam, 1988.

G. A. Maugin. Generalized continuum mechanics: What do we mean by that? In G. A. Maugin and A. V. Metrikine, editors, *Mechanics of Generalized Continua (One hundred years after the Cosserats)*, pages 3–13. Springer, New York, 2010.

G. A. Maugin. A historical perspective of generalized continuum mechanics. In H. Altenbach, G. A. Maugin, and V. Erofeev, editors, *Mechanics of generalized continua – From micromechanical basics to engineering applications*, pages 1–17. Springer, Berlin, 2011a.

G. A. Maugin. *Configurational forces*. CRC/Chapman/Taylor and Francis, Boca Raton, FL., USA, 2011b.

G. A. Maugin and A. C. Eringen. Deformable magnetically saturated media - I - Field equations. *J. Math. Phys.*, 13:143–155, 1972a.

G. A. Maugin and A. C. Eringen. Polarized elastic materials with electronic spin – A relativistic approach. *J. Math. Phys.*, 13:1777–1788, 1972b.

G. A. Maugin and A. C. Eringen. On the equations of the electrodynamics of deformable bodies of finite extent. *J. Mécanique*, 16:101–147, 1977.

G. A. Maugin and A. Hakmi. Magnetoelastic surface waves in elastic ferromagnets - I : Orthogonal setting of the bias magnetic field. *J. Acoust. Soc. Amer.*, 77:1010–1026, 1985.

G. A. Maugin and A. Miled. Solitary waves in elastic ferromagnets. *Phys. Rev. B*, 33:4830–4842, 1986a.

G. A. Maugin and A. Miled. Solitary waves micropolar elastic crystals. *Int. J. Engng. Sci.*, 24:1477–1499, 1986b.

G. A. Maugin and J. Pouget. Electroacoustic equations in one-domain ferroelectric bodies. *J. Acoust. Soc. Amer.*, 68:575–587, 1980.

G. A. Maugin and J. Pouget. A continuum approach to magnon-phonon couplings - III: Numerical results. *Int. J. Engng. Sci.*, 19:479–493, 1981.

G. A. Maugin and J. Sioké-Rainaldy. Magnetoacoustic resonance in antiferromagnetic insulators in weak magnetic fields. *J. Appl. Phys.*, 54: 1507–1518, 1983.

G. A. Maugin and J. Sioké-Rainaldy. Magnetoacoustic resonance in antiferromagnetic insulators in moderate and strong magnetic-fields. *J. Appl. Phys.*, 57:2131–2141, 1985.

G. A. Maugin, J. Pouget, R. Drouot, and B. Collet. *Nonlinear Electromechanical Couplings.* J. Wiley, New York, 1992.

R. D. Mindlin. *Notes: An introduction to the mathematical theory of vibrations of elastic plates.* U.S Army Signal Corps Engineering Laboratories, Fort Monmouth, N.J, 1955. [reprint as a book edited by J. S. Yang, World Scientific, Singapore, 2006].

R. D. Mindlin. Waves and vibrations in isotropic plates. In *Structural Mechanics*, pages 199–232. Pergamon Press, New York, 1960.

R. D. Mindlin. Micro-structure in linear elasticity. *Arch. Rat. Mech. Anal.*, 16:52–78, 1964.

R. D. Mindlin. Elasticity, piezoelectricity and crystal lattice dynamics. *J. Elasticity*, 2:217–282, 1972.

D. F. Nelson. *Electric, optic and acoustic interactions in dielectrics.* J. Wiley, New York, 1979.

D. F. Nelson and B. Chen. Lagrangian treatment of magnetic dielectrics. *Phys. Rev. B*, 50(2):1023–1038, 1994.

R. W. Ogden and D. F. Steigmann, editors. *Mechanics and electrodynamics of magneto- and electro-elastic materials* – Udine, 2009, CISM Courses and Lectures no. 527, 2010. Springer, Wien-New York.

J. Pouget and G. A. Maugin. Coupled acoustic-optic modes in elastic ferroelectrics. *J. Acoust. Soc. Amer.*, 68:588–601, 1980.

J. Pouget and G. A. Maugin. Piezoelectric rayleigh surface modes in elastic ferroelectrics. *J. Acoust. Soc. Amer.*, 69:1319–1325, 1981.

J. Pouget and G. A. Maugin. Solitons and electroelastic interactions in ferroelectric crystals - I - Single solitons and domain walls. *Phys. Rev. B*, 30:5306–5325, 1984.

J. Pouget and G. A. Maugin. Nonlinear dynamics of oriented elastic solids - I - Basic equations. *J. Elasticity*, 22:157–183, 1989.

J. Pouget, A. Askar, and G. A. Maugin. Lattice model for elastic ferroelectric crystals: Microscopic approach. *Phys. Rev. B*, 33:6304–6319, 1986a.

J. Pouget, A. Askar, and G. A. Maugin. Lattice model for elastic ferroelectric crystals: Continuum approximation. *Phys. Rev. B*, 33:6320–6325, 1986b.

L. Restuccia and G. A. Maugin. Thermodynamics of inhomogeneous ferroelectrics. *J. Mech. Mat. & Struct.*, 3(6):1113–1123, 2008.

J. M. Rice. Path-independent integral and the approximate analysis of strain concentrations by notches and cracks. *Trans. ASME J. Appl. Mech.*, 33: 379–385, 1968.

J. Sioké-Rainaldy and G. A. Maugin. Magnetoacoustic equations in antiferromagnetic insulators of the easy-axis type. *J. Appl. Phys.*, 54:1490–1506, 1983.

K. Soumahoro and J. Pouget. Electroacoustic properties for deformable antiferrelectric materials. *J. Acoust. Soc. Amer.*, 96(6):3558–3567, 1994.

E. S. Suhubi. Elastic dielectrics with polarization gradients. *Int. J. Engng. Sci.*, 7:993–997, 1969.

H. F. Tiersten. Coupled magnetomechanical equations for magnetically saturated insulators. *J. Math. Phys.*, 5:1298–1318, 1964.

H. F. Tiersten. Variational principle for saturated magnetoelastic insulators. *J. Math. Phys.*, 6:779–787, 1965.

H. F. Tiersten. On the nonlinear equations of thermoelectroelasticity. *Int. J. Engng. Sci.*, 9:587–604, 1971.

R. A. Toupin. The elastic dielectric. *J. Rat. Mech. Anal.*, 5:840–915, 1956.

R. A. Toupin. A dynamical theory of dielectrics. *Int. J. Engng. Sci.*, 1: 101–126, 1963.

R. A. Toupin. Theory of elasticity with couple stress. *Arch. Rat. Mech. Anal.*, 17:85–112, 1964.

J. H. Van Vleck. *Electric and magnetic susceptibilities.* Clarendon Press, Oxford, UK, 1932.

W. Voigt. *Lehrbuch der Kristallphysik.* Teubner-Verlag, Leipzig, 1928.

Computational Methods for Generalised Continua

René de Borst*
* School of Engineering, University of Glasgow, Glasgow, UK

Abstract Standard continuum models do not incorporate an internal length scale, and therefore suffer from excessive mesh dependence when strain-softening models are used in numerical analyses. In this contribution this phenomenon will be analysed and remedied through the use of higher-order continua. To enable an efficient and robust implementation algorithms based on damage and on plasticity theories will be described for higher-order gradient models and for a Cosserat continuum.

1 Introduction

Localisation of deformation refers to the emergence of narrow regions in a structure where all further deformation tends to concentrate, in spite of the fact that the external actions continue to follow a monotonic loading programme. The remaining parts of the structure usually unload and behave in an almost rigid manner. The phenomenon has a detrimental effect on the integrity of the structure and often acts as a direct precursor to structural failure. It is observed for a wide range of materials, including rocks, concrete, soils, metals, alloys and polymers, although the scale of localisation phenomena in the various materials may differ by some orders of magnitude: the band width is typically less than a millimeter in metals and several meters for crestal faults in rocks.

In this contribution we address the fundamental issue of developing (higher-order) continuum models that admit localisation of deformation while preserving well-posedness of the rate boundary value problem. In a standard (Boltzmann) continuum well-posedness is normally lost when the homogenised constitutive relation exhibits a descending branch, which is commonly referred to as strain softening.

A part of this contribution is hence devoted to uniqueness and stability issues of non-linear boundary value problems in standard continua. Emphasis is placed on the critical conditions which entail a change of type of the differential equation, and the consequences of such a change for the discretisation sensitivity of computations. Subsequently, we discuss simple enhancements of the standard continuum theory within the framework of damage theory and of plasticity theory. Our aim is to obtain an enhanced continuum formulation that does not exhibit a change of type of the differential equation and therefore, does not suffer from excessive mesh sensitivity when strain softening occurs.

For large-scale applications of such higher-order continuum methods it is pivotal that the damage and plasticity models utilised are cast in an algorithmic framework that is robust and efficient. For this reason a major part of this contribution is devoted to the description of such algorithms.

2 Isotropic Elasticity-Based Damage

The basic structure of constitutive models that are set up in the spirit of damage mechanics is simple. We have a total stress-strain relation (Lemaitre and Chaboche, 1990):

$$\boldsymbol{\sigma} = \mathbf{D}^{\mathrm{s}}(\omega, \boldsymbol{\omega}, \boldsymbol{\Omega}) : \boldsymbol{\epsilon} \tag{1}$$

where $\boldsymbol{\sigma}$ is the stress tensor, $\boldsymbol{\epsilon}$ is the strain tensor and $\mathbf{D}^{\mathrm{s}}$ is a secant, fourth-order stiffness tensor, which can depend on a number of internal variables, like scalar-valued variables ω, second-order tensors $\boldsymbol{\omega}$ and fourth-order tensors $\boldsymbol{\Omega}$. Equation (1) differs from non-linear elasticity in the sense that a history dependence is incorporated via a loading-unloading function f. The theory is completed by specifying the appropriate (material-dependent) evolution equations for the internal variables.

For isotropic damage evolution, the secant stiffness tensor of Eq. (1) becomes (in matrix format):

$$\mathbf{D}^{\mathrm{s}} = \frac{E^{\mathrm{s}}}{(1+\nu^{\mathrm{s}})(1-2\nu^{\mathrm{s}})} \begin{bmatrix} 1-\nu^{\mathrm{s}} & \nu^{\mathrm{s}} & \nu^{\mathrm{s}} & 0 & 0 & 0 \\ \nu^{\mathrm{s}} & 1-\nu^{\mathrm{s}} & \nu^{\mathrm{s}} & 0 & 0 & 0 \\ \nu^{\mathrm{s}} & \nu^{\mathrm{s}} & 1-\nu^{\mathrm{s}} & 0 & 0 & 0 \\ 0 & 0 & 0 & \frac{1-2\nu^{\mathrm{s}}}{2} & 0 & 0 \\ 0 & 0 & 0 & 0 & \frac{1-2\nu^{\mathrm{s}}}{2} & 0 \\ 0 & 0 & 0 & 0 & 0 & \frac{1-2\nu^{\mathrm{s}}}{2} \end{bmatrix} \tag{2}$$

with $E^{\mathrm{s}} = E(1-\omega_1)$ the secant stiffness modulus, and $\nu^{\mathrm{s}} = \nu(1-\omega_2)$ the secant value of Poisson's ratio. The scalar-valued damage variables ω_1 and ω_2 grow from zero to one at complete damage. A further simplification can be achieved if it is assumed that Poisson's ratio remains constant during the damage process, which is equivalent to the assumption that the secant shear stiffness and bulk moduli degrade in the same manner during damage evolution. Equation (2) then simplifies to:

$$\mathbf{D}^{\mathrm{s}} = (1 - \omega)\mathbf{D}^{\mathrm{e}} \tag{3}$$

with ω the single damage variable.

The total stress-strain relation (3) is complemented by a damage loading function f, which reads:

$$f = f(\tilde{\epsilon}, \tilde{\sigma}, \kappa) \tag{4}$$

with $\tilde{\epsilon}$ and $\tilde{\sigma}$ scalar-valued functions of the strain and stress tensors, respectively, and κ the internal variable. The internal variable κ starts at an initial level κ_i and is updated by the requirement that during damage growth $f = 0$, whereas at unloading $f < 0$ and $\dot{\kappa} = 0$. Damage growth occurs according to an evolution law such that $\omega = \omega(\kappa)$, which can be determined from a uniaxial test. The loading-unloading conditions of inelastic constitutive models are often formalised using the Karush-Kuhn-Tucker conditions:

$$f \leq 0 \quad , \quad \dot{\kappa} \geq 0 \quad , \quad f\dot{\kappa} = 0 \tag{5}$$

We here limit the treatment to the case that the damage loading function does not depend on $\tilde{\sigma}$. For such a strain-based, or elasticity-based, damage model we have:

$$f(\tilde{\epsilon}, \kappa) = \tilde{\epsilon} - \kappa \tag{6}$$

For metals a common choice for $\tilde{\epsilon}$ is the energy measure:

$$\tilde{\epsilon} = \frac{1}{2}\boldsymbol{\epsilon} : \mathbf{D}^{\mathrm{e}} : \boldsymbol{\epsilon} \tag{7}$$

Equation (7) is less convenient in the sense that it does not reduce to the uniaxial strain for uniaxial stressing. For this reason it is sometimes replaced by the modified expression

$$\tilde{\epsilon} = \sqrt{\frac{1}{E}\,\boldsymbol{\epsilon} : \mathbf{D}^{\mathrm{e}} : \boldsymbol{\epsilon}} \tag{8}$$

Expression (8) is represented graphically in the principal strain space for plane-stress conditions in Figure 1(a). In this figure, a scaling has been applied such that $\tilde{\epsilon} = 1$, while $\nu = 0.2$. The dashed lines are uniaxial stress paths.

The above energy release rate definition for $\tilde{\epsilon}$ gives equal weights to tensile and compressive strain components, which makes it unsuitable to describe the mechanical behaviour of quasi-brittle materials like concrete,

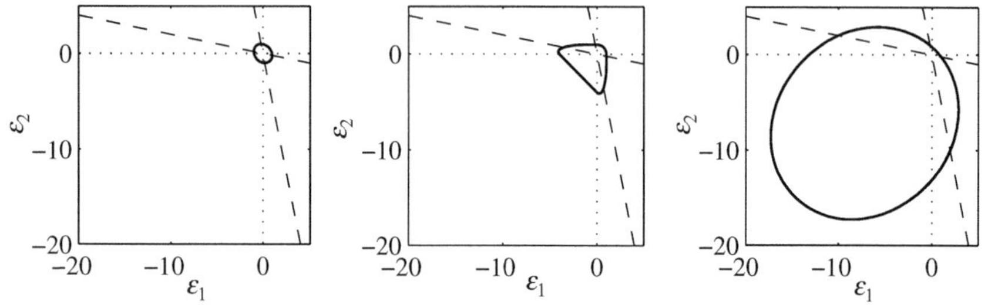

Figure 1. Contour plots for $\tilde{\epsilon}$ for (a) the energy-based concept, (b) the Mazars definition (Mazars and Pijaudier-Cabot, 1989) and (c) the modified Von Mises definition for $k = 10$

rock and ceramics. To remedy this deficiency, Mazars and Pijaudier-Cabot (1989) have suggested the definition

$$\tilde{\epsilon} = \sqrt{\sum_{i=1}^{3} (< \epsilon_i >)^2} \tag{9}$$

with ϵ_i the principal strains, with $< \cdot >$ the MacAulay brackets defined such that $< \epsilon_i >= \epsilon_i$ if $\epsilon_i > 0$ and $< \epsilon_i >= 0$ otherwise. A contour plot for $\tilde{\epsilon} = 1$ is given in Figure 1(b). A third definition for the equivalent strain $\tilde{\epsilon}$ has been proposed by de Vree et al. (1995). This proposition, which has been named a Modified von Mises definition, is given by

$$\tilde{\epsilon} = \frac{k-1}{2k(1-\nu)} I_1^{\epsilon} + \frac{1}{2k} \sqrt{\frac{(k-1)^2}{(1-2\nu)^2} (I_1^{\epsilon})^2 + \frac{12k}{(1+\nu)^2} J_2^{\epsilon}} \tag{10}$$

with I_1^{ϵ} the first invariant of the strain tensor and J_2^{ϵ} the second invariant of the deviatoric strain tensor. The parameter k governs the sensitivity to the compressive strain components relative to the tensile strain components. The definition of $\tilde{\epsilon}$ is such that a compressive uniaxial stress $k\sigma$ has the same effect as a uniaxial tensile stress σ. k is therefore normally set equal to the ratio of the compressive uniaxial strength and the tensile uniaxial strength. A graphical representation of the Modified von Mises definition is given in Figure 1(c).

From a computational point of view the above elasticity-based damage model is cast easily into a simple and robust algorithm. Indeed, in a displacement-based finite element formulation we can directly compute the strains from the given nodal displacements. The equivalent strain follows in a straightforward fashion, since $\tilde{\epsilon} = \tilde{\epsilon}(\boldsymbol{\epsilon})$. After evaluation of the damage loading function, Eq. (6), the damage variable ω can be updated, and the new value for the stress tensor can be computed directly. The simple structure of the algorithm, see Box 1 for details, is due to the fact that the stress-strain relation of Eq. (1) is a total stress-strain relation, in the sense that there exists a bijective relation for unloading, and a surjective, but non-injective relation between the stress and strain tensors for loading.

Box 1. Algorithm for a isotropic elasticity-based damage model.

1. Compute the strain increment: $\Delta\boldsymbol{\epsilon}_{j+1}$
2. Update the total strain: $\boldsymbol{\epsilon}_{j+1} = \boldsymbol{\epsilon}_0 + \Delta\boldsymbol{\epsilon}_{j+1}$
3. Compute the equivalent strain: $\tilde{\epsilon}_{j+1} = \tilde{\epsilon}(\boldsymbol{\epsilon}_{j+1})$
4. Evaluate the damage loading function: $f = \tilde{\epsilon}_{j+1} - \kappa_0$

   if $f \geq 0$, $\kappa_{j+1} = \tilde{\epsilon}_{j+1}$

   else $\kappa_{j+1} = \kappa_0$
5. Update the damage variable: $\omega_{j+1} = \omega(\kappa_{j+1})$
6. Compute the new stresses: $\boldsymbol{\sigma}_{j+1} = (1 - \omega_{j+1})\mathbf{D}^{\mathrm{e}} : \boldsymbol{\epsilon}_{j+1}$

The algorithm described above evaluates the stress from a given strain. To arrive at a computationally efficient procedure that utilises a Newton-Raphson method, it must be complemented by a tangential stiffness tensor, which is derived by a consistent linearisation of the stress-strain relation. Differentiating Eq. (3) gives:

$$\dot{\boldsymbol{\sigma}} = (1 - \omega)\mathbf{D}^{\mathrm{e}} : \dot{\boldsymbol{\epsilon}} - \dot{\omega}\mathbf{D}^{\mathrm{e}} : \boldsymbol{\epsilon} \tag{11}$$

Since $\omega = \omega(\kappa)$, and because the internal variable κ depends on the equivalent strain via $\tilde{\epsilon}$ and the loading function (6), we obtain:

$$\dot{\omega} = \frac{\partial \omega}{\partial \kappa}\frac{\partial \kappa}{\partial \tilde{\epsilon}}\dot{\tilde{\epsilon}} \tag{12}$$

where $\partial\kappa/\partial\tilde{\epsilon} \equiv 1$ for loading and $\partial\kappa/\partial\tilde{\epsilon} \equiv 0$ for unloading. Considering the dependence $\tilde{\epsilon} = \tilde{\epsilon}(\boldsymbol{\epsilon})$, we can elaborate this relation as:

$$\dot{\omega} = \frac{\partial \omega}{\partial \kappa}\frac{\partial \kappa}{\partial \tilde{\epsilon}}\frac{\partial \tilde{\epsilon}}{\partial \boldsymbol{\epsilon}} : \dot{\boldsymbol{\epsilon}} \tag{13}$$

Substitution of Eq. (13) into the expression for the stress rate yields:

$$\dot{\boldsymbol{\sigma}} = \left((1-\omega)\mathbf{D}^{\mathrm{e}} - \frac{\partial \omega}{\partial \kappa}\frac{\partial \kappa}{\partial \tilde{\epsilon}}(\mathbf{D}^{\mathrm{e}} : \boldsymbol{\epsilon}) \otimes \frac{\partial \tilde{\epsilon}}{\partial \boldsymbol{\epsilon}} \right) : \dot{\boldsymbol{\epsilon}} \tag{14}$$

For unloading the second term in Eq. (14) cancels and we retrieve the secant stiffness matrix $(1-\omega)\mathbf{D}^{\mathrm{e}}$ as the tangential stiffness matrix for unloading. It is finally noted, cf. Simo and Ju (1987), that the tangential stiffness matrix as defined in (14) is generally non-symmetric. For the special choice that the equivalent strain is given by Eq. (7), symmetry is restored, since then

$$\dot{\boldsymbol{\sigma}} = \left((1-\omega)\mathbf{D}^{\mathrm{e}} - \frac{\partial \omega}{\partial \kappa}\frac{\partial \kappa}{\partial \tilde{\epsilon}}(\mathbf{D}^{\mathrm{e}} : \boldsymbol{\epsilon}) \otimes (\mathbf{D}^{\mathrm{e}} : \boldsymbol{\epsilon}) \right) : \dot{\boldsymbol{\epsilon}} \tag{15}$$

3 Stability, Ellipticity, and Mesh Sensitivity

A fundamental problem of incorporating damage evolution in standard continuum models is the inherent mesh sensitivity that occurs after reaching a certain damage level. This mesh sensitivity goes beyond the standard discretisation sensitivity of numerical approximation methods for partial differential equations and is not related to deficiencies in the discretisation methods. Instead, the underlying reason for this mesh sensitivity is a local change in character of the governing partial differential equations. This local change of character of the governing set of partial differential equations leads to a loss of well-posedness of the initial boundary value problem and results in an infinite number of possible solutions. After discretisation, a finite number of solutions results. For a finer discretisation, the number of solutions increases, which explains the observed mesh sensitivity.

Since the observed mesh sensitivity is of a fundamental nature, we shall first discuss some basic notions regarding stability and ellipticity. Subsequently, we elucidate the mathematical concepts by simple examples regarding mesh sensitivity.

3.1 Stability and Ellipticity

At the continuum level stable material behaviour is usually defined as the scalar product of the stress rate $\dot{\boldsymbol{\sigma}}$ and the strain rate $\dot{\boldsymbol{\epsilon}}$ being positive (Hill, 1958; Maier and Hueckel, 1979):

$$\dot{\boldsymbol{\epsilon}} : \dot{\boldsymbol{\sigma}} > 0 \tag{16}$$

although it can be linked in a rigorous manner to Lyapunov's mathematical definition of stability only for elastic materials (Koiter, 1969). In Eq. (16)

restriction is made to geometrical linearity. Extension to geometrical non-linearity is straightforward by replacing $\dot{\boldsymbol{\sigma}}$ by the rate of the First Piola-Kirchhoff stress tensor and $\dot{\boldsymbol{\epsilon}}$ by the velocity gradient. Evidently, the scalar product of Eq. (16) becomes negative when, in a uniaxial tension or compression test, the slope of the homogenised axial stress-axial strain curve is negative. This phenomenon is named strain softening and is not restricted to a damage mechanics framework, but can also occur in plasticity.

There is a class of material instabilities that can cause the scalar product of stress rate and strain rate to become negative without the occurrence of strain softening in the sense as defined above. These instabilities can arise when the predominant load-carrying mechanism of the material is due to frictional effects such as in sands, rock joints and in pre-cracked concrete. At a phenomenological level such material behaviour typically results in constitutive models which, in a multiaxial context, have a non-symmetric relation between the stress-rate tensor and the strain-rate tensor, e.g. as in Eq. (14), unless a special choice is made for the equivalent strain $\tilde{\epsilon}$. This lack of symmetry is sufficient to cause loss of material stability, even if the slope of the axial stress-strain curve is still rising (Rudnicki and Rice, 1974).

In the above discussion, the terminology 'homogenised' has been used. Here, we refer to the fact that initial flaws and boundary conditions inevitably induce an inhomogeneous stress state in a specimen. During progressive failure of a specimen these flaws and local stress concentrations cause strongly inhomogeneous deformations of the specimen. The procedure that is normally utilised to derive stress-strain relations, i.e. dividing the force by the virgin load-carrying area and dividing the displacement of the end of the specimen by the original length so as to obtain stress and strain, respectively, then no longer reflects what happens at a lower length scale and loses physical significance.

Limiting the discussion to incrementally-linear stress-strain relations, that is the relation between the stress rate $\dot{\boldsymbol{\sigma}}$ and the strain rate $\dot{\boldsymbol{\epsilon}}$ can be written as

$$\dot{\boldsymbol{\sigma}} = \mathbf{D} : \dot{\boldsymbol{\epsilon}} \tag{17}$$

with $\mathbf{D}$ the material tangential stiffness tensor, inequality (16) can be reformulated as

$$\dot{\boldsymbol{\epsilon}} : \mathbf{D} : \dot{\boldsymbol{\epsilon}} > 0 \tag{18}$$

The limiting case that the inequality (18) is replaced by an equality, marks the onset of unstable material behaviour. Mathematically, this is expressed by the loss of positive definiteness of the material tangential stiffness tensor $\mathbf{D}$:

$$\det(\mathbf{D}^{\mathrm{sym}}) = 0 \tag{19}$$

where the superscript *sym* denotes a symmetrised operator. Material instability can lead to structural instability. For a structure that occupies a volume V, Hill's definition (Hill, 1958) guarantees structural stability if

$$\int_V \dot{\boldsymbol{\epsilon}} : \dot{\boldsymbol{\sigma}} \, \mathrm{d}V > 0 \tag{20}$$

for all kinematically admissible $\dot{\boldsymbol{\epsilon}}$. Obviously, violation of inequality (16), i.e. loss of material stability, can lead to violation of Eq. (20), thus opening the possibility of structural instability. Accordingly, the existence of material instabilities, such as strain softening, can lead to structural instability, even in the absence of geometrically destabilising terms. Of course, there exist many cases where material instabilities and geometrical terms interact and are both (partly) responsible for structural instability.

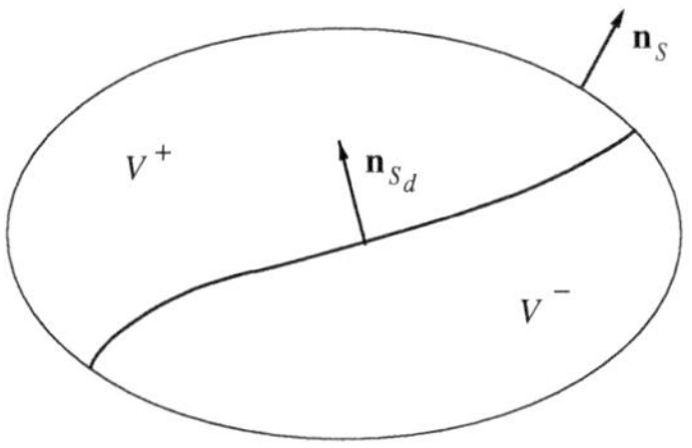

Figure 2. Body composed of continuous displacement fields at each side of the discontinuity S_d

Yet, the occurrence of unstable material behaviour does not explain the frequently observed discretisation-sensitive behaviour of computations of such solids. Indeed, a crucial consequence of the loss of positive definiteness of the material tangential stiffness tensor $\mathbf{D}$ is that it can result in loss of ellipticity of the governing set of rate equations. Considering quasi-static loading conditions, the governing differential equations – equilibrium equations, kinematic equations and constitutive equations – normally have an elliptic character. Mathematically, this implies that discontinuities in the solution are not possible. Now suppose that within the given context of quasi-static loading conditions, a (possibly curved) plane emerges, say S_d (Figure 2), across which the solution can be discontinuous. The difference in the traction rate $\dot{\mathbf{t}}_d$ across this plane reads:

$$[\![\dot{\mathbf{t}}_d]\!] = \mathbf{n}_{S_d} \cdot [\![\dot{\boldsymbol{\sigma}}]\!] \tag{21}$$

with $\mathbf{n}_{S_d}$ the normal vector to the discontinuity S_d. Using the tangential stress-strain relation defined in Eq. (17) we obtain

$$[\![\dot{\mathbf{t}}_d]\!] = \mathbf{n}_{S_d} \cdot \mathbf{D} : [\![\dot{\boldsymbol{\epsilon}}]\!] \tag{22}$$

where the assumption of a linear comparison solid (Hill, 1958) has been introduced, i.e. $\mathbf{D}$ is assumed to have the same value at both sides of the discontinuity S_d. A displacement field $\mathbf{u}$ that is crossed by a single discontinuity can be represented as:

$$\mathbf{u} = \bar{\mathbf{u}} + \mathcal{H}_{S_d}\tilde{\mathbf{u}} \tag{23}$$

with the Heaviside function $\mathcal{H}_{S_d}$ separating the continuous displacement fields $\bar{\mathbf{u}}$ and $\tilde{\mathbf{u}}$. The strain field is subsequently obtained by straightforward differentiation:

$$\boldsymbol{\epsilon} = \nabla^{\mathrm{sym}}\bar{\mathbf{u}} + \mathcal{H}_{S_d}\nabla^{\mathrm{sym}}\tilde{\mathbf{u}} + \delta_{S_d}(\tilde{\mathbf{u}} \otimes \mathbf{n}_{S_d})^{\mathrm{sym}} \tag{24}$$

where δ_{S_d} is the Dirac function placed at the discontinuity S_d. For a stationary discontinuity, so that there is no variation of the Heaviside function $\mathcal{H}_{S_d}$ and the Dirac function δ_{S_d}, the strain rate field follows by differentiation with respect to time:

$$\dot{\boldsymbol{\epsilon}} = \nabla^{\mathrm{sym}}\dot{\bar{\mathbf{u}}} + \mathcal{H}_{S_d}\nabla^{\mathrm{sym}}\dot{\tilde{\mathbf{u}}} + \delta_{S_d}(\dot{\tilde{\mathbf{u}}} \otimes \mathbf{n}_{S_d})^{\mathrm{sym}} \tag{25}$$

The difference in strain rate fields at S_d is proportional to the unbounded term at the interface:

$$[\![\dot{\boldsymbol{\epsilon}}]\!] = \zeta(\dot{\tilde{\mathbf{u}}} \otimes \mathbf{n}_{S_d})^{\mathrm{sym}} \tag{26}$$

also known as the Maxwell compatibility condition and ζ a non-zero scalar. Substitution into Eq. (22) gives:

$$[\![\dot{\mathbf{t}}_d]\!] = \zeta(\mathbf{n}_{S_d} \cdot \mathbf{D} \cdot \mathbf{n}_{S_d}) \cdot \dot{\tilde{\mathbf{u}}} \tag{27}$$

where the minor symmetry of the tangential stiffness tensor has been exploited. A non-trivial solution can exist if and only if the determinant of the acoustic tensor $\mathbf{A} = \mathbf{n}_{S_d} \cdot \mathbf{D} \cdot \mathbf{n}_{S_d}$ vanishes:

$$\det(\mathbf{n}_{S_d} \cdot \mathbf{D} \cdot \mathbf{n}_{S_d}) = 0 \tag{28}$$

Thus, if condition (28) is met, discontinuous solutions can emerge and loss of ellipticity of the governing differential equations occurs. It is noted that condition (28) is coincident with Hill's condition for the propagation of plane acceleration waves in solids (Hill, 1962). Analyses that aim at determining the load level at which the determinant of the acoustic tensor vanishes are also denoted as discontinuous bifurcation analyses, cf. Vardoulakis and Sulem (1995).

Ellipticity is a necessary condition for well-posedness of the rate boundary value problem, in the sense that a finite number of linearly independent solutions are admitted, continuously depending on the data and not involving discontinuities, cf. Benallal et al. (1988). Loss of ellipticity therefore allows an infinite number of solutions to occur, including those which involve discontinuities. A numerical approximation method will try to capture the discontinuity as good as possible and resolve it in the smallest possible volume which the discretisation allows. Accordingly, mesh refinement will result in a smaller and smaller localisation volume, but obviously, a discontinuity cannot be represented exactly unless special approximation methods are used that can capture a discontinuity rigorously.

For small displacement gradients loss of material stability as expressed by Eq. (19) is a necessary condition for loss of ellipticity. We show this by substituting the strain field (26) into the condition for loss of material stability (18):

$$(\tilde{\mathbf{u}} \otimes \mathbf{n}_{S_d}) : \mathbf{D} : (\tilde{\mathbf{u}} \otimes \mathbf{n}_{S_d}) > 0 \tag{29}$$

The left-hand side of this inequality vanishes for arbitrary $\tilde{\mathbf{u}}$ if and only if

$$\det(\mathbf{n}_{S_d} \cdot \mathbf{D}^{\rm sym} \cdot \mathbf{n}_{S_d}) = 0 \tag{30}$$

Because the real-valued eigenspectrum of the acoustic tensor $\mathbf{A}$ is bounded by the minimum and maximum eigenvalues of $\mathbf{n}_{S_d} \cdot \mathbf{D}^{\rm sym} \cdot \mathbf{n}_{S_d}$, Eq. (30) is always met prior to satisfaction of Eq. (28). Since Eq. (30) can only be satisfied if material stability is lost, Eq. (19), it follows that loss of ellipticity can occur only after loss of material stability. However, when geometrically non-linear terms are included, ellipticity can be lost prior to loss of material stability. This, for instance, can occur at low, but positive values of the plastic hardening modulus, in situations where geometrically non-linear terms have a destabilising effect.

3.2 Mesh Sensitivity

Mesh sensitivity in a standard continuum equipped with a strain-softening stress-strain relation is conveniently demonstrated by the example of a simple bar loaded in uniaxial tension, Figure 3. Let the bar be divided into m elements. Prior to reaching the tensile strength f_t a linear relation is assumed between the normal stress σ and the normal strain ϵ:

$$\sigma = E\epsilon$$

After reaching the peak strength a descending slope is defined in this diagram through an affine transformation from the measured load-displacement curve. The result is given in Figure 4(a), where κ_u marks the point where the load-carrying capacity is exhausted. In the post-peak regime the constitutive model thus reads:

$$\sigma = f_t + h(\epsilon - \kappa_i) \tag{31}$$

where, evidently, in case of degrading materials, $h < 0$ and may be termed a softening modulus. For linear strain softening we have

$$h = -\frac{f_t}{\kappa_u - \kappa_i} \tag{32}$$

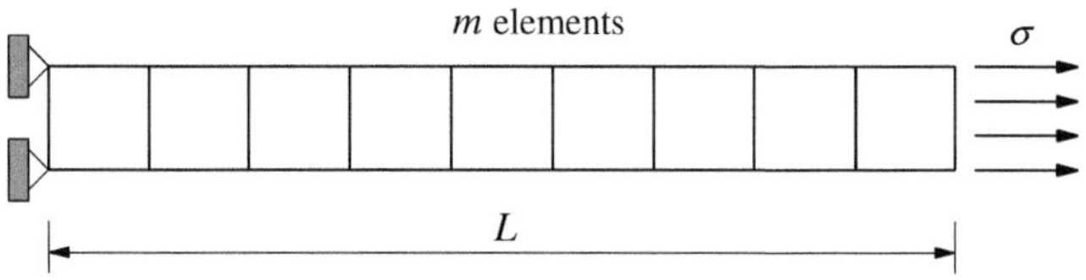

Figure 3. Bar with length L subjected to an axial tensile stress σ

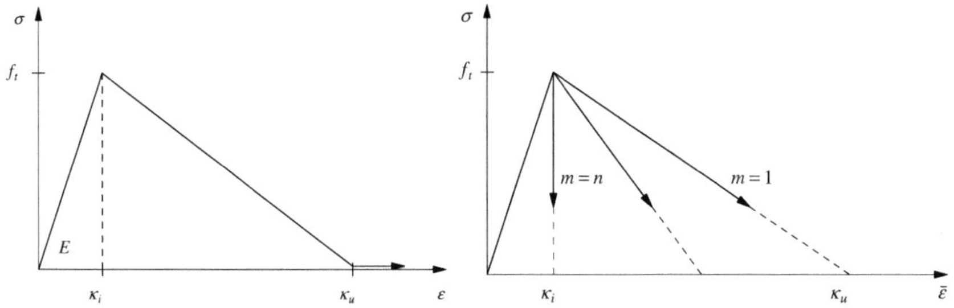

Figure 4. Left: Elastic-linear damaging material behaviour. Right: Response of an imperfect bar in terms of a stress-average strain curve

We next suppose that one element has a tensile strength that is marginally below that of the other $m - 1$ elements. Upon reaching the tensile strength of this element, failure will occur. In the other, neighbouring elements the tensile strength is not exceeded and they will unload elastically. Beyond the peak strength the average strain in the bar is thus given by:

$$\bar{\epsilon} = \frac{\sigma}{E} + \frac{E - h}{Eh} \frac{\sigma - f_t}{m} \tag{33}$$

Substitution of Eq. (32) for the softening modulus h and introduction of n as the ratio between the strain κ_u at which the residual load-carrying capacity is exhausted and the threshold damage level κ_i, $n = \kappa_u/\kappa_i$ and $h = -E/(n-1)$, gives

$$\bar{\epsilon} = \frac{\sigma}{E} + \frac{n(f_t - \sigma)}{mE} \tag{34}$$

This result has been plotted in Figure 4(b) for different values of m for given n. The computed post-peak curves do not seem to converge to a unique curve. In fact, they do, because the governing equations predict the failure mechanism to be a line crack with zero thickness. The numerical solution simply tries to capture this line crack, which results in localisation in one element, irrespective of the width of the element. The impact on the stress-average strain curve is obvious: For an infinite number of elements ($m \to \infty$) the post-peak curve doubles back on the original loading curve. A major problem is now that, since in continuum mechanics the constitutive model is phrased in terms of a stress-strain relation and not as a force-displacement relation, the energy that is dissipated tends to zero upon mesh refinement, simply because the volume in which the failure process occurs also becomes zero. From a physical point of view this is unacceptable.

4 Non-Local and Gradient Damage Models

4.1 Non-Local Damage Models

In a non-local generalisation the equivalent strain $\tilde{\epsilon}$ is normally replaced by a spatially averaged quantity in the damage loading function (Pijaudier-Cabot and Bažant, 1987):

$$f(\bar{\epsilon}, \kappa) = \bar{\epsilon} - \kappa \tag{35}$$

where the non-local strain $\bar{\epsilon}$ is computed from:

$$\bar{\epsilon}(\mathbf{x}) = \frac{1}{\Psi(\mathbf{x})} \int_V \psi(\mathbf{y}, \mathbf{x}) \tilde{\epsilon}(\mathbf{y}) \mathrm{d}V \quad , \quad \Psi(\mathbf{x}) = \int_V \psi(\mathbf{y}, \mathbf{x}) \mathrm{d}V \tag{36}$$

with $\psi(\mathbf{y}, \mathbf{x})$ a weight function. Often, the weight function ψ is assumed to be homogeneous and isotropic, so that it only depends on the norm $s = \parallel \mathbf{x} - \mathbf{y} \parallel$. In this formulation all the other relations remain local: the local stress-strain relation, Eq. (3), the loading-unloading conditions, Eqs. (5), and the dependence of the damage variable ω on the internal variable κ: $\omega = \omega(\kappa)$. As an alternative to Eq. (36), the locally defined internal variable κ can be replaced in the damage loading function f by a spatially averaged quantity $\bar{\kappa}$:

$$\bar{\kappa}(\mathbf{x}) = \frac{1}{\Psi(\mathbf{x})} \int_V \psi(\mathbf{y}, \mathbf{x}) \kappa(\mathbf{y}) \mathrm{d}V \tag{37}$$

The fact that in elasticity-based damage models the stress can be computed directly from the given strain, enables that a straightforward algorithm can be set up for non-local damage models. For the non-local damage model defined by Eq. (36) the algorithm of Box 2 applies. Although conceptually straightforward, the tangential stiffness matrix entails some inconvenient properties. Due to the non-local character of the constitutive relation the tangential stiffness matrix is full, i.e. the bandedness is lost. The introduction of a cut-off on the averaging function partly remedies this disadvantage, but an increased band width will nevertheless result. Secondly, symmetry can be lost (Pijaudier-Cabot and Huerta, 1991).

Box 2. Algorithm for a non-local elasticity-based damage model.

1. Compute the strain increment: $\Delta\boldsymbol{\epsilon}_{j+1}$
2. Update the total strain: $\boldsymbol{\epsilon}_{j+1} = \boldsymbol{\epsilon}_j + \Delta\boldsymbol{\epsilon}_{j+1}$
3. Compute the equivalent strain: $\tilde{\epsilon}_{j+1} = \tilde{\epsilon}(\boldsymbol{\epsilon}_{j+1})$
4. Compute the non-local equivalent strain:

   $$\bar{\epsilon}_{j+1}(\mathbf{x}) = \sum_i w_i \psi(\mathbf{y}_i, \mathbf{x}) \tilde{\epsilon}_{j+1}(\mathbf{y}_i) V_{\mathrm{elem}}$$

5. Evaluate the damage loading function: $f = \bar{\epsilon}_{j+1} - \kappa_0$

   if $f \geq 0$, $\kappa_{j+1} = \bar{\epsilon}_{j+1}$

   else $\kappa_{j+1} = \kappa_0$
6. Update the damage variable: $\omega_{j+1} = \omega(\kappa_{j+1})$
7. Compute the new stresses: $\boldsymbol{\sigma}_{j+1} = (1 - \omega_{j+1})\mathbf{D}^{\mathrm{e}} : \boldsymbol{\epsilon}_{j+1}$

4.2 Gradient Damage Models

Non-local constitutive relations can be considered as a point of departure for constructing gradient models, although we wish to emphasise that the latter class of models can also be defined directly by supplying higher-order gradients in the damage loading function. Yet, we will follow the first-mentioned route to underline the connection between integral and differential type non-local models. This is done either by expanding the kernel

$\tilde{\epsilon}$ of the integral in Eq. (36) in a Taylor series, or by expanding of the internal variable κ in Eq. (37) in a Taylor series. We will first consider the expansion of $\tilde{\epsilon}$ and then we will do the same for κ. If we truncate after the second-order terms and carry out the integration implied in Eq. (36) under the assumption of isotropy, the following relation ensues:

$$\bar{\epsilon} = \tilde{\epsilon} + c\nabla^2\tilde{\epsilon} \tag{38}$$

where c is a gradient parameter of the dimension length squared. It can be related to the averaging volume and then becomes dependent on the precise form of the weight function ψ. For instance, for a one-dimensional continuum and taking

$$\psi(s) = \frac{1}{\sqrt{2\pi}l} \mathrm{e}^{-s^2/2\ell^2} \tag{39}$$

we obtain $c = 1/2\ell^2$. Here, we adopt the phenomenological view that $\ell = \sqrt{2c}$ reflects the internal length scale of the failure process which we wish to describe macroscopically.

Formulation (38), known as the explicit gradient damage model, has a disadvantage when applied in a finite element context, namely that it requires computation of second-order gradients of the local equivalent strain $\tilde{\epsilon}$. Since this quantity is a function of the strain tensor, and since the strain tensor involves first-order derivatives of the displacements, third-order derivatives of the displacements have to be computed, which would necessitate $\mathcal{C}^2$-continuity of the shape functions. To obviate this problem, Eq. (38) is differentiated twice and the result is substituted again into Eq. (38). Again neglecting fourth-order terms this leads to:

$$\bar{\epsilon} - c\nabla^2\bar{\epsilon} = \tilde{\epsilon} \tag{40}$$

In Peerlings et al. (2001) it has been shown that the implicit gradient damage model of Eq. (40) becomes formally identical to a fully non-local formulation for a specific choice of the weighting function ψ in Eq. (36), which underlines that this formulation has a truly non-local character, in contrast to the explicit gradient formulation of Eq. (38).

Higher-order continua require additional boundary conditions. With Eq. (40) governing the damage process, either the averaged equivalent strain $\bar{\epsilon}$ itself or its normal derivative must be specified on the boundary S of the body:

$$\bar{\epsilon} = \bar{\epsilon}_{\mathrm{s}} \quad \text{or} \quad \mathbf{n}_S \cdot \nabla\bar{\epsilon} = \bar{\epsilon}_{\mathrm{ns}} \tag{41}$$

In most example calculations in the literature the natural boundary condition $\mathbf{n}_S \cdot \nabla\bar{\epsilon} = 0$ has been adopted.

In a fashion similar to the derivation of the gradient damage models based on the averaging of the equivalent strain $\tilde{\epsilon}$, we can elaborate a gradient approximation of Eq. (37), i.e. by developing κ into a Taylor series. For an isotropic, infinite medium and truncating after the second term we have (de Borst et al., 1996):

$$\bar{\kappa} = \kappa + c\nabla^2\kappa \tag{42}$$

Since the weight functions for the different gradient formulations may be quite different, also the gradient parameter c may be very different for the various formulations. For instance, the gradient parameter c of Eq. (42) may differ considerably from those in Eqs. (38) or (40). The additional boundary conditions now apply to κ. Although formally similar to those of Eq. (41), namely

$$\kappa = \kappa_{\mathrm{s}} \quad \text{or} \quad \mathbf{n}_S \cdot \nabla\kappa = \kappa_{\mathrm{ns}} \tag{43}$$

they have a different character, since they apply to an internal variable instead of to a kinematic quantity, which seems somewhat suspect. On the other hand, the physical interpretation that can be given to the boundary condition $(43)_2$ is rather clear. Since the damage variable ω is a function of the internal variable κ, and therefore, the differential equation (42) and the boundary conditions (43) can be replaced by (de Borst et al., 1996):

$$\bar{\omega} = \omega + c\nabla^2\omega \tag{44}$$

where $\bar{\omega}$ is a spatially averaged damage field, similar to $\bar{\epsilon}$ or $\bar{\kappa}$, and the corresponding boundary conditions

$$\omega = \omega_{\mathrm{s}} \quad \text{or} \quad \mathbf{n}_S \cdot \nabla\omega = \omega_{\mathrm{ns}} \tag{45}$$

Equation $(45)_2$ with $\omega_{\mathrm{ns}} = 0$ can be identified as a condition of no damage flux through the boundary S of the body.

Numerical schemes for gradient-enhanced continua typically have the character of a coupled problem and depart from the weak form of the balance of momentum,

$$\int_V \delta\boldsymbol{\epsilon}^{\mathrm{T}}\boldsymbol{\sigma}\mathrm{d}V = \int_S \delta\mathbf{u}^{\mathrm{T}}\mathbf{t}\mathrm{d}S \tag{46}$$

and a weak form of the averaging equation, e.g. Eq. (40):

$$\int_V \delta\bar{\epsilon}\;(\bar{\epsilon} - c\nabla^2\bar{\epsilon} - \tilde{\epsilon})\mathrm{d}V = 0 \tag{47}$$

with $\delta\bar{\epsilon}$ the variational field of the non-local strain $\bar{\epsilon}$. Transforming Eq. (47), using the divergence theorem and the natural boundary condition $\mathbf{n}_S \cdot \nabla\bar{\epsilon} =$

0 yields:

$$\int_V (\delta\bar{\epsilon}\ \bar{\epsilon} + c\nabla\delta\bar{\epsilon} \cdot \nabla\bar{\epsilon})\mathrm{d}V = \int_V \delta\bar{\epsilon}\ \tilde{\epsilon}\mathrm{d}V \tag{48}$$

From Eq. (48) it is clear that in this formulation a $\mathcal{C}^0$-interpolation for $\bar{\epsilon}$ suffices. Accordingly, we can discretise the displacements $\mathbf{u}$ and the non-local strains

$$\mathbf{u} = \mathbf{Ha} \ \text{ and } \ \bar{\epsilon} = \bar{\mathbf{H}}\mathbf{e} \tag{49}$$

where $\mathbf{H}$ and $\bar{\mathbf{H}}$ contain $\mathcal{C}^0$-interpolation polynomials which can have a different order. Similarly, for the variations

$$\delta\mathbf{u} = \mathbf{H}\delta\mathbf{a} \ \text{ and } \ \delta\bar{\epsilon} = \bar{\mathbf{H}}\delta\mathbf{e} \tag{50}$$

Substitution into Eqs. (46), (48) and requiring that the result holds for arbitrary $(\delta\mathbf{a}, \delta\mathbf{e})$, yields the discrete format of the equilibrium equation:

$$\int_V \mathbf{B}^{\mathrm{T}}\boldsymbol{\sigma}\mathrm{d}V = \int_S \mathbf{H}^{\mathrm{T}}\mathbf{t}\mathrm{d}S \tag{51}$$

and of the averaging equation:

$$\int_V (\bar{\mathbf{H}}^{\mathrm{T}}\bar{\mathbf{H}} + c\bar{\mathbf{B}}^{\mathrm{T}}\bar{\mathbf{B}})\mathrm{d}V = \int_V \bar{\mathbf{H}}^{\mathrm{T}}\tilde{\boldsymbol{\epsilon}}\mathrm{d}V \tag{52}$$

where $\bar{\mathbf{B}}$ contains the spatial derivatives of $\bar{\mathbf{H}}$. An algorithm for computing the right-hand side of this model is given in Box 3.

Box 3. Algorithm for a second-order implicit gradient damage model.

1. Compute the strain increment: $\Delta\boldsymbol{\epsilon}_{j+1}$ and the non-local strain increment $\Delta\bar{\epsilon}_{j+1}$
2. Update the total strain: $\boldsymbol{\epsilon}_{j+1} = \boldsymbol{\epsilon}_j + \Delta\boldsymbol{\epsilon}_{j+1}$ and the non-local strain $\bar{\epsilon}_{j+1} = \bar{\epsilon}_j + \Delta\bar{\epsilon}_{j+1}$
3. Evaluate the damage loading function: $f = \bar{\epsilon}_{j+1} - \kappa_0$
   if $f \geq 0$, $\kappa_{j+1} = \bar{\epsilon}_{j+1}$
   else $\kappa_{j+1} = \kappa_0$
4. Update the damage variable: $\omega_{j+1} = \omega(\kappa_{j+1})$
5. Compute the new stresses: $\boldsymbol{\sigma}_{j+1} = (1 - \omega_{j+1})\mathbf{D}^{\mathrm{e}} : \boldsymbol{\epsilon}_{j+1}$

The tangential stiffness matrix needed for an iterative solution via the Newton-Raphson method reads (Peerlings et al., 1996):

$$\begin{bmatrix} \mathbf{K}_{\mathrm{aa}} & \mathbf{K}_{\mathrm{ae}} \\ \mathbf{K}_{\mathrm{ea}} & \mathbf{K}_{\mathrm{ee}} \end{bmatrix} \begin{pmatrix} \mathrm{da} \\ \mathrm{de} \end{pmatrix} = \begin{pmatrix} \mathbf{f}^{\mathrm{a}}_{\mathrm{ext}} - \mathbf{f}^{\mathrm{a}}_{\mathrm{int}} \\ \mathbf{f}^{\mathrm{e}}_{\mathrm{int}} - \mathbf{K}_{\mathrm{ee}}\mathbf{e} \end{pmatrix} \tag{53}$$

with $\mathbf{f}^{\mathrm{e}}_{\mathrm{int}}$ given by the right-hand side of Eq. (52). The stiffness matrices are given by:

$$\mathbf{K}_{\mathrm{aa}} = \int_V (1-\omega)\mathbf{B}^{\mathrm{T}}\mathbf{D}^{\mathrm{e}}\mathbf{B}\mathrm{d}V \tag{54}$$

$$\mathbf{K}_{\mathrm{ae}} = \int_V q\mathbf{B}^{\mathrm{T}}\mathbf{D}^{\mathrm{e}}\boldsymbol{\epsilon}\bar{\mathbf{H}}\mathrm{d}V \tag{55}$$

$$\mathbf{K}_{\mathrm{ea}} = \int_V \bar{\mathbf{H}}^{\mathrm{T}}\left(\frac{\partial\tilde{\epsilon}}{\partial\boldsymbol{\epsilon}}\right)\mathbf{B}\mathrm{d}V \tag{56}$$

$$\mathbf{K}_{\mathrm{ee}} = \int_V \left(\bar{\mathbf{H}}^{\mathrm{T}}\bar{\mathbf{H}} + c\bar{\mathbf{B}}^{\mathrm{T}}\bar{\mathbf{B}}\right)\mathrm{d}V \tag{57}$$

where $q = \partial\omega/\partial\kappa$ for loading and vanishes if otherwise. The expressions for $\mathbf{K}_{\mathrm{ae}}$ and $\mathbf{K}_{\mathrm{ea}}$ exhibit a non-symmetry. This non-symmetry is caused by the damage formalism and not by the gradient enhancement, cf. Eq. (14).

5 Cosserat Elasto-Plasticity

5.1 Cosserat Elasticity

In the present treatment we shall limit attention to two-dimensional, planar deformations. In that case, each material point in a micro-polar solid has two translational degrees-of-freedom, namely u_x and u_y and a rotational degree-of-freedom ω_z, the rotation axis of which is orthogonal to the x, y-plane. The normal strains are defined as in a standard continuum, but for the shear strains we have:

$$\epsilon_{xy} = \frac{\partial u_x}{\partial y} + \omega_z \tag{58}$$

and

$$\epsilon_{yx} = \frac{\partial u_y}{\partial x} - \omega_z \tag{59}$$

In addition to the normal strains and the shear strains, the Cosserat theory requires the introduction of micro-curvatures:

$$\kappa_{zx} = \frac{\partial \omega_z}{\partial x} \tag{60}$$

and

$$\kappa_{zy} = \frac{\partial \omega_z}{\partial y} \tag{61}$$

Anticipating the treatment for elasto-plasticity we will rather use the generalised curvatures $\kappa_{zx}\ell$ and $\kappa_{zy}\ell$, where ℓ is a material parameter with the dimension of length. It is this parameter which effectively sets the internal length scale in the continuum, and attains the role of a characteristic length scale.

The strain components introduced sofar may be assembled in a vector,

$$\boldsymbol{\epsilon} = (\epsilon_{xx}, \epsilon_{yy}, \epsilon_{zz}, \epsilon_{xy}, \epsilon_{yx}, \kappa_{zx}\ell, \kappa_{zy}\ell)^{\mathrm{T}} \tag{62}$$

Note that in addition to the strain components, also the normal strain in the z-direction, ϵ_{zz} has been included in the strain vector $\boldsymbol{\epsilon}$. This has been done because, although this strain component remains zero under plane strain conditions during the entire loading process, this is not necessarily the case for the elastic and plastic contributions of this strain component. Also, the normal stress σ_{zz}, which acts in the z-direction, may be non-zero, which necessitates inclusion of ϵ_{zz} and σ_{zz} in the stress-strain relation. It is furthermore noted that by multiplying the micro-curvatures κ_{zx} and κ_{zy} by the length parameter ℓ all the components of the strain vector $\boldsymbol{\epsilon}$ have the same dimension.

We now consider the statics of a Cosserat continuum. While the strain vector $\boldsymbol{\epsilon}$ is comprised of seven components for planar deformations, so is the stress vector $\boldsymbol{\sigma}$. As in a classical continuum we have the normal stresses σ_{xx}, σ_{yy} and σ_{zz}, and the shear stresses σ_{xy}, σ_{yx}. For the Cosserat continuum we also have to introduce stress quantities that are conjugate to the curvatures κ_{zx} and κ_{zy}, namely the couple stresses m_{zx} and m_{zy}. Dividing the couple stresses by the length parameter ℓ, we obtain a stress vector $\boldsymbol{\sigma}$ in which all the entries have the same dimension:

$$\boldsymbol{\sigma} = (\sigma_{xx}, \sigma_{yy}, \sigma_{zz}, \sigma_{xy}, \sigma_{yx}, m_{zx}/\ell, m_{zy}/\ell)^{\mathrm{T}} \tag{63}$$

Omitting body forces and body couples for sake of simplicity, translational equilibrium in the x and the y-directions, respectively, results in the usual balance of momentum:

$$\frac{\partial \sigma_{xx}}{\partial x} + \frac{\partial \sigma_{xy}}{\partial y} = 0$$

$$\frac{\partial \sigma_{yx}}{\partial x} + \frac{\partial \sigma_{yy}}{\partial y} = 0$$

which replicates the results obtained for a classical, non-polar continuum. However, for rotational equilibrium we find that:

$$\frac{\partial m_{zx}}{\partial x} + \frac{\partial m_{zy}}{\partial y} - (\sigma_{xy} - \sigma_{yx}) = 0 \tag{64}$$

which shows that the stress tensor is in general only symmetric – $\sigma_{xy} = \sigma_{yx}$ – if the couple-stresses m_{zx} and m_{zy} vanish, the so-called Boltzmann's Axiom.

Anticipating the treatment of Cosserat plasticity we decompose the strain vector into an elastic contribution $\boldsymbol{\epsilon}^{\mathrm{e}}$ and a plastic part $\boldsymbol{\epsilon}^{\mathrm{p}}$:

$$\boldsymbol{\epsilon} = \boldsymbol{\epsilon}^{\mathrm{e}} + \boldsymbol{\epsilon}^{\mathrm{p}} \tag{65}$$

while we assume that the elastic strains are linearly related to the stresses:

$$\boldsymbol{\sigma} = \mathbf{D}^{\mathrm{e}} : \boldsymbol{\epsilon}^{\mathrm{e}} \tag{66}$$

where $\mathbf{D}^{\mathrm{e}}$ is the stiffness matrix that contains the elastic moduli:

$$\mathbf{D}^{\mathrm{e}} = \begin{bmatrix} 2\mu c_1 & 2\mu c_2 & 2\mu c_2 & 0 & 0 & 0 & 0 \\ 2\mu c_2 & 2\mu c_1 & 2\mu c_2 & 0 & 0 & 0 & 0 \\ 2\mu c_2 & 2\mu c_2 & 2\mu c_1 & 0 & 0 & 0 & 0 \\ 0 & 0 & 0 & \mu + \mu_c & \mu - \mu_c & 0 & 0 \\ 0 & 0 & 0 & \mu - \mu_c & \mu + \mu_c & 0 & 0 \\ 0 & 0 & 0 & 0 & 0 & 2\mu & 0 \\ 0 & 0 & 0 & 0 & 0 & 0 & 2\mu \end{bmatrix} \tag{67}$$

with $c_1 = \frac{1-\nu}{1-2\nu}$ and $c_2 = \frac{\nu}{1-2\nu}$. The elastic constants μ and ν have the classical meaning of the shear modulus and Poisson's ratio, respectively. μ_c is an additional material constant, completing the total of four material constants, viz. μ, ν, ℓ and μ_c that are needed to describe the elastic behaviour of an isotropic Cosserat continuum under planar deformations. The coefficient two has been introduced in the terms D^{e}_{66} and D^{e}_{77} in order to arrive at a convenient form of the elasto-plastic constitutive equations. The total (bending) stiffness that sets the relation between the micro-curvatures and the couple stresses is basically determined by the value of the internal length scale ℓ. All the elastic stiffness moduli in $\mathbf{D}^{\mathrm{e}}$ have the same dimension. This is attributable to the fact that all components of the strain vector $\boldsymbol{\epsilon}$ and the stress vector $\boldsymbol{\sigma}$ have the same dimension.

5.2 Cosserat Plasticity

As an example we use a pressure-dependent J_2-flow theory (Drucker-Prager model). Accordingly, the yield function f can be written as

$$f(\boldsymbol{\sigma}, \gamma) = \sqrt{3J_2} + \alpha p - \bar{\sigma}(\gamma) \tag{68}$$

with $\bar{\sigma}$ a function of the hardening parameter γ and α a friction coefficient. $p = \frac{1}{3}(\sigma_{xx} + \sigma_{yy} + \sigma_{zz})$ and J_2 is the second invariant of the deviatoric stresses, which, for a micro-polar continuum, can be generalised as:

$$J_2 = a_1 \mathbf{s}^{\mathrm{T}} : \mathbf{s} + a_2 \mathbf{s} : \mathbf{s} + a_3 \mathbf{m} : \mathbf{m}/\ell^2 \tag{69}$$

where the summation convention with respect to repeated indices has been adopted. $\mathbf{s}$ is the deviatoric stress tensor and a_1, a_2 and a_3 are material parameters. In the absence of couple-stresses, i.e. $\mathbf{m} = \mathbf{0}$, $\mathbf{s} = \mathbf{s}^{\mathrm{T}}$, so that:

$$J_2 = a_1 \mathbf{s}^{\mathrm{T}} : \mathbf{s} + a_2 \mathbf{s} : \mathbf{s} \tag{70}$$

which implies that the constraint $a_1 + a_2 = \frac{1}{2}$ must be enforced to achieve that the classical expression for J_2 be retrieved. Introduction of the projetion matrix

$$\mathbf{P} = \begin{bmatrix} \frac{2}{3} & -\frac{1}{3} & -\frac{1}{3} & 0 & 0 & 0 & 0 \\ -\frac{1}{3} & \frac{2}{3} & -\frac{1}{3} & 0 & 0 & 0 & 0 \\ -\frac{1}{3} & -\frac{1}{3} & \frac{2}{3} & 0 & 0 & 0 & 0 \\ 0 & 0 & 0 & 2a_1 & 2a_2 & 0 & 0 \\ 0 & 0 & 0 & 2a_2 & 2a_1 & 0 & 0 \\ 0 & 0 & 0 & 0 & 0 & 2a_3 & 0 \\ 0 & 0 & 0 & 0 & 0 & 0 & 2a_3 \end{bmatrix} \tag{71}$$

and the projection vector:

$$\boldsymbol{\pi}^{\mathrm{T}} = \left(\frac{1}{3}, \frac{1}{3}, \frac{1}{3}, 0, 0, 0, 0 \right) \tag{72}$$

for planar deformations, enables a rewriting of the yield function in an appealingly compact format:

$$f(\boldsymbol{\sigma}, \gamma) = \sqrt{\frac{3}{2} \boldsymbol{\sigma}^{\mathrm{T}} \mathbf{P} \boldsymbol{\sigma}} + \alpha \boldsymbol{\pi}^{\mathrm{T}} \boldsymbol{\sigma} - \bar{\sigma}(\gamma) \tag{73}$$

A (non-associated) flow rule is now obtained in an identical fashion to that in a non-polar continuum by defining a resembling plastic potential function:

$$g(\boldsymbol{\sigma}, \gamma) = \sqrt{\frac{3}{2} \boldsymbol{\sigma}^{\mathrm{T}} \mathbf{P} \boldsymbol{\sigma}} + \beta \boldsymbol{\pi}^{\mathrm{T}} \boldsymbol{\sigma} - \bar{\sigma}(\gamma) \tag{74}$$

with β a dilatancy factor, from which the plastic strain rates can be derived:

$$\dot{\boldsymbol{\epsilon}}^{\mathrm{p}} = \dot{\lambda} \frac{\partial g}{\partial \boldsymbol{\sigma}} \tag{75}$$

with $\dot{\lambda}$ the plastic multiplier which, in analogy with standard plasticity theory, is determined from the consistency condition $\dot{f} = 0$. Substitution of the plastic potential for the Drucker-Prager plasticity model, Eq. (74), into the above expression for the plastic strain rate yields

$$\dot{\boldsymbol{\epsilon}}^{\mathrm{p}} = \dot{\lambda} \left(\frac{3\mathbf{P}\boldsymbol{\sigma}}{2\sqrt{\frac{3}{2}\boldsymbol{\sigma}^{\mathrm{T}}\mathbf{P}\boldsymbol{\sigma}}} + \beta\boldsymbol{\pi} \right) \tag{76}$$

It remains to identify the hardening parameter γ in a Cosserat continuum. For this purpose we recall the conventional strain-hardening hypothesis:

$$\dot{\gamma} = \sqrt{\frac{2}{3}\dot{\mathbf{e}}^{\mathrm{p}} : \dot{\mathbf{e}}^{\mathrm{p}}} \tag{77}$$

with $\dot{\mathbf{e}}^{\mathrm{p}}$ the deviatoric plastic strain-rate tensor. Since there are no couple-stress effects in uniaxial loading we require that any modification for a Cosserat continuum does not affect the behaviour for uniaxial loading. A possible generalisation is then:

$$\dot{\gamma} = \sqrt{b_1(\dot{\mathbf{e}}^{\mathrm{p}})^{\mathrm{T}} : \dot{\mathbf{e}}^{\mathrm{p}} + b_2\dot{\mathbf{e}}^{\mathrm{p}} : \dot{\mathbf{e}}^{\mathrm{p}} + b_3\dot{\boldsymbol{\kappa}}^{\mathrm{p}} : \dot{\boldsymbol{\kappa}}^{\mathrm{p}}/\ell^2} \tag{78}$$

with $b_1 + b_2 = \frac{2}{3}$ in order that definition the strain-hardening hypothesis in a non-polar solid be retrieved. Introduction of the matrix

$$\mathbf{Q} = \begin{bmatrix} \frac{2}{3} & -\frac{1}{3} & -\frac{1}{3} & 0 & 0 & 0 & 0 \\ -\frac{1}{3} & \frac{2}{3} & -\frac{1}{3} & 0 & 0 & 0 & 0 \\ -\frac{1}{3} & -\frac{1}{3} & \frac{2}{3} & 0 & 0 & 0 & 0 \\ 0 & 0 & 0 & \frac{3}{2}b_1 & \frac{3}{2}b_2 & 0 & 0 \\ 0 & 0 & 0 & \frac{3}{2}b_2 & \frac{3}{2}b_1 & 0 & 0 \\ 0 & 0 & 0 & 0 & 0 & \frac{3}{2}b_3 & 0 \\ 0 & 0 & 0 & 0 & 0 & 0 & \frac{3}{2}b_3 \end{bmatrix} \tag{79}$$

allows $\dot{\gamma}$ to be written as:

$$\dot{\gamma} = \sqrt{\frac{2}{3}(\dot{\boldsymbol{\epsilon}}^{\mathrm{p}})^{\mathrm{T}}\mathbf{Q}\dot{\boldsymbol{\epsilon}}^{\mathrm{p}}} \tag{80}$$

for planar deformations. We next substitute the flow rule, Eq. (76), into this expression. The result is given by:

$$\dot{\gamma} = \dot{\lambda}\sqrt{\frac{\boldsymbol{\sigma}^{\mathrm{T}}\mathbf{PQP}\boldsymbol{\sigma}}{\boldsymbol{\sigma}^{\mathrm{T}}\mathbf{P}\boldsymbol{\sigma}}} \tag{81}$$

since $\dot{\lambda}$ and $\bar{\sigma}$ are non-negative, and $\mathbf{Q}\boldsymbol{\pi} = \mathbf{0}$. In line with a standard continuum we choose a_1, a_2, a_3 and b_1, b_2, b_3 such, that

$$\mathbf{PQP} = \mathbf{P} \tag{82}$$

and we obtain

$$\dot{\gamma} = \dot{\lambda} \tag{83}$$

which has the same format as in standard Drucker-Prager plasticity theory.

5.3 A Return-Mapping Algorithm

With the governing rate equations for the micro-polar elasto-plastic solid at hand, we can develop an algorithm that determines the stress increment in a finite loading step. Here, a variety of algorithms exist, but we shall only consider the Euler backward algorithm, in which the state parameters are evaluated at the end of the loading step:

$$\boldsymbol{\sigma}_{j+1} = \boldsymbol{\sigma}_0 + \mathbf{D}^{\mathrm{e}}(\Delta\boldsymbol{\epsilon} - \Delta\boldsymbol{\epsilon}^{\mathrm{p}}) \tag{84}$$

The expression for the plastic strain rate, Eq. (76) is now integrated using a single-point Euler backward rule. This results in:

$$\Delta\boldsymbol{\epsilon}^{\mathrm{p}} = \Delta\lambda \left(\frac{3\mathbf{P}\boldsymbol{\sigma}_{j+1}}{2\sqrt{\frac{3}{2}\boldsymbol{\sigma}_{j+1}^{\mathrm{T}}\mathbf{P}\boldsymbol{\sigma}_{j+1}}} + \beta\boldsymbol{\pi} \right) \tag{85}$$

so that the expression for the stress can be elaborated as:

$$\boldsymbol{\sigma}_{j+1} = \boldsymbol{\sigma}_e - \Delta\lambda \left(\frac{3\mathbf{D}^{\mathrm{e}}\mathbf{P}\boldsymbol{\sigma}_{j+1}}{2(\bar{\sigma}(\lambda_{j+1}) - \alpha\boldsymbol{\pi}^{\mathrm{T}}\boldsymbol{\sigma}_{j+1})} + \beta\mathbf{D}^{\mathrm{e}}\boldsymbol{\pi} \right) \tag{86}$$

Unfortunately, $\boldsymbol{\sigma}_{j+1}$ also enters the denominator on the right-hand side. To eliminate $\boldsymbol{\sigma}_{j+1}$ from the right-hand side of the identity we premultiply by the projection vector $\boldsymbol{\pi}$, which results in:

$$\boldsymbol{\pi}^{\mathrm{T}}\boldsymbol{\sigma}_{j+1} = \boldsymbol{\pi}^{\mathrm{T}}\boldsymbol{\sigma}_e - \Delta\lambda\beta K \tag{87}$$

with $K = \boldsymbol{\pi}^{\mathrm{T}}\mathbf{D}^{\mathrm{e}}\boldsymbol{\pi}$. For isotropic elasticity K can be identified as the bulk modulus. Substitution into Eq. (86) yields a formulation in which $\boldsymbol{\sigma}_{j+1}$ is expressed in terms of the trial stress $\boldsymbol{\sigma}_e$ and the elastic parameters:

$$\boldsymbol{\sigma}_{j+1} = \mathbf{A}^{-1}(\boldsymbol{\sigma}_e - \Delta\beta\mathbf{D}^{\mathrm{e}}\boldsymbol{\pi}) \tag{88}$$

where

$$\mathbf{A} = \mathbf{I} + \frac{3\Delta\lambda\mathbf{D}^{\mathrm{e}}\mathbf{P}}{2(\bar{\sigma}(\lambda_{j+1}) + \Delta\lambda\alpha\beta K - \alpha\boldsymbol{\pi}^{\mathrm{T}}\boldsymbol{\sigma}_e)} \tag{89}$$

Substitution in the yield condition $f(\boldsymbol{\sigma}_{j+1}, \gamma_{j+1}) = 0$ gives a non-linear equation in $\Delta\lambda$: $f(\Delta\lambda) = 0$.

5.4 Consistent Tangent Operator

For the derivation of a properly linearised set of tangential moduli we can differentiate the return map, Eq. (88), to give:

$$\dot{\boldsymbol{\sigma}} = \mathbf{H}\left(\dot{\boldsymbol{\epsilon}} - \dot{\boldsymbol{\epsilon}}^{\mathrm{p}}\right) \tag{90}$$

where, for Drucker-Prager plasticity:

$$\mathbf{H}^{-1} = (\mathbf{D}^{\mathrm{e}})^{-1} - \Delta\lambda\sqrt{\frac{3}{2}}\frac{\boldsymbol{\sigma}^{\mathrm{T}}\mathbf{P}\boldsymbol{\sigma}\mathbf{P} - \mathbf{P}\boldsymbol{\sigma}\boldsymbol{\sigma}^{\mathrm{T}}\mathbf{P}}{\sqrt{\boldsymbol{\sigma}^{\mathrm{T}}\mathbf{P}\boldsymbol{\sigma}}} \tag{91}$$

Since $f = f(\boldsymbol{\sigma}, \gamma)$, the consistency condition $\dot{f} = 0$ can be elaborated as:

$$\left(\frac{\partial f}{\partial \boldsymbol{\sigma}}\right)^{\mathrm{T}} \dot{\boldsymbol{\sigma}} - h\dot{\lambda} = 0 \tag{92}$$

with the hardening modulus

$$h = \frac{\partial \bar{\sigma}}{\partial \gamma} \tag{93}$$

Eqs. (90) and (92) can now be combined to give the consistent tangential stiffness relation:

$$\dot{\boldsymbol{\sigma}} = \left(\mathbf{H} - \frac{\mathbf{H}\frac{\partial g}{\partial \boldsymbol{\sigma}}\left(\frac{\partial f}{\partial \boldsymbol{\sigma}}\right)^{\mathrm{T}}\mathbf{H}}{h + \left(\frac{\partial f}{\partial \boldsymbol{\sigma}}\right)^{\mathrm{T}}\mathbf{H}\frac{\partial g}{\partial \boldsymbol{\sigma}}}\right)\dot{\boldsymbol{\epsilon}} \tag{94}$$

A deficiency of the Cosserat plasticity model is that it is only effective when the local rotations are mobilised, i.e. for mode-II failure.

6 Non-Local and Gradient Plasticity

6.1 Non-Local Plasticity

Alternatively, Bažant and Lin (1988) have suggested to average $\dot{\gamma}$ for a standard continuum, such that:

$$\dot{\bar{\gamma}}(\mathbf{x}) = \frac{1}{\Psi(\mathbf{x})}\int_V \psi(\mathbf{y}, \mathbf{x})\dot{\gamma}(\mathbf{y})\mathrm{d}V \tag{95}$$

with $V_r(\mathbf{x}) = \int g(\mathbf{s} - \mathbf{x})\mathrm{d}V$ and $g(\mathbf{s})$ a weighting function, for which the error function is usually substituted, and to make f dependent on $\bar{\gamma}$ instead of on γ:

$$f = f(\boldsymbol{\sigma}, \bar{\gamma}) \tag{96}$$

Alternatively, one can first average the plastic strain rate tensor $\dot{\boldsymbol{\epsilon}}^{\mathrm{p}}$ and then form $\dot{\bar{\gamma}}$. Numerical experience indicates that the differences between both approaches are marginal. Either approach can lead to a set of rate equations that remains elliptic after the onset of localisation. This holds true for mode-I type failure mechanisms (decohesion) and mode-II type failures (slip).

A disadvantage that adheres to non-local plasticity is the fact that the consistency condition, i.e. $\dot{f} = 0$, results in an integro-differential equation instead of in an algebraic equation that can be solved locally:

$$\left(\frac{\partial f}{\partial \boldsymbol{\sigma}}\right)^{\mathrm{T}} \dot{\boldsymbol{\sigma}} - \frac{h}{\Psi} \int_V \psi(\mathbf{y}, \mathbf{x}) \dot{\gamma}(\mathbf{y}) \mathrm{d}V = 0 \tag{97}$$

where it is implied that $\Psi = \Psi(\mathbf{x})$, $\dot{\boldsymbol{\sigma}} = \dot{\boldsymbol{\sigma}}(\mathbf{x})$ etc. Using the elasto-plastic decomposition, Eq. (65), the flow rule, Eq. (75), and the strain-hardening hypothesis, Eq. (77), we can rework this identity as:

$$\dot{\lambda} = \dot{\lambda}_{\mathrm{local}} - \frac{h}{\Psi \left(\frac{\partial f}{\partial \boldsymbol{\sigma}}\right)^{\mathrm{T}} \mathbf{D}^{\mathrm{e}} \frac{\partial g}{\partial \boldsymbol{\sigma}}} \int_V \psi(\mathbf{y}, \mathbf{x}) \dot{\lambda}(\mathbf{y}) \varphi(\mathbf{y}) \mathrm{d}V \tag{98}$$

with

$$\dot{\lambda}_{\mathrm{local}} = \frac{\left(\frac{\partial f}{\partial \boldsymbol{\sigma}}\right)^{\mathrm{T}} \mathbf{D}^{\mathrm{e}} \dot{\boldsymbol{\epsilon}}}{\left(\frac{\partial f}{\partial \boldsymbol{\sigma}}\right)^{\mathrm{T}} \mathbf{D}^{\mathrm{e}} \frac{\partial g}{\partial \boldsymbol{\sigma}}} \tag{99}$$

and

$$\varphi = \sqrt{\frac{2}{3} \left(\frac{\partial g}{\partial \boldsymbol{\sigma}}\right)^{\mathrm{T}} \frac{\partial g}{\partial \boldsymbol{\sigma}}} \tag{100}$$

Next, we consider a one-dimensional continuum for simplicity and we approximate the integral by a finite sum:

$$\dot{\lambda}_i = (\dot{\lambda}_{\mathrm{local}})_i - \frac{h}{\Psi \left(\frac{\partial f}{\partial \boldsymbol{\sigma}}\right)^{\mathrm{T}} \mathbf{D}^{\mathrm{e}} \frac{\partial g}{\partial \boldsymbol{\sigma}}} \sum_{j=0}^{n_i} w_j \psi(y_j, x_i) \dot{\lambda}(y_j) \varphi(y_j) \tag{101}$$

with w_j a weight factor. To obtain a proper solution we must carry out an iterative procedure within each global equilibrium iteration:

$$\dot{\lambda}_i = (\dot{\lambda}_{\mathrm{local}})_i - \frac{h}{\Psi \left(\frac{\partial f}{\partial \boldsymbol{\sigma}}\right)^{\mathrm{T}} \mathbf{D}^{\mathrm{e}} \frac{\partial g}{\partial \boldsymbol{\sigma}}} \sum_{j=0}^{n_i} w_j \psi(y_j, x_i) \dot{\lambda}^{k-1}(y_j) \varphi(y_j) \tag{102}$$

where the superscript k is the iteration counter. A simple alternative would be to carry out no iterations, so that:

$$\dot{\lambda}_i = (\dot{\lambda}_{\mathrm{local}})_i - \frac{h}{\Psi \left(\frac{\partial f}{\partial \boldsymbol{\sigma}}\right)^{\mathrm{T}} \mathbf{D}^{\mathrm{e}} \frac{\partial g}{\partial \boldsymbol{\sigma}}} \sum_{j=0}^{n_i} w_j \psi(y_j, x_i) \dot{\lambda}_{\mathrm{local}}(y_j) \varphi(y_j) \tag{103}$$

Such an averaging procedure has been utilised by Bažant and Lin (1988). Unfortunately, omission of an iteration loop in which $\dot{\lambda}_i$ is computed properly results in loss of satisfaction of the consistency condition, which makes the algorithm defect, especially for large loading steps.

The numerical difficulty discussed above does not occur when total stress-strain relations are employed, that is when the strain is not decomposed into elastic and plastic components. An example is the elasticity-based non-local damage model of Pijaudier-Cabot and Bažant (1987), where the averaging process can be carried out directly with respect to the strains.

6.2 Gradient plasticity

Gradient plasticity models can be derived from fully non-local models by first expanding the weight function $g(\mathbf{s})$ in a Taylor series about $\mathbf{s} = \mathbf{0}$ and then carrying out the integration. The result is given by

$$\dot{\bar{\gamma}} = \dot{\gamma} + c_1 \nabla^2 \dot{\gamma} + c_2 \nabla^4 \dot{\gamma} + \dots \tag{104}$$

where the coefficients c_1, c_2 depend on the form of the weighting function and the dimension considered. Note that the odd derivates cancel because of the implicit assumption of isotropy. Restricting the treatment to second-order derivatives, the functional dependence on the yield function now becomes:

$$f = f(\boldsymbol{\sigma}, \gamma, \nabla^2 \gamma) \tag{105}$$

The numerical problem delineated above, which prevents an efficient use of elasto-plastic non-local models, in principle also applies to gradient plasticity models. However, gradient plasticity has the advantage that the consistency condition yields a partial differential equation instead of an integro-differential equation, namely for the yield function of Eq. (105):

$$\left(\frac{\partial f}{\partial \boldsymbol{\sigma}}\right)^{\mathrm{T}} \dot{\boldsymbol{\sigma}} - h\dot{\lambda} + c\nabla^2 \dot{\gamma} = 0 \tag{106}$$

where h and $\bar{c}$ are defined as:

$$h = -\frac{\dot{\gamma}}{\dot{\lambda}} \frac{\partial f}{\partial \gamma} \tag{107}$$

and

$$c = \frac{\partial f}{\partial \nabla^2 \gamma} \tag{108}$$

If the partial differential equation (106) is, just as the equilibrium condition, cf. Eq. (46), satisfied in a weak sense only, a set of equations results that is suitable as a starting point for large-scale finite element computations in two and three dimensions:

$$\int_V \delta\dot{\boldsymbol{\epsilon}}^{\mathrm{T}}\boldsymbol{\sigma}\mathrm{d}V = \int_S \delta\dot{\mathbf{u}}^{\mathrm{T}}\mathbf{t}\mathrm{d}S \tag{109}$$

and

$$\int_V \delta\dot{\lambda}\left(\left(\frac{\partial f}{\partial \boldsymbol{\sigma}}\right)^{\mathrm{T}}\dot{\boldsymbol{\sigma}} - h\dot{\lambda} + c\nabla^2\dot{\gamma}\right)\mathrm{d}V = 0 \tag{110}$$

Together with the kinematic relations and the elastic stress-strain relation, Eq. (66), which are both satisfied in a pointwise manner, this set of equations define the elasto-plastic rate boundary value problem. The fact that the consistency condition is no longer satisfied in a pointwise manner marks a departure from return-mapping algorithms that are used in standard plasticity and in Cosserat plasticity. Now, the plastic multiplier $\dot{\lambda}$ is considered as a fundamental unknown and has a role similar to that of the displacements. It is solved for at global level together with the displacement degrees-of-freedom.

The displacement field $\mathbf{u}$ and the field of plastic multipliers λ can be discretized to nodal variables $\mathbf{a}$ and $\boldsymbol{\Lambda}$:

$$\mathbf{u} = \mathbf{H}\mathbf{a} \ , \ \ \lambda = \mathbf{h}^{\mathrm{T}}\boldsymbol{\lambda} \tag{111}$$

Use of a Bubnov-Galerkin approach and linearising then leads to the following set of equations:

$$\begin{bmatrix} \mathbf{K}_{\mathrm{aa}} & \mathbf{K}_{\mathrm{a}\boldsymbol{\lambda}} \\ \mathbf{K}_{\mathrm{a}\boldsymbol{\lambda}}^{\mathrm{T}} & \mathbf{K}_{\boldsymbol{\lambda}\boldsymbol{\lambda}} \end{bmatrix} \begin{pmatrix} \mathrm{d}\mathbf{a} \\ \mathrm{d}\boldsymbol{\lambda} \end{pmatrix} = \begin{pmatrix} \mathbf{f}_{\mathrm{ext}}^{\mathrm{a}} - \mathbf{f}_{\mathrm{int}}^{\mathrm{a}} \\ \mathbf{0} \end{pmatrix} \tag{112}$$

where

$$\mathbf{K}_{\mathrm{aa}} = \int_V \mathbf{B}^{\mathrm{T}}\mathbf{D}^{\mathrm{e}}\mathbf{B}\mathrm{d}V \tag{113}$$

$$\mathbf{K}_{\mathrm{a}\boldsymbol{\lambda}} = -\int_V \mathbf{B}^{\mathrm{T}}\mathbf{D}^{\mathrm{e}}\frac{\partial f}{\partial \boldsymbol{\sigma}}\mathbf{h}^{\mathrm{T}}\mathrm{d}V \tag{114}$$

$$\mathbf{K}_{\boldsymbol{\lambda}\boldsymbol{\lambda}} = \int_V \left[\left(h + \left(\frac{\partial f}{\partial \boldsymbol{\sigma}}\right)^{\mathrm{T}}\mathbf{D}^{\mathrm{e}}\frac{\partial f}{\partial \boldsymbol{\sigma}}\right)\mathbf{h}\mathbf{h}^{\mathrm{T}} - c\mathbf{h}\nabla^2\mathbf{h}^{\mathrm{T}}\right]\mathrm{d}V \tag{115}$$

and $\mathbf{f}_{\text{ext}}$ the external force vector. In the preceding $\mathbf{h}^{\text{T}} = (h_1, \ldots, h_n)$ is the vector that contains the interpolation polynomials for the plastic multiplier and $\mathbf{p}^{\text{T}} = (\nabla^2 h_1, \ldots, \nabla^2 h_n)$. The detailed derivation of these equations and the finite elements that have been constructed on the basis of them are described in de Borst and Mühlhaus (1992); de Borst and Pamin (1996).

An unpleasant property of Eq. (112) is the unsymmetry that enters through $\mathbf{K}_{\lambda\lambda}$. For the pure rate problem $\mathbf{K}_{\lambda\lambda}$ can be symmetrised. Introducing $\mathbf{q}^{\text{T}} = (\nabla h_1, \ldots, \nabla h_n)$ and using Green's theorem we obtain

$$-\mathbf{h}\nabla^2\mathbf{h}^{\text{T}} \to \nabla\mathbf{h}\nabla\mathbf{h}^{\text{T}} \tag{116}$$

and the non-standard boundary conditions at the elasto-plastic boundary S_λ: $\delta\dot{\lambda} = 0$ or $(\nabla\dot{\lambda})^{\text{T}}\mathbf{n}_\lambda = 0$, with $\mathbf{n}_\lambda$ the outward normal at S_λ.

Bibliography

Z. P. Bažant and F.B. Lin. Non-local yield limit degradation. *International Journal for Numerical Methods in Engineering*, 26:1805–1823, 1988.

A. Benallal, R. Billardon, and G. Geymonat. Some mathematical aspects of the damage softening rate problem. In J. Mazars and Z. P. Bažant, editors, *Cracking and Damage*, pages 247–258, Amsterdam, 1988. Elsevier.

R. de Borst and H. B. Mühlhaus. Gradient-dependent plasticity: formulation and algorithmic aspects. *International Journal for Numerical Methods in Engineering*, 35:521–539, 1992.

R. de Borst and J. Pamin. Some novel developments in finite element procedures for gradient-dependent plasticity. *International Journal for Numerical Methods in Engineering*, 39:2477–2505, 1996.

R. de Borst, A. Benallal, and O. M. Heeres. A gradient-enhanced damage approach to fracture. *Journal de Physique IV*, C6:491–502, 1996.

H. P. J. de Vree, W. A. M. Brekelmans, and M. A. J. van Gils. Comparison of nonlocal approaches in continuum damage mechanics. *Computers and Structures*, 55: 581–588, 1995.

R. Hill. A general theory of uniqueness and stability in elastic-plastic solids. *Journal of the Mechanics and Physics of Solids*, 6:236–249, 1958.

R. Hill. Acceleration waves in solid. *Journal of the Mechanics and Physics of Solids*, 10:1–16, 1962.

W. T. Koiter. On the thermodynamic background of elastic stability theory. In *Problems of Hydrodynamics and Continuum Mechanics*, pages 423–433. SIAM, 1969.

J. Lemaitre and J. L. Chaboche. *Mechanics of Solid Materials*. Cambridge University Press, Cambridge, UK, 1990.

G. Maier and T. Hueckel. Nonassociated and coupled flow rules of elastoplasticity for rock-like materials. *International Journal of Rock Mechanics and Mining Sciences & Geomechanical Abstracts*, 16: 77–92, 1979.

J. Mazars and G. Pijaudier-Cabot. Continuum damage theory – application to concrete. *ASCE Journal of Engineering Mechanics*, 115:345–365, 1989.

R. H. J. Peerlings, R. de Borst, W. A. M. Brekelmans, and H. P. J. de Vree. Gradient-enhanced damage for quasi-brittle materials. *International Journal for Numerical Methods in Engineering*, 39:3391–3403, 1996.

R. H. J. Peerlings, M. G. D. Geers, R. de Borst, and W. A. M. Brekelmans. A critical comparison of nonlocal and gradient-enhanced softening continua. *International Journal of Solids and Structures*, 38:7723–7746, 2001.

G. Pijaudier-Cabot and Z. P. Bažant. Nonlocal damage theory. *ASCE Journal of Engineering Mechanics*, 113:1512–1533, 1987.

G. Pijaudier-Cabot and A. Huerta. Finite element analysis of bifurcation in nonlocal strain softening solids. *Computer Methods in Applied Mechanics and Engineering*, 90:905–919, 1991.

J. W. Rudnicki and J. R. Rice. Conditions for the localization of deformation in pressure sensitive dilatant materials. *Journal of the Mechanics and Physics of Solids*, 23:371–394, 1974.

J. C. Simo and J. W. Ju. Strain- and stress-based continuum damage formulations – Part I: Formulation – Part II: Computational aspects. *International Journal of Solids and Structures*, 23: 821–869, 1987.

I. Vardoulakis and J. Sulem. *Bifurcation Analysis in Geomechanics*. Blackie, London, UK, 1995.

GPSR Compliance

The European Union's (EU) General Product Safety Regulation (GPSR) is a set of rules that requires consumer products to be safe and our obligations to ensure this.

If you have any concerns about our products, you can contact us on ProductSafety@springernature.com

In case Publisher is established outside the EU, the EU authorized representative is:

Springer Nature Customer Service Center GmbH
Europaplatz 3
69115 Heidelberg, Germany

Batch number: 10726517

Printed by Printforce, the Netherlands